Pabulo H. Rampelotto (Ed.)

Polar Microbiology: Recent Advances and Future Perspectives

MDPI

This book is a reprint of the special issue that appeared in the online open access journal
Biology (ISSN 2079-7737) in 2012 (available at:
http://www.mdpi.com/journal/biology/special_issues/polar-microbio).

Guest Editor
Pabulo H. Rampelotto
Federal University of Rio Grande do Sul
Brazil

Editorial Office
MDPI AG
Klybeckstrasse 64
Basel, Switzerland

Publisher
Shu-Kun Lin

Assistant Editor
Zhenfang Zhao

1. Edition 2016

MDPI • Basel • Beijing • Wuhan

ISBN 978-3-03842-175-7 (Hbk)
ISBN 978-3-03842-176-4 (PDF)

Table of Contents

List of Contributors

Claudio Alimenti: Department of Environmental and Natural Sciences, University of Camerino, Camerino 62032, Italy.

Dale T. Andersen: Carl Sagan Center for the Study of Life in the Universe, SETI Institute, 189 Bernado Avenue, Suite 100, Mountain View, CA 94043, USA.

Justin Beckers: Department of Earth and Atmospheric Sciences, University of Alberta, Edmonton, AB T6G2E9 Canada.

Terrence H. Bell: Department of Natural Resource Sciences, McGill University, Sainte-Anne-de-Bellevue, Quebec H9X 3V9, Canada; National Research Council Canada, Energy, Mining and Environment, 6100 Royalmount Avenue, Montreal, Quebec H4P 2R2, Canada.

Renaud Berlemont: Laboratory of Biological Macromolecules, Centre for Protein Engineering, University of Liège, Institut de Chimie B6a, Liège, Sart-Tilman (4000), Belgium; Department of Earth System Science & Department of Ecology and Evolutionary Biology, University of California Irvine, 3208 Croul Hall, 92697 Irvine CA, USA.

Anna Bramucci: Department of Biological Sciences, University of Alberta, Edmonton, AB T6G2E9 Canada.

Anatoli Brouchkov: Faculty of Geology, Lomonosov Moscow State University, GSP-1,1 Leninskiye Gory, Moscow 119991, Russia.

Katrina L. Callender: Department of Natural Resource Sciences, McGill University, Sainte-Anne-de-Bellevue, Quebec H9X 3V9, Canada; National Research Council Canada, Energy, Mining and Environment, 6100 Royalmount Avenue, Montreal, Quebec H4P 2R2, Canada.

S. Craig Cary: The International Centre for Terrestrial Antarctic Research, Department of Biological Sciences, University of Waikato, Private Bag 3105, Hamilton 3240, New Zealand; College of Earth, Ocean and Environment, University of Delaware, Lewes, DE 19958, USA.

Casper T. Christiansen: Department of Biology, Queen's University, Kingston Ontario, K7L 3N6, Canada.

Brent C. Christner: Department of Biological Sciences, Louisiana State University, Baton Rouge, LA 70803, USA.

Haiyan Chu: Department of Biology, Queen's University, Kingston Ontario, K7L 3N6, Canada; Current address: Institute of Soil Science, Chinese Academy of Sciences, Nanjing 210008, China.

Laurie Connell: School of Marine Sciences, University of Maine, Orono, ME 04496, USA.

Don A. Cowan: Institute for Microbial Biotechnology and Metagenomics, University of the Western Cape, Cape Town, Bellville 7535, South Africa; Centre for Microbial Ecology and Genomics, Department of Genetics, University of Pretoria, Pretoria 0002, South Africa.

Tom D'Elia: Biological Sciences, Indian River State College, 32021 Virginia Avenue, Fort Pierce, FL 34981, USA; Department of Biological Sciences, Bowling Green State University, Bowling Green, OH 43403, USA; Current Address: Biological Sciences, Indian River State College, 32021 Virginia Avenue, Fort Pierce, FL 34981, USA.

Maud Delsaute: Laboratory of Biological Macromolecules, Centre for Protein Engineering, University of Liège, Institut de Chimie B6a, Liège, Sart-Tilman (4000), Belgium.

Jody W. Deming: School of Oceanography, University of Washington, Campus Mailbox 357940, Seattle, WA 98195,USA.

Aurélien Dommergue: Université Joseph Fourier—Grenoble 1/CNRS, LGGE, 54 rue Molière BP56, F-38402 Saint Martin d'Hères, France.

Shawn M. Doyle: Department of Biological Sciences, Louisiana State University, Baton Rouge, LA 70803, USA.

Lisa L. Dreesens: International Centre for Terrestrial Antarctic Research, University of Waikato, Hamilton 3216, New Zealand.

Robyn Edgar: Department of Biological Sciences, Bowling Green State University, Bowling Green, OH 43403, USA.

Marcela Ewert: School of Oceanography, University of Washington, Campus Mailbox 357940, Seattle, WA 98195,USA.

Gerard T. A. Fleming: Microbial Oceanography Research Unit, Microbiology, School of Natural Sciences, National University of Ireland Galway, University Road, Galway, Ireland.

Moreno Galleni: Laboratory of Biological Macromolecules, Centre for Protein Engineering, University of Liège, Institut de Chimie B6a, Liège, Sart-Tilman (4000), Belgium.

Jacques Georis: Puratos Group, Rue Bourrie 12, Andenne, Belgium.

Michael Geralt: Department of Molecular Biology, The Scripps Research Institute, La Jolla, CA 92037, USA.

Charles Gerday: Laboratory of Biochemistry, Institute of Chemistry, University of Liege, Sart-Tilman, B-4000, Liege, Belgium.

Jarishma K. Gokul: Institute for Microbial Biotechnology and Metagenomics, University of the Western Cape, Cape Town, Bellville 7535, South Africa.

Charles W. Greer: National Research Council Canada, Energy, Mining and Environment, 6100 Royalmount Avenue, Montreal, Quebec H4P 2R2, Canada.

Gennady Griva: Tyumen Scientific Center Siberian Branch of Russian Academy of Science, 86 Malygina, Tyumen 625000, Russia.

Paul Grogan: Department of Biology, Queen's University, Kingston Ontario, K7L 3N6, Canada.

Martin Grube: Institute of Plant Sciences, Karl-Franzens-University Graz, Holteigasse 6, A-8010 Graz, Austria.

Christian Haas: Department of Earth and Atmospheric Sciences, University of Alberta, Edmonton, AB T6G2E9 Canada; Current address: Department of Earth and Space Science and Engineering, York University, 105 Petrie Science Building, Toronto, ON M3J 1P3, Canada.

Sukkyun Han: Department of Biological Sciences, University of Alberta, Edmonton, AB T6G2E9 Canada; Current address: IEH Laboratories and Consulting Group, Lake Forest Park, WA 98155, Canada.

Ian Hawes: Gateway Antarctica, University of Canterbury, Private Bag 4800, Christchurch, New Zealand.

Daniela Isola: Department of Ecological and BiologicalSciences (DEB), University of Tuscia, Largo dell'Università snc, Viterbo 01100, Italy.

Olivier Jacquin: Laboratory of Biological Macromolecules, Centre for Protein Engineering, University of Liège, Institut de Chimie B6a, Liège, Sart-Tilman (4000), Belgium.

Swati Joshi: Department of Microbiology, University of Delhi South Campus, New Delhi 110021, India.

Anne D. Jungblut: Department of Life Sciences, The Natural History Museum, Cromwell Road, London, UK.

Caitlin Knowlton: Department of Biological Sciences, Bowling Green State University, Bowling Green, OH 43403, USA.

Zeynep A. Koçer: Department of Biological Sciences, Bowling Green State University, Bowling Green, OH 43403, USA; Current Address: Department of Infectious Diseases, Division of Virology, St. Jude Children's Research Hospital, Memphis, TN 38105, USA.

Eileen Y. Koh: School of Biological Sciences, Victoria University of Wellington, PO Box 600, Wellington 6140, New Zealand; Current address: Department of Microbiology, Yong Loo Lin-School of Medicine, National University of Singapore, Singapore 117579, Singapore.

Niraj Kumar: Department of Biology, Queen's University, Kingston Ontario, K7L 3N6, Canada.

Marcello La Salla: Laboratory of Biological Macromolecules, Centre for Protein Engineering, University of Liège, Institut de Chimie B6a, Liège, Sart-Tilman (4000), Belgium.

Brian Lanoil: Department of Biological Sciences, University of Alberta, Edmonton, AB T6G2E9 Canada.

Catherine Larose: Environmental Microbial Genomics, CNRS, Ecole Centrale de Lyon, Université de Lyon, 36 avenue Guy de Collongue, 69134 Ecully, France.

Charles K. Lee: International Centre for Terrestrial Antarctic Research, University of Waikato, Hamilton 3216, New Zealand.

Pierangelo Luporini: Department of Environmental and Natural Sciences, University of Camerino, Camerino 62032, Italy.

Barbara R. Lyon: School of Environmental Sciences, University of East Anglia, Norwich Research Park, Norwich NR4 7TJ, UK.

Tyler J. Mackey: Department of Geology, University of California, Davis, CA 95616, USA.

Rosa Margesin: Institute of Microbiology, University of Innsbruck, Technikerstrasse 25, A-6020 Innsbruck, Austria.

Andrew R. Martin: Institute for Marine and Antarctic Studies, University of Tasmania, Hobart 7001, Australia.

David M. McCarthy: Microbial Oceanography Research Unit, Microbiology, School of Natural Sciences, National University of Ireland Galway, University Road, Galway, Ireland.

Andrew McMinn: Institute for Marine and Antarctic Studies, University of Tasmania, Hobart 7001, Australia.

Thomas Mock: School of Environmental Sciences, University of East Anglia, Norwich Research Park, Norwich NR4 7TJ, UK.

Scott N. Montross: Department of Earth Sciences, Montana State University, Bozeman, MT 59717, USA.

Silvano Onofri: Department of Ecological and BiologicalSciences (DEB), University of Tuscia, Largo dell'Università snc, Viterbo 01100, Italy.

John W. Patching: Microbial Oceanography Research Unit, Microbiology, School of Natural Sciences, National University of Ireland Galway, University Road, Galway, Ireland.

David A. Pearce: British Antarctic Survey, Natural Environment Research Council, High Cross, Madingley Road, Cambridge, CB3 OET, UK.

Pablo Power: Laboratory of Biological Macromolecules, Centre for Protein Engineering, University of Liège, Institut de Chimie B6a, Liège, Sart-Tilman (4000), Belgium; Department of Microbiology, Immunology and Biotechnology, School of Pharmacy and Biochemistry, University of Buenos Aires, Junin 956 (1113), Buenos Aires, Argentina.

Pabulo H. Rampelotto: Interdisciplinary Center for Biotechnology Research, Federal University of Pampa, AntônioTrilha Avenue, P.O.Box 1847, 97300-000, São Gabriel—RS, Brazil.

Scott O. Rogers: Department of Biological Sciences, Bowling Green State University, Bowling Green, OH 43403, USA.

Ken G. Ryan: School of Biological Sciences, Victoria University of Wellington, PO Box 600, Wellington 6140, New Zealand.

Tulasi Satyanarayana: Department of Microbiology, University of Delhi South Campus, New Delhi 110021, India.

Franz Schinner: Institute of Microbiology, University of Innsbruck, Technikerstrasse 25, A-6020 Innsbruck, Austria.

Laura Selbmann: Department of Ecological and BiologicalSciences (DEB), University of Tuscia, Largo dell'Università snc, Viterbo 01100, Italy.

Yury M. Shtarkman: Department of Biological Sciences, Bowling Green State University, Bowling Green, OH 43403, USA.

Mark L. Skidmore: Department of Earth Sciences, Montana State University, Bozeman, MT 59717, USA.

Hubert Staudigel: Institute of Geophysics and Planetary Physics, Scripps Institution of Oceanography, La Jolla, CA 92093, USA.

Dawn Y. Sumner: Department of Geology, University of California, Davis, CA 95616, USA.

Henry J. Sun: Division of Earth and Ecosystem Sciences, Desert Research Institute, Las Vegas, NV 89119, USA.

Marla Tuffin: Institute for Microbial Biotechnology and Metagenomics, University of the Western Cape, Cape Town, Bellville 7535, South Africa.

Adriana Vallesi: Department of Environmental and Natural Sciences, University of Camerino, Camerino 62032, Italy.

Angel Valverde: Centre for Microbial Ecology and Genomics, Department of Genetics, University of Pretoria, Pretoria 0002, South Africa.

Ram Veerapaneni: Department of Biological Sciences, Bowling Green State University, Firelands Campus, Huron, OH 44839, USA; Current Address: Department of Biological Sciences, Bowling Green State University, Firelands Campus, Huron, OH 44839, USA.

Fabienne Verté: Puratos Group, Industrielaan 25, Groot-Bijgarden, Belgium.

Timothy M. Vogel: Environmental Microbial Genomics, CNRS, Ecole Centrale de Lyon, Université de Lyon, 36 avenue Guy de Collongue, 69134 Ecully, France.

Virginia K. Walker: Department of Biology, Queen's University, Kingston Ontario, K7L 3N6, Canada; Department of Biomedical and Molecular Sciences, School of Environmental Studies, Queen's University, Kingston Ontario, K7L 3N6, Canada.

Lyle G. Whyte: Department of Natural Resource Sciences, McGill University, Sainte-Anne-de-Bellevue, Quebec H9X 3V9, Canada.

Kurt Wüthrich: Department of Molecular Biology, The Scripps Research Institute, La Jolla, CA 92037, USA; Skaggs Institute for Chemical Biology, The Scripps Research Institute, La Jolla, CA 92037, USA.

De-Chao Zhang: Institute of Microbiology, University of Innsbruck, Technikerstrasse 25, A-6020 Innsbruck, Austria.

Laura Zucconi: Department of Ecological and BiologicalSciences (DEB), University of Tuscia, Largo dell'Università snc, Viterbo 01100, Italy.

About the Guest Editor

Pabulo Henrique Rampelotto is Editor-in-Chief of the Book Series *Grand Challenges in Biology and Biotechnology* (Springer) and *Astrobiology: Exploring Life on Earth and Beyond* (Imperial College Press). In addition, he serves as Editor-in-Chief of **Current Biotechnology** as well as Associate Editor, Guest Editor and member of the editorial board of several scientific journals in the field of Life Sciences and Biotechnology. Prof. Rampelotto is also member of four Scientific Advisory Boards (Astrobiology/SETI Board, Biotech/Medical Board, Policy Board, and Space Settlement Board) of the Lifeboat Foundation, alongside several Nobel Laureates and other distinguished scientists, philosophers, educators, engineers, and economists. In his books and special issues, some of the most distinguished team leaders in the field have published their work, ideas, and findings.

Preface

Polar microbiology is a promising field of research that can tell us much about the fundamental features of life. The microorganisms that inhabit Arctic and Antarctic environments are important not only because of the unique species they represent, but also because of their diverse and unusual physiological and biochemical properties. Furthermore, microorganisms living in Polar Regions provide useful models for general questions in ecology and evolutionary biology given the reduced complexity of their ecosystems, the relative absence of confounding effects associated with higher plants or animals, and the severe biological constraints imposed by the polar environment. In terms of applied science, the unique cold-adapted enzymes and other molecules of polar microorganisms provide numerous opportunities for biotechnological development. Another compelling reason to study polar microbial ecosystems is the fact that they are likely to be among the ecosystems most strongly affected by global change. For these reasons, polar microbiology is a thriving branch of science with the potential to provide new insights into a wide range of basic and applied issues in biological science. In this context, it is timely to review and highlight the progress so far and discuss exciting future perspectives. In this special issue, some of the leaders in the field describe their work, ideas and findings.

Pabulo H. Rampelotto
Guest Editor

The Distribution and Identity of Edaphic Fungi in the McMurdo Dry Valleys

Lisa L. Dreesens, Charles K. Lee and S. Craig Cary

Abstract: Contrary to earlier assumptions, molecular evidence has demonstrated the presence of diverse and localized soil bacterial communities in the McMurdo Dry Valleys of Antarctica. Meanwhile, it remains unclear whether fungal signals so far detected in Dry Valley soils using both culture-based and molecular techniques represent adapted and ecologically active biomass or spores transported by wind. Through a systematic and quantitative molecular survey, we identified significant heterogeneities in soil fungal communities across the Dry Valleys that robustly correlate with heterogeneities in soil physicochemical properties. Community fingerprinting analysis and 454 pyrosequencing of the fungal ribosomal intergenic spacer region revealed different levels of heterogeneity in fungal diversity within individual Dry Valleys and a surprising abundance of Chytridiomycota species, whereas previous studies suggested that Dry Valley soils were dominated by Ascomycota and Basidiomycota. Critically, we identified significant differences in fungal community composition and structure of adjacent sites with no obvious barrier to aeolian transport between them. These findings suggest that edaphic fungi of the Antarctic Dry Valleys are adapted to local environments and represent an ecologically relevant (and possibly important) heterotrophic component of the ecosystem.

Reprinted from *Biology*. Cite as: Dreesens, L.L.; Lee, C.K.; Cary, S.C. The Distribution and Identity of Edaphic Fungi in the McMurdo Dry Valleys. *Biology* **2014**, *3*, 466-483.

1. Introduction

Located between the Polar Plateau and Ross Sea in Southern Victoria Land, the McMurdo Dry Valleys (hereinafter the Dry Valleys) are the largest contiguous ice-free area on the Antarctic continent. Dry Valley soils are known as some of the oldest, coldest, driest, and most oligotrophic soils on Earth [1]; consequently, the Dry Valley ecosystem is characterized by a lack of nutrients [2], low precipitation levels and biologically available water [3–5], high levels of salinity [6–8], large temperature fluctuations [5,9,10], steep chemical and biological gradients [11], and high incidence of UV-solar radiation [12–14]. Early studies suggested that Dry Valley soils contained very little microbial biota [1], but recent molecular evidence has demonstrated the presence of diverse and heterogeneous bacterial communities potentially driven by steep physicochemical gradients [1,10,15–19]. In contrast, comparatively limited molecular evidence exists on the distribution and drivers of fungal communities in Dry Valley soils [20–23].

Fungal identification in Dry Valley soils by means of a combination of culturing and molecular tools (*i.e.*, denaturing gradient gel electrophoresis and DNA sequencing) has detected primarily members of Dikarya (*i.e.*, Ascomycota and Basidiomycota), including both filamentous and non-filamentous species [24–27]. A survey of Dry Valley sites including Mt Flemming, Allan Hills, New Harbor, and Ross Island revealed the dominant free-living fungal genera in Dry Valley soils as *Cadophora* (Ascomycota), *Cryptococcus* (Basidiomycota), *Geomyces* (Ascomycota), and *Cladosporium* (Ascomycota) [22]. A study of cultivable fungi in Taylor Valley showed that filamentous fungi appeared to be associated with high soil pH and moisture, whereas yeasts and yeast-like fungi had wider distribution across habitats examined [23]. Basidiomycetous *Cryptocococcus* and *Leucosporidium* species were the most frequently isolated genera in a regional survey of yeasts and yeast-like fungi in the Dry Valleys [20]. The diversity of yeasts and yeast-like fungi was positively correlated with soil pH and negatively with conductivity [20]. The same study also revealed apparent segregation of *Cryptococcus* clades found in Taylor Valley and the Labyrinths of Wright Valley [20], hinting at the presence of localized communities adapted to environmental conditions, as has been reported for soil bacteria in the Dry Valleys [15]. A culture-based study of soils taken from McKelvey Valley detected no fungal colony-forming units (CFUs) in most of the samples [21], and a molecular survey of McKelvey Valley also detected no fungal signals in the soils [18]. However, sequences affiliated with genera *Dothideomycetes* (Ascomycota), *Sordariomycetes* (Ascomycota), and *Cystobasidiomycetes* (Basidiomycota) were found in endolithic and chasmolithic communities in McKelvey Valley [18]. The evidence so far suggests that the cultivable components of Dry Valley fungal communities are dominated by ascomycetous and basidiomycetous species, although their biogeography and factors that shape their distribution in the Dry Valleys remain unclear due to the lack of systematic and culture-independent evidence. Furthermore, the ecological relevance of fungi in Dry Valley soils remains unknown since neither cultivation nor molecular techniques can effectively distinguish active fungal cells from dormant spores.

For this study, we carried out a molecular survey of Dry Valley soil fungi at six study sites (Battleship Promontory, Upper Wright Valley, Beacon Valley, Miers Valley, Alatna Valley, and University Valley) using terminal restriction fragment length polymorphism (tRFLP) and 454 pyrosequencing analyses of the fungal ribosomal intergenic spacer. Soil physicochemical properties were also characterized to examine potential environmental drivers of fungal diversity.

2. Experimental

2.1. Sample Collection

Soil was collected at six different sites in the McMurdo Dry Valleys (Table 1 and Figure 1) as described previously [15]. Briefly, sampling sites were all located on a south facing, 0–20° slope. An intersection was made by two 50 m transects, with the intersection in the middle being the central sampling point (X or C). Four sampling points around the central point were marked (A–D with A being the southernmost point and the remaining points in an anti-clockwise order, or N, E, S, W). Five scoops of the top 2 cm of soil were collected and homogenized at each identified (1 m²) sampling point after pavement pebbles were removed. Samples were stored in sterile Whirl-Pak (Nasco International, Fort Atkinson, WI, USA) at −20 °C until returned to New Zealand, where they were stored at −80 °C until analysis.

Table 1. List of sampling sites.

Valley	Coordinates	Elevation	Sampling Date
Miers Valley	78°05.486'S, 163°48.539'E	171 m	December 2006
Beacon Valley	77°52.321'S, 160°29.725'E	1376 m	December 2006
Upper Wright Valley	77°31.122'S, 160°45.813'E	947 m	January 2008
Battleship Promontory	76°54.694'S, 160°55.676'E	1028 m	January 2008
Alatna Valley	76°54.816'S, 161°02.213'E	1057 m	November 2010
University Valley	77°51.668'S, 160°42.736'E	1680 m	November 2010

Figure 1. Antarctica is presented in the lower right corner, with the McMurdo Dry Valleys marked in a blue rectangle. The locations of the sampling sites within the McMurdo Dry Valleys are displayed by red dots.

2.2. Soil Chemistry

Soil moisture content was determined by drying 6 g of soil at 35 °C until its weight stabilized and then at 105 °C until the sample reached constant weight. Soil pH and electrical conductivity were determined using the slurry technique, which is based on a 2:5 unground dried soil:de-ionized water mixture rehydrated overnight before measurement, using a Thermo Scientific Orion 4 STAR pH/Conductivity meter (Thermo Scientific, Beverly, MA, USA). For total and organic carbon and nitrogen contents, dried soils were ground to fine powders using an agate mortar and pestle and precisely weighed out to 100 mg. Samples were analyzed with an Elementar Isoprime 100 analyzer (Elementar Analysensysteme, Hanau, Germany). Sample preparation for elemental analysis was adapted from US EPA Analytical Methods 200.2 (Revision 2.8, 1994) and Lee *et al.* [15], in which ground dried soil samples were acid digested and analyzed using an E2 Instruments Inductively

Coupled Plasma Mass Spectrometer (ICP-MS) (Perkin-Elmer, Shelton, CT, USA) at the Waikato Mass Spectrometry Facility following manufacturer protocols [15]. For soil grain size, 0.3–0.4 g of 2-mm-sieved dried soil was incubated overnight with 10% hydrogen peroxide. A second excess of hydrogen peroxide was then added to the sample and heated on a hotplate. Finally, 10 mL of 10% Calgon was added to the sample and left overnight before being placed in an ultrasonic bath for 5 min. Measurements were taken on a Mastersizer 2000 (Malvern, Taren Point, NSW, Australia).

2.3. DNA Extraction

DNA was extracted from soils using a modified version of a previously published cetyl trimethylammonium bromide (CTAB) bead beating protocol designed for maximum recovery of DNA from low biomass soils [15,28] (Supplementary Material Text). DNA quantification was done using the QuBit-IT dsDNA HS Assay Kit (Invitrogen, Carlsbad, CA, USA).

2.4. Terminal Restriction Fragment Length Polymorphism Analysis

Terminal restriction fragment length polymorphism analysis (tRFLP) was utilized to identify fungal community structure and relative diversity by amplifying the intergenic spacer (ITS) between the 18S and the 28S genes of the fungal *rrn* operon. PCR was performed in triplicate and pooled together to reduce stochastic inter-reaction variability. PCR master mix included 1x PCR buffer (with 1.5 mM Mg^{2+}) (Invitrogen, Carlsbad, CA, USA), 0.2 mM dNTPs (Roche Applied Science, Branford, CT, USA), 0.02 U Platinum Taq (Invitrogen, Carlsbad, CA, USA), 0.25 µM of both forward and reverse primer (Custom Science, Auckland, New Zealand) (ITS1-F and 3126R; Table S1), and 0.02 mg/mL bovine serum albumin (Sigma Aldrich, St. Louis, MO, USA) and was treated with ethidium monoazide at a final concentration of 25 pg/µL to inhibit contaminating DNA in the reagents [29]. PCR was carried out using the following thermal cycling conditions: 94 °C for 3 min; 35 cycles of 94 °C for 20 s, 52 °C for 20 s, 72 °C for 1 min 15 s; and 72 °C for 5 min on a DNA Engine thermal cycler (Bio-Rad Laboratories, Hercules, CA, USA). Successful PCR was confirmed with 1% Tris-acetate-EDTA (TAE) agarose gels, and PCR products were cleaned using the Ultraclean 15 DNA Purification kit (MOBIO Laboratories, Carlsbad, CA, USA) according to manufacturer instructions. DNA was quantified using the QuBit-IT dsDNA HS Assay Kit. 40 ng of DNA was digested with 2 U of MspI and 1× restriction enzyme buffer (Roche Applied Science, Branford, CT, USA) according to manufacturer instructions and purified with Ultraclean 15 DNA Purification kit. Lengths of fluorescent-labeled PCR amplicons (*i.e.*, tRFLP fragments) were determined by capillary electrophoresis at the Waikato DNA Sequencing Facility using an ABI 3130 Genetic Analyzer (Life Technologies, Carlsbad, CA, USA) at 10 kV, a separation temperature of 44 °C for 2 h, and the GeneScan 1200 LIZ dye Size Standard (Life Technologies, Carlsbad, CA, USA).

2.5. 454 Pyrosequencing

PCR protocol for preparing amplicons for pyrosequencing was identical to that for tRFLP, except a different reverse primer (ITS4, Table S1) was used. PCR products were purified using gel extraction and the QuickClean 5M PCR Purification Kit (GenScript, Piscataway, NJ, USA). A second round of PCR using fusion primers containing adapters for 454 pyrosequencing was performed (Table S1). These products were purified using Agencourt AMPure XP Beads (Beckman Coulter, Inc., Brea, CA, USA) for PCR amplicon recovery and removal of unincorporated dNTPs, primers, primer dimmers, salts and other contaminants (Beckman Coulter, Beverly, MA, USA) according to manufacturer instructions. Quality of PCR amplicon libraries was checked using the Agilent High Sensitivity DNA Kit with a BioAnalyzer (Agilent 2100, Agilent Technologies, Santa Clara, CA, USA) and the Kapa Library Quantification Kit—454 Titanium (Kapa Biosystems, Wilmington, MA, USA). 454 pyrosequensing was performed using a Roche 454 Junior sequencer at the Waikato DNA Sequencing Facility following manufacturer protocols.

2.6. Data Analysis

Environmental variables were $\log(x + c)$ transformed, where c is the 1st percentile value for the variable (except [Ag] where c is the mean due to low values), prior to analysis; pH values were not transformed. A Euclidean distance matrix was calculated in PRIMER 6 (PRIMER-E Ltd., Ivybridge, UK) from the transformed environmental variables and used for downstream analyses. tRFLP traces were first processed using PeakScanner 1.0 (Life Technologies, Carlsbad, CA, USA) to export all peaks above 5 relative fluorescence units (RFU). The resulting profiles were further processed using an in-house collection of python and R scripts (available from authors upon request) to identify true signal peaks as well as binning peaks based on their sizes. Briefly, peaks outside the size range of 50–1200 bp were excluded from analysis, and only peaks whose heights are greater than the 99% confidence threshold (*i.e.*, alpha value of 0.01) within a log-normal distribution were considered to be non-noise. Additionally, peaks had to be greater than 50 RFU to be considered non-noise, and all peaks above 200 RFU were by default designated as non-noise peaks. Peaks were then binned to the nearest 1 bp, and only peaks whose relative abundance was greater than 0.1% were retained. The resulting matrix of peaks expressed as relative abundances was imported into PRIMER 6, and a Bray-Curtis similarity matrix was calculated for downstream analyses. Using these distance matrices, PRIMER 6 was used to generate non-metric multidimensional scaling (MDS) plots, perform group-average hierarchical clustering, and carry out one-way analysis of similarities (ANOSIM) and biota-environmental stepwise (BEST) analyses.

454 pyrosquencing flowgrams were denoised using AmpliconNoise v1.24 [30], including a SeqNoise step to remove PCR errors and a Perseus step to remove PCR chimeras [30]. Denoised reads were aligned pair-wise using ESPIRIT [31], which directly generated a distance matrix. Mothur 1.26 was used to cluster the sequences at 0.15 distance with nearest neighbor clustering [32], and the representative sequences for the resulting operational taxonomic units (OTUs) were checked (blastn with word size of 7) against the GenBank *nr* database to allow manual identification of fungal ITS sequences (>250 bp and >80% similarity to known fungal ITS sequences). The curated sequences were then re-clustered using average neighbor at 0.05 distance. OTUs with fewer than 9 reads were excluded from downstream analysis as an aggressive filter against spurious OTUs that arose from non-specific PCR amplification and sequencing errors.

3. Results and Discussion

3.1. Soil Geochemistry

Soils from six Dry Valleys were characterized as loamy sand or sand due to their low clay (<2%) and silt (<13%) contents (Table S2), which is congruent with Antarctica's known slow and primarily physical weathering processes [7]. The coarse soil texture likely resulted from low erosivity of cold-based glaciers and salt weathering, which causes comminution of coarse fragments and provides a steady supply of sandy grains to the soils [7]. Consequently, these soils lack significant aggregation and have poor moisture retention capacity, which is consistent with their low gravimetric water content (Table S2). Water availability has been suggested to be a major factor controlling biomass and diversity of Antarctic vegetation [33,34]. Among the six study sites, Miers Valley soils contained the lowest average moisture content (0.53%, ANOVA p-value = 0.002; Table S2). But due to its low elevation (elev. 171 m) and variable wind direction, temperatures in Miers Valley can reach above 0 °C in austral summers [35]. This likely leads to increased water availability from melt streams of Miers and Adams Glaciers, which can trigger rapid responses from local microorganisms [16,34]. Water availability in austral summers is also elevated in Alatna Valley and Battleship Promontory, where transient ponds are formed from snow melt. This is in contrast with the low moisture content and water availability in higher (elev. >1500 m) and more inland valleys (e.g., University Valley). The high altitude of University Valley results in colder air temperatures all year round, leading to a lower net ice loss rate when compared to Beacon Valley (*ca.* 450 m below University Valley) [36]. Soil salt content is a proxy for water availability [37], and Miers Valley, Alatna Valley, and Battleship Promontory soils showed relatively low conductivity. Soil physicochemical properties (Table S2) were significantly different among the sampling sites (ANOSIM global R = 0.963, p-value = 0.001) with each valley clearly forming its own clade. In a broader view, distinct grouping patterns emerged for Miers Valley in the MDS plot (Figure 2), possibly due to its alkaline pH reflective of greater influence from salts of marine origin [38] and its higher C/N ratio. Overall, geochemical analysis revealed a wide range of soil salinity (107–3920 µS), low moisture content (1%–3% w/v), low levels of organic carbon (<0.46%) and nitrogen (<0.12%).

3.2. Community Fingerprinting with tRFLP

DNA extractions from soils proved difficult, and DNA samples from Beacon, University, and Upper Wright Valleys were mostly below the detection limit of 0.05 ng/µL. The highest recovery yields were obtained from Miers Valley samples, followed by those from Battleship Promontory and Alatna Valley (Table 2). Fungal tRFLP analysis of extracted DNA returned positive signals for 12 of the 30 soil samples, with no polymorphic fragments (PFs) detected in any of the samples from University Valley. A total of 33 PFs were obtained (Table 3), whose lengths varied between 145 and 781 bp. Samples from Battleship Promontory collectively returned the highest diversity (13 PFs), followed by Alatna Valley (11 PFs) and Miers Valley (5 PFs). ANOSIM analysis of PF profiles demonstrated statistically significant differences among valleys (ANOSIM global

$R = 0.731$, p-value $= 0.001$), and there was no robust correlation between diversity (PF count) and biomass (averaged DNA yield from 1 gram of soil) ($R = 0.35$, p-value $= 0.06$).

Figure 2. Nonmetric multidimensional scaling (MDS) plot based on Euclidean distances between soil physicochemical profiles. Significant correlations (Pearson $R > 0.25$) between plot ordinations and soil physicochemical properties are represented as vectors in gray.

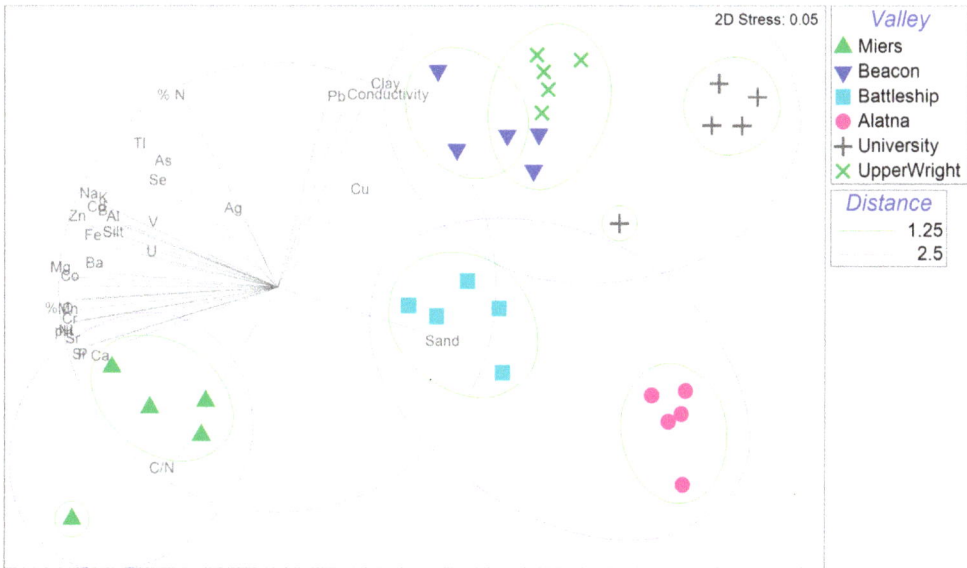

Table 2. Average concentrations of DNA extracted from 1 g of soil.

Valley	Average Concentration ± S.D.
Miers Valley	48.60 ± 27.79 ng/μL
Beacon Valley	0.48 ± 0.55 ng/μL
Battleship Promontory	20.87 ± 5.61 ng/μL
Upper Wright Valley	3.68 ± 7.57 ng/μL
Alatna Valley	15.84 ± 13.49 ng/μL
University Valley	0.05 ± 0.09 ng/μL

Table 3. Summary of terminal restriction fragment length polymorphism (tRFLP) polymorphic fragments (PF).

Valley	Total PF	Average PF ± S.D.
Miers Valley	5	1.0 ± 1.2
Beacon Valley	2	0.4 *
Battleship Valley	13	2.6 ± 1.5
Wright Valley	2	0.4 *
Alatna Valley	11	2.2 ± 3.2
University Valley	0	0

* S.D. not calculated.

Interestingly, a MDS plot of tRFLP data showed a clear separation of samples from Battleship Promontory and Alatna Valley (Figure 3), despite the fact that the two sampling sites are less than 5 km apart and within line-of-sight. This suggests that aeolian dispersal between these sites is very limited or outweighed by other environmental drivers that shape edaphic fungal diversity at these locations. There was only one sample each from Beacon and Upper Wright Valleys, but they were >50% similar to each other. Samples from Miers Valley were widely dispersed in the MDS plot, making Miers Valley a clear outlier.

Figure 3. Nonmetric multidimensional scaling (MDS) plot based on Bray-Curtis similarities of tRFLP profiles. Samples used for 454 PCR amplicon pyrosequencing are labeled by name. Significant correlations (Pearson $R > 0.25$) between plot ordinations and soil physicochemical properties are represented as vectors in gray.

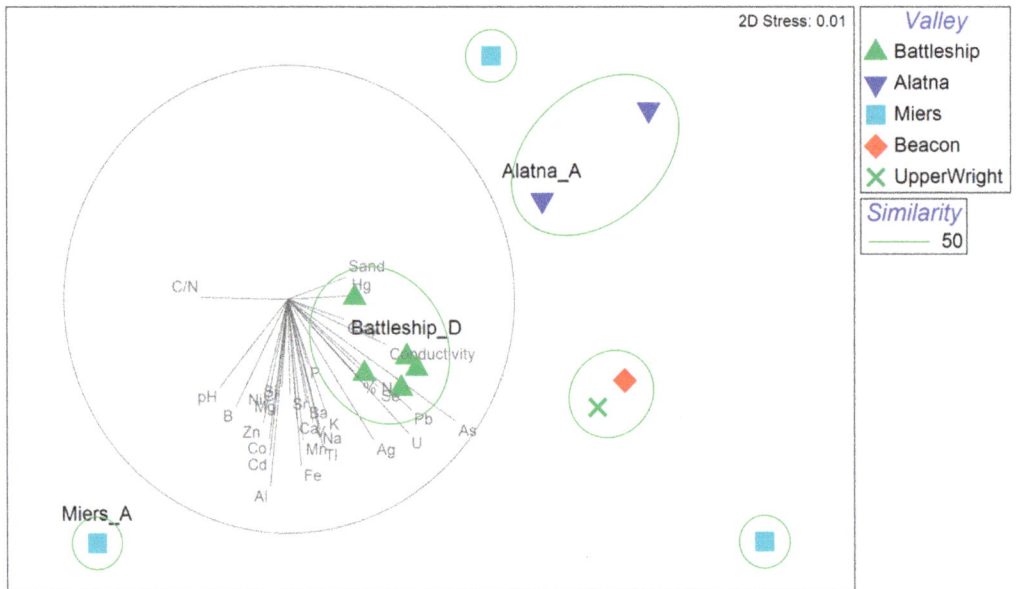

3.3. 454 Pyrosequencing

To identify the fungal species present, three samples that represented the greatest diversity based on results from tRFLP analysis were chosen for 454 PCR amplicon pyrosequencing. DNA extracted from Battleship Promontory sample D, referred to as Battleship_D, Alatna Valley sample N (Alatna_N), and Miers Valley sample A (Miers_A) appeared to be most representative of each major cluster (Figure 3). Fungal signals in Beacon and Upper Wright Valley were considered unsequenceable due to very low DNA extraction and amplification yields and therefore excluded from pyrosequencing. After filtering, denoising, chimera removal, and quality control, 262 fungal OTUs (from 21,101 reads) were obtained, of which 37 contained more than 9 reads (*i.e.*, >0.2% of the sample with fewest reads) and were used for downstream analysis. Species richness (Table 4) was highest in Miers Valley (31 OTUs from 1771 reads), followed by Battleship Promontory

(18 OTUs from 2091 reads), and Alatna Valley (17 OTUs from 5081 reads). A Venn diagram illustrates the distribution of OTUs among the three samples (Figure 4). Nine OTUs (representing 8943 reads) were found in all three Valleys (Figure 4), including the five most abundant OTUs.

Figure 4. Venn diagram of fungal OTUs.

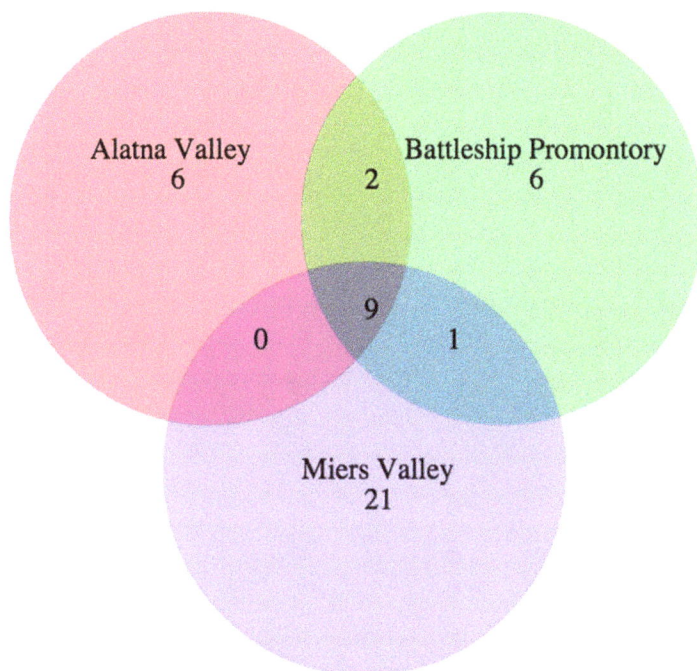

A significant number of OTUs were annotated as unclassified (Table 4 and Figure 5), which is likely reflective of the comparative lack of high quality annotated fungal ITS sequences in the GenBank *nr* database. Therefore, results that rely on classification of fungal sequences must be interpreted carefully. However, multiple studies identified Ascomycota and Basidiomycota as the dominant fungal phyla in the Dry Valleys [22,25,27,39], whereas our results showed an unexpected prominence of Chytridiomycota among all three valleys (Figure 5). It should be noted that Chytridiomycota were reported in a molecular survey on west Antarctic sites [40], including Signy Island, Mars Oasis, and Coal Nunatak, at significant abundances but not in the Dry Valleys.

Table 4. Overview of fungal OTUs from PCR amplicon pyrosequencing.

OTU #	Read Count				Best Match in GenBank *nr* Database			
	AV_N	BP_D	MV_A	Total	GenBank ID	Identity (%)	Phylum	Organism
3	1852	407	1	2283	AB032673	99	Basidiomycota	*Cryptococcus consortionis*
4	841	728	191	1760	EF432821	93	Chytridiomycota	*Lobulomycetales* sp. AF017
6	1058	122	369	1542	EF060799	99	Ascomycota	*Herpotrichiellaceae* sp. LM500
7	505	68	233	806	JF747078	99	Ascomycota	*Exophiala equina*
10	129	351	61	541	EU480339	93	Unknown	Uncultured clone
11	0	0	372	372	GQ250013	92	Ascomycota	*Cordyceps* sp. BCC22921
14	246	0	0	246	EF535204	90	Ascomycota	*Candelaria crawfordii* strain CHN265
16	179	0	0	179	FJ827708	90	Chytridiomycota	*Powellomyces* sp. PL 142
20	0	109	0	109	EU352772	93	Chytridiomycota	*Chytridiales* sp. JEL178
22	0	0	109	109	DQ457086	85	Unknown	Uncultured clone
24	0	83	0	83	AM901700	97	Ascomycota	*Ascomycete* sp. BF104
25	0	0	81	81	FJ827708	94	Chytridiomycota	*Powellomyces* sp. PL 142
26	80	0	0	80	GU184116	96	Ascomycota	*Acarospora rosulata* isolate ACABUL_USA2
28	36	30	0	66	KC222134	83	Ascomycota	*Trichoglossum octopartitum*
29	0	0	61	61	EF585664	83	Chytridiomycota	*Betamyces americaemeridionalis*
35	0	0	54	54	EU352770	92	Chytridiomycota	*Lobulomyces poculatus*
39	0	47	0	47	AF106527	91	Ascomycota	*Arthrobotrys arcuata* strain CBS 174.89
40	8	33	1	42	DQ494379	94	Ascomycota	*Vermispora fusarina*
41	12	27	3	42	JX171180	94	Basidiomycota	*Meira* sp. ANTCW08-165
45	34	1	5	40	FJ827741	96	Chytridiomycota	*Gaertneriomyces* sp. JEL 550
48	29	1	0	30	HQ634632	97	Ascomycota	*Chaetothyriales* sp. M-Cre1-2
49	29	0	0	29	JX124723	98	Ascomycota	*Taphrina* sp. CCFEE 5198
51	0	0	28	28	JX036093	93	Ascomycota	*Polysporina frigida*
54	0	10	17	27	EU352770	92	Chytridiomycota	*Lobulomyces poculatus*
56	0	0	25	25	JF809853	99	Chytridiomycota	*Betamyces* sp. PL 173
59	0	0	23	23	AY373015	91	Unknown	*Olpidium brassicae*
60	0	22	0	22	JQ936330	99	Unknown	*Phaeosphaeriopsis* sp. CBP21E

Table 4. *Cont.*

OTU #	Read Count				Best Match in GenBank *nr* Database			
	AV_N	BP_D	MV_A	Total	GenBank ID	Identity (%)	Phylum	Organism
61	0	0	22	22	JX219783	91	Ascomycota	*Cortinarius callisteus*
62	0	0	22	22	JN416510	89	Basidiomycota	*Basidiobolus* sp. BCU1
64	1	19	1	21	JX173100	99	Ascomycota	*Cladosporium* sp. AF13
67	18	0	0	18	AY781244	89	Unknown	*Ascomycete* sp. olrim401
68	0	18	0	18	AY394892	94	Ascomycota	*Mycorrhizal* sp. pkc11
72	0	0	17	17	EF634250	80	Chytridiomycota	*Coralloidiomyces digitatus*
78	0	15	0	15	EU480016	90	Unknown	Uncultured clone
101	0	0	11	11	JN882333	94	Chytridiomycota	*Monoblepharis hypogyna*
102	0	0	11	11	DQ485612	93	Chytridiomycota	*Rhizophydium carpophilum*
105	0	0	10	10	JQ711836	99	Basidiomycota	*Russula nigricans*

Abbreviations: OTU, operational taxonomic unit; AV_N, Alatna Valley sample N; BP_D, Battleship Promontory sample D; MV_A, Miers Valley sample A.

Figure 5. Phylum-level distribution of fungal OTUs.

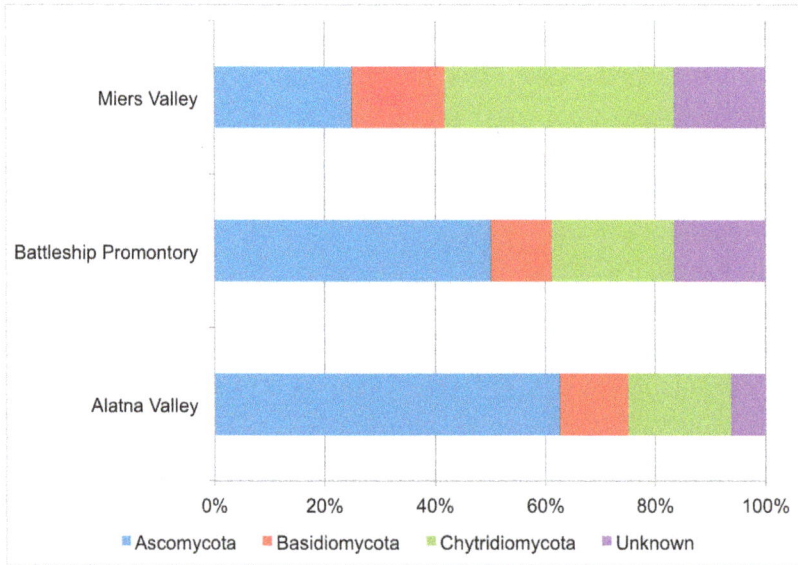

Contrary to fungal tRFLP results, PCR amplicon pyrosequencing analysis of the fungal ITS region identified Miers Valley as having the highest level of diversity of the three valleys (Figure 4), despite the lowest sequencing depth. In particular, Miers Valley appeared to harbor a limited presence of Ascomycota compared to the other two valleys, but also the highest number of Chytridiomycota OTUs (Figure 5).

The most abundant OTU (#3) was found in Alatna Valley (1875 reads), Battleship Promontory (407 reads), and Miers Valley (1 read) (Table 4). Its best match in GenBank (99% identity) was the psychrotolerant species *Cryptococcus consortionis* (Basidiomycota), which was previously observed and commonly found in Dry Valley soils [22,41]. *Cryptococcus consortionis* is characterized by the combination of amylase production and inability to utilize nitrate, cellobiose, D-galactose, *myo*-inositol, and mannitol [41]. The second most abundant OTU (#4) was also found in all three Dry Valleys (Table 4). Its best match in GenBank (93% similarity) was *Lobulomycetales* sp. AF017 (Chytridiomycota), which has been reported to occur in barren alpine soil in Peru [42]. Two other OTUs (#35 and #54) appeared to be affiliated with this genus as well.

Other abundant OTUs found in all three valleys (Table 4) were 99% similar to the species *Herpotrichiellaceae* sp. LM500 (Ascomycota) and 99.9% similar to *Exophiala equine* (Ascomycota), which was curiously reported to occur exclusively in waterborne cold-blooded animals [43]. Less abundant OTUs show similarity to fungal taxa described as Dry Valley lichen *Polysporina frigida* [44], *Meira* sp. ANTCW08-165 [45], and *Tetracladium* sp. ANTCW08-156 [45] which were previously detected in Antarctica. The genus *Cladosporium* has been reported as a dominant group by multiple studies [24,46,47] of pristine areas with little biotic influence [24,46], likely because of its prolific production of spores and high abundance in the air [24,47]. This is in contrast to our study, where *Cladosporium* species appear to be very rare (21 reads total). Notably, these fungi have been reported to survive repeated inoculations [24] and form spores, which can remain dormant for considerable periods of time [26]. It should be stressed that no conclusions can be drawn as to whether these fungi are active based on PCR amplicon pyrosequencing, as the method only detects the presence of DNA and does not indicate the viability of the organism [48,49].

3.4. Biogeography and Local Adaptation

The most important dispersal mechanisms for biomass in Antarctica have been suggested as aeolian transport [4,50,51]. If, as hypothesized previously [52], fungal species in the Dry Valleys are inactive spores that only respond to cultivation efforts and do not exhibit localized adaptations, neighboring valleys would be expected to harbor very similar fungal communities; for example, between Battleship Promontory and Alatna Valley and between Beacon and University Valley, which are located next to each other (<1 km) without any physical barrier. The tRFLP results indicated highly localized community structures, with Battleship Promontory and Alatna Valley forming statistically distinct clades (Figure 3). In addition, no fungal signals were detected in samples from University Valley while some were detected in Beacon Valley samples. Rao *et al.* previously hypothesized that the biogeography may be important for fungi in the Dry Valleys [52] and that fungal tolerance to saline conditions could confer selective advantage in high-elevation Dry Valleys [52]. Although the five most abundant OTUs reported here were found in all three samples sequenced, the relative abundances of individual OTUs were highly divergent. Since each of the sequenced samples can be considered representative of distinct diversity patterns found in the three Dry Valleys (Figure 3), the relative abundance patterns suggest that distinct fungal communities exist in each of these locations (Table 4). It should be noted that the limited spatial coverage in each Dry Valley and lack of replicates for sequencing analysis preclude definitive conclusions from being drawn, but these observations could indicate that aeolian transport plays a less important role than previous believed, or that Dry Valley fungal communities exhibit adaption to local conditions and thus are ecologically relevant.

3.5. Environmental Drivers of Fungal Distribution

Whether and how environmental factors shape fungal communities in Dry Valleys soils remains largely unexplored, but it has been suggested that both contemporary environmental conditions and historical contingencies play important roles in the distribution of fungal taxa in general [53]. It has been shown that abiotic factors play the most dominant role in extremely simplified food webs [5,11,54,55]. This makes the Dry Valleys soil ecosystem, with its extreme environmental stress, an excellent model for resolving the influence of abiotic factors on soil microbiota [19,55,56]. Miers Valley and Battleship Promontory, whose soils generally have a lower salinity, were reported to harbor greater bacterial and cyanobacterial diversity [15]. This study reveals similar trends for edaphic fungal diversity in these Dry Valleys as well as Alatna Valley; compared with Beacon Valley, University Valley, and Upper Wright Valley, where the lack of amplifiable fungal signal in extracted DNA could indicate potential limits of fungal growth and distribution. Importantly, soil C/N ratios are higher in all three coastal and lower elevation valleys, which potentially indicate higher levels of primary productivity that can in turn sustain diverse populations of heterotrophic fungi [4,16,57]. Rao *et al.* suggested that substrate availability could limit diversity [52], since Dry Valley soils with higher carbon content harbored greater species richness [22,52]. Biota-environmental stepwise (BEST) analysis of soil physicochemical properties and tRFLP results supported this view, identifying C/N ratio as the most consistent differentiator of fungal community structure, followed by As and Ca (Supplementary Table S3). Calcium can be considered as a proxy for the mineral composition of underlying soils. The influence of arsenic on fungal populations is not clear since its concentrations are very low in our samples (Supplementary Table S2). The complete/near absence of detectable fungal signal in samples from University Valley and Beacon Valley is intriguing. Compared with other valleys, Beacon Valley and University Valley have higher elevations, resulting in lower average temperature and possibly less ice melting [36]. Therefore, contrary to an earlier hypothesis [52], lower temperature and water availability, combined with lower C/N ratio and higher salinity, may create conditions in these inland Dry Valleys that restrict fungal growth while permitting bacterial presence [15]. However, given that our samples were taken within comparatively small areas (2500 m^2) on south-facing slopes, the possibility that our observations are reflective of specific geographic features of the sampling sites cannot be ruled out. South-facing slopes of the Dry Valleys are generally colder due to the lack of solar radiation input [1] and possibly more oligotrophic (compared with north-facing slopes) [16], and as such may restrict the colonization and growth of fungi.

4. Conclusions

Soil physicochemical properties among the Dry Valley sites showed distinct grouping patterns, with each valley forming its own clade. tRFLP results revealed similar grouping patterns, with significant variations in relative abundances of fungal signals between sites. Miers Valley was identified as a clear outlier by geochemical and tRFLP analyses, which were corroborated by pyrosequencing results, showing that Miers Valley harbored the highest level of fungal diversity and an unexpected abundance of Chytridiomycota. This is in contrast with the relatively low abundance of Basidiomycota, which was previously reported as the most dominant fungal phyla in the Dry Valleys. In total, nine OTUs were found in all three valleys, including the five most abundant ones, indicating that a set of core fungal species is present throughout the Dry Valleys. However, the relative abundances of these dominant OTUs are notably different among the three sites, suggesting that there is significant biogeography for Dry Valley edaphic fungi and that they likely respond and adapt to local environmental conditions. This in turn implies that much of the fungal biomass in the Dry Valleys is biological active and ecologically relevant, rather than spores whose distribution pattern is largely dictated by aeolian transport. The comparative lack of fungal signals in the inland high elevation Dry Valleys suggests that environmental conditions at those locations may represent limits of fungal growth.

Acknowledgments

This research was supported by grants from the New Zealand Foundation for Research, Science and Technology (FRST) (UOWX0710) and the United States National Science Foundation (ANT-0944556, ANT-0944560) to S. Craig Cary. Charles K. Lee and S. Craig Cary were also supported by the New Zealand Marsden Fund (UOW0802 and UOW1003) and the New Zealand Antarctic Research Institute (NZARI2013-7). We would like to thank John Longmore of Waikato DNA Sequencing Facility, Anjana Rejendram of Waikato Stable Isotope Unit, Steve Cameron of Waikato Mass Spectrometry Facility, and Roanna Richards-Babbage and Eric Bottos of Thermopile Research Unit at University of Waikato for their support and assistance.

Author Contributions

Lisa L. Dreesens carried out DNA extraction and analysis. Lisa L. Dreesens and Charles K. Lee carried out data analyses and wrote the manuscript. Charles K. Lee and S. Craig Cary designed the study and carried out field sampling. Charles K. Lee and S. Craig Cary provided funding for the study and coordinated field expeditions.

Conflict of Interest

The authors declare no conflict of interest.

References

1. Cary, S.C.; McDonald, I.R.; Barrett, J.E.; Cowan, D.A. On the rocks: The microbiology of Antarctic Dry Valley soils. *Nat. Rev. Microbiol.* **2010**, *8*, 129–138.
2. Vishniac, H.S. The microbiology of Antarctic soils. In *Antarctic Microbiology*; Friedmann, E.I., Ed.; Wiley-Liss: New York, NY, USA, 1993; pp. 297–341.
3. Horowitz, N.H.; Cameron, R.E.; Hubbard, J.S. Microbiology of the Dry Valleys of Antarctica. *Science* **1972**, *176*, 242–245.
4. Wynn-Williams, D.D. Ecological aspects of Antarctic microbiology. In *Advances in Microbial Ecology*; Marshall, K.C., Ed.; Springer US: New York, NY, USA, 1990; Volume 11, pp. 71–146.
5. Doran, P.T.; Priscu, J.C.; Lyons, W.B.; Walsh, J.E.; Fountain, A.G.; McKnight, D.M.; Moorhead, D.L.; Virginia, R.A.; Wall, D.H.; Clow, G.D.; *et al.* Antarctic climate cooling and terrestrial ecosystem response. *Nature* **2002**, *415*, 517–520.
6. Claridge, G.G.C.; Campbell, I.B. The salts in Antarctic soils, their distribution and relationship to soil processes. *Soil Sci.* **1977**, *123*, 377–384.
7. Bockheim, J.G. Properties and classification of cold desert soils from Antarctica. *Soil Sci. Soc. Am. J.* **1997**, *61*, 224–231.
8. Treonis, A.M.; Wall, D.H.; Virginia, R.A. The use of anhydrobiosis by soil nematodes in the Antarctic Dry Valleys. *Funct. Ecol.* **2000**, *14*, 460–467.
9. Vincent, W.F. *Microbial Ecosystems of Antarctica*; Cambridge University Press: Cambridge, UK, 1988; p. 59.

10. Aislabie, J.M.; Chhour, K.L.; Saul, D.J.; Miyauchi, S.; Ayton, J.; Paetzold, R.F.; Balks, M. Dominant bacteria in soils of Marble Point and Wright Valley, Victoria Land, Antarctica. *Soil Biol. Biochem.* **2006**, *38*, 3041–3056.

11. Poage, M.A.; Barrett, J.E.; Virginia, R.A.; Wall, D.H. The influence of soil geochemistry on nematode distribution, McMurdo Dry Valleys, Antarctica. *Arct. Antarct. Alp. Res.* **2008**, *40*, 119–128.

12. Priscu, J.C. *Ecosystem Dynamics in A Polar Desert: The McMurdo Dry Valleys, Antarctica*, 1st ed.; American Geophysical Union: Washington, DC, USA, 1998; Volume 72, p. 369.

13. Smith, R.C.; Prezelin, B.B.; Baker, K.S.; Bidigare, R.R.; Boucher, N.P.; Coley, T.; Karentz, D.; MacIntyre, S.; Matlick, H.A.; Menzies, D.; *et al.* Ozone depletion: Ultraviolet radiation and phytoplankton biology in Antarctic waters. *Science* **1992**, *255*, 952–959.

14. Tosi, S.; Brusoni, M.; Zucconi, L.; Vishniac, H. Response of Antarctic soil fungal assemblages to experimental warming and reduction of UV radiation. *Polar Biol.* **2005**, *28*, 470–482.

15. Lee, C.K.; Barbier, B.A.; Bottos, E.M.; McDonald, I.R.; Cary, S.C. The inter-valley soil comparative survey: The ecology of Dry Valley edaphic microbial communities. *ISME J.* **2012**, *6*, 1046–1057.

16. Wood, S.A.; Rueckert, A.; Cowan, D.A.; Cary, S.C. Sources of edaphic cyanobacterial diversity in the Dry Valleys of eastern Antarctica. *ISME J.* **2008**, *2*, 308–320.

17. Niederberger, T.D.; McDonald, I.R.; Hacker, A.L.; Soo, R.M.; Barrett, J.E.; Wall, D.H.; Cary, S.C. Microbial community composition in soils of Northern Victoria Land, Antarctica. *Environ. Microbiol.* **2008**, *10*, 1713–1724.

18. Pointing, S.B.; Chan, Y.; Lacap, D.C.; Lau, M.C.; Jurgens, J.A.; Farrell, R.L. Highly specialized microbial diversity in hyper-arid polar desert. *Proc. Natl. Acad. Sci. USA* **2009**, *106*, 19964–19969.

19. Wall, D.H.; Virginia, R.A. Controls on soil biodiversity: Insights from extreme environments. *Appl. Soil. Ecol.* **1999**, *13*, 137–150.

20. Connell, L.; Redman, R.; Craig, S.; Scorzetti, G.; Iszard, M.; Rodriguez, R. Diversity of soil yeasts isolated from South Victoria Land, Antarctica. *Microb. Ecol.* **2008**, *56*, 448–459.

21. Arenz, B.E.; Blanchette, R.A. Distribution and abundance of soil fungi in Antarctica at sites on the Peninsula, Ross Sea Region and McMurdo Dry Valleys. *Soil Biol. Biochem.* **2011**, *43*, 308–315.

22. Arenz, B.E.; Held, B.W.; Jurgens, J.A.; Farrell, R.L.; Blanchette, R.A. Fungal diversity in soils and historic wood from the Ross Sea region of Antarctica. *Soil Biol. Biochem.* **2006**, *38*, 3057–3064.

23. Connell, L.; Redman, R.; Craig, S.; Rodriguez, R. Distribution and abundance of fungi in the soils of Taylor Valley, Antarctica. *Soil Biol. Biochem.* **2006**, *38*, 3083–3094.

24. Farrell, R.L.; Arenz, B.E.; Duncan, S.M.; Held, B.W.; Jurgens, J.A.; Blanchette, R.A. Introduced and indigenous fungi of the Ross Island historic huts and pristine areas of Antarctica. *Polar Biol.* **2011**, *34*, 1669–1677.

25. Blanchette, R.A.; Held, B.W.; Arenz, B.E.; Jurgens, J.A.; Baltes, N.J.; Duncan, S.M.; Farrell, R.L. An Antarctic hot spot for fungi at Shackleton's historic hut on Cape Royds. *Microb. Ecol.* **2010**, *60*, 29–38.

26. Duncan, S.M.; Farrell, R.L.; Jordan, N.; Jurgens, J.A.; Blanchette, R.A. Monitoring and identification of airborne fungi at historic locations on Ross Island, Antarctica. *Polar Sci.* **2010**, *4*, 275–283.

27. Selbmann, L.; de Hoog, G.S.; Mazzaglia, A.; Friedmann, E.I.; Onofri, S. Fungi at the edge of life: Cryptoendolithic black fungi from Antarctic desert. *Stud. Mycol.* **2005**, *51*, 1–32.

28. Coyne, K.J.; Hutchins, D.A.; Hare, C.E.; Cary, S.C. Assessing temporal and spatial variability in *Pfiesteria piscicida* distributions using molecular probing techniques. *Aquat. Microb. Ecol.* **2001**, *24*, 275–285.

29. Rueckert, A.; Morgan, H.W. Removal of contaminating DNA from polymerase chain reaction using ethidium monoazide. *J. Microbiol. Methods* **2007**, *68*, 596–600.

30. Quince, C.; Lanzen, A.; Davenport, R.J.; Turnbaugh, P.J. Removing noise from pyrosequenced amplicons. *BMC Bioinform.* **2011**, *12*, 1–18.

31. Sun, Y.; Cai, Y.; Liu, L.; Yu, F.; Farrell, M.L.; McKendree, W.; Farmerie, W. Esprit: Estimating species richness using large collections of 16s rRNA pyrosequences. *Nucleic Acids Res.* **2009**, *37*, e76.

32. Schloss, P.D.; Westcott, S.L.; Ryabin, T.; Hall, J.R.; Hartmann, M.; Hollister, E.B.; Lesniewski, R.A.; Oakley, B.B.; Parks, D.H.; Robinson, C.J.; *et al.* Introducing mothur: Open-source, platform-independent, community-supported software for describing and comparing microbial communities. *Appl. Environ. Microbiol.* **2009**, *75*, 7537–7541.

33. Kennedy, A.D. Water as a limiting factor in the Antarctic terrestrial environment: A biogeographical synthesis. *Arct. Alp. Res.* **1993**, *25*, 308–315.

34. McKnight, D.M.; Tate, C.M.; Andrews, E.D.; Niyogi, D.K.; Cozzetto, K.; Welch, K.; Lyons, W.B.; Capone, D.G. Reactivation of a cryptobiotic stream ecosystem in the McMurdo Dry Valleys, Antarctica: A long-term geomorphological experiment. *Geomorphology* **2007**, *89*, 186–204.

35. Katurji, M.; Zawar-Reza, P.; Zhong, S. Surface layer response to topographic solar shading in Antarctica's Dry Valleys. *J. Geophys. Res. Atmos.* **2013**, *118*, 12332–12344.

36. Pollard, W.H.; Lacelle, D.; Davila, A.F.; Andersen, D.; McKay, C.P.; Marinova, M.; Heldmann, J. Ground ice conditions in University Valley, McMurdo Dry Valleys, Antarctica. In Proceedings of the Tenth International Conference on Permafrost (TICOP), Salekhard, Russia, 25–29 June 2012; Volume 1, pp. 305–310.

37. Lamsal, K.; Paudyal, G.N.; Saeed, M. Model for assessing impact of salinity on soil water availability and crop yield. *Agric. Water Manag.* **1999**, *41*, 57–70.

38. Campbell, I.B.; Claridge, G.G.C. *Antarctica: Soils, Weathering Processes and Environment*; Elsevier Science Publishers: Amsterdam, The Netherlands, 1987; Volume 16, p. 406.

39. Duncan, S.M.; Farrell, R.L.; Thwaites, J.M.; Held, B.W.; Arenz, B.E.; Jurgens, J.A.; Blanchette, R.A. Endoglucanase-producing fungi isolated from Cape Evans historic expedition hut on Ross Island, Antarctica. *Environ. Microbiol.* **2006**, *8*, 1212–1219.

40. Lawley, B.; Ripley, S.; Bridge, P.; Convey, P. Molecular analysis of geographic patterns of eukaryotic diversity in Antarctic soils. *Appl. Environ. Microbiol.* **2004**, *70*, 5963–5972.

41. Vishniac, H.S. *Cryptococcus socialis* sp. nov. and *Cryptococcus consortionis* sp. nov., Antarctic Basidioblastomycetes. *Int. J. Syst. Bacteriol.* **1985**, *35*, 119–122.

42. Simmons, D.R.; James, T.Y.; Meyer, A.F.; Longcore, J.E. *Lobulomycetales*, a new order in the *Chytridiomycota*. *Mycol. Res.* **2009**, *113*, 450–460.

43. De Hoog, G.S.; Vicente, V.A.; Najafzadeh, M.J.; Harrak, M.J.; Badali, H.; Seyedmousavi, S. Waterborne *Exophiala* species causing disease in cold-blooded animals. *Persoonia* **2011**, *27*, 46–72.

44. Kantvilas, G.; Seppelt, R.D. *Polysporina frigida* sp. Nov. from Antarctica. *Lichenologist* **2006**, *38*, 109–113.

45. Slemmons, C.; Johnson, G.; Connell, L.B. Application of an automated ribosomal intergenic spacer analysis database for identification of cultured Antarctic fungi. *Antarct. Sci.* **2013**, *25*, 44–50.

46. Kerry, E. Effects of temperature on growth rates of fungi from Subantarctic Macquarie Island and Casey, Antarctica. *Polar Biol.* **1990**, *10*, 293–299.

47. Marshall, W.A. Seasonality in Antarctic airborne fungal spores. *Appl. Environ. Microbiol.* **1997**, *63*, 2240–2245.

48. Adams, B.J.; Bargett, R.D.; Ayres, E.; Wall, D.H.; Aislabie, J.; Bamforth, S.; Bargagli, R.; Cary, C.; Cavacini, P.; Conell, L.; *et al.* Diversity and distribution of Victoria Land biota. *Soil. Biol. Biochem.* **2006**, *38*, 3003–3018.

49. Fletcher, L.D.; Kerry, E.J.; Weste, G.M. Microfungi of MacRobertson and Enderby lands, Antarctica. *Polar Biol.* **1985**, *4*, 81–88.

50. Vincent, W.F. Evolutionary origins of Antarctic microbiota: Invasion, selection and endemism. *Antarct. Sci.* **2000**, *12*, 374–385.

51. Marshall, W.A. Biological particles over Antarctica. *Nature* **1996**, *383*, 680.

52. Rao, S.; Chan, Y.; Lacap, D.C.; Hyde, K.D.; Pointing, S.B.; Farrell, R.L. Low-diversity fungal assemblage in an Antarctic Dry Valleys soil. *Polar Biol.* **2012**, *35*, 567–574.

53. Green, J.L.; Holmes, A.J.; Westoby, M.; Oliver, I.; Briscoe, D.; Dangerfield, M.; Gillings, M.; Beattie, A.J. Spatial scaling of microbial eukaryote diversity. *Nature* **2004**, *432*, 747–750.

54. Convey, P. The influence of environmental characteristics on life history attributes of Antarctic terrestrial biota. *Biol. Rev.* **1996**, *71*, 191–225.

55. Hogg, I.D.; Cary, C.S.; Convey, P.; Newsham, K.K.; O'Donnell, A.G.; Adams, B.J.; Aislabie, J.; Frati, F.; Stevens, M.I.; Wall, D.H. Biotic interactions in Antarctic terrestrial ecosystems: Are they a factor? *Soil. Biol. Biochem.* **2006**, *38*, 3035–3040.

56. Hopkins, D.W.; Sparrow, A.D.; Novis, P.M.; Gregorich, E.G.; Elberling, B.; Greenfield, L.G. Controls on the distribution of productivity and organic resources in Antarctic Dry Valley soils. *Proc. R. Soc. Sci. B* **2006**, *273*, 2687–2695.

57. Barrett, J.E.; Virginia, R.A.; Wall, D.H.; Cary, S.C.; Adams, B.J.; Hacker, A.L. Co-variation in soil biodiversity and biogeochemistry in Northern and Southern Victoria Land, Antarctica. *Antarct. Sci.* **2006**, *18*, 535–548.

Recent Advances and Future Perspectives in Microbial Phototrophy in Antarctic Sea Ice

Eileen Y. Koh, Andrew R. Martin, Andrew McMinn and Ken G. Ryan

Abstract: Bacteria that utilize sunlight to supplement metabolic activity are now being described in a range of ecosystems. While it is likely that phototrophy provides an important competitive advantage, the contribution that these microorganisms make to the bioenergetics of polar marine ecosystems is unknown. In this minireview, we discuss recent advances in our understanding of phototrophic bacteria and highlight the need for future research.

Reprinted from *Biology*. Cite as: Koh, E.Y.; Martin, A.R.; McMinn, A.; Ryan, K.G. Recent Advances and Future Perspectives in Microbial Phototrophy in Antarctic Sea Ice. *Biology* **2012**, *1*, 542-556.

1. Introduction

Microorganisms have been fundamentally important to the history and function of life on Earth. They have played a central role in the climatic, geological, and biological evolution of the planet [1]. They are found in every conceivable ecological niche, from the tropics to the poles, from underground mines and oil fields to the stratosphere and mountain ranges, from deserts to the Dead Sea and from hot springs to underwater hydrothermal vents [2–5]. Microbes dominate the flux of energy and biologically important chemical elements in the world's oceans and, as a result, are estimated to be five to ten times the mass of all multicellular marine organisms [6]. Bacteria harbor a potential reservoir of useful genes for medicine and biotechnology, and unraveling their complex taxonomic diversity is considered the key to understanding the process of evolution [7,8].

Sea ice is one of the most seasonally dynamic ecosystems on Earth. An important driver of the global climate system, annual sea ice at polar latitudes influences both physical and biological processes; particularly in modulating the exchange of heat and moisture between the atmosphere and ocean, and restricting the penetration of solar radiation. Importantly, sea ice also provides a stable platform for the colonization and growth of marine microbes [9,10]. Although a range of microbial taxa are initially scavenged from the water column during ice formation, only some are able to adapt to the physicochemical variability that characterizes the brine inclusions and interstices of the ice matrix. The most conspicuous ice-bound organisms are microalgae and research efforts have historically focused on the composition, physiology, and ecology of the diatoms that dominate sea ice assemblages [11–15]. Sea ice algae contribute between 10%–28% of the total primary production in ice-covered regions of the Southern Ocean [10,16] and over 90% of this biogenic carbon is produced within first-year ice and approximately 60% during the austral spring (November-December) when the algal cells typically discolor the bottom 1–20 cm of the ice [16] (Figure 1). Microalgae provide a crucial source of winter nutrition for juvenile zooplankton such as the Antarctic krill *Euphausia superba* [17,18], and may provide inocula for bloom events at the receding ice edge in the austral summer [11,16,19,20].

Figure 1. Cross-section of sea ice. A distinct brown coloration is present at the bottom 20 cm of a 1 m diameter section of sea ice. This is due to the high concentration of bacteria and microalgae within the sea ice.

While bacteria are now recognized as a major biological component of the oceanic carbon cycle and ecosystem structure [21,22], an understanding of the phylogenetic diversity and functional capabilities of ice-associated bacteria remains fragmentary [23]. Evidence that bacteria actively grow within the ice dates back to only the 1980's when Sullivan and Palmisano [24] observed large and morphologically distinct bacteria undergoing cell division in fast-ice within McMurdo Sound, Antarctica. This initial observation indicated an active heterotrophic community, and the subsequent microautoradiographic uptake of radiolabeled compounds such as ^{14}C-L-serine, ^{3}H-serine, ^{3}H-glucose and ^{3}H-thymidine confirmed community-level activity in the form of DNA synthesis [24,25]. More recent single-cell analyses, including the use of tetrazolium chloride (CTC) and fluorescence *in situ* hybridization (FISH), have shown that ~80% of the bacteria present in the bottom of Antarctic sea ice have a probe-positive cellular rRNA content and >30% of the cells have an electron transport system that is capable of reducing CTC [26,27]. Most of these cells appear to be heterotrophic bacteria, which either live freely or attached to microalgae or detritus [28,29]. Molecular-based surveys of SSU rRNA gene diversity typically reveal psychrophilic and halotolerant members of the Proteobacteria, Bacteroidetes (previously known as the Cytophaga-Flavobacteria-Bacteroides (CFB) cluster) and Gram-positive bacteria [23,28,30].

Following a decade of seminal research conducted within McMurdo Sound, Antarctica, Sullivan [25] suggested that sea ice bacteria might play an important role in secondary microbial production mediated through the microbial loop and remineralisation and recycling of ice-associated organic matter (Figure 2). Sullivan [25] also postulated that these bacteria maintain a balance of oxygen concentration in the ice microenvironment through their respiration and may be involved in ice nucleation and early stages of sea ice formation although these ideas remain largely unsubstantiated.

Figure 2. Sea ice food web and the microbial loop. The microbial loop re-drawn and abridged from Azam *et al.* (1983) and Fenchel (2008). Only the bacteria discussed in this review are presented; the other bacteria are grouped as heterotrophs. AAnP = aerobic anaerobic phototroph, DOC = dissolved organic carbon, DOM = dissolved organic matter, POC = particulate organic carbon, PR = proteorhodopsins.

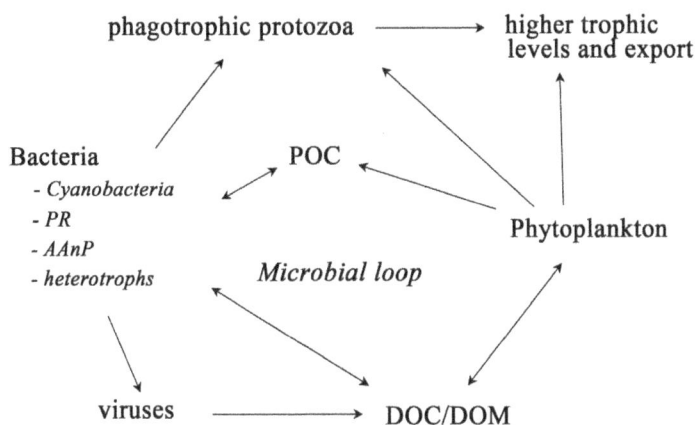

2. Bacteria with Light-Harvesting Capabilities

The energy to support life in the sea is ultimately derived from phototrophy in the euphotic zone [31]. The most significant contribution from prokaryotic life forms is from cyanobacteria, which utilize chlorophyll-based phototrophy and contribute 30% of all globally fixed carbon [32]. However, in recent years, non-cultivation-based studies of bacteria have led to the discovery of novel genes, proteins and phototrophic mechanisms that are rapidly gaining scientific interest [33–37]. In particular, widespread reports of bacteriochlorophyll (bchl) and proteorhodopsin (PR) in planktonic marine prokaryotes are challenging the assumption that chl-*a* is the only light-capturing pigment of ecological importance. It remains to be seen whether these metabolic pathways will require a significant revision of oceanic energy budgets [22,38], but alternative light-based metabolic strategies are now being described in aquatic ecosystems that range from the deep-sea biosphere to high-altitude glaciers [39,40].

2.1. Cyanobacteria

Cyanobacteria colonize a variety of polar terrestrial ecospheres including rocks, glaciers, ice shelves, streams, ponds and lakes [41–45]. Phototrophs in these environments are generally psychrotolerant and exhibit an assortment of cold-protection mechanisms and slow growth rates to endure freeze-thaw cycles. The intracellular accumulation of salts to sustain osmotic balance, variation in DNA repair mechanisms and the use of photo-complexes are additional strategies that terrestrial cyanobacteria employ in extreme cold environments [44,46,47]. Picocyanobacteria such as *Synechococcus* and *Prochlorococcus* are the most abundant phototropic cells in the World's oceans [48]. Despite this significant contribution to primary production, these cells are small, 0.5–1.5 μm and, in the case of *Prochlorococcus*, remained undetected until 1986 when they were discovered using flow cytometry [49]. Importantly, the abundance of marine cyanobacteria decreases rapidly south of latitude 40° [50–52] and this has been attributed to eco-physiological factors such as temperature, salinity and nutrient requirements [44].

Considering their prevalence in cold terrestrial environments, the apparent absence of cyanobacteria within sea ice is, however, unexpected. Interestingly, *Synechococcus* was detected from coastal waters off East Antarctica in 1989 using microscopy and pigment chemistry [53] and a decade later, cyanobacterial-like pigments (phycoerythrin and phycocyanin) were detected for the first time within the ice matrix using flow cytometry [54]. Pigment-based confirmation is questionable however as phycoerythrin and phycocyanin are also present in other algae including Cryptophytes which are common during Antarctic coastal blooms [54]. To potentially validate these earlier findings, a multi-method molecular analysis was recently carried out on fast-ice cores extracted from sites spanning 300 km in the Ross Sea region of Antarctica [55]. Clone libraries were constructed from the 16S rDNA gene, the internal transcribed sequence (ITS) region and the cyanobacterial core RNA polymerase (*rpoC*). Analysis of all sections of extracted ice did not reveal the presence of *Synechococcus* sp., *Prochlorococcus* sp., or any other marine cyanobacteria-related species. Additional screening was carried out using ligation detection reaction-based microarray, which can detect as little as 1 *f*mol of DNA [56]. Data from the ITS and microarray analysis of these sea ice samples showed close affiliation to the freshwater cyanobacteria *Phormidium* sp. and *Cylindrospermopsis* sp., respectively [55], but the closest Antarctic relative was an uncultured cyanobacteria clone from the nearby meromictic Lake Fryxell [42]. Aerobiology studies conducted in the Antarctic [57,58] and the Arctic [59] suggested that much of the biological material present in the air originates locally. Harding *et al.* [59] observed that >47% of the operational taxonomic units (OTUs) in Arctic snow samples were from previously reported local cyanobacteria [45,60]. Given that Terra Nova Bay is situated in a katabatic wind cross-zone [61], it is likely that the cyanobacterial propagules identified by Koh *et al.* [55] were wind-transported from nearby freshwater ponds or terrestrial soil and incorporated into the ice during seasonal formation. Despite earlier anecdotal findings, molecular-based evidence now confirms that cyanobacteria do not play a significant role in sea ice ecosystem dynamics.

2.2. Bacteriochlorophyll

Phototropic metabolism is a feature of four other eubacterial phyla (*i.e.*, Proteobacteria, Chlorobi, Chloroflexi and Firmicutes). Unlike cyanobacteria, these phototrophs utilize the most ancient form of photosynthesis: anoxygenic photosynthesis [62]. This pathway is important for nitrogen-fixation, and cells with bacteriochlorophyll (bchl) also play an important role in the microbial loop [63,64]. The presence of highly diverse anoxygenic phototrophic bacterial communities in marine environments now suggests that non-chlorophyll-*a* phototrophy may be a more common life history strategy than previously realized. For example, the Proteobacteria contain the largest group of anoxyphototrophs [65], which were thought to be strictly anaerobic until three decades ago when the first aerobic representative was identified [66]. In both aerobic and anaerobic taxa, bchl-*a* is the primary light-harvesting pigment and absorbs red light at 770 to 880 nm and blue light at ~385nm [67–69]. This provides a useful contrast to the chlorophyll-*a* present in cyanobacteria and algae, which absorbs at 430 nm and 665 nm (Figure 3).

Figure 3. Schematic diagram of light pigments/proteins of sea ice phototrophic bacteria. Bchl = bacteriochlorophyll; Chl-*a* = Chlorophyll-*a*; BPR = Blue proteorhodopsin; GPR = Green proteorhodopsin. Diagram not drawn to scale.

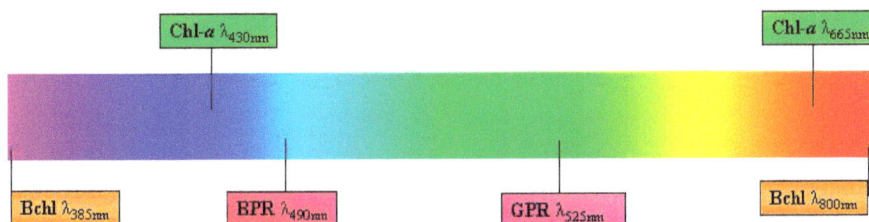

The aerobic anoxygenic phototrophic (AAnP) bacteria are a diverse group of prokaryotes with respect to their functionality, physiology, and morphology [69]. These are obligate aerobes with unusually high concentrations of carotenoids, low cellular contents of bchl-*a* and a distinct lack of the light-harvesting complex II [38,39]. In the anaerobic phototrophic bacteria (AnPB), the *puf* operon coding for the bchl is repressed by both oxygen and high light [70]. However, in the AAnP, the expression of the *puf* operon is not limited by oxygen, but it is still repressed by strong light [37,71,72]. Despite these physiological differences, both AnPB and AAnPs have a similar photosynthetic apparatus and similar electron carriers and structural polypeptides [73]. The close correlation between AAnPs and oxygenic phototrophs in the euphotic zone may indicate that these cells contribute to a light-controlled component of the microbial redox cycle [38]. Several molecular studies based on the genes of the *puf* operon have been carried out in tropical, temperate and polar marine environments [74–79] and these organisms have been estimated to account for up to 10% of the energy production in the upper layers of the water column [38,80] in most temperate and tropical oceans.

There is some evidence to suggest that AAnP bacteria are absent at high latitudes [81] and are genetically distinct from their freshwater counterparts [82,83]. However, positive *puf*M [64,73] clonal sequences were detected from extracted DNA and messenger RNA transcripts from the lower sections of Antarctic annual fast-ice as well as the underlying water column [84]. All clones grouped with the cultured α-Proteobacteria [39,74]. No β- and δ-Proteobacteria AAnPs were detected in the sea ice, matching the observations of Karr *et al.* [82] at Lake Fryxell. In fact, all the sea ice and seawater clones were likely α-Proteobacteria *Roseobacter*-clade affiliated [84], which could constitute ~20% of the Southern Ocean bacterial community [85,86], given that *Roseobacter denitrificans* is able to illicit specific defence systems against photo-oxidative stress [87]. More importantly, their presence in RNA extracts indicates that bacteria within the sea ice are actively expressing the gene for bchl synthesis.

AAnPs may constitute only ~0.05% of the prokaryotic abundance in the Western Antarctic waters [80], however the ease with which AAnPs were found within sea ice suggests that their relative proportion may be higher in ice-associated microbial communities. Results obtained by quantitative PCR suggest that Bchl OTUs may comprise up to 10% of the sea ice bacterial community [88], although further work is clearly needed to ascertain the ecological significance of this metabolic pathway.

2.3. Proteorhodopsin

The discovery of phototrophic energy generated via proteorhodopsin (PR) was a major finding in microbial ecology [89]. PRs are retinal binding bacterial integral membrane proteins that belong to the microbial rhodopsin super-family of proteins and function as light-driven proton pumps [89,90]. Unlike Bchl, PR cells do not generate cellular reducing power through NADPH, however ATP is produced upon light stimulation without the evolution of oxygen or fixation of inorganic carbon. Since the first reported PR sequence was obtained in 2000 [89], many other PR-bearing bacteria have been identified in environments ranging from freshwater lakes to the deep marine biosphere [91–96]. PR genes appear to be abundant in the genomes of oceanic bacteria [95], accounting for 13% of the prokaryotic community in the Mediterranean Sea and Red Sea and 70% in the Sargasso Sea [94,95,97,98]. Importantly, *in vitro* studies have demonstrated proton pumping and increased growth rates of PR-bearing bacteria under illuminated conditions [99–102]. Recently, PR in *Candidatus* Pelagibacter ubique was reported to play a critical role in a cellular response that maintains cell function during periods of carbon starvation [102]. These observations suggest that harvesting light energy via PR may be important in marine environments [103], but again the ecological significance of this metabolic pathway is currently unknown.

Sea ice bacteria that express the PR gene were described for the first time in 2010 [104]. PR-bearing representatives from the classes α-Proteobacteria, γ-Proteobacteria and Flavobacteria were present throughout the fast-ice in the Ross Sea region of Antarctica. Complementary DNA (cDNA) generated from RNA samples suggested that PR bacteria were metabolically active at the time of sampling. The bulk of the positive cDNA samples were collected from the middle and bottom part of the ice matrix, which possibly indicates that PR bacteria favor the lower light intensity and relatively stable temperatures found in the bottom half of the ice. Essentially, as light penetrates deeper into the ice matrix, the more energetic blue light predominates [105]. A stratified distribution of different forms of PR-bacteria in marine waters has been observed previously [106,107], and this has been attributed to a single-residue switch mechanism whereby the presence of leucine or glutamine at amino acid position 105 determines whether the protein absorbs in the green or blue wavelength [106]. Koh *et al.* [104] found both blue-absorbing (BPR) and green-absorbing (GPR) forms, but BPR were found primarily in the middle of the ice where red and green wavelengths of the solar spectrum are relatively low [105]. Conversely, GPR appear to be distributed throughout the ice, but their highest concentrations were at the ice/water interface [104], where, due to the presence of eukaryotic chl-*a*, the only available light is green.

3. Future Research and the Significance of Light-Harvesting Pigments for Antarctic Sea Ice

Research-to-date has confirmed that some of the bacteria present in Antarctic sea ice are capable of phototrophic metabolism, most likely as a supplement to an otherwise heterotrophic lifestyle. A mechanistic understanding of the diversity, ecophysiology, and functionality of marine photoheterotrophs is therefore a worthy goal, but one that is extremely challenging [108].

Logistic and weather constraints are the primary reasons that polar studies are conducted during the summer months. As a result, insight into ice-associated light-harvesting bacteria has thus far come from cores extracted during the austral summer. In the future, it will be particularly important to contrast the light-driven energy flux with the metabolic processes and activity level that take place during the dark polar winter. Only a handful of over-winter studies have been reported in the more accessible Arctic [78,109,110], however microorganisms are more active during summer months compared with winter.

Next generation pyrosequencing [111,112], microfluidics [113] and microarray analysis [114–116] are rapidly changing the way microbial communities are studied. These high-throughput methods could be employed to elucidate more phototrophic bacteria from Antarctic sea ice and accurately determine their *in situ* distribution and abundance. Techniques such as catalyzed reporter deposition (CARD)-FISH [117] and quantitative PCR [80,118] will enable functional gene expression to be quantified at the single-cell level of resolution. In addition, metatranscriptomics and proteomics would provide a valuable tool with which to link *in situ* expression dynamics with environmental stress [99,119,120]. Coupled with the chromatin immune-precipitation (ChIP) procedure, it is now possible to characterize both the genome-wide location and function of novel energy-binding proteins [120–122].

4. Concluding Remarks

Sea ice represents one of the most ephemeral habitats on Earth and the ice-associated microbial communities are integral to the energy base of the Southern Ocean ecosystem. The specific physiological roles and adaptive strategies of phototrophic bacteria within this ecosystem have yet to be elucidated; however, the future looks promising given the expanding range of technologies that may be used to explore the bioenergetics of light-harvesting pathways. There is also a growing need to quantify the resilience of sea ice microbes to increased environmental stress and to provide a real-time biological response to climate change. Considering the variety of genetic, physiological and environmental contexts in which light-harvesting bacteria are found, the diversity observed to date may reflect only a subset of the organisms present and more light-dependent adaptive strategies are likely to exist in the microbial world. The combined sequencing of cultivated and uncultivated organisms will undoubtedly reveal more microbial groups with known, or even novel, photosynthetic abilities.

Acknowledgments

This work was supported by grant VICX0706 from the New Zealand Foundation of Research Science and Technology and Victoria University of Wellington Faculty Strategic Research Grants 103167/2585 and 96429/2661.

References

1. Xu, J. Microbial ecology in the age of genomics and metagenomics: Concepts, tools, and recent advances. *Mol. Ecol.* **2006**, *15*, 1713–1731.
2. Junge, K.; Imhoff, F.; Staley, T.; Deming, J.W. Phylogenetic diversity of numerically important Arctic sea-ice bacteria cultured at subzero temperature. *Microb. Ecol.* **2002**, *43*, 315–328.
3. Nagy, M.L.; Pérez, A.; Garcia-Pichel, F. The prokaryotic diversity of biological soil crusts in the Sonoran Desert (Organ Pipe Cactus National Monument, AZ). *FEMS Microb. Ecol.* **2005**, *54*, 233–245.
4. McCliment, E.A.; Voglesonger, K.M.; O'Day, P.A.; Dunn, E.E.; Holloway, J.R.; Cary, S.C. Colonization of nascent, deep-sea hydrothermal vents by a novel Archaeal and Nanoarchaeal assemblage. *Environ. Microbiol.* **2006**, *8*, 114–125.
5. Soo, R.M.; Wood, S.A.; Grzymski, J.J.; McDonald, I.R.; Cary, S.C. Microbial biodiversity of thermophilic communities in hot mineral soils of Tramway Ridge, Mount Erebus, Antarctica. *Environ. Microbiol.* **2009**, *11*, 715–728.
6. Pomeroy, L.R.; Williams, P.J.I.; Azam, F.; Hobbie, J.E. The microbial loop. *Oceanography* **2007**, *20*, 28–33.
7. Pace, N.R. A molecular view of microbial diversity and the biosphere. *Science* **1997**, *276*, 734–740.
8. Pedrós-Alió, C. Marine microbial diversity: Can it be determined? *Trends Microbiol.* **2006**, *14*, 257–263.
9. Delille, D. Marine bacterioplankton at the Weddell Sea ice edge, distribution of psychrophilic and psychrotrophic populations. *Polar Biol.* **1992**, *12*, 205–210.
10. Legendre, L.; Ackley, S.F.; Dieckmann, G.S.; Gulliksen, B.; Horner, R.; Hoshiai, T.; Melnikov, I.A.; Reeburgh, W.S.; Spindler, M.; Sullivan, C.W. Ecology of sea ice biota. 2. Global significance. *Polar Biol.* **1992**, *12*, 429–444.
11. Kottmeier, S.T.; Sullivan, C.W. Sea ice microbial communities (SIMCO) 9. Effects of temperature and salinity on rates of metabolism and growth of autotrophs and heterotrophs. *Polar Biol.* **1988**, *8*, 293–304.
12. McMinn, A.; Ashworth, C.; Ryan, K.G. *In situ* net primary productivity of an Antarctic fast ice bottom algal community. *Aquat. Microb. Ecol.* **2000**, *21*, 177–185.
13. Ryan, K.G.; McMinn, A.; Mitchell, K.A.; Trenerry, L. Mycosporine-like amino acids in Antarctic sea ice algae, and their response to UVB radiation. *Z. Naturforsch. C* **2002**, *57*, 1–7.

14. Ryan, K.G.; Cowie, R.O.M.; Liggins, E.; McNaughtan, D.; Martin, A.; Davy, S.K. The short-term effect of irradiance on the photosynthetic properties of Antarctic fast-ice microalgal communities. *J. Phycol.* **2009**, *45*, 1290–1298.

15. Ryan, K.G.; Tay, M.L.; Martin, A.; McMinn, A.; Davy, S.K. Chlorophyll fluorescence imaging analysis of the responses of Antarctic bottom-ice algae to light and salinity during melting. *J. Exp. Mar. Biol. Ecol.* **2011**, *399*, 156–161.

16. Arrigo, K.R.; Thomas, D.N. Large scale importance of sea ice biology in the Southern Ocean. *Antarct. Sci.* **2004**, *16*, 471–486.

17. Daly, K.L. Overwintering development, growth, and feeding of larval *Euphausia superba* in the Antarctic marginal ice zone. *Limnol. Oceanogr.* **1990**, *35*, 1564–1576.

18. Kottmeier, S.T.; Sullivan, C.W. Bacterial biomass and production in pack ice of Antarctic marginal ice edge zones. *Deep-Sea Res.* **1990**, *37*, 1311–1330.

19. Giesenhagen, H.C.; Detmer, A.E.; de Wall, J.; Weber, A.; Gradinger, R.R.; Jochem, F.J. How are Antarctic planktonic microbial food webs and algal blooms affected by melting of sea ice? Microcosm simulations. *Aquat. Microb. Ecol.* **1999**, *20*, 183–201.

20. Lizotte, M.P. The contribution of sea ice algae to Antarctic marine primary production. *Am. Zool.* **2001**, *41*, 57–73.

21. Delille, D.; Rosiers, C. Seasonal changes of Antarctic marine bacterioplankton and sea ice bacterial assemblages. *Polar Biol.* **1996**, *16*, 27–34.

22. Azam, F.; Worden, A.Z. Microbes, molecules, and marine ecosystems. *Science* **2004**, *303*, 1622–1624.

23. Murray, A.E.; Grzymski, J.J. Diversity and genomics of Antarctic marine micro-organisms. *Phil. Trans. R. Soc. B* **2007**, *362*, 2259–2271.

24. Sullivan, C.W.; Palmisano, A.C. Sea ice microbial communities: distribution, abundance, and diversity of ice bacteria in McMurdo Sound, Antarctica, in 1980. *Appl. Environ. Microbiol.* **1984**, *47*, 788–795.

25. Sullivan, C.W. Reciprocal interactions of the organisms and their environment. In *Sea Ice Biota*; Horner, R., Ed.; Boca Raton Chemical Rubber Company: Boca Raton, FL, USA, 1985; pp. 160–171.

26. Martin, A.; Hall, J.A.; O'Toole, R.; Davy, S.K.; Ryan, K.G. High single-cell metabolic activity in Antarctic sea ice bacteria. *Aquat. Microb. Ecol.* **2008**, *52*, 25–31.

27. Martin, A.; Hall, J.A.; Ryan, K.G. Low salinity and high-level UV-B radiation reduce single-cell activity in Antarctic sea ice bacteria. *Appl. Environ. Microbiol.* **2009**, *75*, 7570–7573.

28. Brown, M.V.; Bowman, J.P. A molecular phylogenetic survey of sea-ice microbial communities. *FEMS Microbiol. Ecol.* **2001**, *35*, 267–275.

29. Junge, K.; Krembs, C.; Deming, J.W.; Stierle, A.; Eicken, H. A microscopic approach to investigate bacteria under in situ conditions in sea-ice samples. *Ann. Glaciol.* **2001**, *33*, 304–310.

30. Gentile, G.; Giuliano, L.; D'Auria, G.; Smedile, F.; Azzaro, M.; de Domenico, M.; Yakimov, M.M. Study of bacterial communities in Antarctic coastal waters by a combination of 16S rRNA and 16S rDNA sequencing. *Environ. Microbiol.* **2006**, *8*, 2150–2161.

31. Karl, D.M. Microbial oceanography: Paradigms, processes and promise. *Nat. Rev. Microbiol.* **2007**, *5*, 759–769.

32. Kumari, N.; Srivastava, A.K.; Bhargava, P.; Rai, L.C. Molecular approaches towards assessment of cyanobacterial biodiversity. *Afr. J. Biotechnol.* **2009**, *8*, 4284–4298.

33. Madigan, M.T. Anoxygenic phototrophic bacteria from extreme environments. *Photosynth. Res.* **2003**, *76*, 157–171.

34. Lanyi, J.K. Bacteriorhodopsin. *Annu. Rev. Physiol.* **2004**, *66*, 665–688.

35. Bryant, D.A.; Frigaard, N.-U. Prokaryotic photosynthesis and phototrophy illuminated. *Trends Microbiol.* **2006**, *14*, 488–496.

36. Fuhrman, J.A.; Schwalbach, M.S.; Stingl, U. Proteorhodopsins: An array of physiological roles? *Nat. Rev. Microbiol.* **2008**, *6*, 488–494.

37. Yurkov, V.; Csotonyi, J.T. New light on aerobic anoxygenic phototrophs. In *The Purple Phototrophic Bacteria*; Hunter, C.N., Daldal, F., Thurnauer, M.C., Beatty, J.T., Eds.; Springer Science: Dordrecht, The Netherlands, 2009; pp. 31–55.

38. Kolber, Z.S.; Gerald, F.; Lang, A.S.; Beatty, J.T.; Blankenship, R.E.; VanDover, C.L.; Vetriani, C.; Koblizek, M.; Rathgeber, C.; Falkowski, P.G. Contribution of aerobic photoheterotrophic bacteria to the carbon cycle in the ocean. *Science* **2001**, *292*, 2492–2495.

39. Rathgeber, C.; Beattyc, J.T.; Yurkov, V. Aerobic phototrophic bacteria: New evidence for the diversity, ecological importance and applied potential of this previously overlooked group. *Photosynth. Res.* **2004**, *81*, 113–128.

40. Jiang, H.; Dong, H.; Yu, B.; Lv, G.; Deng, S.; Wu, Y.; Dai, M.; Jiao, N. Abundance and diversity of aerobic anoxygenic phototrophic bacteria in saline lakes on the Tibetan plateau. *FEMS Microbiol. Ecol.* **2009**, *67*, 268–278.

41. Hawes, I.; Schwarz, A.-M.J. Absorption and utilization of irradiance by cyanobacterial mats in two ice-covered antarctic lakes with contrasting light climates. *J. Phycol.* **2001**, *37*, 5–15.

42. Taton, A.; Grubisic, S.; Brambilla, E.; de Wit, R.; Wilmotte, A. Cyanobacterial diversity in natural and artificial microbial mats of Lake Fryxell (McMurdo Dry Valleys, Antarctica): A morphological and molecular approach. *Appl. Environ. Microbiol.* **2003**, *69*, 5157–5169.

43. Jungblut, A.D.; Hawes, I.; Mountfort, D.; Hitzfeld, B.; Dietrich, D.R.; Burns, B.P.; Neilan, B.A. Diversity within cyanobacterial mat communities in variable salinity meltwater ponds of McMurdo Ice Shelf, Antarctica. *Environ. Microbiol.* **2005**, *7*, 519–529.

44. Vincent, W.F. Cold tolerance in cyanobacteria and life in the cryosphere. In *Cellular Origin and Life in Extreme Habitats and Astrobiology*; Seckbach, J., Ed.; Springer Science: Dordrecht, The Netherlands, 2007; pp. 289–301.

45. Jungblut, A.D.; Lovejoy, C.; Vincent, W.F. Global distribution of cyanobacterial ecotypes in the cold biosphere. *ISME J.* **2010**, *4*, 191–202.

46. Tang, E.P.Y.; Vincent, W.F. Strategies of thermal adaptation by high-latitude cyanobacteria. *New Phytol.* **1999**, *142*, 315–323.

47. Zakhia, F.; Jungblut, A.-D.; Taton, A.; Vincent, W.F.; Wilmotte, A. Cyanobacteria in cold ecosystems. In *Psychrophiles: From Biodiversity to Biotechnology*; Margesin, R., Schinner, F., Marx, J.-C., Gerday, C., Eds.; Springer Science: Heidelberg, Germany, 2008; pp. 121–135.

48. Partensky, F.; Hess, W.R.; Vaulot, D. *Prochlorococcus*, a marine photosynthetic prokaryote of global significance. *Microbiol. Mol. Biol. Rev.* **1999**, *63*, 106–127.

49. Chisholm, S.W.; Olson, R.J.; Zettler, E.R.; Goericke, R.; Waterbury, J.B.; Welschmeyer, N.A. A novel free-living prochlorophyte abundant in the oceanic euphotic zone. *Nature* **1988**, *334*, 340–343.

50. Murphy, L.S.; Haugen, E.M. The distribution and abundance of phototrophic ultraplankton in the North-Atlantic. *Limnol. Oceanogr.* **1985**, *30*, 47–58.

51. Marchant, H.J.; Davidson, A.T.; Wright, S.W. The distribution and abundance of chroococcoid cyanobacteria in the Southern Ocean. *Proc. NIPR Symp. Polar Biol.* **1987**, *1*, 1–9.

52. Vincent, W.F. Cyanobacterial dominance in the polar regions. In *The Ecology of Cyanobacteria*; Whitton, B.A., Potts, M., Eds.; Kluwer Academic Publishers: Dordrecht, the Netherlands, 2000; pp. 321–340.

53. Walker, T.D.; Marchant, H.J. The seasonal occurrence of chroococcoid cyanobacteria at an Antarctic coastal site. *Polar Biol.* **1989**, *9*, 193–196.

54. Andreoli, C.; Moro, I.; La Rocca, N.; Dalla Valle, L.; Masiero, L.; Rascio, N.; Dalla Vecchia, F. Ecological, physiological and biomolecular surveys on microalgae from Ross Sea (Antarctica). *Ital. J. Zool.* **2000**, *67*, 147–156.

55. Koh, E.Y.; Cowie, R.O.M.; Simpson, A.M.; O'Toole, R.; Ryan, K.G. The origin of cyanobacteria in Antarctic sea ice: Marine or freshwater? *Environ. Microbiol. Rep.* **2012**, *4*, 479–483.

56. Castiglioni, B.; Rizzi, E.; Frosini, A.; Sivonen, K.; Rajaniemi, P.; Rantala, A.; Mugnai, M.A.; Ventura, S.; Wilmotte, A.; Boutte, C. Development of a universal microarray based on the ligation detection reaction and 16S rRNA gene polymorphism to target diversity of cyanobacteria. *Appl. Environ. Microbiol.* **2004**, *70*, 7161–7172.

57. Hughes, K.A.; McCartney, H.A.; Lachlan-Cope, T.A.; Pearce, D.A. A preliminary study of airborne microbial biodiversity over peninsular Antarctica. *Cell. Mol. Biol.* **2004**, *50*, 537–542.

58. Pearce, D.A.; Bridge, P.D.; Hughes, K.A.; Sattler, B.; Psenner, R.; Russell, N.J. Microorganisms in the atmosphere over Antarctica. *FEMS Microbiol. Ecol.* **2009**, *69*, 143–157.

59. Harding, T.; Jungblut, A.D.; Lovejoy, C.; Vincent, W.F. Microbes in high Arctic snow and implications for the cold biosphere. *Appl. Environ. Microbiol.* **2011**, *77*, 3234–3243.

60. Bottos, E.M.; Vincent, W.F.; Greer, C.W.; Whyte, L.G. Prokaryotic diversity of arctic ice shelf microbial mats. *Environ. Microbiol.* **2008**, *10*, 950–966.

61. Marshall, G.J.; Turner, J. Katabatic wind propagation over the western Ross Sea observed using ERS-1 scatterometer data. *Antarct. Sci.* **1997**, *9*, 221–226.

62. Raymond, J.; Blankenship, R.E. The evolutionary development of the protein complement of photosystem 2. *Biochim. Biophys. Acta.* **2004**, *1655*, 133–139.

63. Azam, F.; Smith, D.C.; Hollibaugh, J.T. The role of the microbial loop in Antarctic pelagic ecosystems. *Polar Res.* **1991**, *10*, 239–243.

64. Achenbach, L.A.; Carey, J.; Madigan, M.T. Photosynthetic and phylogenetic primers for detection of anoxygenic phototrophs in natural environments. *Appl. Environ. Microbiol.* **2001**, *67*, 2922–2926.

65. Stackebrandt, E.; Rainey, F.A.; Ward-Rainey, N. Anoxygenic phototrophy across the phylogenetic spectrum: Current understanding and future perspectives. *Arch. Microbiol.* **1996**, *166*, 211–223.

66. Shiba, T.; Simidu, U.; Taga, N. Distribution of aerobic bacteria which contain bacteriochlorophyll *a*. *Appl. Environ. Microbiol.* **1979**, *38*, 43–45.

67. Kolber, Z.S.; van Dover, C.L.; Niederman, R.A.; Falkowski, P.G. Bacterial photosynthesis in surface waters of the open ocean. *Nature* **2000**, *407*, 177–179.

68. Yurkov, W.; Beatty, J.T. Aerobic anoxygenic phototrophic bacteria. *Microbiol. Mol. Biol. Rev.* **1998**, *62*, 695–724.

69. Goericke, R. Bacteriochlorophyll *a* in the ocean: is anoxygenic bacterial photosynthesis important? *Limnol. Oceanogr.* **2002**, *47*, 290–295.

70. Gregor, J.; Klug, G. Regulation of bacterial photosynthesis genes by oxygen and light. *FEMS Microbiol. Lett.* **1999**, *179*, 1–9.

71. Nishimura, K.; Shimada, H.; Ohta, H.; Masuda, T.; Shioi, Y.Z.; Takamiya, K. Expression of the puf operon in an aerobic photosynthetic bacterium, *Roseobacter denitrificans*. *Plant Cell Physiol.* **1996**, *37*, 153–159.

72. Masuda, S.; Nagashima, K.V.P.; Shimada, K.; Matsuura, K. Transcriptional control of expression of genes for photosynthetic reaction center and light-harvesting proteins in the purple bacterium *Rhodovulum sulfidophilum*. *J. Bacteriol.* **2000**, *182*, 2778–2786.

73. Yutin, N.; Beja, O. Putative novel photosynthetic reaction centre organizations in marine aerobic anoxygenic photosynthetic bacteria: Insights from metagenomics and environmental genomics. *Environ. Microbiol.* **2005**, *7*, 2027–2033.

74. Béjà, O.; Suzuki, M.T.; Heidelberg, J.F.; Nelson, W.C.; Preston, C.M.; Hamada, T.; Eisen, J.A.; Fraser, C.M.; DeLong, E.F. Unsuspected diversity among marine aerobic anoxygenic phototrophs. *Nature* **2002**, *415*, 630–633.

75. Lambeck, K.; Esat, T.M.; Potter, E.K. Links between climate and sea levels for the past three million years. *Nature* **2002**, *419*, 199–206.

76. Oz, A.; Sabehi, G.; Koblizek, M.; Massana, R.; Beja, O. *Roseobacter*-like bacteria in Red and Mediterranean Sea aerobic anoxygenic photosynthetic populations. *Appl. Environ. Microbiol.* **2005**, *71*, 344–353.

77. Yutin, N.; Suzuki, M.T.; Beja, O. Novel primers reveal wider diversity among marine aerobic anoxygenic phototrophs. *Appl. Environ. Microbiol.* **2005**, *71*, 8958–8962.

78. Cottrell, M.T.; Kirchman, D.L. Photoheterotrophic microbes in the Arctic Ocean in summer and winter. *Appl. Environ. Microbiol.* **2009**, *75*, 4958–4966.

79. Jiao, N.Z.; Zhang, F.; Hong, N. Significant roles of bacteriochlorophylla supplemental to chlorophylla in the ocean. *ISME J.* **2010**, *4*, 595–597.

80. Schwalbach, M.S.; Fuhrman, J.A. Wide-ranging abundances of aerobic anoxygenic phototrophic bacteria in the world ocean revealed by epifluorescence microscopy and quantitative PCR. *Limnol. Oceanogr.* **2005**, *50*, 620–628.

81. Cottrell, M.T.; Michelou, V.K.; Nemcek, N.; DiTullio, G.; Kirchman, D.L. Carbon cycling by microbes influenced by light in the Northeast Atlantic Ocean. *Aquat. Microb. Ecol.* **2008**, *50*, 239–250.

82. Karr, E.A.; Sattley, W.M.; Jung, D.O.; Madigan, M.T.; Achenbach, L.A. Remarkable diversity of phototrophic purple bacteria in a permanently frozen Antarctic lake. *Appl. Environ. Microbiol.* **2003**, *69*, 4910–4914.

83. Perreault, N.N.; Greer, C.W.; Andersen, D.T.; Tille, S.; Lacrampe-Couloume, G.; Lollar, B.S.; Whyte, L.G. Heterotrophic and autotrophic microbial populations in cold perennial springs of the high Arctic. *Appl. Environ. Microbiol.* **2008**, *74*, 6898–6907.

84. Koh, E.Y.; Phua, W.; Ryan, K.G. Aerobic anoxygenic phototrophic bacteria in Antarctic sea ice and seawater. *Environ. Microbiol. Rep.* **2011**, *3*, 710–716.

85. Giebel, H.A.; Brinkhoff, T.; Zwisler, W.; Selje, N.; Simon, M. Distribution of *Roseobacter* RCA and SAR11 lineages and distinct bacterial communities from the subtropics to the Southern Ocean. *Environ. Microbiol.* **2009**, *11*, 2164–2178.

86. Giebel, H.A.; Kalhoefer, D.; Lemke, A.; Thole, S.; Gahl-Janssen, R.; Simon, M.; Brinkhoff, T. Distribution of *Roseobacter* RCA and SAR11 lineages in the North Sea and characteristics of an abundant RCA isolate. *ISME J.* **2011**, *5*, 8–19.

87. Berghoff, B.A.; Glaeser, J.; Nuss, A.M.; Zobawa, M.; Lottspeich, F.; Klug, G. Anoxygenic photosynthesis and photooxidative stress: A particular challenge for *Roseobacter*. *Environ. Microbiol.* **2011**, *13*, 775–791.

88. Cowie, R.O.M.; Maas, E.W.; Ryan, K.G. Archaeal diversity revealed in Antarctic sea ice. *Antarct. Sci.* **2011**, *23*, 531–536.

89. Béja, O.; Aravind, L.; Koonin, E.V.; Suzuki, M.T.; Hadd, A.; Nguyen, L.P.; Jovanovich, S.B.; Gates, C.M.; Feldman, R.A.; Spudich, J.L. Bacterial rhodopsin: Evidence for a new type of phototrophy in the sea. *Science* **2000**, *289*, 1902–1906.

90. Beja, O.; Spudich, E.N.; Spudich, J.L.; Leclerc, M.; Delong, E.F. Proteorhodopsin phototrophy in the ocean. *Nature* **2001**, *411*, 786–789.

91. Atamna-Ismaeel, N.; Sabehi, G.; Sharon, I.; Witzel, K.P.; Labrenz, M.; Jürgens, K.; Barkay, T.; Stomp, M.; Huisman, J.; Beja, O. Widespread distribution of proteorhodopsins in freshwater and brackish ecosystems. *ISME J.* **2008**, *2*, 656–662.

92. De La Torre, J.R.; Christianson, L.M.; Béjà, O.; Suzuki, M.T.; Karl, D.M.; Heidelberg, J.; DeLong, E.F. Proteorhodopsin genes are distributed among divergent marine bacterial taxa. *Proc. Natl. Acad. Sci.* **2003**, *100*, 12830–12835.

93. Giovannoni, S.J.; Bibbs, L.; Cho, J.C.; Stapels, M.D.; Desiderio, R.; Vergin, K.L.; Rappé, M.S.; Laney, S.; Wilhelm, L.J.; Tripp, H.J. Proteorhodopsin in the ubiquitous marine bacterium SAR11. *Nature* **2005**, *438*, 82–85.

94. Campbell, B.J.; Waidner, L.A.; Cottrell, M.T.; Kirchman, D.L. Abundant proteorhodopsin genes in the North Atlantic Ocean. *Environ. Microbiol.* **2007**, *10*, 99–109.

95. Rusch, D.B.; Halpern, A.L.; Sutton, G.; Heidelberg, K.B.; Williamson, S.; Yooseph, S.; Wu, D.; Eisen, J.A.; Hoffman, J.M.; Remington, K. The *Sorcerer II* global ocean sampling expedition: Northwest Atlantic through Eastern Tropical Pacific. *PLoS Biol.* **2007**, *5*, e77.

96. Zhao, M.; Chen, F.; Jiao, N. Genetic diversity and abundance of Flavobacterial proteorhodopsin in China Seas. *Appl. Environ. Microbiol.* **2009**, *75*, 529–533.

97. Sabehi, G.; Loy, A.; Jung, K.H.; Partha, R.; Spudich, J.L.; Isaacson, T.; Hirschberg, J.; Wagner, M.; Béjà, O. New insights into metabolic properties of marine bacteria encoding proteorhodopsins. *PLoS Biol.* **2005**, *3*, e273.

98. Venter, J.C.; Remington, K.; Heidelberg, J.F.; Halpern, A.L.; Rusch, D.; Eisen, J.A.; Wu, D.; Paulsen, I.; Nelson, K.E.; Nelson, W. Environmental genome shotgun sequencing of the Sargasso Sea. *Science* **2004**, *304*, 66–74.

99. Poretsky, R.S.; Bano, N.; Buchan, A.; LeCleir, G.; Kleikemper, J.; Pickering, M.; Pate, W.M.; Moran, M.A.; Hollibaugh, J.T. Analysis of microbial gene transcripts in environmental samples. *Appl. Environ. Microbiol.* **2005**, *71*, 4121–4126.

100. Gómez-Consarnau, L.; González, J.M.; Coll-Lladó, M.; Gourdon, P.; Pascher, T.; Neutze, R.; Pedrós-Alió, C.; Pinhassi, J. Light stimulates growth of proteorhodopsin-containing marine Flavobacteria. *Nature* **2007**, *445*, 210–213.

101. Michelou, V.K.; Cottrell, M.T.; Kirchman, D.L. Light-stimulated bacterial production and amino acid assimilation by cyanobacteria and other microbes in the North Atlantic Ocean. *Appl. Environ. Microbiol.* **2007**, *73*, 5539–5546.

102. Steindler, L.; Schwalbach, M.S.; Smith, D.P.; Chan, F.; Giovannoni, S.J. Energy starved *Candidatus* Pelagibacter ubique substitutes light-mediated ATP production for endogenous carbon respiration. *PLoS One* **2011**, *6*, e19725.

103. Eiler, A. Evidence for the ubiquity of mixotrophic bacteria in the upper ocean: Implications and consequence. *Appl. Environ. Microbiol.* **2006**, *72*, 7431–7437.

104. Koh, E.Y.; Atamna-Ismaeel, N.; Martin, A.; Cowie, R.O.M.; Beja, O.; Davy, S.K.; Maas, E.W.; Ryan, K.G. Proteorhodopsin-bearing bacteria in Antarctic sea ice. *Appl. Environ. Microbiol.* **2010**, *76*, 5918–5925.

105. Buckley, R.G.; Trodahl, H.J. Scattering and absorption of visible light by sea ice. *Nature* **1987**, *326*, 867–869.

106. Man, D.; Wang, W.; Sabehi, G.; Aravind, L.; Post, A.F.; Massana, R.; Spudich, E.N.; Spudich, J.L.; Béjà, O. Diversification and spectral tuning in marine proteorhodopsins. *EMBO J.* **2003**, *22*, 1725–1731.

107. Sabehi, G.; Kirkup, B.C.; Rozenberg, M.; Stambler, N.; Polz, M.F.; Beja, O. Adaptation and spectral tuning in divergent marine proteorhodopsins from the eastern Mediterranean and the Sargasso Seas. *ISME J.* **2007**, *1*, 48–55.

108. DeLong, E.F.; Beja, O. The light-driven proton pump proteorhodopsin enhances bacterial survival during tough times. *PLoS Biol.* **2010**, *8*, e1000359.

109. Terrado, R.; Lovejoy, C.; Massana, R.; Vincent, W.F. Microbial food web responses to light and nutrients beneath the coastal Arctic Ocean sea ice during the winter-spring transition. *J. Mar. Sys.* **2008**, *74*, 964–977.

110. Collins, R.E.; Rocap, G.; Deming, J.W. Persistence of bacterial and archaeal communities in sea ice through an Arctic winter. *Environ. Microbiol.* **2010**, *12*, 1828–1841.

111. Goldberg, S.M.D.; Johnson, J.; Busam, D.; Feldblyum, T.; Ferriera, S.; Friedman, R.; Halpern, A.; Khouri, H.; Kravitz, S.A.; Lauro, F.M. A Sanger/pyrosequencing hybrid approach for the generation of high-quality draft assemblies of marine microbial genomes. *Proc. Natl. Acad. Sci.* **2006**, *103*, 11240–11245.

112. Rothberg, J.M.; Leamon, J.H. The development and impact of 454 sequencing. *Nat. Biotechnol.* **2008**, *26*, 1117–1124.

113. Stepanauskas, R.; Sieracki, M.E. Matching phylogeny and metabolism in the uncultured marine bacteria, one cell at a time. *Proc. Natl. Acad. Sci.* **2007**, *104*, 9052–9057.

114. Gentry, T.J.; Wickham, G.S.; Schadt, C.W.; He, Z.; Zhou, J. Microarray applications in microbial ecology research. *Microb. Ecol.* **2006**, *52*, 159–175.

115. Palmer, C.; Bik, E.M.; Eisen, M.B.; Eckburg, P.B.; Sana, T.R.; Wolber, P.K.; Relman, D.A.; Brown, P.O. Rapid quantitative profiling of complex microbial populations. *Nucleic Acids Res.* **2006**, *34*, e5.

116. DeSantis, T.; Brodie, E.; Moberg, J.; Zubieta, I.; Piceno, Y.; Andersen, G. High-density universal 16S rRNA microarray analysis reveals broader diversity than typical clone library when sampling the environment. *Microb. Ecol.* **2007**, *53*, 371–383.

117. Pernthaler, A.; Amann, R. Simultaneous fluorescence in situ hybridization of mRNA and rRNA in environmental bacteria. *Appl. Environ. Microbiol.* **2004**, *70*, 5426–5433.

118. Smith, C.J.; Osborn, A.M. Advantages and limitations of quantitative PCR (Q-PCR)-based approaches in microbial ecology. *FEMS Microbiol. Ecol.* **2009**, *67*, 6–20.

119. Frias-Lopez, J.; Shi, Y.; Tyson, G.W.; Coleman, M.L.; Schuster, S.C.; Chisholm, S.W.; DeLong, E.F. Microbial community gene expression in ocean surface waters. *Proc. Natl. Acad. Sci.* **2008**, *105*, 3805–3810.

120. Zhang, W.; Li, F.; Nie, L. Integrating multiple "omics" analysis for microbial biology: Application and methodologies. *Microbiology* **2010**, *156*, 287–301.

121. Herring, C.D.; Raffaelle, M.; Allen, T.E.; Kanin, E.I.; Landick, R.; Ansari, A.Z.; Palsson, B.Ø. Immobilization of *Escherichia coli* RNA polymerase and location of binding sites by use of chromatin immunoprecipitation and microarrays. *J. Bacteriol.* **2005**, *187*, 6166–6174.

122. Uyar, E.; Kurokawa, K.; Yoshimura, M.; Ishikawa, S.; Ogasawara, N.; Oshima, T. Differential binding profiles of StpA in wild-type and *hns* mutant cells: A comparative analysis of cooperative partners by chromatin immunoprecipitation-microarray analysis. *J. Bacteriol.* **2009**, *191*, 2388–2391.

Isolation and Characterization of Bacteria from Ancient Siberian Permafrost Sediment

De-Chao Zhang, Anatoli Brouchkov, Gennady Griva, Franz Schinner and Rosa Margesin

Abstract: In this study, we isolated and characterized bacterial strains from ancient (Neogene) permafrost sediment that was permanently frozen for 3.5 million years. The sampling site was located at Mammoth Mountain in the Aldan river valley in Central Yakutia in Eastern Siberia. Analysis of phospolipid fatty acids (PLFA) demonstrated the dominance of bacteria over fungi; the analysis of fatty acids specific for Gram-positive and Gram-negative bacteria revealed an approximately twofold higher amount of Gram-negative bacteria compared to Gram-positive bacteria. Direct microbial counts after natural permafrost enrichment showed the presence of $(4.7 \pm 1.5) \times 10^8$ cells g^{-1} sediment dry mass. Viable heterotrophic bacteria were found at 0 °C, 10 °C and 25 °C, but not at 37 °C. Spore-forming bacteria were not detected. Numbers of viable fungi were low and were only detected at 0 °C and 10 °C. Selected culturable bacterial isolates were identified as representatives of *Arthrobacter phenanthrenivorans*, *Subtercola frigoramans* and *Glaciimonas immobilis*. Representatives of each of these species were characterized with regard to their growth temperature range, their ability to grow on different media, to produce enzymes, to grow in the presence of NaCl, antibiotics, and heavy metals, and to degrade hydrocarbons. All strains could grow at −5 °C; the upper temperature limit for growth in liquid culture was 25 °C or 30 °C. Sensitivity to rich media, antibiotics, heavy metals, and salt increased when temperature decreased (20 °C > 10 °C > 1 °C). In spite of the ligninolytic activity of some strains, no biodegradation activity was detected.

Reprinted from *Biology*. Cite as: Zhang, D-C.; Brouchkov, A.; Griva, G.; Schinner, F.; Margesin, R. Isolation and Characterization of Bacteria from Ancient Siberian Permafrost Sediment. *Biology* **2013**, *2*, 85-106.

1. Introduction

Permafrost is one of the most extreme environments on earth and covers more than 20% of the earth's land surface; it has been defined as lithosphere material (soil, sediment or rock) that is permanently exposed to temperatures ≤0 °C and remains frozen for at least two consecutive years, and can extend down to more than 1,500 m in the subsurface [1]. Regions with permafrost occur at high latitudes, but also at high elevations; a significant part of the global permafrost is represented by mountains [2].

The microbial long-term survival in permafrost has been questioned; however, there is evidence that bacteria are able to survive in 500,000-year-old permafrost [3]. Considerable abundance and diversity of microorganisms, including bacteria, archaea, phototrophic cyanobacteria and green algae, fungi and protozoa, are present in permafrost [4–6]. The characteristics of these microorganisms reflect the unique and extreme conditions of the permafrost environment. Permafrost soils may contain up to 20% or more unfrozen water in the form of salt solutions with a low water activity (a_w = 0.8–0.85) [7]. Microorganisms in this environment have additionally to thrive under permanently frozen conditions, oligotrophic conditions, complete darkness, constant gamma radiation and extremely low rates of nutrient and metabolite transfer [4,5]. Substantial growth and metabolic activity (respiration and biosynthesis) of permafrost microorganisms at temperatures down to −20 °C and even −35 °C have been demonstrated [8–10].

Relict microorganisms from ancient permafrost are not only of interest from an ecological point of view, recent studies pointed to their significance as objects of gerontology. *Bacillus* sp. isolated from permafrost sands of the Mammoth Mountain in Central Yakutia was characterized by an extraordinary viability (about 3.5 million years old) and enhanced longevity, immunity and resistance to heat shock and UV irradiation in *Drosophila melanogaster* and mice [11–14]; probiotic activity by a *Bacillus* sp. strain isolated from the same sample has been recently reported [15].

Frozen soils consisting of mineral particles and ice of different ages contain live microorganisms [16]. It has been shown that microbial cells, even showing features of aging [17,18], are able to live or stay viable for a long time. Despite the fact that it is unknown whether these cells are individually surviving or growing, *Bacillus anthracis* remains viable for about 10^5 years [19]. Colonies of bacteria from amber have been reported to survive for 40 million or more years [20].

Viability of bacteria below 0 °C has been investigated [21]. Unfrozen water, held tightly by electrochemical forces onto the surfaces of mineral particles, occurs even in hard-frozen permafrost. Bacterial cells are not frozen at temperatures of −2 °C and −4 °C [22,23]. The thin liquid layers provide a route for water flow, carrying solutes and small particles, possibly nutrients or metabolites, but movement is extremely slow. A bacterium of greater size (0.3–1.4 μm) than the thickness of the water layer (0.01–0.1 μm at temperatures of −2 °C and −4 °C) is unlikely to move, at least in ice [2]. Therefore, microorganisms trapped among mineral particles and ice in permafrost have been isolated [16]. In some cases, their age can be proved by geological conditions, the history of freezing, and radioisotope dating [21].

The nature of extreme longevity of permafrost microorganisms has no comprehensive explanation. Cell structures are far from being stable [24]. The genome is subject to destruction, and the reparation mechanisms of the majority of organisms are not effective enough to prevent accumulation of damages [25]. The half-life of cytosine does not exceed a few hundred years [26]. Ancient DNA of mummies, mammoths, insects in amber and other organisms appears destroyed [20,27,28].

Microorganisms in permafrost have been studied by culture-dependent and culture-independent methods [4–6]. Microbial abundance is often based on culture-based methods. However, culturable cells may only represent less than 1% of the total microbial community in an environment [29] and numerous bacteria enter a viable but non-culturable (VBNC) state in response to environmental stress [30]. Therefore, culture-independent, molecular assays, such as profiling soil DNA, rRNA, or phospholipid fatty acids, are increasingly used in environmental microbiology. Direct recovery of bacterial 16S rDNA theoretically represents the entire microbial population from environmental samples [31]. However, molecular methods also have their limitations, such as variable efficiency of lysis and DNA extraction, and differential amplification of target genes [32]. Only through isolation can microorganisms be fully characterized at the physiological and functional level.

Although major advances have been made in the last decade, our knowledge on the genetics, biochemistry and ecology of microorganisms in permafrost is still limited.

In this study, we investigated the culturable heterotrophic microbial population in ancient (Neogene) permafrost collected from one of the oldest permafrost areas on earth, located in Siberia and permanently frozen for 3.5 million years. We analyzed the bacterial and fungal population by using a combination of culture-dependent and culture-independent techniques. Selected bacterial isolates were characterized with regard to their growth characteristics, their ability to grow on different media, to produce enzymes and to degrade hydrocarbons, and their sensitivity to NaCl, antibiotics, and heavy metals.

2. Materials and Methods

2.1. Sampling Site

The sampling site was located at Mammoth Mountain in the Aldan river valley in Central Yakutia in Eastern Siberia. The site is an exposure located on the left bank of the Aldan river, 325 km upstream from the mouth of the River Lena (N62°56' E134°0.1'). The exposure is a consequence of recent river erosion of a few cm, up to 0.7 m per year. Prior to the erosion, the sampling site would have been considerably deeper.

Annual mean temperature of the deposits is presently about −4 °C near the surface; the temperature is constantly below 0 °C. Alluvial deposits consisting of fine-grained sands and aleurolites with interlayers of plant remains (trunks, branches, leaves) are exposed. The systematic composition of seeds, pollen and leafs is related to Middle Miocene [33], about 11–16 million years ago. This is the northernmost part of the known Eurasian localities of Neogenic leaf and trunk remains of *Salix*, *Populus*, *Alnus* and other species. The deposits in the area became frozen at least 1.8–1.9 million years ago [34], and probably earlier than 3.5 million years ago, and never thawed until now because of the cold climate of the Pleistocene [2]. Recent studies showed that an intensive cooling began there in Late Pliocene 3–3.5 million years ago. The temperature was estimated as ranging from −12 °C to −32 °C in January and from about +12 °C to +15.6 °C in July [35], thus the age of the permafrost at Mammoth Mountain likely reaches up to 3.5 million years. Geological data indicated the absence of thawing of deposited sediment for millions of years, which assures the ancient age of the sample [15].

The exposure has three visible major ancient layers that can be attributed to Late Pleistocene (about 15,000–40,000 years old Ice Complex), Middle and Early Pleistocene (sands and clay; 0.1–1 million years old, frozen at the time of formation) and Neogene (Miocene and Pliocene), mostly sand formations, frozen probably at the end of Neogene about 3.5 million years ago (Figures 1 and 2). A slightly decomposed frozen trunk (Figure 3) was found about 15 m above river level in the Middle Miocene deposits. The topsoil on the exposure, an active layer of about 0.9–1 m, is of modern age and consisted of (acidic) raw humus on frozen siliceous sand and silt and was covered with vegetation that consisted of birch, alder, conifers (spruce, larch and pine tree) and shrubs.

2.2. Sampling

Samples were collected in July 2009 at an altitude of 83 m above sea level, exposition north, and at a depth of 1.5 m from the surface of the Neogene formation (Figure 4). A deep hole of approximately 100 cm was horizontally dug into the frozen Neogene horizon. After sterilizing the surface of this sampling hole by flame, pieces of frozen sediment (icy sand) were collected from a horizontal depth of 75–100 cm, cleaved with a sterilized axe, and collected in sterile 50 mL vials by using sterile spatulas. The mean temperature of the icy sand at the time of sampling was −4 °C.

Figure 1. Sampling site at Mammoth Mountain in the Aldan river valley (Central Yakutia, Eastern Siberia). Formations of Late Pleistocene (Ice Complex) (red), Pleistocene (green) and Neogene (yellow) are visible.

Figure 2. Profile of the exposure of the Mammoth Mountain: **1**, Neogene sands; **2**, Pleistocene sediment: **a**, pebbles in the ferrous sands; **b**, sands; **c**, lacustrine silt; **d**, silt; **3**, ice wedge; **4**, active layer (after data from Markov [36]).

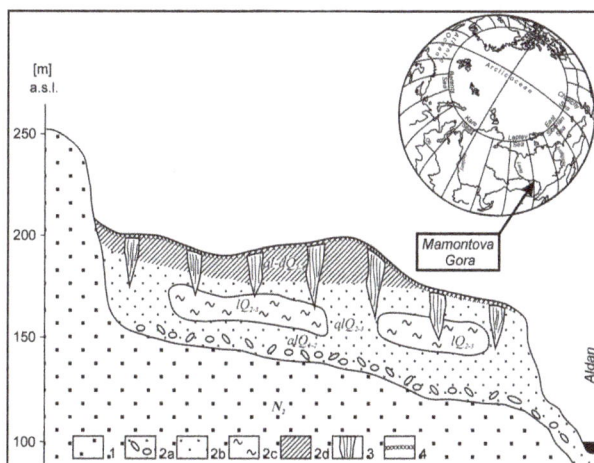

Figure 3. A frozen trunk slightly decomposed was found about 15 m above the river level in the Middle Miocene (Neogene) deposits (3.5 million years old).

Figure 4. Permafrost immediately before sampling. Neogenic deposits consisting of fine-grained sands and aleurolites with interlayers of plant remains characterize the sediment at the site.

Samples were immediately embedded in frozen natural permafrost material, then stored in a cryogenic mixture of NaCl and water to keep the material constantly frozen. The samples were kept frozen during transport from Yakutia to the laboratory in Innsbruck where samples were stored at −20 °C. Thus, the collected material was constantly kept frozen and never subjected to thawing. A composite sample was produced under sterile conditions immediately before analysis.

2.3. Enrichment of Microorganisms

The composite sample was kept for one month at 0 °C for natural permafrost enrichment (NPE) [37]. Afterwards, a number of analyses (physical and chemical soil properties, PLFA, direct and viable microbial counts) were performed.

After NPE, a liquid enrichment (LE) culture was produced by preparing a 1:20 dilution of the NPE with 1/10 strength R2A broth. This enrichment culture was kept at 1 °C on a shaker at 100 rpm. After two and four weeks, samples were analyzed again for direct and viable microbial counts.

2.4. Physical and Chemical Sediment Properties

Dry mass content was determined from mass loss after 24 h at 105 °C. Soil organic matter (SOM) was determined from loss on ignition (LOI) after heating dried soil for 3 h at 430 °C [38]. Soil pH was determined in 10 mM $CaCl_2$ [39]. Contents of nitrate, nitrite and phosphorus were determined spectrophotometrically [39].

2.5. Phospholipid Fatty Acids (PLFA)

Phospholipids were determined after NPE as described [40] and were extracted from 6 g (fresh mass) of sediment, fractionated and quantified using the procedures described [41,42]. Separated fatty acid methyl-esters were identified using gas chromatography and a flame ionization detector. Fatty acid nomenclature was used as described [41]. The fatty acids i15:0, a15:0, 15:0, i16:0, 16:1ω7c, 17:0, i17:0, cy17:0, 18:1ω7c and cy19:0 were chosen to represent bacterial biomass (bacterial PLFA), and 18:2ω6,9c (fungal PLFA) was taken to indicate fungal biomass [43,44]. The ratio of bacterial PLFA to fungal PLFA was calculated to indicate shifts in the ratio between bacterial and fungal biomass. The Gram-positive specific fatty acids i15:0, a15:0, i16:0 and i17:0 and the Gram-negative specific fatty acids cy17:0 and cy19:0 [45] were taken as a measure of the ratio between Gram-positive and Gram-negative bacteria. The fatty acid 20:5ω3c was used as an indicator for soil algae [46]. PLFA concentrations (nmol g^{-1} sediment) were calculated on a dry mass basis and were determined with three replicates.

2.6. Direct Microbial Counts

Total microbial counts were determined by using acridine orange staining and Calcofluor-white staining and epifluorescence microscopy [47,48]. After NPE, 1 mL 10^{-2} diluted sediment extract (the same that was also used for the determination of viable microbial counts) was stained with 1 mL of 0.01% acridine orange or with 1 mL of Calcofluor white M2R (15 µg mL^{-1}) for 3 min. To remove excess staining, the stained suspension was filtered through a 0.4 µm pore size filter (Millipore HTBP02500 Isopore black) held on a 25 mm diameter filter holder. The filter was air dried, cleared in immersion oil and covered by a cover glass. Slides were examined with a Nikon Microphot-SA epifluorescent microscope equipped with a high intensity mercury light source. A Nikon B-2A filter cube was used for examination of acridine orange stained slides. Ten randomly-chosen fields of view were photographed with an 8-bit digital color camera (Nikon Digital sight DS U1) and cells were counted.

2.7. Enumeration of Culturable Heterotrophic Aerobic Sediment Microorganisms

Culturable microorganisms in the sediment sample were enumerated with three replicates by the plate-count method for viable cells. Pre-chilled glassware and solutions were used. Sediment suspensions were prepared by shaking sediment after NPE (10 g fresh mass) for 15 min at 150 rpm with 90 mL of ice-cold sodium pyrophosphate solution (0.28%). Dilutions of this sediment suspension prepared in ice-cold pyrophosphate solution were surface spread onto agar plates.

Similarly, the liquid enrichment culture was diluted in ice-cold pyrophosphate solution and surface spread onto agar plates.

R2A agar and 1/10 strength R2A agar (prepared as R2A broth diluted 1:10 with sterile distilled water and supplemented with agar) were used to determine numbers of viable heterotrophic bacteria. To determine numbers of spore-forming bacteria, the dilutions of the sediment suspension were kept for 15 min at 80 °C in a water bath and afterwards spread on R2A and 1/10 strength R2A agar. Saboraud agar, 1/10 strength Saboraud agar, and malt extract agar (each of these media was supplemented with chlorampenicol (100 µg mL^{-1}) and tetracyclin (100 µg mL^{-1}) to inhibit bacterial growth) were used to determine numbers of viable fungi. Sterile controls were incubated under the same conditions as inoculated plates.

All plates were incubated at −5 °C, 0 °C, 10 °C, 25 °C and 37 °C. Colonies were incubated up to 42 days (−5 °C and 0 °C), 28 days (25 °C), 5–14 days (25 °C) or 7 days (37 °C) until no growth of new colonies was detected. Colony-forming units (CFU) were calculated on a sediment dry mass basis.

2.8. Phylogenetic Analysis of Culturable Bacteria and Restriction Fragment Length Polymorphism (RFLP)

Genomic DNA of 32 culturable bacterial strains (collected from plates incubated at 0 °C) differing in phenotypic characteristics (colony morphology, pigmentation, growth characteristics) was extracted using the UltraClean Microbial DNA isolation kit (Mo Bio Laboratories). The 16S rRNA genes were amplified as described earlier [49].

Restriction fragment length polymorphism (RFLP) was carried out as described [50]. Amplified 16S rRNA genes were restricted using the enzymes RsaI and HhaI (Invitrogen) at 37 °C overnight. Restriction digests were analyzed by agarose gel electrophoresis (2% agarose, 0.5× TBE buffer). Unique restriction patterns were identified visually and two representatives of each restriction pattern were used as a template for 16S rRNA gene sequencing. Sequencing reactions were carried out by Eurofins MWG Operon (Ebersberg, Germany). The 16S rRNA gene sequences were submitted for comparison and identification to the GenBank databases using the NCBI Blastn algorithm and to the EMBL databases using the Fasta algorithm.

2.9. Characterization of Culturable Bacteria

2.9.1. Growth Temperature Range

Growth at −5, 1, 5, 10, 15, 20, 25, 30 and 35 °C was assessed on R2A agar and in R2A broth at 150 rpm (except for cultures at −5 °C where shaking was not possible), using two replicates per strain and temperature. Growth on agar plates was regularly monitored up to an incubation time of 28 days; growth in liquid cultures was monitored by measuring regularly OD600.

2.9.2. Growth on Different Media

Growth on different media was assessed on 1/10 strength R2A agar, R2A agar, nutrient agar (NA, 0.5% peptone, 0.3% meat extract, 1.5% agar; pH 7), TSA (trypticase soy agar; 1.5% casein peptone, 0.5% soy peptone, 0.5% sodium chloride, 1.5% agar; pH 7) and LB agar (Luria Bertani agar; 1% tryptone, 0.5% yeast extract, 0.5% NaCl). Plates were incubated at 1 °C, 10 °C and 20 °C and growth was monitored up to an incubation time of 21 days.

2.9.3. Facultatively Anaerobic Growth

Facultative growth under anaerobic conditions was determined as described [51] on R2A agar, on half-concentrated nutrient agar and on nutrient agar supplemented with 10 mM KNO_3 after incubation at 1 °C, 10 °C and 20 °C in an anaerobic jar (containing Anaerocult A (Merck) to produce anaerobic conditions).

2.9.4. Salt Tolerance

Growth in the presence of various salt concentrations was determined on R2A agar supplemented with 0, 1, 2, 3, 5, 7 and 10% (w/v) NaCl. Two replicates per strain and NaCl concentration were tested. Plates were incubated at 1 °C, 10 °C and 20 °C and growth was monitored up to an incubation time of 21 days.

2.9.5. Resistance to Antibiotics

Susceptibility to antibiotics was determined on R2A agar supplemented with penicillin, ampicillin, kanamycin, streptomycin, rifampicin, tetracyclin, chloramphenicol and cyclosporin. Two concentrations (20 and 100 μg mL^{-1}) were tested for each antibiotic. Growth was tested with two replicates per strain, antibiotic and temperature at 1 °C, 10 °C and 20 °C. Growth was regularly monitored up to an incubation time of 21 days. Strain N1-17 was additionally tested for its susceptibility to antibiotics (20 μg mL^{-1}) in R2A broth at 1 °C, 10 °C and 20 °C; growth in liquid cultures was monitored by measuring regularly OD600.

2.9.6. Resistance to Heavy Metals

Resistance to heavy metals was determined on a mineral salts medium in order to avoid complexation of heavy metals in a complex medium [52]. The used medium was pH-neutral and Tris-buffered [53] and contained 0.1% glucose and 0.1% gluconate as carbon sources, and 1.5% purified agar. The medium was supplemented with the heavy metals Zn^{2+} (1, 3, 5 mM; supplied as $Zn(NO_3)_2 \times 6H_2O$), Pb^{2+} (1, 3, 5 mM; supplied as $Pb(NO_3)_2$) or Cu^{2+} (0.1, 1, 2, 3 mM; supplied as $CuSO_4$). All metals were provided in a soluble, bioavailable form. Plates were incubated up to 21 days at 1, 10 and 20 °C.

2.9.7. Biodegradation of Hydrocarbons

Biodegradation of hydrocarbons was tested as described [54] on mineral medium agar plates amended with the following hydrocarbons as the sole carbon source: n-hexadecane (2,000 mg L^{-1}), diesel oil (2,500 mg L^{-1}), phenol (2.5 mM), naphthalene, phenanthrene, anthracene (2 and 10 mg per plate). Plates were incubated up to 28 days at 1 °C, 10 °C and 20 °C.

2.9.8. Enzyme Activities

Amylase, protease, cellulase and esterase-lipase activities were tested with two replicates as described [54,55] on R2A agar supplemented with starch, skim milk (each compound 0.4% w/v), carboxymethylcellulose and trypan blue (0.4% and 0.01% w/v, respectively) or Tween 80 and $CaCl_2$ (0.4% v/v and 0.01% w/v, respectively). Ligninolytic activity was evaluated on MM agar plates containing 0.4% (w/v) lignosulfonic acid sodium salt [56]. Plates were incubated up to 21 days at 1, 10 and 20 °C.

3. Results

3.1. Sediment Properties

The alluvial sediment material at the sampling site contained a mixture of pale sand, silt, clay and plant debris (litter). The predominant minerals were quartz and feldspar. Multiple stratifications occurred at intervals of 30–300 cm and contained dark-colored, sparsely silicified plant debris; sometimes fragments of stems of monocotyledoneus and dicotyledoneus trees and shrubs were visible. Fruits of conifers and walnut were found; walnut appeared in the sampling area in the warm period shortly before ice formation.

The investigated composite sediment sample had a dry mass content of 88%, a SOM content of 3.6% and a pH ($CaCl_2$) of 4.5. Nutrient contents (nitrate, nitrite, phosphorus) were below the detection limit (<20 mg/kg dry sediment).

3.2. PLFA

The biomass estimate based on PLFA was 0.76 nmol g^{-1} dry sediment. Analysis of PLFA specific for bacteria, fungi and algae demonstrated the dominance of bacteria. Bacterial PLFA were detected to a (9.0 ± 1.2)-fold higher amount compared to fungi. Among bacterial PLFA, the analysis of fatty acids specific for Gram-positive and Gram-negative bacteria revealed a (1.8 ± 0.3)-fold higher amount of Gram-negative bacteria compared to Gram-positive bacteria. PLFA related to algae were not detected.

3.3. Direct Microbial Counts

Direct microbial counts after NPE revealed the presence of (4.7 ± 1.5) × 10^8 cells g^{-1} sediment dry mass (mainly rods and occasionally fungal hyphae), which corresponded to approximately 0.02%–0.5% and 0.01%–0.6% of viable numbers obtained on R2A and 1/10 strength R2A agar, respectively.

When counting after staining with acridine orange, only green fluorescent cells that are often attributed to living cells, (2.5 ± 1.2) ×10^7 cells g^{-1} sediment dry mass were counted after NPE (corresponding to 0.3%–9% of viable counts on R2A, see below), this number further increased after LE in the presence of nutrients and paralleled the increase in viable counts.

3.4. Enumeration of Culturable Bacteria and Fungi

Independent of the enrichment period and of the culture medium, viable heterotrophic bacteria were found at 0 °C, 10 °C and 25 °C, but not at 37 °C (detection limit 100 cfu g^{-1} sediment) (Figure 5). However, the relation between bacteria able to grow at the different incubation temperatures was influenced by the period of enrichment. After NPE, viable bacterial numbers determined at 0 °C were 7.4 × 10^4 (R2A agar) and 5.8 × 10^4 g^{-1} sediment dry mass (1/10 strength R2A agar). They were 30-fold (R2A) or 50-fold (1/10 strength R2A) higher at 10 °C and 10-fold (R2A) or 18-fold (1/10 strength R2A) higher at 25 °C. Thus, only 3% and 10% of the viable bacterial numbers obtained on R2A and able to grow at 0 °C could also grow at 10 °C and 25 °C, respectively. An additional LE in the presence of nutrients after NPE resulted in an increase in viable numbers; this was also confirmed by counts of green fluorescent cells after staining with acridine orange. After two weeks of LE at 2 °C, the fraction of bacteria able to grow at 0 °C had increased to 20% of the fraction able to grow at 10 °C, but had decreased to 7% of the fraction able to grow at 25 °C. This trend was also observed after four weeks of LE when even 90% of the fraction able to grow at 10 °C could grow at 0 °C. An almost identical trend was observed for bacteria cultured on 1/10 strength R2A agar. Thus, enrichment in the presence of nutrients favored the enrichment of bacteria able to grow at 0 °C, while the opposite was observed for bacteria able to grow at 25 °C. Spore-forming bacteria were not detected at any of the tested incubation temperatures neither on R2A nor on 1/10 strength R2A agar (detection limit 100 cfu g^{-1} sediment).

In contrast to numbers of viable bacteria, numbers of viable fungi were very low and were only detected on media incubated at 0 and 10 °C, but not at higher incubation temperatures. After NPE, 2–5 fungal colonies appeared on all three media used to detect viable fungi. Colonies with the same appearance (color, size) also appeared after two and four weeks of LE in the presence of nutrients. The techniques applied in this study for the recovery of viable fungi might be limited since NPE was originally developed for bacteria, and subsequent LE might have favored bacteria rather than fungi.

Since these colonies did not differ in their visible appearance or growth behavior, only one of them was subjected to identification by CBS (Delft, The Netherlands) based on the rRNA gene sequence of the Internal Transcribed Spacer 1 and 2 (ITS). A sequence identity of 97% with a fungal endophyte associated with Antarctic mosses was detected. The strain was able to grow between 0 °C and 20 °C with fastest growth rates at 20 °C; sporulation was not detected.

Figure 5. Effect of enrichment on numbers of culturable heterotrophic bacteria on R2A agar and 1/10 strength R2A agar incubated at 0, 10 and 25 °C. No growth was observed at 37 °C. **NPE**, natural permafrost enrichment (1 month at 0 °C in undisturbed sediment); **NPE + LE** 2 weeks, 2 weeks of liquid enrichment at 1 °C after NPE; **NPE + LE** 4 weeks, 4 weeks of liquid enrichment at 1 °C after NPE.

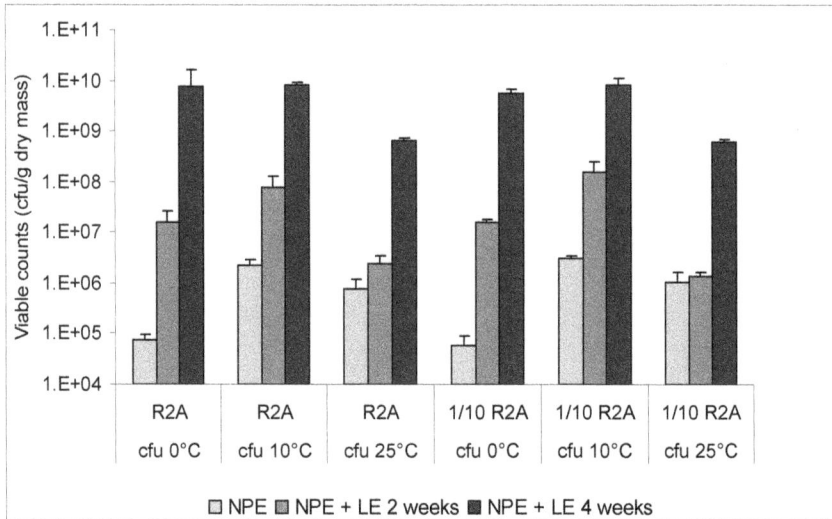

3.5. Characterization of Culturable Bacterial Isolates

According to RFLP, the 32 analyzed bacterial isolates could be divided into three groups and the 16s rRNA gene sequence of two representatives of each restriction pattern were determined. 14 bacterial strains were identified as *Arthrobacter phenanthrenivorans* (98.1% sequence identity; GenBank accession number JX545208), 13 strains were identified as *Subtercola frigoramans* (99.4% 16S rDNA sequence identity; GenBank accession number JX545207), while only five strains could be affiliated to *Glaciimonas immobilis* (98.8% sequence identity; GenBank accession number JX545209). All *Arthrobacter phenanthrenivorans* and *Subtercola frigoramans* strains could be isolated after NPE and after additional LE. In contrast, none of the *Glaciimonas immobilis* strains were isolated after NPE, while all of them appeared after two weeks of LE, which indicates the selective enrichment of this species as already previously stated [37].

Six strains (two representatives of each of the three identified bacterial species) were characterized with regard to their growth temperature range. Their ability to grow on different media, to produce enzymes, to grow in the presence of NaCl, antibiotics and heavy metals and to degrade hydrocarbons was evaluated at 1 °C, 10 °C and 20 °C. Characteristics of one representative for each bacterial species are shown in Table 1. All strains were able to grow at −5 °C, the upper temperature limit for growth in liquid culture was 30 °C for *Arthrobacter* and *Subtercola* strains, while *Glaciimonas* strains could grow up to 25 °C. All strains exhibited fastest growth rates in R2A broth at the maximum temperature for growth, biomass yield (cell density) was highest at 1–5 °C for *Glaciimonas*, while *Arthrobacter* and *Subtercola* strains produced the highest amount of biomass at 30 °C and their cell yields were *ca.* 20% lower at 1–25 °C than at 30 °C (Figure 6). *Arthrobacter* strains produced an approx. threefold higher amount of biomass than *Subtercola* and *Glaciimonas* strains. Growth at −5 °C is not shown in Figure 6 since the strains were cultured without shaking at this temperature.

All strains were initially exposed to oxygen and thus able to grow under aerobic conditions, but none of them was facultatively anaerobic. Growth on different media was temperature-dependent. All strains could grow at 1 °C, 10 °C and 20 °C on 1/10 strength R2A and R2A. *Arthrobacter* strains preferred rich media (NA, TSA, LB) at all three temperatures tested. *Subtercola* grew on NA at 10 °C and 20 °C but not at 1 °C, growth on LB was good at 20 °C, week at 10 °C and absent at 1 °C. *Glaciimonas* did not grow on rich media (NA, LB and TSA).

Table 1. Properties of bacterial strains isolated from ancient permafrost sediment.

Strain properties	*Subtercola frigoramans* N1-13	*Arthrobacter phenanthrenivorans* N1-17	*Glaciimonas immobilis* N1-38
Tmin/Tmax			
R2A broth	−5 °C/30 °C	−5 °C/30 °C	−5 °C/25 °C
R2A agar	1 °C/30 °C	1 °C/30 °C	1 °C/20 °C
Growth on various media (R2A, NA, LB, TSA)			
1 °C	R2A	R2A NA LB TSA	R2A
10 °C	R2A NA	R2A NA LB TSA	R2A
20 °C	R2A NA LB	R2A NA LB TSA	R2A
Growth in presence of NaCl (% w/v)			
1 °C	0%	1% weak	0%
10 °C	0% (1% weak)	2%	0%
20 °C	1%	3% (5% weak)	0%
Growth in presence of cyclosporin A (100 µg mL^{-1}) *			
1 °C	Weak	+	weak
10 °C	+	+	+
20 °C	+	+	+
Utilization of lignosulfonic acid			
1 °C	−	+	−
10 °C	Weak	+	−
20 °C	Weak	+	−
Resistance to heavy metals (0.1 mM Cu^{2+}, 1 mM Pb^{2+}) *			
1 °C	−	−	−
10 °C	(+)	+	−
20 °C	+	++	−

* All strains were sensitive to rifampicin, kanamycin, tetracyclin, streptomycin, chloramphenicol (20 µg mL^{-1} and 100 µg mL^{-1}) at 1 °C, 10 °C and 20 °C. ** All strains were sensitive to 1–5 mM Zn^{2+}, 3–5 mM Pb^{2+} and 1–3 mM Cu^{2+} at 1 °C, 10 °C and 20 °C.

Figure 6. Effect of temperature on growth of bacterial strains isolated from ancient permafrost sediment (data at −5 °C were obtained without shaking and are therefore not shown).

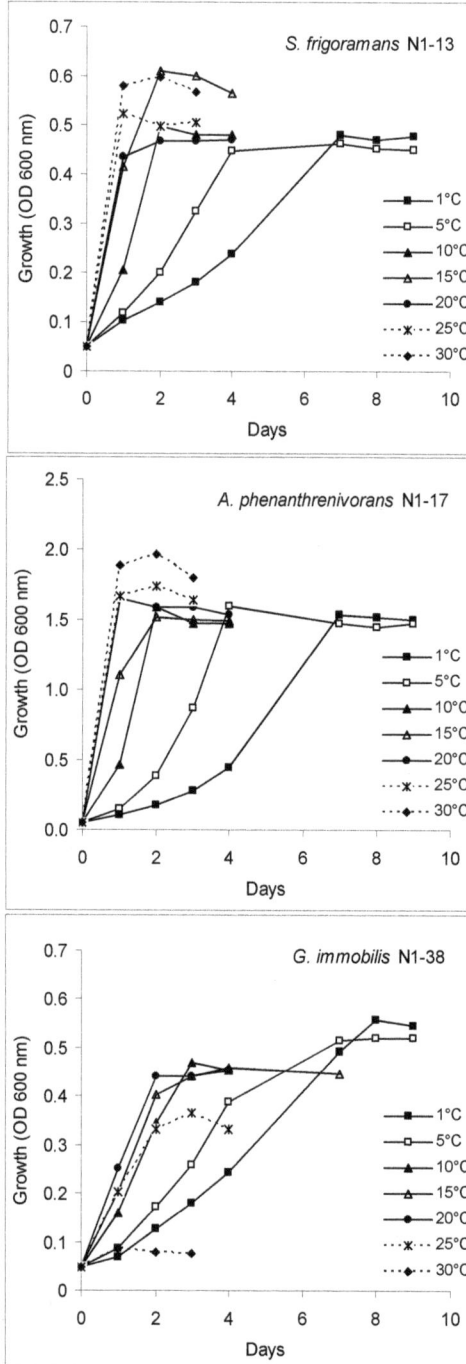

Arthrobacter exhibited a higher salt tolerance than all other strains. Generally, an increased sensitivity to NaCl was noted at decreased temperatures. All strains could grow in the presence of cyclosporin A and at 10 °C and 20 °C, growth at 1 °C in the presence of this antibiotic was weak for *Subtercola*. *Arthrobacter* strains showed weak resistance towards low amounts (20 µg mL^{-1}) of penicillin and ampicillin at 20 °C but behaved sensitive at 10 °C and 1 °C. None of the strains could grow in the presence of other antibiotics tested in this study (rifampicin, kanamycin, tetracyclin, streptomycin, chloramphenicol; 20 or 100 µg mL^{-1}) at 1 °C, 10 °C or 20 °C on agar plates. However, strain *Arthrobacter phenanthrenivorans* N1-17 showed growth in the presence of a number of antibiotics in liquid culture, whereby resistance was clearly influenced by the growth temperature and was generally highest at 20 °C and lowest at 1 °C (Table 2). All tested strains in this study were sensitive to 1–3 mM Cu^{2+} and 1–3 mM Zn^{2+}. *Subtercola* and *Arthrobacter* strains were resistant to 0.1 mM Cu^{2+} and to 1 mM Pb^{2+} at 10 °C and 20 °C, with a more pronounced resistance at 20 °C than at 10 °C.

Table 2. Effect of temperature on sensitivity of *Arthrobacter phenanthrenivorans* N1-17 to antibiotics in R2A broth.

Antibiotic (20 µg mL^{-1})	Relative growth (%)		
	1 °C	10 °C	20 °C
Without antibiotic	100	100	100
Chloramphenicol	24	36	45
Kanamycin	28	41	52
Rifampicin	25	38	46
Streptomycin	26	36	50
Tetracyclin	28	37	48

None of the strains produced protease, amylase, lipase (Tween 80 hydrolysis) or CM-cellulase. Ligninolytic activity was noted for *Arthrobacter* at 10 °C and 20 °C and was weak at 1 °C. Since the substrate used for this activity test, lignosulfonic acid, contains a high amount of phenolic compounds, these strains were expected to degrade phenol [56], however, none of the strains was able to utilize n-hexadecane, diesel oil, phenol, naphthalene, phenanthrene or anthracene at any of the temperatures tested.

4. Discussion

Microbial abundance in permafrost varies depending on the environment. Microbial permafrost communities contain culturable, viable-but-non-culturable, non-culturable and dead cells [57]. Due to constant subzero temperatures in permafrost, dead or compromised microbial cells may remain well preserved and contribute to total microbial counts [6]. The presence of partially degraded bacterial cells and empty "ghost" cells has been demonstrated in Siberian permafrost [58]. Siberian permafrost is dominated by very small (≤1 µm) cells or ultramicroforms of cells (≤0.4 µm) [57], which are typical of the viable but non-culturable state [59]. The microscopic investigation in this study also revealed the dominance of small-sized cells.

Data from direct and microbial counts reported in this study are within the range of data described in other studies on permafrost. Viable counts of aerobic heterotrophs in Siberian permafrost range from 0 to 10^8 cfu g^{-1} material [6,37,60]. Viable counts obtained in our study after NPE were $(6-7) \times 10^4$ cfu g^{-1} dry sediment. Direct counts by epifluorescence microscopy, which is frequently used for the enumeration of bacteria in environmental samples [61,62], in Siberian permafrost range from 10^3–10^8 cells g^{-1} investigated material [5,6,57,60]; ancient Siberian permafrost sediments (100,000 years old, from late Pleistocene) contained 2×10^7 to 1.2×10^8 cells g^{-1} sediment [37]. The percentage of viable counts in relation to direct microscopic counts (DTAF staining) ranged from 0.02% [63] to <0.01%–0.3% [37]. Higher fractions of viable cells (0.1%–10%) were counted with acridine orange staining [39]; we obtained very similar values (0.3%–9% of acridine orange-stained cells were culturable after NPE) in our study, these values are also in agreement with others [64] who reported that 1%–10% of cells stained with acridine orange are culturable.

The successful recovery of viable cells from permafrost depends on a number of factors. The occurrence of viable microorganisms was independent of the depth of permafrost sampling and sometimes even increased with depth [65]. The number of bacterial isolates has been reported to decrease with increasing permafrost age, while species diversity remained almost unaffected [65]. Viable microbial cells could be recovered from 3-million-year-old Siberian permafrost [7,11,66]. Long-term survival of bacteria in 500,000-year-old permafrost samples was closely tied to cellular metabolic activity and DNA repair that, over time, may be superior to dormancy as a strategy to sustain viability [3].

Other important factors for the successful recovery of permafrost microorganisms are low-temperature enrichment strategies and media composition. NPE of unthawed (undisturbed) permafrost soil at 4 °C for up to 12 weeks resulted in enhanced recovery of permafrost bacteria and led to the isolation of genotypes that could not be recovered by means of low-temperature liquid enrichments, since diverse soil microbial communities can better develop independently in various soil microenvironments than in liquid culture [37]. Therefore, we applied this enrichment strategy in our study. We additionally enriched permafrost microorganisms after NPE in LE in the presence of nutrients, which resulted in increased viable counts and in the isolation of representatives of the genus *Glaciimonas* that could not be isolated after NPE, while all representatives of *Arthrobacter phenanthrenivorans* and *Subtercola frigoramans* could be isolated after NPE and after additional LE.

Rich media favor morphological diversity, while diluted media (with low nutrient contents) enhance the quantitative recovery of viable microorganisms [37]. In our study, we did not observe significant differences between viable counts in R2A and 1/10 strength R2A medium, which demonstrates that R2A is a suitable medium for the isolation of oligotrophic microorganisms from environmental habitats such as permafrost.

Both Gram-positive and Gram-negative bacteria have been described in Siberian and other permafrost samples. *Firmicutes* and *Actinobacteria* generally represent a high proportion of the bacterial permafrost community and accounted for 45% of Siberian isolates [67]; *Arthrobacter* (*Actinobacteria*) and *Planococcus* (*Firmicutes*) accounted for 85% of cultured isolates from a northeast Siberian permafrost sample [68]. In our study, the majority of the identified bacterial isolates could be attributed to *Actinobacteria*: *Arthrobacter phenanthrenivorans*, previously isolated from creosote oil-polluted soil [69], and *Subtercola frigoramans*, so far only found in cold groundwater [70]. Only a small fraction belonged to the species *Glaciimonas immobilis* (*Betaproteobacteria*) previously found in alpine glacier cryoconite [71]. PLFA analysis (a culture-independent approach), however, demonstrated an approximately two-fold higher amount of Gram-negative compared to Gram-positive bacteria. Unfortunately no PLFA data are available for permafrost, which makes it impossible to compare our values. Microbial community analysis of subalpine and alpine soils demonstrated a general decrease of PLFA representing bacteria and fungi, as well as a shift of the bacterial population towards the increase of the Gram-negative population with altitude [40].

The high percentage of high G + C Gram-positive, non-spore-forming bacteria (such as *Actinobacteria*) among ancient permafrost isolates has been attributed to their adaptation to frozen environments, to their metabolic activity at low temperatures, to their ability to form dormant cells, to their efficient DNA repair mechanisms, and to the fact that they are more easily cultured [3,66,68]. In contrast, the dominance of spore-forming bacterial genera in Canadian permafrost was attributed to the ability of this microbial group to survive as spores rather than vegetative cells [63]. The first metagenomic analysis of permafrost samples confirmed that *Actinobacteria* are well adapted to the conditions prevailing in permafrost habitats [72].

The abundance of spore-forming bacteria varies between geographically isolated permafrost samples [5,6]. In Siberian permafrost samples, they represented 1%–30% of viable isolates [67] or were not at all detected [68]. This was also the case in our study. However, Brushkov *et al.* [11] isolated *Bacillus* sp. strains from 3-million-year-old permafrost.

There is only little information on fungi in permafrost habitats. Viable fungi can be isolated from permafrost, however, fungal abundance is low, while species diversity is high [73].

Permafrost microorganisms are primarily cold-adapted and only few representatives are mesophilic or thermophilic [4,6,7]. Growth at 37 °C has been rarely reported [68]. Steven *et al.* [63] reported a three-fold higher amount of viable cells growing at 5 °C compared to viable counts at 25 °C; in our study, only 10% of culturable cells growing at 0 °C could also grow at 25 °C. A number of cold-adapted permafrost microorganisms are able to grow at subzero temperatures down to −10 °C [4,63,66,68]. In our study, all strains investigated could grow at −5 °C, and the maximum temperature for growth ranged from 20 to 30 °C.

Permafrost microorganisms tend to be more halotolerant than organisms from the overlaying active layer [6,74], which is seen as a microbial survival strategy in environments with low water activity, such as permafrost, where little water is bioavailable [75]. Tolerance to 7% NaCl [63] or to 8% NaCl [68] has been reported. *Arthrobacter* strains in our study tolerated up to 3% NaCl and showed weak growth in the presence of 5% (w/v) NaCl at 20 °C; however, sensitivity to salt increased when temperature decreased, which has not yet been described before.

Permafrost bacteria are resistant to a wide range of antibiotics combined with the presence of mobile genetic elements [76], which might be part of a generalized bacterial response to stress conditions [77]. Metagenomic analysis of ancient DNA from 30,000-year-old Beringian permafrost sediments demonstrated the presence of a highly diverse collection of genes encoding resistance to ß-lactam, tetracycline and glycopeptide antibiotics [78]. The strains investigated in our study were resistant to cyclosporin A (an immunosuppressive cyclopeptide) and sensitivity towards this compound increased when temperature decreased. The same tendency was noted when growth was tested in the presence of other antibiotics in liquid culture. Similarly, a trend of increased antibiotic sensitivity at 4 °C *versus* 24 °C with all classes of antibiotics except erythromycin was described for permafrost bacteria [66,79].

5. Conclusion

In conclusion, our data demonstrate the presence of viable bacteria in ancient Siberian permafrost. The sample was collected from an ancient Neogene deposit that was permanently frozen for 3.5 million years. The low diversity of viable microorganisms observed in the studied sample may be attributed to a number of factors, such as strong selection pressure due to harsh conditions, nutrient deficiency, presence of inorganic and organic inhibitors, permanently freezing conditions for 3.5 million years combined with the lack of contamination by percolating water from surface, groundwater, lakes and rivers. Since this is the first study of permafrost microbial diversity in the area of Central Yakutia, we cannot compare our data with those of others. Bacterial isolates were able to grow at subzero temperatures and some were halotolerant. In spite of the ligninolytic activity of some strains, no biodegradation activity was detected. In general, sensitivity to rich media, antibiotics, heavy metals, and salt increased when temperature decreased (20 °C > 10 °C > 1 °C). This could be explained as the reaction to an increased stress situation at low temperatures. However, further studies are needed to elucidate the mechanisms behind this process.

Acknowledgements

We thank P. Thurnbichler and R. Kuhnert (University of Innsbruck) for technical assistance and S. Rudolph (University of Hohenheim) for PLFA analysis.

References and Notes

1. Ershov, E.D. *Foundations of Geocryology (in Russian)*; Moscow State University: Moscow, Russia, 1998; pp. 1–575.
2. Margesin, R. *Permafrost Soils*; Springer Verlag: Berlin/Heidelberg, Germany, 2009; pp. 1–348.
3. Johnson, S.S.; Hebsgaard, M.B.; Christensen, T.R.; Mastepanov, M.; Nielsen, R.; Munch, K.; Brand, T.B.; Gilbert, M.T.P.; Zuber, M.T.; Bunce, M.; *et al.* Ancient bacteria show evidence of DNA repair. *Proc. Natl. Acad. Sci. USA* **2007**, *104*, 14401–14405.
4. Steven, B.; Leveille, R.; Pollard, W.H.; Whyte, L.G. Microbial ecology and biodiversity in permafrost. *Extremophiles* **2006**, *10*, 259–267.
5. Gilichinsky, D.; Vishnivetskaya, T.; Petrova, M.; Spirina, E.; Mamykin, V.; Rivkina, E. In *Psychrophiles: From Biodiversity to Biotechnology*; Margesin, R., Schinner, F., Marx, J.C., Gerday, C., Eds.; Springer Verlag: Berlin/Heidelberg, Germany, 2008; pp. 83–102.
6. Steven, B.; Niederberger, T.D.; Whyte, L.G. *Permafrost Soils*; Margesin, R., Ed.; Springer Verlag: Berlin/Heidelberg, Germany, 2009; pp. 59–72.

7. Gilichinsky, D. *Encyclopedia of Environmental Microbiology*; Bitton, G., Ed.; Wiley: New York, NY, USA, 2002; pp. 932–956.

8. Bakermans, C. *Psychrophiles: From Biodiversity to Biotechnology*; Margesin, R., Schinner, F., Marx, J.C., Gerday, C., Eds.; Springer Verlag: Berlin/Heidelberg, Germany, 2008; pp. 17–28.

9. Panikov, N.S.; Sizova, M.V. Growth kinetics of microorganisms isolated from Alaskan soil and permafrost in solid media frozen down to −5 degrees C. *FEMS Microbiol. Ecol.* **2007**, *59*, 500–512.

10. Amato, P.; Doyle, S.M.; Battista, J.R.; Christner, B.C. Implications of subzero metabolic activity on long-term microbial survival in terrestrial and extraterrestrial permafrost. *Astrobiology* **2010**, *10*, 789–798.

11. Brushkov, A.V.; Melnikov, V.P.; Sukhovei, I.G.; Griva, G.I.; Repin, V.E.; Kalenova, L.F.; Brenner, E.V.; Subbotin, A.M.; Trofimova, I.B.; Tanaka, M.; *et al.* Relict microorganisms of cryolithozone as possible objects of gerontology (in Russian). *Adv. Gerontol.* **2009**, *22*, 253–258.

12. Brushkov, A.V.; Bezrukov, V.V.; Griva, G.I.; Muradyan, K.K. The effects of the relict microorganism *B.* sp. on development, gas exchange, spontaneous motor activity, stress resistance, and survival of *Drosophila melanogaster*. *Adv. Gerontol.* **2012**, *2*, 19–26.

13. Kalenova, L.F.; Suhovey, U.G.; Broushkov, A.V.; Melnikov, V.P.; Fisher, T.A.; Besedin, I.M.; Novikova, M.A.; Efimova, J.A.; Subbotin, A.M. Experimental study of the effects of permafrost microorganisms on the morphofunctional activity of the immune system. *Bull. Exp. Biol. Med.* **2011**, *151*, 201–204.

14. Kalenova, L.F.; Sukhovei, U.G.; Brushkov, A.V.; Melnikov, V.P.; Fisher, T.A.; Besedin, I.M.; Novikova, M.A.; Efimova, J.A. Effects of permafrost microorganisms on the quality and duration of life of laboratory animals. *Neurosci. Behav. Physiol.* **2011**, *41*, 484–490.

15. Fursova, O.; Potapov, V.; Brouchkov, A.; Pogorelko, G.; Griva, G.; Fursova, N.; Ignatov, S. Probiotic activity of bacterial strain isolated from ancient permafrost against *Salmonella* infection in mice. *Probiotics Antimicrob. Proteins* **2012**, *403*, 145–153.

16. Friedmann, E.I. *Viable Microorganisms in Permafrost*; Gilichinsky, D., Ed.; Russian Academy of Sciences: Pushchino, Russia, 1994; pp. 21–26.

17. Stewart, E.J.; Madden, R.; Paul, G.; Taddei, F. Ageing and death in an organism that reproduces by morphologically symmetric division. *PLoS Biol.* **2005**, *3*, e45.

18. Johnson, L.R.; Mangel, M. Life histories and the evolution of aging in bacteria and other single-celled organisms. *Mech. Ageing Dev.* **2006**, *127*, 786–793.

19. Nicholson, W.L.; Munakata, N.; Horneck, G.; Melosh, H.J.; Setlow, P. Resistance of *Bacillus* endospores to extreme terrestrial and extraterrestrial environments. *Microbiol. Mol. Biol.* **2000**, *64*, 548–572.

20. Greenblatt, C.L.; Davis, A.; Clement, B.G.; Kitts, C.L.; Cox, T.; Cano, R.J. Diversity of microorganisms isolated from amber. *Microb. Ecol.* **1999**, *38*, 58–68.

21. Katayama, T.; Tanaka, M.; Moriizumi, J.; Nakamura, T.; Brouchkov, A.; Douglas, T.; Fukuda, M.; Tomita, M.; Asano, K. Phylogenetic analysis of bacteria preserved in a permafrost ice wedge for 25,000 Years. *Appl. Environ. Microbiol.* **2007**, *73*, 2360–2363.

22. Clein, J.S.; Schimel, J.P. Microbial activity of tundra and taiga soils at sub-zero temperatures. *Soil Biol. Biochem.* **1995**, *27*, 1231–1234.

23. Ashcroft, F. *Life at the Extremes*; HarperCollins: London, UK, 2000; pp. 1–326.

24. Jaenicke, R. Stability and folding of ultrastable proteins: Eye lens crystallins and enzymes from thermophiles. *FASEB J.* **1996**, *10*, 84–92.

25. Cairns, J.; Overbaugh, J.; Miller, S. The origin of mutations. *Nature* **1994**, *335*, 142–145.

26. Levy, M.; Miller, S.L. The stability of the RNA bases: Implications for the origin of life. *Biochemistry* **1998**, *95*, 7933–7938.

27. Rauser, C.L.; Mueller, L.D.; Rose, M.R. Evolution of late life. *Ageing Res. Rev.* **2005**, *5*, 14–32.

28. Willerslev, E.; Cooper, A. Ancient DNA. *Proc. Roy. Soc. B* **2005**, *272*, 3–16.

29. Amann, R.; Ludwig, W.; Schleifer, K.H. Phylogenetic identification and *in situ* detection of individual microbial cells without cultivation. *Microbiol. Rev.* **1995**, *59*, 143–169.

30. McDougald, D.; Rice, S.A.; Kjelleberg, S. New perspectives on the viable but nonculturable response. *Biologia* **2009**, *54*, 617–623.

31. Spiegelman, D.; Whissell, G.; Greer, C.W. A survey of the methods for the characterization of microbial consortia and communities. *Can. J. Microbiol.* **2005**, *51*, 355–386.

32. Kirk, L.J.; Beaudette, L.A.; Hart, M.; Moutoglis, P.; Klironomas, J.N.; Lee, H.; Trevors, J.T. Methods of studying soil microbiol diversity. *J. Microbiol. Meth.* **2004**, *58*, 169–188.

33. Baranova, U.P.; Il'inskay, I.A.; Nikitin, V.P.; Pneva, G.P.; Fradkina, A.F.; Shvareva, N.Y. *Works of Geological Institute of Russian Academy of Sciences (in Russian)*; Nauka: Moscow, Russia, 1976; pp. 1–284.

34. Romanovsky, N.N. *Basics of Croygenesis of Lithosphere (in Russian)*; Moscow State University: Moscow, Russia, 1993; pp. 1–336.

35. Bakulina, N.T.; Spector, V.B. *Climate and Permafrost (in Russian)*; Maksimov, G.N., Fedorov, A.N., Eds.; Permafrost Institute: Yakutsk, Russia, 2000; pp. 21–32.

36. Markov, K.K. *Cross-Section of the Newest Sediments*; Moscow University Press: Moscow, Russia, 1973; pp. 1–198.

37. Vishnivetskaya, T.A.; Kathariou, S.; McGrath, J.; Gilichinsky, D.; Tiedje, J.M. Low-temperature recovery strategies for the isolation of bacteria from ancient permafrost sediments. *Extremophiles* **2000**, *4*, 165–173.

38. Schlichting, E.; Blume, H.P.; Stahr, K. *Bodenkundliches Praktikum, 2. Auflage (in German)*; Blackwell: Wissenschafts-Verlag, Berlin, Germany, 1995; pp. 1–296.

39. Schinner, F.; Öhlinger, R.; Kandeler, E.; Margesin, R. *Methods in Soil Biology*; Springer Lab Manual: Berlin/Heidelberg, Germany, 1996; pp. 1–426.

40. Margesin, R.; Jud, M.; Tscherko, D.; Schinner, F. Microbial communities and activities in alpine and subalpine soils. *FEMS Microbiol. Ecol.* **2009**, *67*, 208–218.

41. Frostegard, A.; Baath, E.; Tunlid, A. Shifts in the structure of soil microbial communities in limed forests as revealed by phospholipid fatty acid analysis. *Soil Biol. Biochem.* **1993**, *25*, 723–730.

42. Bardgett, R.D.; Hobbs, P.J.; Frostegard, A. Changes in soil fungal:bacterial biomass ratios following reductions in the intensity of management of an upland grassland. *Biol. Fertil. Soils* **1996**, *22*, 261–264.

43. Federle, T.W. *Perspectives in Microbial Ecology*; Megusar, F., Gantar, M., Eds.; Slovene Society for Microbiology: Ljubljana, Slovenia, 1986; pp. 493–498.

44. Zelles, L. Fatty acid patterns of phospholipids and lipopolysaccharides in the characterisation of microbial communities in soil: A review. *Biol. Fertil. Soils* **1999**, *29*, 111–129.

45. Haubert, D.; Häggblom, M.M.; Langel, R.; Scheu, S.; Ruess, L. Trophic shift of stable isotopes and fatty acids in Collembola on bacterial diets. *Soil Biol. Biochem.* **2006**, *38*, 2004–2007.

46. Khotimchenko, S.V.; Vaskovsky, V.E.; Titlyanova, T.V. Fatty acids of marine algae from the pacific coast of North California. *Bot. Mar.* **2002**, *45*, 17–22.

47. Parkinson, D.; Gray, T.R.G.; Williams, S.T. *Methods of Studying the Ecology of Soil Microorganisms*; Handbooks International Biological Programme: Blackwell Sci. Publ.: Oxford/Edinburgh, UK, 1971; pp. 1–128.

48. Hansen, J.F.; Thingstad, T.F.; Godsoyr, J. Evaluation of fungal lengths and hyphal biomass in soil by a membrane filter technique. *Oikos* **1974**, *25*, 102–107.

49. Zhang, D.C.; Liu, H.C.; Xin, Y.H.; Zhou, Y.G.; Schinner, F.; Margesin, R. *Sphingopyxis bauzanensis* sp. nov., a novel psychrophilic bacterium isolated from soil. *Int. J. Syst. Evol. Microbiol.* **2010**, *60*, 2618–2622.

50. Zhang, D.C.; Moertelmaier, C.; Margesin, R. Characterization of the bacterial and archaeal diversity in hydrocarbon-contaminated soil. *Sci. Total Environ.* **2012**, *421–422*, 184–196.

51. Zhang, D.C.; Schumann, P.; Redzic, M.; Zhou, Y.G.; Liu, H.C.; Schinner, F.; Margesin, R. *Nocardioides alpinus* sp. nov., a psychrophilic actinomycete isolated from alpine glacier cryoconite. *Int. J. Syst. Evol. Microbiol.* **2012**, *62*, 445–450.

52. Margesin, R.; Plaza, G.A.; Kasenbacher, S. Characterization of bacterial communities at heavy-metal-contaminated sites. *Chemosphere* **2011**, *82*, 1583–1588.

53. Mergeay, M.; Nies, D.; Schlegel, H.G.; Gerits, J.; Charles, P.; van Gijsegem, F. *Alcaligenes eutrophus* CH34 is a facultative chemolithotroph with plasmid-bound resistance to heavy metals. *J. Bacteriol.* **1985**, *162*, 328–334.

54. Margesin, R.; Gander, S.; Zacke, G.; Gounot, A.M.; Schinner, F. Hydrocarbon degradation and enzyme activities of cold-adapted bacteria and yeasts. *Extremophiles* **2003**, *7*, 451–458.

55. Gratia, E.; Weekers, F.; Margesin, R.; D'Amico, S.; Thonart, P.; Feller, G. Selection of a cold-adapted bacterium for bioremediation of wastewater at low temperatures. *Extremophiles* **2009**, *13*, 763–768.

56. Margesin, R.; Moertelmaier, C.; Mair, J. Low-temperature biodegradation of petroleum hydrocarbons (n-alkanes, phenol, anthracene, pyrene) by four actinobacterial strains. *Int. Biodeterior. Biodegradation* **2012**, doi:10.1016/j.ibiod.2012.05.004.

57. Vorobyova, E.; Soina, V.; Gorlenko, M.; Minkovskaya, N.; Zalinova, N.; Mamukelashvili, A.; Gilichinsky, D.A.; Rivkina, E.; Vishnivetskaya, T. The deep cold biosphere: Facts and hypothesis. *FEMS Microbiol. Rev.* **1997**, *20*, 277–290.

58. Dmitriev, V.V.; Suzina, N.E.; Rusakova, T.G.; Gilichinsky, D.A.; Duda, V.I. Ultrastructural characteristics of natural forms of microorganisms isolated from permafrost grounds of Eastern Siberia by the method of low-temperature fractionation. *Dokl. Biol. Sci.* **2001**, *378*, 304–306.

59. Vorobyova, E.; Minkovsky, N.; Mamukelashvili, A.; Zvyagintsev, D.; Soina, V.; Polanskaya, L.; Gilichinsky, D. *Permafrost Response on Economic Development, Environmental Security and Natural Resources*; Paepe, R., Melnikov, V.P., Eds.; Kluwer Acedemic Publishers: Norwell, MA, USA, 2001; pp. 527–541.

60. Rivkina, E.; Gilichinsky, D.; Wagener, S.; Tiedje, J.; McGrath, J. Biogeochemical activity of anaerobic microorganisms from buried permafrost sediments. *Geomicrobiology* **1998**, *15*, 87–193.

61. Kepner, R.L.; Pratt, J.R. Use of fluorochromes for direct enumeration of total bacteria in environmental samples: Past and present. *Microbiol. Rev.* **1994**, *58*, 603–615.

62. Nadeau, J.L.; Perreault, N.N.; Niederberger, T.D.; Whyte, L.G.; Sun, H.J.; Leon, R. Fluorescence microscopy as a tool for *in situ* life detection. *Astrobiology* **2008**, *8*, 859–874.

63. Steven, B.; Briggs, G.; McKay, C.P.; Pollard, W.H.; Greer, C.W.; Whyte, L.G. Characterization of the microbial diversity in a permafrost sample from the Canadian high Arctic using culture-dependent and culture-independent methods. *FEMS Microbiol. Ecol.* **2007**, *59*, 513–523.

64. Trolldenier, G. *Methods in Soil Biology*; Schinner, F., Öhlinger, R., Kandeler, E., Margesin, R., Eds.; Springer Lab Manual: Berlin/Heidelberg, Germany, 1996; pp. 15–19.

65. Gilichinsky, D.A.; Wilson, G.S.; Friedmann, E.I.; McKay, C.P.; Sletten, R.S.; Rivkina, E.M.; Vishnivetskaya, T.A.; Erokhina, L.G.; Ivanushkina, N.E.; Kochkina, G.A.; *et al.* Microbial populations in Antarctic permafrost: Biodiversity, state, age, and implication for astrobiology. *Astrobiology* **2007**, *7*, 275–311.

66. Vishnivetskaya, T.A.; Petrova, M.A.; Urbance, J.; Ponder, M.; Moyer, C.L.; Gilichinsky, D.A.; Tiedje, J.M. Bacterial community in ancient Siberian permafrost as characterized by culture and culture-independent methods. *Astrobiology* **2006**, *6*, 400–414.

67. Shi, T.; Reeves, R.H.; Gilichinsky, D.A.; Friedmann, E.I. Characterization of viable bacteria from Siberian permafrost by 16S rDNA sequencing. *Microb. Ecol.* **1997**, *33*, 169–179.

68. Hinsa-Leisure, S.M.; Bhavaraju, L.; Rodrigues, J.L.M.; Bakermans, C.; Gilichinsky, D.A.; Tiedje, J.A. Characterization of a bacterial community from a northeastern Siberian seacost permafrost sample. *FEMS Microbiol. Ecol.* **2010**, *74*, 103–113.

69. Vandera, E.; Kavakiotis, K.; Kallimanis, A.; Kyrpides, N.C.; Drainas, C.; Koukkou, A.I. Heterologous expression and characterization of two 1-hydroxy-2-naphthoic acid dioxygenases from *Arthrobacter phenanthrenivorans*. *Appl. Environ. Microbiol.* **2012**, *78*, 621–627.

70. Mannisto, M.K.; Schumann, P.; Rainey, F.A.; Kampfer, P.; Tsitko, I.; Tiirola, M.A.; Salkinoja-Salonen, M.S. *Subtercola boreus* gen. nov., sp. nov. and *Subtercola frigoramans* sp. nov., two new psychrophilic actinobacteria isolated from boreal groundwater. *Int. J. Syst. Evol. Microbiol.* **2000**, *50*, 1731–1739.

71. Zhang, D.C.; Redzic, M.; Schinner, F.; Margesin, R. *Glaciimonas immobilis* gen. nov., sp. nov., a novel member of the family *Oxalobacteraceae* isolated from alpine glacier cryoconite. *Int. J. Syst. Evol. Microbiol.* **2011**, *61*, 2186–2190.

72. Yergeau, E.; Hogues, H.; Whyte, L.G.; Greer, C.W. The functional potential of high Arctic permafrost revealed by metagenomic sequencing, qPCR and microarray analyses. *ISME J.* **2010**, *4*, 1206–1214.

73. Ozerskaya, S.; Kochkina, G.; Ivanushhkina, N.; Gilichinsky, D.A. *Permafrost Soils*; Margesin, R., Ed.; Springer Verlag: Berlin/Heidelberg, Germany, 2009; pp. 85–95.

74. Steven, B.; Pollard, W.H.; Greer, C.W.; Whyte, L.G. Microbial diversity and activity through a permafrost/ground ice core profile from the Canadian high Arctic. *Environ. Microbiol.* **2008**, *10*, 3388–3403.

75. Franks, F. Nucleation of ice and its management in ecosystems. *Phil. Trans. Roy. Soc. Lond. A* **2003**, *361*, 557–574.

76. Petrova, M.; Gorlenko, Z.; Mindlin, S. Tn5045, a novel integron-containing antibiotic and chromate resistance transposon isolated from a permafrost bacterium. *Res. Microbiol.* **2011**, *162*, 337–345.

77. Margesin, R.; Miteva, V. Diversity and ecology of psychrophilic microorganisms. *Res. Microbiol.* **2011**, *162*, 346–361.

78. D'Costa, V.M.; King, C.E.; Kalan, L.; Morar, M.; Sung, W.W.L.; Schwarz, C.; Froese, D.; Zazula, G.; Calmels, F.; Debruyne, R.; *et al.* Antibiotic resistance is ancient. *Nature* **2011**, *477*, 457–461.

79. Ponder, M.A.; Gilmour, S.J.; Bergholz, P.W.; Mindock, C.A.; Hollingsworth, R.; Thomashow, M.F.; Tiedje, J.M. Characterization of potential stress responses in ancient Siberian permafrost psychroactive bacteria. *FEMS Microbiol. Ecol.* **2005**, *53*, 103–115.

Thermodynamic Stability of Psychrophilic and Mesophilic Pheromones of the Protozoan Ciliate *Euplotes*

Michael Geralt, Claudio Alimenti, Adriana Vallesi, Pierangelo Luporini and Kurt Wüthrich

Abstract: Three psychrophilic protein pheromones (E*n*-1, E*n*-2 and E*n*-6) from the polar ciliate, *Euplotes nobilii*, and six mesophilic pheromones (E*r*-1, E*r*-2, E*r*-10, E*r*-11, E*r*-22 and E*r*-23) from the temperate-water sister species, *Euplotes raikovi*, were studied in aqueous solution for their thermal unfolding and refolding based on the temperature dependence of their circular dichroism (CD) spectra. The three psychrophilic proteins showed thermal unfolding with mid points in the temperature range 55–70 °C. In contrast, no unfolding was observed for any of the six mesophilic proteins and their regular secondary structures were maintained up to 95 °C. Possible causes of these differences are discussed based on comparisons of the NMR structures of the nine proteins.

Reprinted from *Biology*. Cite as: Geralt, M.; Alimenti, C.; Vallesi, A.; Luporini, P.; Wüthrich, K. Thermodynamic Stability of Psychrophilic and Mesophilic Pheromones of the Protozoan Ciliate *Euplotes*. *Biology* **2013**, *2*, 142-150.

1. Introduction

Protozoan ciliates represent a major micro-eukaryotic component of the polar ecosystem [1,2], which can readily be collected from every aquatic habitat for use in stable laboratory cultures [3]. Strains of *Euplotes* species such as *E. patella*, *E. raikovi*, *E. octocarinatus* and *E. crassus* inhabiting non-polar temperate waters, and of *E. nobilii* inhabiting Arctic and Antarctic waters are capable of secreting cell type-specific signaling proteins genetically specified at a single multi-allelic locus (designated as *mating-type*, or *mat* locus) [4,5]. These water-borne "pheromones" are functionally associated with the genetic mechanism of the mating types and act as prototypic autocrine (autologous) growth factors and as paracrine (heterologous) inducers of mating pair formation [6,7]. In addition to the full-length coding gene sequences [8–10], the three-dimensional molecular structures of a significant number of pheromones were determined by NMR spectroscopy in solution, firstly, from the temperate-water species, *E. raikovi* [11–17], and subsequently from the polar-water species, *E. nobilii* [18–20]. These mesophilic (*E. raikovi*) and psychrophilic (*E. nobilii*) pheromone families, both characterized by small, helical and disulfide-rich proteins of 37 to 63 amino acids, thus represent an interesting source of material for structure based comparative studies of protein adaptation to cold.

Here we present data on the thermal denaturation of the three pheromones E*n*-1, E*n*-2 and E*n*-6 from the psychrophilic pheromone family of *E. nobilii*, and the six pheromones E*r*-1, E*r*-2, E*r*-10, E*r*-11, E*r*-22 and E*r*-23 from the mesophilic pheromone family of *E. raikovi* (Figure 1). Considering that NMR solution structures are available for all the nine proteins, and for one (*i.e.*, E*r*-1) is available also the crystallographic structure [21], we expect that these data will be of interest for in-depth studies of correlations between molecular protein structure, thermodynamic stability, and cold adaptation.

Figure 1. Amino acid sequences of *E. nobilii* and *E. raikovi* pheromones. The sequence alignment was maximized by insertion of gaps. Cysteines are marked in red and by progressive Roman numerals, and their pairing into disulfide bonds is indicated by brackets. The sequence regions involved in the formation of helical structures are shadowed. The PDB codes of the pheromone NMR and crystal (E*r*-1) structures are the following: E*r*-1, 1ERC, 1ERI; E*r*-2, 1ERD; E*r*-10, 1ERP; E*r*-11, 1ERY; E*r*-22, 1HD6; E*r*-23, 1HA8; E*n*-1, 2NSV; E*n*-2, 2NSW; E*n*-6, 2JMS.

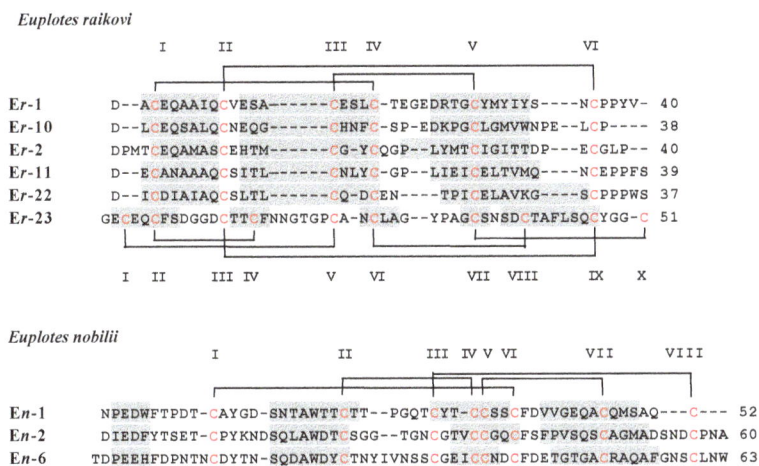

Euplotes raikovi

```
                    I        II           III  IV        V              VI
Er-1     D--ACEQAAIQCVESA-------CESLC-TEGEDRTGCYMYIYS----NCPPYV-  40
Er-10    D--LCEQSALQCNEQG-------CHNFC-SP-EDKPGCLGMVWNPE--LCP----  38
Er-2     DPMTCEQAMASCEHTM-------CG-YCQGP--LYMTCIGITTDP---ECGLP--  40
Er-11    D--ECANAAAQCSITL-------CNLYC-GP--LIEICELTVMQ----NCEPPFS  39
Er-22    D--ICDIAIAQCSLTL-------CQ-DCEN----TPICELAVKG----SCPPPWS  37
Er-23    GECEQCFSDGGDCTTCFNNGTGPCA-NCLAG--YPAGCSNSDCTAFLSQCYGG-C  51
         I  II     III IV      V    VI          VII VIII       IX   X
```

Euplotes nobilii

```
                 I               II        III IV V VI      VII       VIII
En-1     NPEDWFTPDT-CAYGD-SNTAWTTCTT--PGQTCYT-CCSSCFDVVGEQACQMSAQ---C---  52
En-2     DIEDFYTSET-CPYKNDSQLAWDTCSGG--TGNCGTVCCGQCFSFPVSQSCAGMADSNDCPNA  60
En-6     TDPEEHFDPNTNCDYTN-SQDAWDYCTNYIVNSSCGEICCNDCFDETGTGACRAQAFGNSCLNW  63
```

2. Materials and Methods

The isolation and purification of the nine proteins investigated in this paper has previously been described [22–25]. Lyophilized samples of each pheromone were dissolved in 20 mM sodium phosphate buffer at pH 6.0 and diluted to a protein concentration of 20 μM before aliquoting into a 0.1 cm path length quartz cuvette. CD experiments were recorded using the Temperature/Wavelength Scan software supplied with the Jasco 815 CD spectrophotometer. Melting curves over the range 20–95 °C were measured at a constant wavelength of 220 nm by increasing the temperature at a rate of 1.0 or 0.5 °C/min. Wavelength scans from 260–190 nm were measured at 5 °C intervals.

In additional exploratory studies, the chemical denaturants guanidine-HCl and urea were added to solutions of the mesophilic pheromones E*r*-1, E*r*-10, E*r*-22 and E*r*-23 to further test their stabilities. In particular melting curves were recorded for E*r*-1 in 7.8 M urea and 20 mM sodium phosphate at pH 6.0, 6 M guanidine-HCl and 20 mM sodium phosphate at pH 6.0, 4 M guanidine-HCl and 20 mM sodium phosphate at pH 6.0, 4.5 and 3.0, and 6 M urea at pH 3.0. Similar analyses on E*r*-10, E*r*-22 and E*r*-23 were performed using 4 M guanidine-HCl in 20 mM formic acid at pH 2.0.

3. Results

For all the three psychrophilic pheromones E*n*-1, E*n*-2, and E*n*-6 of *E. nobilii*, the CD spectra (Figure 2) show that the regular secondary structures are unfolded at 95 °C, and that this unfolding is reversible upon cooling of the solutions to the starting temperature at 20 °C. Nevertheless, among the individual proteins there are appreciable variations with regard to the shape of the thermal unfolding curves. For E*n*-6 a nearly symmetrical sigmoidal denaturation curve was observed with a midpoint near 65 °C, and a sigmoidal curve was also obtained for the refolding upon cooling of the solution; in addition, only a very small loss of protein was recorded during the unfolding/refolding procedure. On the other hand, E*n*-1 and E*n*-2 showed more sluggish unfolding transitions, and for E*n*-2 there was an indication that the unfolding and refolding processes involve equilibria between more than two states. For the present qualitative survey of the stability of these three proteins we retain that, similar to E*n*-6, the regular secondary structures in E*n*-1 and E*n*-2 are unfolded at 95 °C and refolded upon cooling of the solutions to 20 °C.

In the family of the six mesophilic pheromones E*r*-1, E*r*-2, E*r*-10, E*r*-11, E*r*-22, and E*r*-23 from *E. raikovi*, the regular secondary structures manifested in the CD spectra at 20 °C were found to be maintained up to the highest temperature studied (Figure 2). The temperature dependence of the signal intensity at 220 nm did not provide evidence for unfolding, and for E*r*-22 the small reduction of the signal intensity at the higher temperatures appeared to be fully reversible upon solution cooling. The data recorded for this pheromone are representative of the observations made with the other pheromones E*r*-1, E*r*-2, E*r*-10, and E*r*-11. Also the pheromone E*r*-23 showed qualitatively similar behavior, but it was clearly the most stable member of the *E. raikovi* mesophilic protein family since it did not undergo denaturation (Figure 3).

In view of the high thermal stability of the *E. raikovi* pheromones in neutral aqueous solution, we also performed exploratory experiments with the addition of chemical denaturants (see Materials and Methods). These experiments provided further indications of the remarkably high stability of these proteins. We did not observe their full unfolding in the temperature range 20–95 °C with the solution conditions listed in the Materials and Methods section, although partial melting within this temperature range was observed for some of them.

Figure 2. Temperature-induced denaturation of the *E. nobilii* psychrophilic pheromones En-6, En-1 and En-2 monitored by CD spectroscopy. (**a**) Temperature variation of the signal intensity at 220 nm during heating and cooling over the range from 20 °C to 95 °C. (**b**) CD spectra at different temperatures, as indicated by the color code in the figure. The protein concentration was 20 µM in 20 mM sodium phosphate at pH 6.0, and the cell length 0.1 cm. The temperature variation of the signal intensity at 220 nm in panels (**a**) was recorded at a rate of 1.0 °C/min. Scans were recorded with a speed of 100 nm/min, in 5 °C intervals over the temperature range of panels (**a**); for improved clarity, only four traces are shown in panels (**b**).

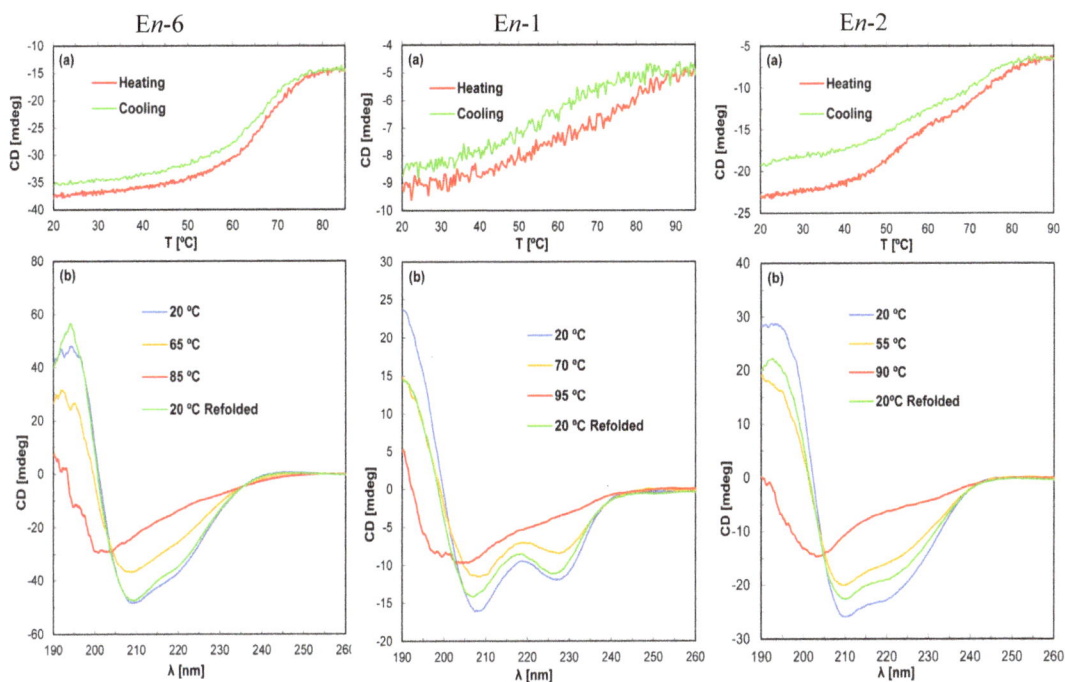

4. Discussion

The psychrophilic and mesophilic pheromone families of the protozoan ciliate *Euplotes* studied here are both represented by single-domain small disulfide-rich proteins, which have usually been shown to be outstandingly stable with regard to thermal denaturation in aqueous solution [26–28]. However, despite their extensive homology on the level of the amino acid sequences and three-dimensional structures, the psychrophilic *E. nobilii* pheromones showed a significantly lower thermal stability than their mesophilic *E. raikovi* counterparts. This finding coincides with observations derived from comparisons between other psychrophilic and mesophilic homologous proteins [29–33]. However, while these comparisons are essentially based on individual proteins from distantly related organisms, our finding involves families of proteins with known NMR solution structures [11–21] from two closely related species [34,35]. It therefore provides data which, at least in principle, are more reliable (being unaffected by the evolutionary noise which is intrinsic to comparisons between distantly related systems) for further detailed analyses by other groups of researchers who are interested in studying the correlations between protein structure, thermodynamic stability, and cold-adaptation.

Figure 3. Temperature-induced denaturation of the *E. raikovi* mesophilic pheromones Er-22 and Er-23 monitored by CD spectroscopy. Er-22 has been taken as representative of the other *E. raikovi* pheromones Er-1, Er-2, Er-10 and Er-11. Same experimental conditions and presentations as in Figure 2, except that all the spectra recorded in 5 °C temperature intervals are shown in panels (**b**).

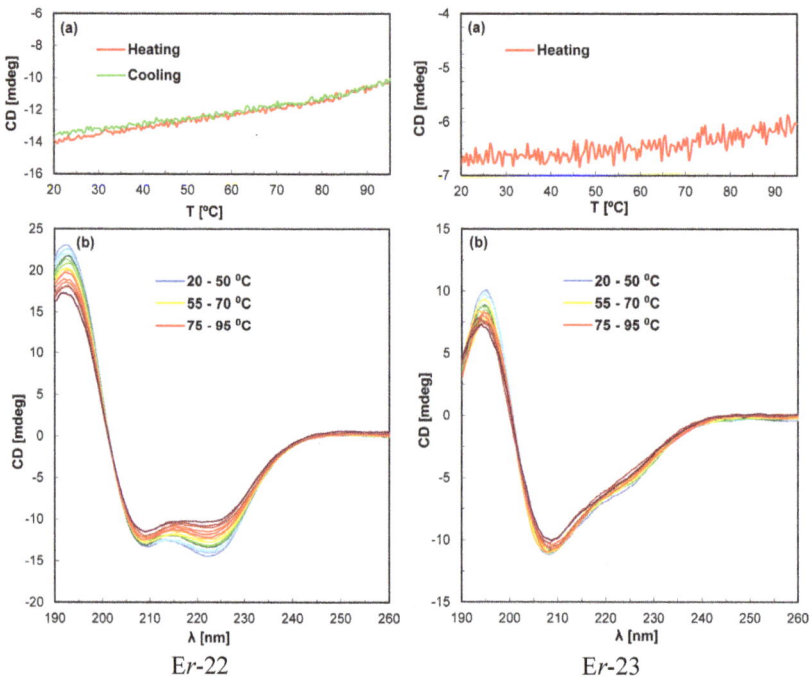

Er-22   Er-23

As previously discussed [3,10,21], the two homologous protein families of *E. nobilii* psychrophilic pheromones and *E. raikovi* mesophilic pheromones differ significantly in the composition of polar, hydrophobic and aromatic amino acids, in global hydrophilicity and hydrophobicity, as well as in various aspects of their three-dimensional structures. It will now be interesting to explore the extent to which these physico-chemical and structural variations can be related to variations of the stability of the folded proteins. Two structural features distinctive of the two pheromone families appear to be particularly promising starting points for further investigations into the structural basis of the observed differences in the thermal stability of these proteins. One is the *N*-terminal elongation of 10 to 12 residues, which is common to all the *E. nobilii* pheromones and has no counterparts in any of the *E. raikovi* pheromones. It includes only a 3_{10}-helical turn as regular secondary structure and is devoid of connections through disulfide bonds to other parts of the proteins. Secondly, there is a higher density of disulfide bonds in the mesophilic *E. raikovi* pheromones than in the psychrophilic ones of *E. nobilii*. On average, the *E. raikovi* pheromones contain one disulfide bond per 13 amino acid residues, while in the *E. nobilii* pheromones there is one disulfide bond per 15 residues. In particular, the high density of one disulfide bond per 10 residues appears to be the dominant feature that makes Er-23 most stable among the *E. raikovi* pheromones, notwithstanding that Er-23 mimics the *E. nobilii* psychrophilic pheromones with regard to the contents of polar and hydrophobic residues as well as the aliphatic index.

5. Conclusions

The key observation reported in this paper is that the *E. nobilii* and *E. raikovi* pheromone families show a thermal denaturation behavior that is uniform among their members and significantly divergent between them. The mesophilic *E. raikovi* pheromones have higher stability by at least 30 °C when compared to the psychrophilic *E. nobilii* pheromones. Given the small size of these water-borne signaling proteins and the availability of their three-dimensional structures, the data presented in this communication should be of interest for systematic investigations of protein adaptation to cold environments.

Acknowledgments

The authors are grateful to R. Glockshuber (ETH, Zürich) for helpful discussions. This research was financially supported by the National Program for Antarctic Research (PNRA).

References

1. Petz, W. Ciliates. In *Antarctic Marine Protists*; Scott, F.J., Marchant, H.J., Eds.; Australian Biological Resources Study: Canberra and Australian Antarctic Division, Hobart, Australia, 2005; pp. 347–448.
2. Petz, W.; Valbonesi, A.; Schiftner, U.; Quesada, A.; Cynan Ellis-Evans, J. Ciliate biogeography in Antarctic and Arctic freshwater ecosystems: endemism or global distribution of species? *FEMS Microbiol. Ecol.* **2007**, *59*, 396–408.
3. Di Giuseppe, G.; Erra, F.; Dini, F.; Alimenti, C.; Vallesi, A.; Pedrini, B.; Wüthrich, K.; Luporini, P. Antarctic and Arctic populations of the ciliate *Euplotes nobilii* show common pheromone-mediated cell-cell signaling and cross-mating. *Proc. Natl. Acad. Sci. USA* **2011**, *108*, 3181–3186.
4. Luporini, P.; Alimenti, C.; Ortenzi, C.; Vallesi, A. Ciliate mating type and their specific protein pheromones. *Acta Protozool.* **2005**, *44*, 89–101.
5. Alimenti, C.; Vallesi, A.; Federici, S.; di Giuseppe, G.; Dini, F.; Carratore, V.; Luporini, P. Isolation and structural characterization of two water-borne pheromones from *Euplotes crassus*, a ciliate commonly known to carry membrane-bound pheromones. *J. Eukaryot. Microbiol.* **2011**, *58*, 234–241.

6. Vallesi, A.; Giuli, G.; Bradshaw, R.A.; Luporini, P. Autocrine mitogenic activity of pheromone produced by the protozoan ciliate *Euplotes raikovi*. *Nature* **1995**, *376*, 522–524.

7. Vallesi, A.; Ballarini, P.; di Pretoro, B.; Alimenti, C.; Miceli, C.; Luporini, P. Autocrine, mitogenic pheromone receptor loop of the ciliate *Euplotes raikovi*: Pheromone-induced receptor internalization. *Eukaryot. Cell* **2005**, *4*, 1221–1227.

8. La Terza, A.; Dobri, N.; Alimenti, C.; Vallesi, A.; Luporini, P. The water-borne protein signals (pheromones) of the Antarctic ciliated protozoan *Euplotes nobilii*: Structure of the gene coding for the E*n*-6 pheromone. *Can. J. Microbiol.* **2009**, *5*, 57–62.

9. Vallesi, A.; Alimenti, C.; La Terza, A.; di Giuseppe, G.; Dini, F.; Luporini, P. Characterization of the pheromone gene family of an Antarctic and Arctic protozoan ciliate, *Euplotes nobilii*. *Mar. Genomics* **2009**, *2*, 27–32.

10. Vallesi, A.; Alimenti, C.; Pedrini, B.; di Giuseppe, G.; Dini, F.; Wüthrich, K.; Luporini, P. Coding genes and molecular structures of the diffusible signalling proteins (pheromones) of the polar ciliate, *Euplotes nobilii*. *Mar. Genomics* **2012**, *8*, 9–13.

11. Brown, L.R.; Mronga, S.; Bradshaw, R.A.; Ortenzi, C.; Luporini, P.; Wüthrich, K. Nuclear magnetic resonance solution structure of the pheromone E*r*-10 from the ciliated protozoan *Euplotes raikovi*. *J. Mol. Biol.* **1993**, *231*, 800–816.

12. Mronga, S.; Luginbühl, P.; Brown, L.R.; Ortenzi, C.; Luporini, P.; Bradshaw, R.A.; Wüthrich, K. The NMR solution structure of the pheromone E*r*-1 from the ciliated protozoan *Euplotes raikovi*. *Protein Sci.* **1994**, *3*, 1527–1536.

13. Ottiger, M.; Szyperski, T.; Luginbühl, P.; Ortenzi, C.; Luporini, P.; Bradshaw, R.A.; Wüthrich, K. The NMR solution structure of the pheromone E*r*-2 from the ciliated protozoan *Euplotes raikovi*. *Protein Sci.* **1994**, *3*, 1517–1526.

14. Luginbühl, P.; Ottiger, M.; Mronga, S.; Wüthrich, K. Structure comparison of the NMR structures of the pheromones E*r*-1, E*r*-10, and E*r*-2 from *Euplotes raikovi*. *Protein Sci.* **1994**, *3*, 1537–1546.

15. Luginbühl, P.; Wu, J.; Zerbe, O.; Ortenzi, C.; Luporini, P.; Wüthrich, K. The NMR solution structure of the pheromone E*r*-11 from the ciliated protozoan *Euplotes raikovi*. *Protein Sci.* **1996**, *5*, 1512–1522.

16. Liu, A.; Luginbühl, P.; Zerbe, O.; Ortenzi, C.; Luporini, P.; Wüthrich, K. NMR structure of the pheromone E*r*-22 from *Euplotes raikovi*. *J. Biomol. NMR* **2001**, *19*, 75–78.

17. Zahn, R.; Damberger, F.; Ortenzi, C.; Luporini, P.; Wüthrich, K. NMR structure of the *Euplotes raikovi* pheromone E*r*-23 and identification of its five disulfide bonds. *J. Mol. Biol.* **2001**, *313*, 923–931.

18. Placzek, W.J.; Etezady-Esfarjani, T.; Herrmann, T.; Pedrini, B.; Peti, W.; Alimenti, C.; Luporini, P.; Wüthrich, K. Cold-adapted signal proteins: NMR structures of pheromones from the Antarctic ciliate *Euplotes nobilii*. *IUBMB Life* **2007**, *59*, 578–585.

19. Pedrini, B.; Placzek, W.J.; Koculi, E.; Alimenti, C.; La Terza, A.; Luporini, P.; Wüthrich, K. Cold-adaptation in sea-waterborne signal proteins: Sequence and NMR structure of the pheromone E*n*-6 from the Antarctic ciliate *Euplotes nobilii*. *J. Mol. Biol.* **2007**, *372*, 277–286.

20. Alimenti, C.; Vallesi, A.; Pedrini, B.; Wüthrich, K.; Luporini, P. Molecular cold-adaptation: Comparative analysis of two homologous families of psychrophilic and mesophilic signal proteins of the protozoan ciliate *Euplotes*. *IUBMB Life* **2009**, *61*, 838–845.

21. Weiss, M.S.; Anderson, D.H.; Raffioni, S.; Bradshaw, R.A.; Ortenzi, C.; Luporini, P.; Eisenberg, D.A. Cooperative model for ligand recognition and cell adhesion: Evidence from the molecular packing in the 1.6 Å crystal structure of the pheromone Er-1 from the ciliate protozoan *Euplotes raikovi*. *Proc. Natl. Acad. Sci. USA* **1995**, *92*, 10172–10176.

22. Concetti, A.; Raffioni, S.; Miceli, C.; Barra, D.; Luporini, P. Purification to apparent homogeneity of the mating pheromone of *mat-1 Euplotes raikovi*. *J. Biol. Chem.* **1986**, *61*, 10582–10587.

23. Raffioni, S.; Miceli, C.; Vallesi, A.; Chowdhury, S.K.; Chait, B.T.; Luporini, P.; Bradshaw, R.A. Primary structure od the *Euplotes raikovi* pheromones: Comparison of five sequences of pheromones from cell with variable mating interactions. *Proc. Natl. Acad. Sci. USA* **1992**, *89*, 2071–2075.

24. Alimenti, C.; Ortenzi, C.; Carratore, V.; Luporini, P. Structural characterization of protein pheromone from a cold-adapted (Antarctic) single-cell eukaryote, the ciliate *Euplotes nobilii*. *FEBS Lett.* **2002**, *514*, 329–332.

25. Alimenti, C.; Ortenzi, C.; Carratore, V.; Luporini, P. Structural characterization of En-1, a cold-adapted protein pheromone isolated from the Antarctic ciliate *Euplotes nobilii*. *Biochem. Biophys. Acta* **2003**, *1621*, 17–21.

26. Devi, V.S.; Sprecher, C.B.; Hunziker, P.; Mittl, P.R.; Bosshard, H.R.; Jelesarov, I. Disulfide formation and stability of a cysteine-rich repeat protein from *Helicobacter pylori*. *Biochemistry* **2006**, *45*, 1599–1607.

27. Trivedi, M.V.; Laurence, J.S.; Siahaan, T.J. The Role of thiols and disulfides on protein stability. *Curr. Protein Pept. Sci.* **2009**, *10*, 614–625.

28. Fass, D. Disulfide bonding in protein biophysics. In *Annual Review of Biophysics*; Rees, D.C., Ed.; Annual Reviews: Palo Alto, CA, USA, 2012; Volume 41, pp. 63–79.

29. Aghajari, N.; Feller, G.; Gerday, C.; Haser, R. Structures of the psychrophilic *Alteromonas haloplanctis* α-amylase give insights into cold adaptation at a molecular level. *Structure* **1998**, *6*, 1503–1516.

30. Bae, E.; Phillips, G.N., Jr. Structure and analysis of highly homologous psychrophilic, mesophilic, and thermophilic adenylate kinases. *J. Biol. Chem.* **2004**, *279*, 28202–28208.

31. Fedoy, A.E.; Yang, N.; Martinez, A.; Leiros, H.K.; Steen, I.H. Structural and functional properties of isocitrate dehydrogenase from the psychrophilic bacterium *Desulfotalea psychrophila* reveal a cold-active enzyme with an unusual high thermal stability. *J. Mol. Biol.* **2007**, *372*, 130–149.

32. Garcia-Arribas, O.; Mateo, R.; Melanie, M.; Tomczak, M.M.; Davies, P.L.; Mateu, M.G. Thermodynamic stability of a cold-adapted protein, type III antifreeze protein, and energetic contribution of salt bridges. *Protein Sci.* **2007**, *16*, 227–238.

33. Lockwood, B.L.; Somero, G.N. Functional determinants of temperature adaptation in enzymes of cold- *versus* warm-adapted mussels (genus *Mytilus*). *Mol. Biol. Evol.* **2012**, *29*, 3061–3070.

34. Vallesi, A.; di Giuseppe, G.; Dini, F.; Luporini, P. Pheromone evolution in the protozoan ciliate, *Euplotes*: The ability to synthesize diffusibile forms is ancestral and secondarily lost. *Mol. Phylogenet. Evol.* **2008**, *47*, 439–442.
35. Jiang, J.; Zhang, Q.; Warren, A.; Al-Rasheid, K.A.; Song, W. Morphology and SSUrRNA gene-based phylogeny of two marine *Euplotes* species, *E. orientalis* spec. nov. and *E. raikovi* (Ciliophora, Euplotida). *Eur. J. Protisotol.* **2010**, *46*, 121–132.

Microbial Analyses of Ancient Ice Core Sections from Greenland and Antarctica

Caitlin Knowlton, Ram Veerapaneni, Tom D'Elia and Scott O. Rogers

Abstract: Ice deposited in Greenland and Antarctica entraps viable and nonviable microbes, as well as biomolecules, that become temporal atmospheric records. Five sections (estimated to be 500, 10,500, 57,000, 105,000 and 157,000 years before present, ybp) from the GISP2D (Greenland) ice core, three sections (500, 30,000 and 70,000 ybp) from the Byrd ice core, and four sections from the Vostok 5G (Antarctica) ice core (10,500, 57,000, 105,000 and 105,000 ybp) were studied by scanning electron microscopy, cultivation and rRNA gene sequencing. Bacterial and fungal isolates were recovered from 10 of the 12 sections. The highest numbers of isolates were found in ice core sections that were deposited during times of low atmospheric CO_2, low global temperatures and low levels of atmospheric dust. Two of the sections (GISP2D at 10,500 and 157,000 ybp) also were examined using metagenomic/metatranscriptomic methods. These results indicated that sequences from microbes common to arid and saline soils were deposited in the ice during a time of low temperature, low atmospheric CO_2 and high dust levels. Members of Firmicutes and Cyanobacteria were the most prevalent bacteria, while *Rhodotorula* species were the most common eukaryotic representatives. Isolates of *Bacillus*, *Rhodotorula*, *Alternaria* and members of the Davidiellaceae were isolated from both Greenland and Antarctica sections of the same age, although the sequences differed between the two polar regions.

Reprinted from *Biology*. Cite as: Knowlton, C.; Veerapaneni, R.; D'Elia, T.; Rogers, S.O. Microbial Analyses of Ancient Ice Core Sections from Greenland and Antarctica. *Biology* **2013**, *2*, 206-232.

1. Introduction

Microorganisms have been identified and isolated from glacial ice samples from many different regions of the world. The global distribution of these microorganisms in snow and glacial ice is the result of wind and atmospheric circulation. Bacteria and fungi have been detected and isolated from many different frozen environments such as permafrost, cryopegs, sea ice, glacial ice, and accretion ice from subglacial lakes [1–17]. Bacteria and fungi have been isolated from both Arctic and Antarctic regions [4,5,7–9]. Most of the bacteria and fungi isolated from these permanently cold environments were psychrotolerant as opposed to psychrophilic, having optimal growth temperatures well above freezing [4,5]. Glacial ice has constant temperatures, making it ideal for long-term preservation of microorganisms and biomolecules [16–19]. Environmental ice acts as a protective matrix for microorganisms over extended periods of time and provides a record of microbial evolution and ancient biodiversity [8].

While the numbers of viable organisms and intact nucleic acid sequences decline with depth, they have been recovered from ice that is more than a million years old [4,5,14,16–18,20]. Study of the organisms isolated from environmental ice has yielded insights into microbial longevity. However, studies have not been performed to compare microorganisms belonging to similar timescales entrapped in glacial ice from geographically distant sites. The Arctic and the Antarctic are geographically distant regions, but also are distinct for other reasons. Much of Antarctica is a desert with little precipitation, and the continent is far from other large landmasses. In general, Greenland receives more precipitation, and is geographically less isolated, receiving winds from Europe, Asia and North America [21]. Study of microbes isolated from these two distinct locations, dating back to similar time periods may provide insights into microbial community composition through time and transportation of microbes in the atmosphere, as well as deposition and preservation of microbes and biomolecules in ice.

Sections from the GISP2D (Greenland Ice Sheet Project 2), Vostok 5G (Antarctica) and Byrd (Antarctica) ice cores were selected for comparison (Figure 1). Sections were chosen that were representative of times of high and low atmospheric carbon dioxide, dust and mean global temperature [22–26]. Scanning electron microscopy, cultivation, sequencing and phylogenetic analyses were used to assess the influences of atmospheric characteristics on microbe deposition and survival in ice. Metagenomic/metatranscriptomic analyses were used to determine and compare the microbial communities at times when carbon dioxide levels and temperatures were at minima (157,000 ybp) [22,23] and at a time when carbon dioxide levels and temperatures were rising nearly to modern levels (10,500 ybp) [22–26]. During the past 420,000 years, the concentration of atmospheric carbon dioxide has ranged from a low of approximately 200 ppm 150,000 years ago [22,23] to a high of >390 ppm currently [27], with another high peak of approximately 280 ppm occurring approximately 125,000 years ago [23]. Carbon dioxide levels vary directly with changes in global temperature. Low CO_2 levels are normally associated with low temperatures, low precipitation and high dust levels, while high CO_2 levels generally indicate high temperatures, high precipitation rates, and low dust levels, factors that may influence transportation of microbes and overall community composition. Here we report research that was designed to test whether the same species are concurrently deposited in ice at both poles, and whether changes in atmospheric carbon dioxide, dust and global temperature affect deposition of microbes in glacial ice.

2. Results and Discussion

2.1. Results

2.1.1. Scanning Electron Microscopy (SEM)

Most of the cells in the GISP2D and Vostok 5G core sections that were examined were rod-shaped bacteria (Figures 2 and 3). A few were coccoid, including diplococcoid forms (Figure 2, panels 16 and 22; Figure 3, panel 23). Two were spiral shaped (Figure 2, panels 18 and 19). Only three appeared to be fungi (Figure 3, panels 27–29), more complex shapes and larger cells, some of which may be conidia.

Figure 1. Locations of ice core sections used in this study. Sections from the Greenland Ice Sheet Project 2 (GISP2) (Greenland), Byrd (Antarctica) and Vostok 5G (Antarctica) ice cores were examined. See Experimental Section for more details about the core sections.

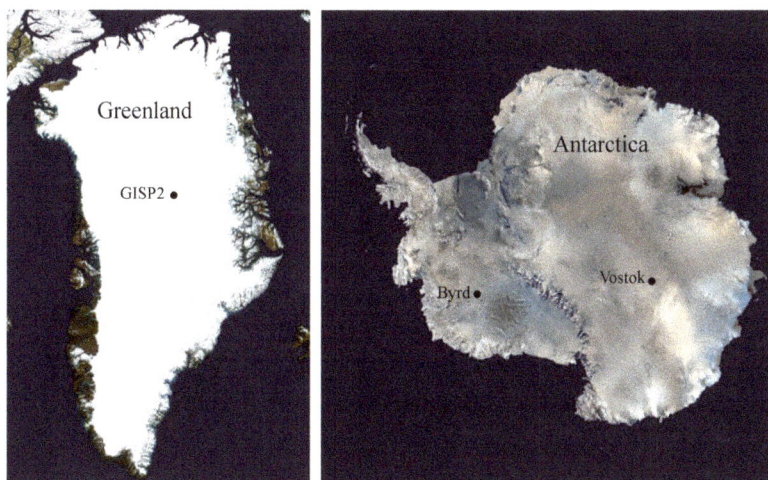

2.1.2. Revival and Molecular Identification of Glacial Isolates

There are geographical and temporal differences in the microbial composition of ice core samples (Table 1). Of the 27 cultures recovered from the 10 ice core sections, 19 were from Greenland ice and only eight were from Antarctica ice. Also, in general, the number of viable microbes, as indicated by number of cultures, was inversely proportional to the age of the ice core (Figure 4). The only exceptions to this were for the youngest ice sections that were analyzed, which were less than 500 years old (from Byrd and GISP2D). In both cases, few cultures were obtained. The ice core sections that were 10,500 years old from the GISP2D (1,601 m) and Vostok 5G (316 m) cores yielded higher numbers of cultures (seven fungi and two bacteria; Table 1). The number of cultures decreased with increasing age of the ice core sections (Table 1, Figure 4). There was a correlation between the number of cultures obtained and the atmospheric conditions at the time of ice deposition (Figure 5). Higher numbers of fungi and bacteria were isolated when atmospheric CO_2 levels were below 240 ppmv (parts per million per volume), temperatures were at least 3 °C lower than the present mean global temperature, and dust levels were <0.4 ppm (atmospheric measurements are from [23]).

Figure 2. Scanning electron micrographs (SEM) of bacteria from the GISP2D ice core sections (panels 1–6, 1,601 m; 7–11, 2,501 m; 12–18, 2,777 m; 19–22, 3,014 m). Microbes in micrographs 1, 3, 5 and 6 are rod shaped bacteria. The microbes in micrographs 1 and 6 have a sheath-like cover over the exterior, similar to the microbes previously observed in Vostok accretion ice (red arrow; [4]). The organisms in micrograph 3 are similar to the rod shaped bacteria observed in previous GISP2D sediment ice studies (white arrow; [28]). Microbes in micrographs 7 and 11 are similar to long rod shaped bacteria observed in the Vostok ice cores (green arrow; [4]). Organisms in micrograph 14 and 15 are rod shaped bacteria, similar to those previously observed from Vostok ice cores (blue arrow; [4]). The organism in micrograph 16 has a diplococcoid form (yellow arrow) and has been observed in other studies of the Vostok and GISP2D ice cores and from several cold environments including the Siberian permafrost [4,14,17]. This microbe is similar to *Psychrobacter*, a diplococcoid bacterium commonly found in the cold environments. The microbe in micrograph 18 has an unusual spiral structure (orange arrow) similar to organisms isolated from the Vostok core [4]. These may be related to the order Spirochaetales, which have long helically coiled cells. The microbes in micrographs 8 and 13 resemble cells of *Caulobacter* spp. while the curving form in micrograph 21 cannot be identified.

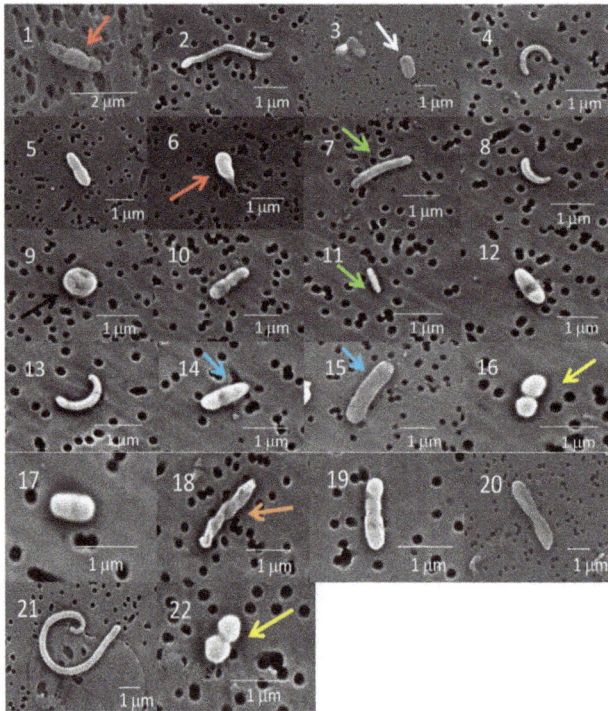

Figure 3. Scanning electron micrographs of bacteria from the Vostok 5G ice core section (micrographs 23–25, 316 m; 26–27, 900 m; 28–30, 1,529 m). No organisms were observed in the Vostok 5G 2,149 m ice core. Also, no cultures were obtained from this core section (Table 1). Diplococcoid organisms similar to the ones observed in the GISP2D cores were evident in micrograph 23 (yellow arrow). The microbe in micrograph 24 has a spiral structure and may be related to Spirochaetales. Spore-like structures were observed commonly (red arrow). A fungal hyphae-like structure was observed in the Vostok 5G 900 m ice core (panel 27). The spherical form in micrograph 29 appears to be a fungal coniduim. The microbes in Panels 25 and 30 are very small, similar to ultramicrobacteria found previously in the GISP2D ice core [28,29].

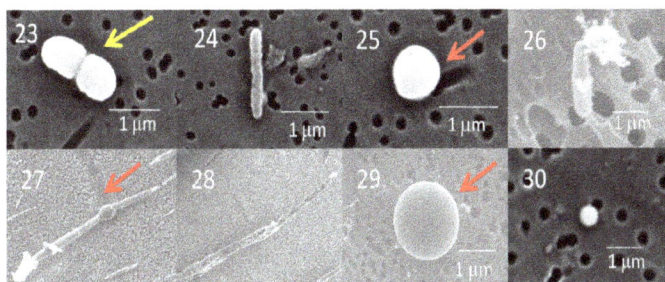

Table 1. Summary of BLAST search results. Included are sequences from cultures and metagenomic/metatranscriptomic analyses.

Core	Depth (m)	Age (ybp)	Sequence number [a]	Accession number	Closest BLASTn match [b]	Phylum [c]	Identity (%)
Byrd	99	<500	**CULT C-2**	KC146580	*Alternaria* sp.	As	99 [d]
	1,593	30,000	**CULT C-8**	KC146581	*Alternaria alternata*	As	99 [d]
			CULT A-1	KC146555	*Bacillus lichniformis*	Fi	99 [d]
	2,131	70,000	**CULT D-7**	KC146582	*Alternaria alternata*	As	99 [d]
GISP 2D	104	<500	**CULT B-1**	KC146556	*Bacillus thioparans*	Fi	100 [e]
	1,601	10,500	**GI860**	KC206482	*Alternaria alternata*	As	100 [d]
			c2581	KC155352	*Lactobacillus helveticus*	Fi	100 [d]
			s3382	KC146557	*Microcoleus vaginatus*	Cy	99 [e]
			c797	KC146560	*Microcoleus vaginatus*	Cy	99 [e]
			c1832	KC146558	Uncultured bacterium	Cy	99 [e]
			GI859	KC206491	*Fusarium culmorum*	As	98 [d]
			s1867	KC155351	*Lactobacillus crispatus*	Fi	98 [e]
			GI862	KC206492	*Penicillium corylophilum*	As	98 [d]
			GI867	KC206485	*Rhodotorula mucilacinosa*	Ba	98 [e]
			GI861	KC206488	*Cladosporium tennuissimum*	As	97 [d]
			GI866	KC206494	*Rhodotorula mucilacinosa*	Ba	97 [d]
			c2097	KC146554	Uncultured cyanobacterium	Cy	95 [e]
			GI865	KC206477	*Caulobacter crescentus*	Ap	94 [e]
			c4392	KC146571	*Escherichia coli*	Gp	93 [e]
			s3934	KC146559	Uncultured bacterium	Ac	93 [e]
			c3088	KC146584	*Halomonas neptunia*	Gp	92 [e]
	2,501	57,000	**GI868**	KC206495	*Aspergillus restrictus*	As	99 [d]

Table 1. *Cont.*

Core	Depth (m)	Age (ybp)	Sequence number [a]	Accession number	Closest BLASTn match [b]	Phylum [c]	Identity (%)
			GI869	KC206484	*Aureobasidium pullulans*	Ba	99 [d]
			GI871	KC206489	*Aspergillus conicus*	As	98 [d]
			GI873	KC206481	*Rhodotorula mucilacinosa*	Ba	98 [d]
			GI872	KC206478	*Caulobacter crescentus*	Ap	86 [e]
	2,777	105,000	**GI875**	KC206483	*Cryptococcus magnus*	Ba	99 [d]
			GI876	KC206487	*Rhodotorula mucilacinosa*	Ba	99 [d]
			GI874	KC209502	*Bacillus subtilis*	Fi	98 [e]
	3,014	157,000	c2575	KC146576	*Lactobacillus helveticus*	Fi	100 [f]
			c4301	KC146578	*Lactobacillus helveticus*	Fi	100 [f]
			c1729	KC146569	*Micrococcus luteus*	Ac	100 [f]
			CULT E-5	KC146583	*Cladosporium tenuissimum*	As	99 [d]
			s3361	KC146561	*Microcoleus vaginatus*	Cy	99 [e]
			c1654	KC146563	*Microcoleus vaginatus*	Cy	99 [e]
			c552	KC146564	Uncultured bacterium	Cy	99 [e]
			c1812	KC146565	Uncultured bacterium	Cy	99 [e]
			c2330	KC146567	Uncultured bacterium	Fi	99 [e]
			c785	KC146568	Uncultured bacterium	Cy	99 [e]
			c3738	KC146577	*Lactobacillus helveticus*	Fi	98 [f]
			c833	KC146573	*Lactobacillus helveticus*	Fi	98 [e]
			c1018	KC146574	*Lactobacillus helveticus*	Fi	98 [f]
			GI855	KC206493	*Penicillium chrysognum*	As	98 [d]
			GI858	KC206480	*Rhodotorula mucilacinosa*	Ba	98 [d]
			c1826	KC146566	Uncultured bacterium	Fi	98 [e]
			c3135	KC146552	Uncultured cyanobacterium	Cy	98 [e]
			c3856	KC146553	Uncultured cyanobacterium	Cy	98 [e]
			s1137	KC146575	*Lactobacillus helveticus*	Fi	97 [e]
			c2005	KC146570	*Micrococcus luteus*	Ac	97 [e]
			c67	KC146562	*Microcoleus vaginatus*	Cy	97 [e]
			c3509	KC146585	*Brevudimonas diminuta*	Ap	95 [f]
			c3833	KC146587	Uncultured bacterium	Bp	93 [e]
			c2683	KC146579	*Pseudomonas stutzeri*	Gp	91 [f]
			c4709	KC146572	*Bordetella petrii*	Bp	89 [f]
			s3724	KC146586	*Bordetella pertussis*	Bp	88 [f]
			c2094	KC146588	*Geobacillus* sp.	Fi	84 [g]
Vostok 5G	316	10,500	**GI878**		*Bacillus amyloliquifaciens*	Fi	100 [e]
			GI877		*Davidiella tassiana*	As	99 [d]
	900	57,000	None		----------	--	----------
	1,529	105,000	**GI879**		*Alternaria tenuissimum*	As	100 [d]
			GI880		*Rhodotorula mucilacinosa*	Ba	99 [d]
	2,149	157,000	None		----------	--	----------

[a] Cultures are indicated in bold font. Metagenomic/metatranscriptomic sequences are in standard font. [b] Fungi are in red font. [c] Phyla: Ac—Actinobacteria; Ap—Alphaproteobacteria; As—Ascomycota; Ba—Basidiomycota; Bp—Betaproteobacteria; Cy—Cyanobacteria; Fi—Firmicutes; Gp—Gammaproteobacteria. [d] Ribosomal RNA internal transcribed spacer (ITS1 and ITS2). [e] Ribosomal RNA small subunit gene. [f] Ribosomal RNA large subunit gene. [g] Glycerol kinase gene.

Figure 4. Number of cultures from GISP2D, Byrd and Vostok 5G ice core sections, based on age of ice. Black bars indicate total number of cultures, red bars indicate fungal cultures and blue bars indicate bacterial cultures.

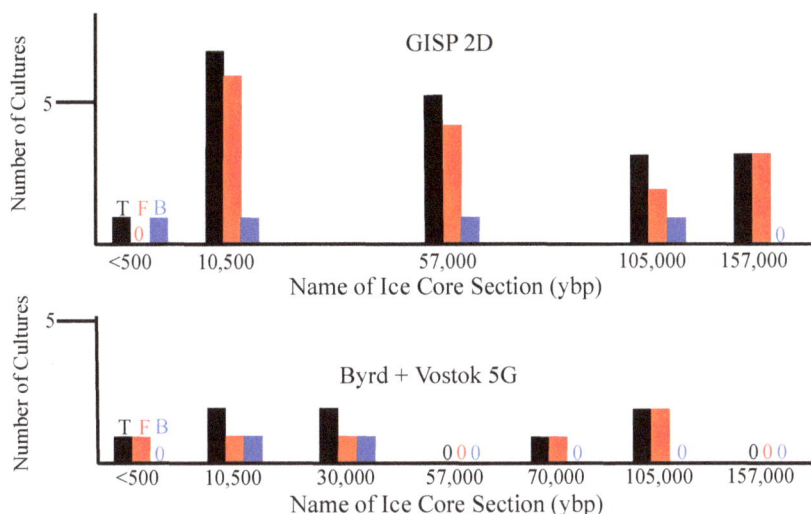

Figure 5. Cultures obtained from ice core sections based on atmospheric conditions at the time of deposition. Black bars indicate total number of cultures, red bars indicate fungal cultures and blue bars indicate bacterial cultures. Values for atmospheric measurements are from reference [23].

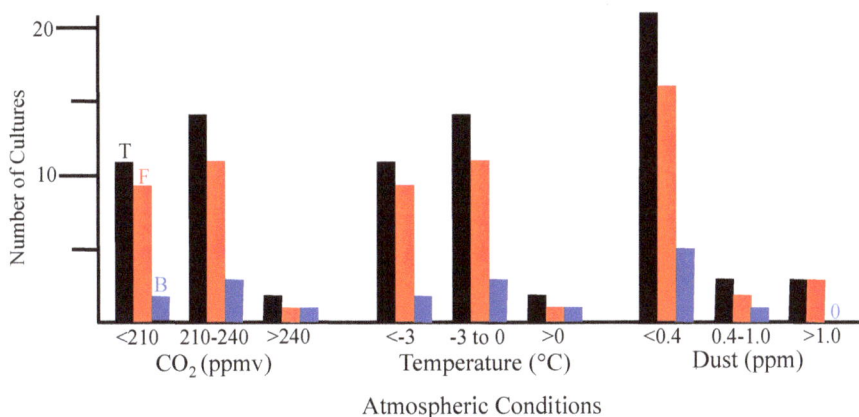

2.1.3. Metagenomic/metatranscriptomic Analyses

A total of 107,700 high quality 454 sequence reads were obtained from the sequencing facility. After categorization by the MID primers, there were 25,748 high quality reads associated with the MID4 (GISP2D, 1,601 m) primer, 29,506 associated with the MID5 (GISP2D, 3,014 m) primer. 18,015 high quality reads associated with the MID6 (control 1) primer, and 34,063 high quality reads associated with the MID10 (control 2) primer. After assembly, a total of 15,891 contigs resulted. Sequences that were in common with the control samples were removed, leaving 33 sequences that were unique to the ice for analysis (Table 1). Nine unique Basic Local Alignment Search Tool (BLAST) hits were present in the 10,500 ybp GISP2D core section and 24 unique hits were present in the 157,000 ybp core section. All of the metagenomic/metatranscriptomic sequences were from bacteria (Table 1). There were five species common to both samples. All others were unique to one sample or the other. In the 10,500 ybp sample, all nine sequences were unique, while in the 157,000 ybp sample, 15 species were indicated among the 24 sequences.

In general, the taxa in the cores were similar to organisms found in polar ice (both Arctic and Antarctic), soil, dust, water, hot springs, marine environments, deep-sea thermal vents and associated with animals. A number of isolates and sequences recovered from the ice in this study were those that have yet to be scientifically described. In the National Center for Biotechnology Information (NCBI) database, they are termed "environmental taxa" or "unidentified" or "uncultured." In the 10,500 ybp ice, five of the sequences were closest to unidentified environmental taxa that were similar to *Microcoleus* and uncultured bacteria. Also, the 10,500 ybp GISP2D ice sample contained one sequence closest to a sequence of a *Halomonas* isolate that lives in marine environments, including deep-sea thermal vents. In the 157,000 ybp ice section sample, fifteen sequences were closest to those from unidentified environmental species. They were closest to members of *Geobacillus*, *Micrococcus*, *Microcoleus* and *Pseudomonas*. Five were associated with soil, three specifically with arid land soils, and two sequences were closest to *Micrococcus*, known to be saprotrophic.

2.1.4. Phylogenetics

For bacteria, phylogenetic trees were produced using SSU rRNA (small subunit ribosomal RNA; Figure 6) and LSU rRNA (large subunit rRNA; Figure 7) sequences. For fungi, the rRNA internal transcribed spacers (ITS1 and ITS2) were used, due to their increased precision for species determination for Eukarya. Phylogenetics was used to confirm the BLAST search results, as well as to refine species identities. The bacteria grouped into four clades, corresponding to phyla: Actinobacteria, Cyanobacteria, Firmicutes, and Proteobacteria (including Alphaproteobacteria, Betaproteobacteria and Gammaproteobacteria), the majority being in either the Cyanobacteria or the Firmicutes. Three sequences from the 157,000 year old ice core were closest to sequences from *Microcoleus vaginatus* and *Phormidium autumnale* that were isolated from soils, desert soils or deglaciated soil (including one from Antarctica). The other six sequences formed a separate Cyanobacteria clade, which was within the larger clade that included sequences from other Antarctic and arid soil taxa. The Actinobacteria clade included three sequences, all closely allied with sequences from environmental samples from soil or dust. Two were closest to sequences from *Micrococcus luteus*. Four sequences grouped within the Firmicutes clade. Two were from cultures whose sequences were within the genus *Bacillus*. In previous studies of Greenland ice cores, several *Bacillus* spp. were isolated [3]. Two metagenomic/metatranscriptomic sequences from the 157,000 year old ice core section most closely matched sequences from *Marinococcus* sp. and *Sinococcus beijingensis*, both of which had been isolated from saline soils. This is consistent with the low temperature and high atmospheric dust concentration at that time [23].

Figure 6. Phylogram of bacterial small subunit ribosomal RNA (SSU rRNA) sequences (maximum parsimony using phylogenetic analysis using parsimony (PAUP) [30]) from the GISP2D and Byrd ice cores. Sequences determined in this study are indicated in bold font. Sequences that start with CULT are isolates. All other sequences in bold were from the metagenomic/metatranscriptomic analysis. Ice core location and depth (meters below the glacial surface) are indicated. Phyla are on the right. The closest National Center for Biotechnology Information (NCBI) sequences to the queried sequences were selected from BLAST search results for use in the phylogenetic analysis. NCBI accession numbers (in parentheses) and sources of isolates (square brackets) are provided. Bootstrap values (1,000 replications) are given for branches with support greater than 50%.

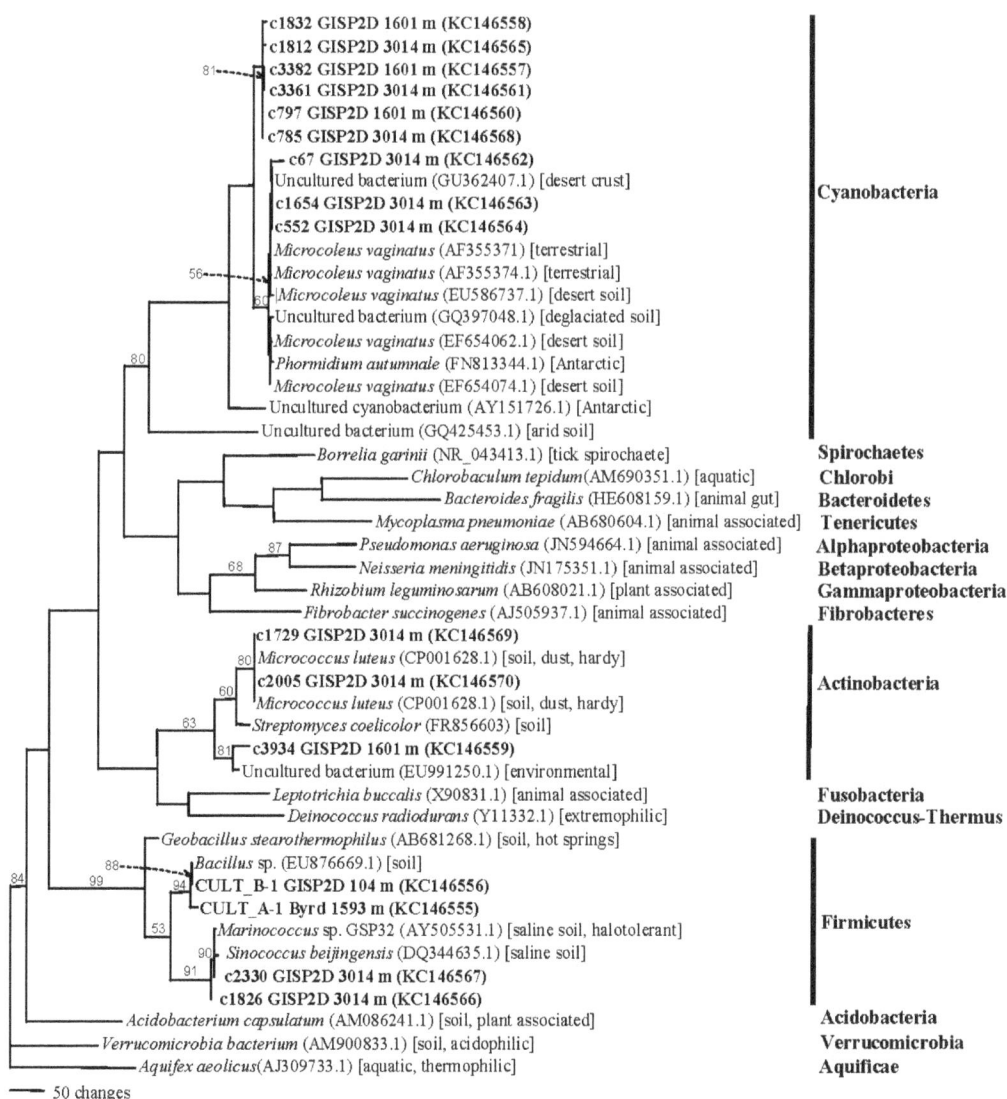

Figure 7. Phylogram of bacterial large subunit rRNA (LSU rRNA) sequences (maximum parsimony using PAUP [30]) from the GISP2D ice core. Sequences determined in this study are indicated in bold font. Ice core location and depth (meters below the glacial surface) are indicated for each. Phyla are on the right. The closest NCBI sequences to the queried sequences were selected from BLAST search results for use in the phylogenetic analysis. NCBI accession numbers (in parentheses) and sources of isolates (square brackets) are provided. Bootstrap values (1,000 replications) are given for branches with support greater than 50%.

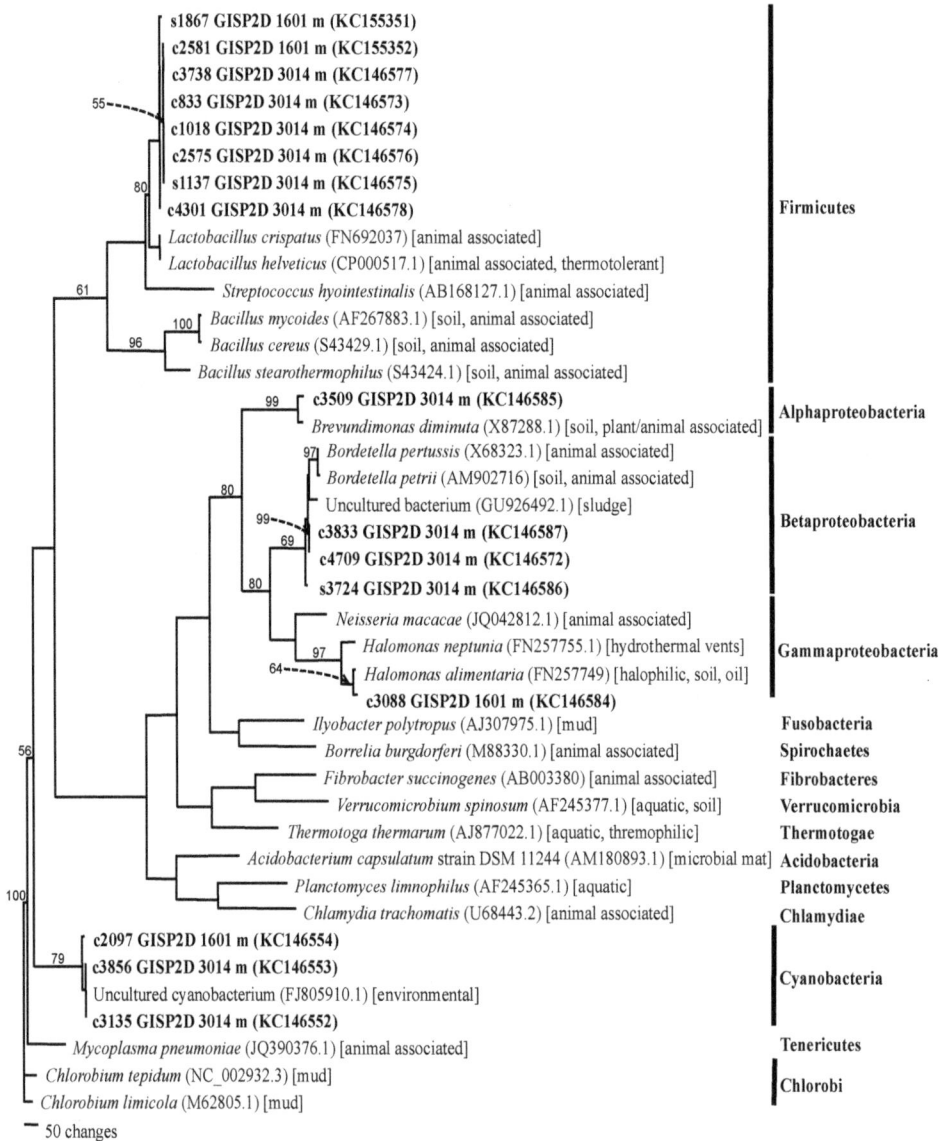

Fewer sequences from Cyanobacteria were found in the LSU rRNA gene data set. However, they all were within a clade that included an uncultured cyanobacterial sequence from an unidentified environmental sample. Firmicute sequences predominated in the LSU rRNA data set. All were most closely allied with sequences from *Lactobacillus crispatus* and *L. helveticus*. Both are associated with animals, but similar taxa can be found in soils. Similarly, one Alphaproteobacterium and three Betaproteobacteria were most closely related to sequences from bacteria associated with animals, but which can also be found in soil and sludge. One sequence most closely matched sequences from Gammaproteobacteria. It was closest to members of the halophilic and thermotolerant genus *Halomonas*. Species of *Halomonas* and *Lactobacillus* are thermotolerant. Coincidentally, dark granular material was found in the 10,500-year-old ice core section that resembles volcanic ash (data not shown).

All of the fungi that were isolated and sequenced (internal transcribed spacer (ITS) regions) were either in the Basidiomycota or the Ascomycota. *Rhodotorula* was the most common genus (Figure 8). Several isolates (GI858, GI867, GI873 and GI876) were recovered from sections of the GISP2D ice core that were 10,500, 57,000 and 157,000 years old, as well as from a 105,000-year-old Vosok 5G ice core section (GI880). Most grouped closest to *R. mucilaginosa, R. glutius* and related isolates that had previously been recovered from marine, deep-sea or Lake Vostok ice samples. However, isolate GI866 was distant from all other isolates, as well as from sequences on NCBI. While it appears that it could be within the Sporidiobolales, it might be outside of the genus *Rhodotorula*.

Isolates of *Penicillium* and *Aspergillus* also were common. The ITS sequence from isolate GI855 (from GISP2D, 3,014 m) was closest to *P. chrysogenum* isolates from marine and deep-sea sediments. Another isolate (GI862, from GISP2D, 1,601 m) also was close to a deep-sea isolate. Two isolates (GI868 and GI871, both from Vostok 5G, 2,501 m) were closest to an isolate of *Aspergillus* from Lake Vostok accretion ice. All of these isolates, as well as the isolates described in the NCBI database originated from cold high-pressure environments. This is likely a common trait for organisms that are able to survive in deep ice. The sequence of isolate GI869 (from Vostok 5G, 2,501 m) is closest to marine isolate within the Dothioraceae, which includes the genus *Aureobasidium*. However, the sequence differs from all other such sequences in this family on the NCBI database, as indicted by the branch length (Figure 8). Sequences from two isolates (GI861, from GISP2D, and GI877, from Vosok 5G; both at 10,500 ybp) are within the *Cladosporium/Davidiella* complex. They were closest to marine and Greenland ice taxa [8].

One isolate of *Alternaria* was recovered from the 105,000-year-old section from Vostok 5G. The sequence was closest to *A. tenuissima*, as well as to isolates from Greenland and from soil. One isolate (GI859) was closest to sequences from *Fusarium* isolates recovered from Greenland and northern Spain. Isolate GI860 (from GISP2D, 1,601 m) could not be placed within a specific taxon. It appears to be approximately midway between *Alternaria* and *Fusarium* (Figure 8). One additional basidiomycete was isolate GI875 (from GIDP2D, 2,777 m), whose sequence is closest to *Cryptococcus magnus*, a species isolated from soil that is also an opportunistic pathogen of humans and other mammals.

Figure 8. Phylogram of fungal rRNA internal transcribed spacer (ITS) sequences (with maximum parsimony on PAUP, [30]) from the GISP2D and Vostok 5G ice cores. Sequences that start with GI (in bold font) are the isolates from the ice sections. Ice core location and depth (meters below the glacial surface) are also given. Genera and phyla designations are on the right. Organisms belonging to the genera *Rhodotorula*, *Alternaria* and *Cladosporium* have previously been isolated from the GISP2D and the Vostok 5G ice cores. The closest NCBI sequences to the queried sequences were selected from BLAST search results for use in the phylogenetic analysis. NCBI accession numbers (in parentheses) and sources of isolates (square brackets) are provided. Bootstrap values (1,000 replications) are given for branches with support greater than 50%.

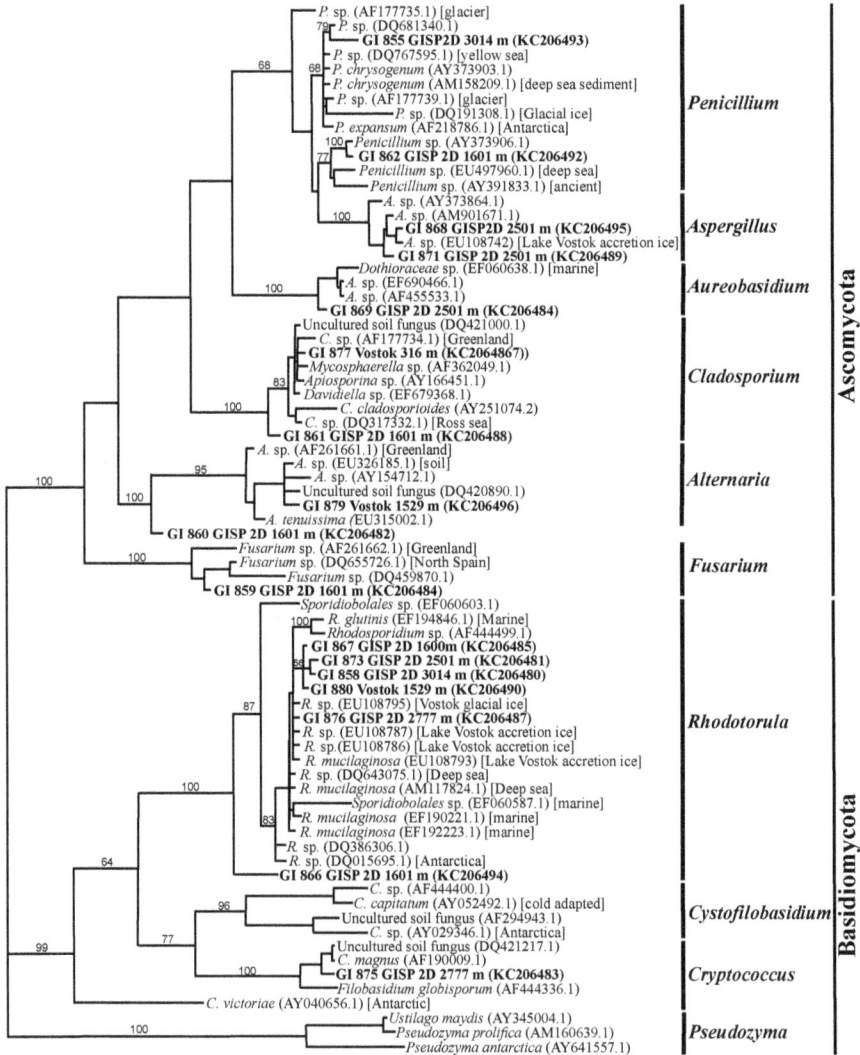

P. sp. (AF177735.1) [glacier]
79 *P.* sp. (DQ681340.1)
GI 855 GISP2D 3014 m (KC206493)
P. sp. (DQ767595.1) [yellow sea]
68 — 68 *P. chrysogenum* (AY373903.1)
P. chrysogenum (AM158209.1) [deep sea sediment]
P. sp. (AF177739.1) [glacier]
P. sp. (DQ191308.1) [Glacial ice]
P. expansum (AF218786.1) [Antarctica]
100 *Penicillium* sp. (AY373906.1)
77 **GI 862 GISP 2D 1601 m (KC206492)**
Penicillium sp. (EU497960.1) [deep sea]
Penicillium sp. (AY391833.1) [ancient]   **Penicillium**

A. sp. (AY373864.1)
100 *A.* sp. (AM901671.1)
GI 868 GISP2D 2501 m (KC206495)
A. sp. (EU108742) [Lake Vostok accretion ice]
GI 871 GISP 2D 2501 m (KC206489)   **Aspergillus**

Dothioraceae sp. (EF060638.1) [marine]
100 *A.* sp. (EF690466.1)
A. sp. (AF455533.1)   **Aureobasidium**
GI 869 GISP 2D 2501 m (KC206484)

Uncultured soil fungus (DQ421000.1)
C. sp. (AF177734.1) [Greenland]
GI 877 Vostok 316 m (KC2064867))
Mycosphaerella sp. (AF362049.1)
83 *Apiosporina* sp. (AY166451.1)
Davidiella sp. (EF679368.1)
100 *C. cladosporioides* (AY251074.2)
C. sp. (DQ317332.1) [Ross sea]   **Cladosporium**
GI 861 GISP 2D 1601 m (KC206488)

A. sp. (AF261661.1) [Greenland]
95 *A.* sp. (EU326185.1) [soil]
A. sp. (AY154712.1)
100 — 100 Uncultured soil fungus (DQ420890.1)
GI 879 Vostok 1529 m (KC206496)   **Alternaria**
A. tenuissima (EU315002.1)
GI 860 GISP 2D 1601 m (KC206482)

100 *Fusarium* sp. (AF261662.1) [Greenland]
Fusarium sp. (DQ655726.1) [North Spain]
Fusarium sp. (DQ459870.1)   **Fusarium**
GI 859 GISP 2D 1601 m (KC206484)

Sporidiobolales sp. (EF060603.1)
100 *R. glutinis* (EF194846.1) [Marine]
Rhodosporidium sp. (AF444499.1)
GI 867 GISP 2D 1600m (KC206485)
GI 873 GISP 2D 2501 m (KC206481)
GI 858 GISP 2D 3014 m (KC206480)
GI 880 Vostok 1529 m (KC206490)
87 *R.* sp. (EU108795) [Vostok glacial ice]
GI 876 GISP 2D 2777 m (KC206487)
R. sp. (EU108787) [Lake Vostok accretion ice]
R. sp.(EU108786) [Lake Vostok accretion ice]
R. mucilaginosa (EU108793) [Lake Vostok accretion ice]
R. sp. (DQ643075.1) [Deep sea]   **Rhodotorula**
100 *R. mucilaginosa* (AM117824.1) [Deep sea]
Sporidiobolales sp. (EF060587.1) [marine]
83 *R. mucilaginosa* (EF190221.1) [marine]
R. mucilaginosa (EF192223.1) [marine]
R. sp. (DQ386306.1)
R. sp. (DQ015695.1) [Antarctica]
GI 866 GISP 2D 1601 m (KC206494)

C. sp. (AF444400.1)
96 *C. capitatum* (AY052492.1) [cold adapted]
Uncultured soil fungus (AF294943.1)   **Cystofilobasidium**
C. sp. (AY029346.1) [Antarctica]
77 Uncultured soil fungus (DQ421217.1)
100 *C. magnus* (AF190009.1)
GI 875 GISP 2D 2777 m (KC206483)   **Cryptococcus**
Filobasidium globisporum (AF444336.1)

C. victoriae (AY040656.1) [Antarctic]
100 *Ustilago maydis* (AY345004.1)
Pseudozyma prolifica (AM160639.1)   **Pseudozyma**
Pseudozyma antarctica (AY641557.1)

99
64

Ascomycota
Basidiomycota

— 10 changes

2.2. Discussion

2.2.1. Atmospheric Conditions and Microbial Deposition

Microbial cells are transported by wind, clouds and precipitation [12,31]. Some become entrapped in glaciers, becoming records of the atmosphere at specific points in time. In this study, two major questions are addressed: (1) Are the same species of microbes concurrently deposited in ice in the Arctic and Antarctic?, and (2) Do atmospheric conditions affect the deposition of microbes in polar ice? The answer to Question 1 is, no. The sequences found in the Arctic differed from those found during the same time periods in the Antarctic, although very similar sequences were sometimes found in ice from both polar regions (Table 1, Figures 6–8). The answer to Question 2 is, yes. The highest numbers of isolates were from ice core sections that contained ice deposited during times of moderate to low atmospheric CO_2 (<240 ppmv), low temperature (<−3 °C below current global mean) and low dust (<0.4 ppm). These correlations to the cultivation results may indicate increased preservation of the microbes during times of colder global mean temperatures. The metagenomic/metatranscriptomic data indicated that sequences from microbes common to arid and saline soils were deposited in the ice during a time of low temperature, low atmospheric CO_2 and high dust levels. The presence of nucleic acids does not indicate whether viable cells are present, and therefore, some of the organisms represented in these samples may have been nonviable. The metagenomic/metatranscriptomic data might be more indicative of the nucleic acids that are in the highest concentrations in the ice, rather than an indication of intact microbes.

The lower number of microbes in the Vostok sections could partly be due to the high elevation of the site and the fact that the Vostok site lies within a cold desert that receives much less snowfall compared to Greenland. Conversely, the higher number of microbes isolated from the Greenland ice cores may be related to a much warmer climate in the recent history compared to the Antarctic cores. During the Eemian interglacial (130,000 to 114,000 ybp) the northern hemisphere had warm temperatures comparable or higher to the temperatures of the Holocene period [32,33]. During this period, part of Greenland (southernmost part of Greenland) was covered with forests including a diverse array of conifers [34]. The warmer temperatures along with the higher precipitation and the geographically closer location to forests may be responsible for the higher number of organisms isolated from the Greenland ice cores. Distances to the nearest land masses with temperate climates likely contributed to the differences observed. The GISP2D site lies relatively close to North America, Europe and Asia, and winds often intersect the GISP2D site from these regions. On the other hand, most of the winds that reach the Vostok and Byrd sites originate in Antarctica, and the closest land mass is the narrowest part of South America. Therefore, the deposition of organisms from land masses onto Antarctica glaciers would be expected to be much less than onto Greenland glaciers. The numbers of cosmopolitan species isolated in Arctic and Antarctic regions along with spore trap data add credence to the point [35].

2.2.2. Isolated Fungi and Bacteria

All of the bacterial and fungal isolates grew at 22 °C. These isolates do not fit the profile of psychrophiles, whose optimal growth temperature is below 15 °C and do not grow at temperatures above 20 °C. The isolates can be classified as psychrotolerant, as they can survive and grow at low temperatures while having the optimal and maximum growth temperatures above 20 °C. Earlier studies indicated that most of the polar ice habitats are dominated by psychrotolerant organisms [4,5,31,36]. Psychrophiles thrive and flourish in continuously cold environments like the sea ice where long periods of sustained cold temperatures act as the major selective pressure. However, in the polar regions, sharp and rapid changes in seasonal temperatures act as the main selective pressure. Significant environmental changes may kill many true psychrophiles, while generalists such as psychrotolerant species have better chances to survive. The organisms in these environments should be able to survive freeze thaw cycles that occur during the seasonal changes. Sharp changes in the surface temperatures have been observed in the Antarctic soils and have been recorded to go as high as 30 °C [37]. If fungi and bacteria from these surrounding environments are deposited in the glacial ice, we would expect to find psychrotolerant species among our isolates. It is important to note that the highest number of cultures were from ice core sections that were deposited during times of cold global temperature (e.g., 30,000, 70,000 and 157,000 ybp; Table 1, Figure 5). This might be caused by increased preservation of the microbes during transport and deposition due to the cold temperatures in the atmosphere.

Species from three genera were isolated from both Greenland and Antarctic ice: *Alternaria*, *Bacillus* and *Rhodotorula*. However, in all cases, the species, as indicated by the rRNA sequences, differed from one pole to the other (Table 1, Figures 6–8). The taxon that was most often isolated was the genus *Rhodotorula* (Table 1, Figure 8), as was the case in our previous studies of Greenland and Antarctic ice core sections [5,12]. Members of this genus are hardy and adaptable, and have been isolated frequently from polar regions [12]. All of the *Rhodotorula* isolates were pigmented (shades of yellow, orange, pink and red), which might be important to survival in cold environments, and might explain their abundance in the cores analyzed [38]. *Rhodotorula* spp. remain viable after many freeze thaw cycles, which is another explanation for the abundance in the ice cores analyzed [12].

Isolates from the Greenland core sections included sequences closely related to *Aspergillus* and *Penicillium*. Isolate KC206495 (99% similar to *Aspergillus restrictus*) and KC206489 (99% similarity to *Aspergillus conicus*) were isolated from the 57,000-year-old ice core. Isolate KC206492 (98% similarity with *Penicillium corylophilum*) was isolated from the 10,500-year-old ice core and KC206493 (98% sequence similarity with *Penicillium chrysogenum*) was isolated from a 157,000-year-old ice core. Several species of *Penicillium* are tolerant to cold conditions and are known to survive for long periods of time [39]. Four of the isolates from Greenland ice cores were closely related to the species of *Cladosporium*, *Cryptococcus* and *Aureobasidium*. All have been frequently isolated from cold environments, such as Antarctic moss, which is one of the richest microhabitats for fungi [40]. *Cladosporium* spores have been found in high numbers in the air, especially in the temperate and tropical environments; the total fungal spore load in the Arctic region air is only about 5% [38]. Spore counts in the Antarctic atmosphere varied with the seasons, and highest spore counts were observed during the summer months, attributed to winds from northern regions [38]. One of the isolates from the Vostok ice core, one from GISP2D and three from the Byrd ice core were closely related to *Alternaria* sp. isolated from cryopegs in Siberia [15]. All of the cultured fungal sequences from this study had a 97% similarity or higher to NCBI sequences.

Several bacterial species isolated from the ice cores had high sequence similarities to the genera *Bacillus* and *Caulobacter*. Organisms belonging to these genera are tolerant to low temperatures and have been isolated from glacial and accretion ice samples from Lake Vostok [4]. Castello *et al.* [3] performed an extensive study of Greenland glacial ice (GISP2D and Dye 3) to report isolation of *Bacillus* through culturing meltwater. In a very similar method to theirs, we obtained our sequences related to *Bacillus*. All of the organisms isolated from the Arctic and Antarctic ice cores were closely related to species that are known to survive and thrive in cold environments and have been isolated from similar environments, indicating that the isolates probably are from the glacial ice and not contaminants.

2.2.3. Metagenomic/Metatranscriptomic Analysis

While we saw a decrease in the number of viable microbes cultured from older ice (Figure 4), the metagenomic/metatranscriptomic results suggest that more nucleic acid (probably carried within living and dead microbes) deposition in glaciers occurred during periods of low CO_2, lowered temperatures and high dust (e.g., 157,000 ybp). The cultivation results indicated that more viable microbes were present in ice that had been deposited in the glacier during periods of moderate to low CO_2, moderate to low temperatures and low dust (e.g., 500, 10,500 and 105,000 ybp). Together, the results indicate that cold temperatures are needed to retain cell viability, and to lessen degradation of nucleic acids. While decreases in temperature and CO_2 were positively correlated with the recovery of viable microbes, increases in the amount of dust were negatively correlated. There have been reports of a variety of organisms present in Greenland ice (e.g., Ma *et al.* [8,9]). It has been shown that more long-range transport and deposition of microbes occurs during times of lower temperature, but it has been assumed that dust particles are responsible for the increases in transport. The results presented here argue against this assumption. Rather, temperature might be more important than the presence of dust particles. Microorganisms can be deposited through precipitation. Reports of Actinobacteria, Firmicutes, Proteobacteria (primarily Alphaproteobacteria, Betaproteobacteria and Gammaproteobacteria) and Bacteroidetes, and genera such as, *Pseudomonas*, *Sphingomonas*, *Staphylococcus*, *Streptomyces* and *Arthrobacter* were found in tropospheric clouds [31]. These phyla are often found in cold environments, freshwater, marine water, soil or vegetation. Also, fog water bacteria consisted of *Pseudomonas*, *Bacillus*, *Actinetobacter* and several fungi [31]. It may be that most of the organisms isolated in this study were preserved and deposited in snow, not dust.

Another important component to long-term survival is the ability to survive cold and desiccation. An example of a microbe found in arid conditions that was found in the ice cores was *Microcoleus vaginatus* (99% identity). This bacterium is found most often in biocrusts of arid land [41–43]. The phylogenetic analyses based on the SSU rRNA gene sequences (Figure 6) indicated that the majority of microbes that were in the 157,000 year old ice originated from soils, especially those from deserts and saline soils, and that they were hardy and desiccation resistant. This is consistent with conditions at the time, which were dry and cold [23]. Cyanobacteria and Firmicutes were present in the Greenland (GISP2D) ice core sections (Figures 5 and 6), including many that are similar to species from soil, arid soil and polar regions. Microorganisms in both hot and cold arid environments have mechanisms for adaptation to extremes in climate (temperature, salt levels and lack available water), such as accumulation of organic compounds that aid in resistance to desiccation and freezing [41]. Cyanobacteria have been reported to enrich the soil stability due to the protection and adaptability of the polysaccharide sheath that surrounds the filaments [42]. *Microcoleus vaginatus* is capable of synthesizing and degrading trehalose [41] from maltose and is highly motile [43]. Trehalose is a cryoprotectant that can help this organism survive in the ice [44,45]. Firmicutes surround themselves with sheaths that protect against desiccation, osmotic shock and temperature changes.

The results presented here confirm that an overwhelming majority of sequences in ancient ice are derived from bacteria [1,4,5,46], while the greatest number of isolates were fungi. This might be due to the relative numbers of total microbes to the total viable microbes in each core section, although rigorous testing of this supposition will have to be performed in subsequent research. There was a diverse assemblage of bacterial phyla in the two ice samples analyzed by metagenomics/metatranscriptomics. In general, it is consistent with previous research, which indicates that Proteobacteria and Actinobacteria are commonly present in glacial ice [47,48]. Castello *et al.* [3] reported finding a sequence related to *Rhodococcus erythreus*, similar to our two sequences closely related to *Micrococcus luteus*, which are also in the Actinobacteria. In one study by Christner *et al.* [49], molecular analysis of glacial ice from a variety of global locations indicated approximately 50% of organisms to be Gram-positive, spore forming organisms. This could be due to the thick layer of peptidoglycan that provides strength and protection of the cell. Many of the microbes found in this study also form spores. Studies have shown Actinobacteria, Firmicutes, Proteobacteria and Fungi to be in the highest proportions in ice cores [48]. In our analyses, the same taxa were also found, with the addition of Cyanobacteria. The number of members of the Cyanobacteria was lower in the 10,500 ybp core section than in the 157,000 ybp section. The reason for this remains unclear.

2.2.4. Global Mixing, Gene Flow and Genome Recycling

Organisms were isolated from 10 of the 12 ice cores used. However, there were differences among the ice core sections. Ice cores from the Arctic region consistently contained more viable organisms than those from Antarctica. Furthermore, species differed in the Arctic and Antarctic, suggesting that atmospheric transport, gene flow and genome recycling [50] between the poles occur infrequently. It appears that the two polar regions entrap microbes only from geographically local regions. Therefore, it is likely that latitudinal global mixing, leading to the global distribution of single genotypes, is inefficient over the long distance between the polar regions. Additionally, there appears to be a temporal component to preservation and deposition in polar ice. Far more isolates and sequences were found in ice core sections that had been deposited during periods of global cooling, when dust levels also were low. Thus, local and global conditions appear to be influential in the patterns of preservation of microbes and nucleic acids in Arctic and Antarctic ice.

3. Experimental Section

3.1. Ice Core Selection and Samples

Three ice core sections (104 m, 1,593 m, and 2,131 m; approximately 500, 30,000, and 70,000 ybp, respectively; [51]) from Byrd Station (80°1' S, 119°31' W, elevation 1,553 m), Antarctica; five core sections (99 m, 1,601 m, 2,501 m, 2,777 m and 3,014 m; approximately 500, 10,500, 57,000, 105,000 and 157,000 ybp, respectively [23]) from GISP 2D (Greenland Ice Sheet Project, core 2D; 72°36' N, 38°30' W, elevation 3,203 m); and four sections (316 m, 900 m, 1,529 m and 2,149 m; approximately 10,500, 57,000, 105,000 and 157,000 ybp, respectively [22,24,26]) from the Vostok 5G ice core (Vostok Station, Antarctica, 78°52' S, 106°50' W, elevation 3,488 m) were selected (Figures 1 and 9). The ice cores sections were chosen based on atmospheric CO_2, dust, and global temperature at the time of ice deposition [23].

3.2. Surface Sterilization and Melting

The outer surfaces of the ice core sections were decontaminated before melting. Clorox (a 5.25% sodium hypochlorite solution) was used as the decontaminating agent. The decontamination protocol developed by Rogers et al. [52] was used for this procedure.

3.3. Scanning Electron Microscopy

An aliquot (meltwater from ice cores) of 5 mL was filtered through a 0.2 μm polycarbonate filter using a sterile syringe. The filter was transferred to a Petri dish with 2.5% glutaraldehyde in 0.1 M phosphate buffer (pH 7.2). The filter was fixed in this solution for 1 h, followed by three 10 min rinses in 0.1 M phosphate buffer (pH 7.2). The filter was dehydrated with 40%, 60%, 80% and 95% ethanol solutions, sequentially for 10 min each and finally with 100% ethanol three times (10 min each). After dehydration, the filter was dried using a Samdri 780A critical point dryer. Then, the filter was cut into four pieces to be mounted into a coating unit. A Polaron E500 SEM coating unit was used to sputter coat the filters with a 5 nm thick gold-palladium coat. The filters then were observed in a scanning electron microscope (Hitachi S-2700, SEM). Control filters were prepared for SEM using sterilized water, taken through the fixation and dehydration procedure and observed.

82

Figure 9. Depths and estimated ages of ice core sections used in this study. Atmospheric CO_2 levels (all in ppmv): 500 ybp = 250; 10,500 and 105,000 ybp = 230; 70,000 ybp = 210; 30,000 = 200; 57,000 and 157,000 = 180. Global temperature: (relative to present temperature, in °C): 500 ybp = +2; 10,500 and 105,000 = −3; 57,000 = −4; 30,000, 70,000 and 157,000 = −6. Atmospheric dust (ppm): 500 = <0.1; 105,000 = 0.1; 10,500 and 57,000 = 0.2; 70,000 = 0.5; 30,000 = 0.7; 157,000 = 1.2. Values for atmospheric measurements are from reference [23]. All values are from reference [23].

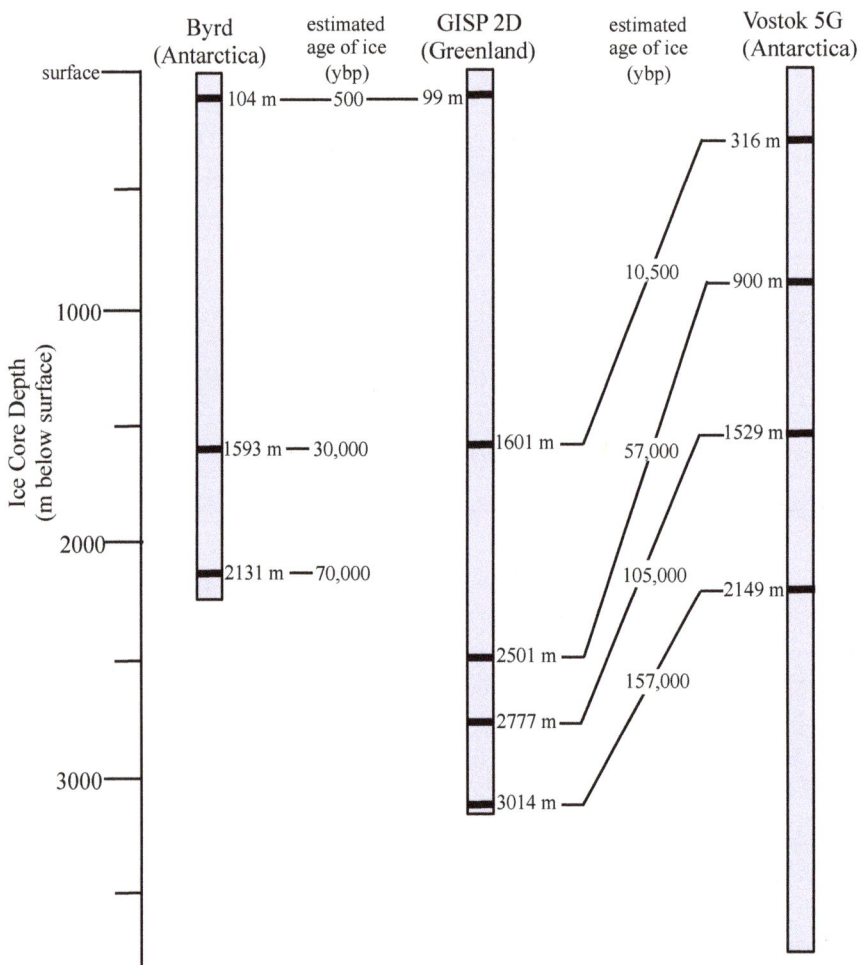

3.4. Culturing

The meltwater (200 μL) from each shell of the ice cores was spread onto several types of solid media in duplicate, and incubated at 8 °C to test for the presence of viable microorganisms. The following media were used: malt extract agar [1.28% maltose, 0.27% dextrin, 0.24% glycerol,

0.08% peptone, 1.5% agar (pH 4.7)], potato dextrose agar [0.4% potato starch, 2% dextrose, 1.5% agar (pH 5.6)], rose bengal agar [0.5% soytone, 1% dextrose, 0.1% monopotassium phosphate, 0.005% rose bengal, 1.5% agar (pH 7.2)], nutrient agar [0.3% beef extract, 0.5% peptone, 1.5% agar (pH 6.8)], oatmeal agar [6% oatmeal, 1.25% agar (pH 6.0)], Sabouraud dextrose agar [1% enzymatic digest of casein, 2% dextrose, 2% agar (pH 7.0)], yeast extract agar [3% yeast extract, 3% malt extract, 0.5% peptone, 1% dextrose, 2% agar (pH 6.2)], acidic yeast extract agar [3% yeast extract, 3% malt extract, 0.5% peptone, 1% dextrose, 2% agar (pH 4.5)], meat-liver agar [2% meat liver base, 0.075% D(+)-glucose, 0.075% starch, 0.12% sodium sulfite, 0.05% ammonium ferric citrate, 1.1% agar (pH 7.6)], blood agar [1.5% pancreatic digest of casein, 0.5% papaic digest of soybean meal, 0.5% sodium chloride, 5% sheep's blood, 1.5% agar (pH 7.3)], R2A [0.05% yeast extract, 0.05% proteose peptone No. 3, 0.05% casamino acids, 0.05% dextrose, 0.05% soluble starch, 0.03% sodium pyruvate, 0.03% dipotassium phosphate, 0.005% magnesium sulfate, 1.5% agar (pH 7.2)], Luria-Bertani agar [1% tryptone, 0.5% yeast extract, 0.5% sodium chloride, 1.5% agar (pH 7)], water agar [2% agar]. They were incubated for 2 weeks, monitored for any growth and transferred to 15 °C for 2 weeks. The plates then were incubated at 22 °C and monitored periodically for microbial growth. The cultures obtained were recorded and subcultured for future use. Throughout the decontamination and culturing procedures, control plates (two malt extract agar (MEA) and two lysogeny broth agar (LBA)) were employed in the laminar flow hood. Each plate was placed at one corner and exposed to the environment in the hood. They were incubated along with the culture plates and observed regularly for any growth.

3.5. PCR Amplification

DNA from the fungal and bacterial isolates obtained by culturing was subjected to polymerase chain reaction (PCR) amplification. DNA was extracted using a CTAB (cetyltrimethylammonium bromide) extraction method [53–55]. ITS4 and ITS5 primers were used to amplify the ribosomal DNA (rDNA) internal transcribed spacers (ITS1 and ITS2) and the 5.8S gene in fungi [56]. For bacterial isolates, primers 16S-2 and 23S-7 were used to amplify the rDNA intergenic spacer region (ITS1), along with portions of the small subunit and large subunit genes [57]. A GeneAmp PCR Reagent kit (Applied Biosystems, Carlsbad, CA, USA) was used for amplification. The composition of each 50 µL reaction was as follows: five microliters of cell suspension, 50 pmol of each primer, 10 pmol of each dNTP, 2U Taq DNA polymerase, 50 mM KCl and 1.5 mM MgCl$_2$. The program cycle used for amplification was: 95 °C for 8 min, 40 cycles of 94 °C for 1 min, 54 °C for 1 min 30 s and 72 °C for 2 min, followed by 72 °C for 8 min, and finally cooled to 4 °C. PCR reactions were subjected to electrophoresis on 1% agarose gels with TBE (90 mM Tris-Borate, 2 mM EDTA, pH 8.0) and 0.5 µg/mL ethidium bromide was used to view the amplification using UV light. For PCR amplification, sterilized water was used as a negative control. *Rhodotorula mucilaginosa* was used as the positive control for PCR with fungal primers and *Bacillus subtilis* as positive control for bacterial PCR.

3.6. Cloning and Sequencing

A TOPO TA cloning kit (Invitrogen, Grand Island, NY, USA) was used for cloning the amplified products. The amplified DNA from the fungal and bacterial isolates was ligated into the PCR 4-TOPO vector. The ligation reaction included: two and a half microliters of PCR product, 1.0 μL of salt solution (200 mM NaCl, 10 mM MgCl₂) and 1.5 μL of vector (10 ng μL⁻¹). The ligation reaction was carried out following the manufacturer's directions. One Shot® TOP10 Competent *E. coli* cells were transformed with the ligation reaction. The plasmid DNA was isolated from transformed cells using the Cyclo-Prep Plasmid DNA isolation kit (Amresco, Solon, OH, USA). The plasmids then were tested for the presence of inserts using *Eco*RI digestion. The products were subjected to electrophoresis on 1% agarose gels with TBE and 0.5 μg/mL ethidium bromide to analyze the digestion products. Plasmids containing the inserts then were sent to Gene Gateway LLC (Hayward, CA, USA) for sequencing.

3.7. Direct Sequencing of Cultures

After subculturing, DNA was extracted from the growing cultures of meltwater. The extracted DNA was amplified using a GeneAmp (with Ampli*Taq*) PCR Reagent Kit (Applied Biosystems, Carlsbad, CA, USA). A final concentration of each reaction was 5 pmol of dATP, dCTP, dGTP, dTTP, 1× PCR Buffer [10 mM Tris-HCl, pH 8.3, 50 mM KCl, 1.5 mM MgCl₂, 0.001% (w/v) gelatin], 1 unit Ampli*Taq* DNA polymerase and fungal rRNA ITS and bacterial rRNA SSU primers, in 25 μL total volume. Each reaction had 25 pmol of a forward and 25 pmol of a reverse primer. The reaction was carried in a thermocycler and the program was: 94 °C for 4 min; then 40 cycles of 94 °C for 1 min, 55 °C for 3 min, 72 °C for 3 min; followed by an incubation for 10 min at 72 °C. After amplification the samples were quantified on 1% agarose gels in TBE (as above). Samples were cleaned using a Exo-SAP IT kit (Affymetrix, Inc., USB Products, Cleveland, OH, USA). Exo-SAP IT is composed of Exonuclease I and Shrimp Alkaline Phosphatase. It is a special buffer that removes any remaining primer and dNTPs from the PCR reaction. Samples were sent to GENEWIZ (South Plainfield, NJ, USA) for Sanger sequencing. They were examined and aligned using Clustal X or MAFFT [58,59], and then were subjected to phylogenetic analysis using maximum parsimony with PAUP [30].

3.8. Metagenomic/Metatranscriptomic Analyses

3.8.1. Ultracentrifugation

Two samples were selected for metagenomic/metatranscriptomic analysis, GISP2D 1,601 m and 3,014 m. The former sample corresponds to 10,500 ybp [22,24,26], a time of moderate and rising atmospheric CO₂ and rising temperatures, while the latter corresponds to 157,000 ybp, a time of low atmospheric CO₂ and low temperatures [50]. Starting from Shell 2 inward, the meltwater was subjected to ultracentrifugation at 32,500 rpm in a Beckman type 60 Ti rotor (100,000 rcf) at 4 °C for 15–17 h to concentrate cells, viruses and nucleic acids. The supernatants were placed into sterile BD Falcon 50 mL conical tubes and stored at −20 °C. Each pellet was resuspended in 50 μL of 0.1× TE (1 mM Tris HCl, pH 8.0; 0.1 mM EDTA, pH 8.0) and then the suspensions from each tube (for a given meltwater sample) were pooled (~500 μL total) and stored at −20 °C. Two additional samples were used as controls. The first control was autoclaved Nanopure water (18.2 MΩ, <1 ppb TOC) that was subjected to ultracentrifugation, as above. Nanopure water (18.2 MΩ, <1 ppb TOC) was the second control.

3.8.2. Isolation of Nucleic Acids

Each concentrated meltwater sample (as well as the two controls) was divided into two 200 μL fractions. One fraction was used for RNA extraction using TRIzol LS reagent (Life Technologies, Grand Island, NY, USA). To carry out homogenization, three volumes of TRIzol LS reagent were added, mixed and incubated to allow complete dissociation of proteins. Next, an equal volume of chloroform was added, followed by vigorous mixing. After incubation at room temperature, the samples were centrifuged in a microfuge at 16,100 rcf, which resulted in three layers. The top aqueous layer, containing the RNA, was removed and the RNA was precipitated by adding 400 μL of isopropanol. This was placed at −20 °C for 2 h, followed by centrifugation in a microfuge to pellet the RNA. This pellet was washed with 800 μL 80% ethanol, air-dried and stored at −80 °C until needed. Directly before use, the RNA pellet was resuspended in 20 μL of sterile RNAse free water. The second 200 μL fraction of pooled ultracentrifuged meltwater was used for DNA extraction using a CTAB method, as described above.

3.8.3. Preparation of Samples for Sequencing

Complementary DNAs (cDNAs) were synthesized from the extracted RNAs. The procedure was performed using SuperScript Choice cDNA kit (Invitrogen, Grand Island, NY, USA), according to the manufacturer's instructions, using 10 μL of the extracted RNA and 80 pmol of random hexamer primers. The cDNA was then mixed with 10 μL of extracted DNA (less than 1 ng/μL) from the same meltwater sample, and *Eco*RI (*Not*I) adapters (AATTCGCGGCCGCGTCGAC, dsDNA) were added using T4 DNA ligase. The final concentration of components in each reaction for *Eco*RI adapter addition was: 66 mM Tris-HCl (pH 7.6), 10 mM $MgCl_2$, 1 mM ATP, 14 mM DTT, 100 pmols *Eco*RI (*Not*I) adapters and 0.5 units of T4 DNA ligase, in 50 μL total volume. The reaction was incubated at 15 °C for 20 h. Then, the reaction was heated to 70 °C for 10 min to inactive the ligase. [Note: The cDNA were mixed in order to maximize the biomass of nucleic acids, necessary for successful pyrosequencing. Thus, the cDNA comprises a metatranscriptomic fraction and the DNA comprises the metagenomic fraction of each sample].

The products were size fractionated by column chromatography. Each 2 mL plastic column contained 1 mL of Sephacryl® S-500 HR resin. TEN buffer (10 mM Tris-HCl [pH 7.5], 0.1 mM EDTA, 25 mM NaCl; autoclaved) was utilized in washing the columns and eluting the samples through the columns. Fractions of approximately 40 μL were collected by chromatography. After measurement of the volume of each fraction, fractions 6–18 were precipitated to concentrate the DNA. Concentration consisted of adding 0.5 volumes (of the fraction size) of 1 M NaCl, and two volumes of −20 °C absolute ethanol. After gentle mixing, each was left to precipitate at −20 °C overnight. The fractions were centrifuged in a microfuge for 20 min at room temperature and decanted. Then, the pellets were washed with 0.5 mL of −20 °C 80% ethanol and centrifuged for 5 min. Finally, each of the DNA pellets was dried under vacuum and rehydrated in 20 μL of 0.1× TE buffer.

After resuspension, fractions 6–18 were subjected to PCR amplification using *Eco*RI (*Not*I) adapter primers (AATTCGCGGCCGCGCTCGAC). The samples were amplified using a GeneAmp PCR Reagent Kit (as described above). The thermal cycling program was 94 °C for 4 min; then 40 cycles of 94 °C for 1 min, 55 °C for 2 min, 72 °C for 2 min; followed by an incubation for 10 min at 72 °C. Each fraction was subjected to gel electrophoresis on 1% agarose gels in TBE (as above) to confirm amplification, and to determine the size distributions.

454-specific primers were used to amplify the fragments, which also added the specific 454 sequences to the ends of the amplified products. All of the primers included the A-tag or the B-tag sequence (CGTATCGCCTCCCTCGCGCCA, CTATGCGCCTTGCCAGCCCGC, respectively), as well as the short four-base tag sequence (TCAG), unique multiplex identifiers (MID sequences (MID4: AGCACTGTAG, MID5: ATCAGACACG, MID6: ATATCGCGAG, MID10: TCTCTATGCG)), and *Eco*RI (*Not*I) adapter sequences (AATTCGCGGCCGCGTCGAC). The samples were amplified using a GeneAmp PCR Reagent Kit (as described above). All PCR products were cleaned with a PCR purification kit (Qiagen, Valencia, CA, USA). Once the sizes to be used for sequencing were confirmed by gel electrophoresis, the fractions containing distributions from approximately 200 bp to 1.5 kb were pooled. After concentration by precipitation and rehydration, the samples were quantified on 1% agarose gels in TBE (as above), using serial dilutions of plasmid (pGEM-3Z, from Promega, Madison, WI, USA) for quantitation.

3.8.4. Sequencing

DNA concentrations were estimated to be ~150 ng/μL for GISP 2D 3,014 m (sample 1); ~200 ng/μL for GISP 2D 1,601 m (sample 2); ~150 ng/μL for the ultracentrifuge sterile Nanopure water (sample 3); and finally the ~250 ng/μL for the unconcentrated Nanopure water. These were mixed in equimolar amounts in the proportion of 4:3:4:2, respectively, based on their estimated concentrations. Eight μL of Sample 1 (GISP 2D 3,014 m), 6 μL of Sample 2 (GISP 2D 1,601 m), 8 μL of Sample 3 (control ultracentrifuged water), and 4 μL of Sample 4 (control water) were mixed to produce a composite sample for sequencing. To this 26-μL mixture, 28 μL of RNase free water was added. A total of 54 μL (4.6 mg) of the amplified DNA at an approximate concentration of 85 ng/mL was sent to the University of Pennsylvania Medical School Sequencing Facility for Roche/454 GS FLX service.

3.8.5. Analysis of the Metagenomic/Metatranscriptomic Sequences

The sequences were assembled using MIRA 3.0.5 [60] and then were separated into four files, based on their multiplex identifier (MID) sequences (corresponding to the four ice core and control samples). Then, the 454 primers were clipped off (in silico) and sequence assembly was performed. The assembled sequences were used in Megablast searches for determination of sequence, taxon and gene similarities. The Megablast results (with the cut off parameter 1e-10) were retrieved for further analysis. Database sequences were reorganized based on gene regions. Multiple sequence alignments for ribosomal RNA small subunit (rRNA SSU) and large subunit (rRNA LSU) genes, separately, were performed with MAFFT (Multiple Sequence Alignment based on Fast Fourier Transform, Kyoto University, Kyoto, Japan) [58,59,61] global and local alignment tools. Alignment files were converted into NEXUS format with SeaView (Université de Lyon, Villeurbanne, France) [62,63]. Sequences from bacterial isolates CULT A-1 and CULT B-1 were included in the phylogenetic analyses. PAUP (Phylogenetic Analysis Using Parsimony, [30]) maximum parsimony phylogenetic reconstructions were performed on the aligned SSU and LSU rRNA gene sequences. Bootstrap analyses (1,000 replications) were used to examine the support for the branches. Results from the phylogenetic analyses of the sequences from uncultured samples allowed a reevaluation of the microbial compositions of the two ice samples.

3.9. Phylogenetic Analysis for Fungal and Bacterial Sequences

The sequences obtained from cultures were subjected to BLAST searches and phylogenetic analysis in the same manner as the metagenomic sequences. The rRNA gene sequences obtained from both the bacterial and fungal isolates were used in BLAST searches of the GenBank NCBI-database to identify sequences of related taxa. Sequence alignments were created using ClustalX 2.0 for bacteria and fungi using the isolate sequences and related NCBI sequences. Alignment files were used to generate maximum parsimony phylogenetic trees using the program PAUP [30]. The phylogenetic trees were created using the heuristic search option, and gaps were treated as a fifth base. Bootstrap support using 1,000 replications also was determined using the same criteria.

4. Conclusions

A variety of microbes and their nucleic acids were recovered from Greenland and Antarctica ice core sections. In general, the viability of microbes decreased with increasing ice core age (Figure 4). While similar species were found in Arctic and Antarctic ice of the same age, the species differed, and in all cases the species composition within each ice core section was unique (Table 1, Figures 6–8). Among the bacteria, species of *Bacillus* were common, while among the fungi, species of *Rhodotorula* were the most common. Cultivation and metagenomic/metatranscriptomic studies indicated that viability was dependent on cold temperatures at the time of deposition (Figure 5). Low concentrations of atmospheric dust and CO_2 also were related to increased viability.

Acknowledgments

We thank NICL-SMO and the Ice Core Working group for allowing us access to the ice core sections used in this research. This research was partially funded by Bowling Green State University. We thank Marilyn Cayer for her valuable help with scanning electron microscopy.

References

1. Abyzov, S.S.; Poglazova, M.N.; Mitskevich, J.N.; Ivanov, M.V. Common features of microorganisms in ancient layers of the Antarctic ice sheet. In *Life in Ancient Ice*; Castello, J.D., Rogers, S.O., Eds.; Princeton University Press: Princeton, NJ, USA, 2005; pp. 240–250.

2. Bell, R.; Studinger, M.; Tikku, A.; Castello, J.D. Comparative biological analyses of accretion ice from subglacial Lake Vostok. In *Life in Ancient Ice*; Castello, J.D., Rogers, S.O., Eds.; Princeton University Press: Princeton, NJ, USA, 2005; pp. 251–267.

3. Castello, J.D.; Rogers, S.O.; Smith, J.E.; Starmer, W.T.; Zhao, Y. Plant and bacterial viruses in the Greenland ice sheet. In *Life in Ancient Ice*; Castello, J.D., Rogers, S.O., Eds.; Princeton University Press: Princeton, NJ, USA, 2005; pp. 196–207.

4. D'Elia, T.; Veerappaneni, R.; Rogers, S.O. Isolation of microbes from Lake Vostok accretion ice. *Appl. Environ. Microb.* **2008**, *74*, 4962–4965.

5. D'Elia, T.; Veerappaneni, R.; Theraisnathan, V.; Rogers, S.O. Isolation of fungi from Lake Vostok accretion ice. *Mycologia* **2009**, *101*, 751–763.

6. Kellogg, D.E.; Kellogg, T.B. Frozen in time: The diatom record in ice cores from remote drilling sites on the Antarctic ice sheets. In *Life in Ancient Ice*; Castello, J.D., Rogers, S.O., Eds.; Princeton University Press: Princeton, NJ, USA, 2005; pp. 69–93.

7. Ma, L.J.; Catranis, C.; Starmer, W.T.; Rogers, S.O. Revival and characterization of fungi from ancient polar ice. *Mycologist* **1999**, *13*, 70–73.

8. Ma, L.J.; Rogers, S.O.; Catranis, C.; Starmer, W.T. Detection and characterization of ancient fungi entrapped in glacial ice. *Mycologia* **2000**, *92*, 286–295.

9. Ma, L.J.; Catranis, C.M.; Starmer, W.T.; Rogers, S.O. The significance and implications of the discovery of filamentous fungi in glacial ice. In *Life in Ancient Ice*; Castello, J.D., Rogers, S.O., Eds.; Princeton University Press: Princeton, NJ, USA, 2005; pp. 159–180.

10. Nichols, D.S. The growth of prokaryotes in Antarctic sea ice: Implications for ancient ice communities. In *Life in Ancient Ice*; Castello, J.D., Rogers, S.O., Eds.; Princeton University Press: Princeton, NJ, USA, 2005; pp. 50–68.

11. Rivkina, E.; Laurinavichyus, K.; Gilichinsky, D.A. Microbial life below the freezing point within permafrost. In *Life in Ancient Ice*; Castello, J.D., Rogers, S.O., Eds.; Princeton University Press: Princeton, NJ, USA, 2005; pp. 106–117.

12. Starmer, W.T.; Fell, J.W.; Catranis, C.M.; Aberdeen, V.; Ma, L.J.; Zhou, S.; Rogers, S.O. Yeasts in the genus *Rhodotorula* recovered from the Greenland ice sheet. In *Life in Ancient Ice*; Castello, J.D., Rogers, S.O., Eds.; Princeton University Press: Princeton, NJ, USA, 2005; pp. 181–195.

13. Vishnivetskaya, T.A.; Erokhina, L.G.; Spirina, E.V.; Shatilovich, A.V.; Vorobyova, E.A.; Tsapin, A.I.; Gilichinsky, D.A. Viable phototrophs: Cyanobacteria and green algae from the permafrost darkness. In *Life in Ancient Ice*; Castello, J.D., Rogers, S.O., Eds.; Princeton University Press: Princeton, NJ, USA, 2005; pp. 140–158.

14. Gilichinsky, D.A.; Khlebnikova, G.M.; Zvyagintsev, D.G.; Fedorov-Davydov, D.G.; Kudryavtseva, N.N. Microbiology of sedimentary materials in the permafrost zone. *Int. Geol. Rev.* **1989**, *31*, 847–858.

15. Gilichinsky, D.; Rivkina, E.; Bakermans, C.; Shcherbakova, V.; Petrovskaya, L.; Ozerskaya, S.; Ivanushkina, N.; Kochkina, G.; Laurinavichuis, K.; Pecheritsina, S.; *et al.* Biodiversity of cryopegs in permafrost. *FEMS Microbiol. Ecol.* **2005**, *53*, 117–128.

16. Abyzov, S.S. Microorganisms in Antarctic ice. In *Antarctic Microbiology*; Friedmann, E.I., Ed.; Princeton University Press: Princeton, NJ, USA, 1993; pp. 265–295.

17. Gilichinsky, D.A.; Vorobyova, E.; Erokhina, L.G.; Fyordorov-Dayvdov, D.G.; Chaikovskaya, N.R. Long-term preservation of microbial ecosystems in permafrost. *Adv. Space Res.* **1992**, *12*, 255–263.

18. Rogers, S.O.; Ma, L.J.; Zhao, Y.; Theraisnathan, V.; Shin, S.G.; Zhang, G.; Catranis, C.M.; Starmer, W.T.; Castello, J.D. Recommendations for elimination of contaminants and authentication of isolates in ancient ice cores. In *Life in Ancient Ice*; Castello, J.D., Rogers, S.O., Eds.; Princeton University Press: Princeton, NJ, USA, 2005; pp. 5–21.

19. Willerslev, E.; Hansen, E.J.; Poinar, H.N. Isolation of nucleic acids and cultures from fossil ice and permafrost. *Trends Ecol. Evol.* **2004**, *19*, 141–147.

20. Bidle K.J.; Lee, S.; Marchant, D.R.; Falkowski, P.G. Fossil genes and microbes in the oldest ice on Earth. *Proc. Natl. Acad. Sci. USA* **2007**, *104*, 13455–13460.

21. Lamb, H.H.; Woodroffe, A. Atmospheric circulation during the last ice age. *Quat. Res.* **1970**, *1*, 29–58.

22. Lüthi, D.; Floch, M.L.; Bereiter, B.; Blunier, T.; Barnola, J.M.; Siegenthaler, U.; Raynaud, D.; Jouzel, J.; Fischer, H.; Kawamura, K.; *et al.* High-resolution carbon dioxide concentration record 650,000–800,000 years before present. *Nature* **2008**, *453*, 379–382.

23. Petit, J.R.; Jouzel, J.; Raynaud, D.; Barkov, N.I.; Barnola, J.M.; Basile, I.; Bender, M.; Chappellaz, J.; Davis, M.; Delaygue, G.; *et al.* Climate and atmospheric history of the past 420,000 years from the Vostok ice core, Antarctica. *Nature* **1999**, *399*, 429–436.

24. Ahn, J.; Brook, E. Atmospheric CO_2 and climate on millennial time scales during the last glacial period. *Science* **2008**, *322*, 83–85.

25. Fischer, H.; Siggaard-Andersen, M.; Ruth, U.; Röthlisberger, R.; Wolff, E. Glacial/interglacial changes in mineral dust and sea-salt records in polar ice cores: Sources, transport, and deposition. *Rev. Geophys.* **2007**, *45*, 1–26.

26. Suwa, M.; Fischer, J.C.; Bender, M.L.; Landais, A.; Brook, E.J. Chronology reconstruction for the disturbed bottom section of the GISP2 and the GRIP ice cores: Implications for termination II in Greenland. *J. Geophys. Res.* **2006**, *111*, D02101.

27. Glikson, A. Milestones in the evolution of the atmosphere with reference to climate change. *Aust. J. Earth Sci.* **2008**, *55*, 125–139.

28. Miteva, V.I.; Sheridan, P.P.; Brenchley, J.E. Phylogenetic and physiological diversity of microorganisms isolated from a deep Greenland ice core. *Appl. Environ. Microb.* **2004**, *70*, 202–213.

29. Loveland-Curtze, J.; Miteva, V.I.; Brenchley, J.E. *Herminiimonas glaciei* sp. nov., a novel ultramicrobacterium from 3,042 m deep Greenland glacial ice. *Int. J. Syst. Evol. Microbiol.* **2009**, *59*, 1272–1277.

30. Swofford, D. *PAUP: Phylogenetic Analysis Using Parsimony, Version 4.0b2a (PPC)*; Sinauer and Associates: Sunderland, MA, USA, 1999.

31. Amato, P.; Parazols, M.; Sancelme, M.; Laj, P.; Mailhot, G.; Delort, A. Microorganisms isolated from the water phase of tropospheric clouds at the Puy de Dôme: Major groups and growth abilities at low temperatures. *FEMS Microbiol. Ecol.* **2007**, *59*, 242–254.

32. Zachos, J.; Pagani, M.; Sloan, L.; Thomas E.; Billups, K. Trends, rhythms, and aberrations in global climate change 65 Ma to present. *Science* **2001**, *292*, 686–693.

33. Utescher, T.; Bruch, A.A.; Micheels, A.; Mosbrugger, V.; Popova, S. Cenozoic climate gradients in Eurasia—A paleo-perspective on future climate change? *Paleogeog. Paleoclim. Paleoecol.* **2011**, *304*, 351–358.

34. Willerslev, E.; Cappellini, E.; Boomsma, W.; Nielsen, R.; Hebsgaard, M.B.; Brand, T.B.; Hofreiter, M.; Bunce, M.; Poinar, H.N.; Dahl-Jensen, D.; *et al.* Ancient biomolecules from deep ice cores reveal a forested southern Greenland. *Science* **2007**, *317*, 111–114.

35. Vincent, W.F. Evolutionary origins of Antarctic microbiota: Invasion, selection and endemism. *Antarct. Sci.* **2000**, *12*, 374–385.

36. Franzmann, P.D. Examination of Antarctic prokaryotic diversity through molecular comparisons. *Biodivers. Conserv.* **1996**, *5*, 1295–1305.

37. Marshall, W.A. Seasonality in Antarctic airborne fungal spores. *Appl. Environ. Microb.* **1997**, *63*, 2240–2245.

38. Sonjak, S.; Frisvad, J.C.; Gunde-Cimerman, N. *Penicillium* mycobiota in arctic subglacial ice. *Microb. Ecol.* **2006**, *52*, 207–216.

39. Tosi, S.; Casado, B.; Gerdol, R.; Caretta, G. Fungi isolated from Antarctic mosses. *Polar Biol.* **2002**, *25*, 262–268.

40. Starkenburg, S.R.; Reitenga, K.G.; Freitas, T.; Johnson, S.; Chain, P.S.G.; Garcia-Pichel, F.; Kuske, C.R. Genome of the Cyanobacterium *Microcoleus vaginatus* FGP-2, a photosynthetic ecosystem engineer of arid land soil biocrusts worldwide. *J. Bacteriol.* **2011**, *193*, 4569–4570.

41. Belnap, J.; Gardner, J.S. Soil microstructure in soils of the Colorado plateau: The role of the cyanobacterium *Microcoleus vaginatus*. *Great Basin Nat.* **1993**, *53*, 40–47.

42. Boyer, S.L.; Johansen, J.R.; Flechtner, V.R.; Howard, G.L. Phylogeny and genetic variance in terrestrial *Microcoleus* species based on the sequence analysis of the 16S rRNA gene and associated 16S-23S ITS region. *J. Phycol.* **2002**, *38*, 1222–1235.

43. Goodrich, R.P.; Handel, T.M.; Baldeschwieler, J.D. Modification of lipid phase behaviour with membrane-bound cryoprotectants. *Biochim. Biophys. Acta* **1998**, *938*, 143–154.

44. Snider, C.S.; Hsiang, T.; Zhao, G.; Griffith, M. Role of ice nucleation and antifreeze activities in pathogenesis and growth of snow molds. *Phytopathology* **2000**, *90*, 354–361.

45. Shtarkman, Y.M.; Koçer, Z.A.; Edgar, R.; Veerapaneni, R.; D'Elia, T.; Morris, P.F.; Rogers, S.O. Subglacial Lake Vostok (Antarctica) accretion ice contains a diverse set of sequences from aquatic and marine Bacteria and Eukarya. *PLoS One* **2012**, in press.

46. Simon, C.; Wiezer, A.; Strittmatter, A.W.; Daniel, R. Phylogenetic diversity and metabolic potential revealed in a glacier ice metagenome. *Appl. Environ. Microb.* **2009**, *75*, 7519–7526.

47. Miteva, V.; Teacher, C.; Sowers, T.; Brenchley, J. Comparison of the microbial diversity at different depths of the GISP2 Greenland ice core in relationship to deposition climates. *Environ. Microbiol.* **2009**, *11*, 640–656.

48. Christner, B.C.; Mosley-Thompson, E.; Thompson, L.G.; Reeve, J.N. Classification of bacteria from polar and nonpolar glacial ice. In *Life in Ancient Ice*; Castello, J.D., Rogers, S.O., Eds.; Princeton University Press: Princeton, NJ, USA, 2005; pp. 227–239.

49. Rogers, S.O.; Starmer, W.T.; Castello, J.D. Recycling of pathogenic microbes through survival in ice. *Med. Hypotheses* **2004**, *63*, 773–777.

50. Smith, A.W.; Skilling, D.E.; Castello, J.D.; Rogers, S.O. Ice as a reservoir for pathogenic human viruses: Specifically, caliciviruses, influenza viruses, and enteroviruses. *Med. Hypotheses* **2004**, *63*, 560–566.

51. Hamer, C.U.; Clausen, H.B.; Langway, C.C. Electrical conductivity method (ECM) stratigraphic dating of the Byrd Station ice core, Antarctica. *Ann. Glaciol.* **1994**, *20*, 115–120.

52. Rogers, S.O.; Theraisnathan, V.; Ma, L.J.; Zhao, Y.; Zhang, G.; Shin, S.G.; Castello, J.D.; Starmer, W.T. Comparisons of protocols for decontamination of environmental ice samples for biological and molecular examinations. *Appl. Environ. Microb.* **2004**, *70*, 2540–2544.

53. Rogers, S.O.; Bendich, A.J. Extraction of DNA from milligram amounts of fresh, herbarium and mummified plant tissues. *Plant Mol. Biol.* **1985**, *5*, 69–76.

54. Rogers, S.O.; Bendich, A.J. Extraction of total cellular DNA from plants, algae and fungi. In *Plant Molecular Biology Manual*, 2nd ed.; Gelvin, S.B., Schilperoort, R.A., Eds.; Kluwer Academic Press: Dordrecht, The Netherlands, 1994; pp. D1:1–D1:8.

55. Rogers, S.O.; Rehner, S.; Bledsoe, C.; Mueller, G.J.; Ammirati, J.F. Extraction of DNA from Basidiomycetes for ribosomal DNA hybridizations. *Can. J. Bot.* **1989**, *67*, 1235–1243.

56. White T.J.; Bruns, T.; Lee, S.; Taylor, J. Amplification and direct sequencing of fungal ribosomal RNA genes for phylogenteics. In *PCR Protocols, a Guide to Methods and Applications*; Innis, M.A., Gelfand, D.H., Sninsky, J.J., White, T.J., Eds.; Academic Press, Inc. Harcourt Brace Janovich Publishers: New York, NY, USA, 1990; pp. 315–322.

57. Kabadjova, P.; Dousset, X.; Le Cam, V.; Prevost, H. Differentiation of closely related *Carnobacterium* food isolates based on 16S-23S ribosomal DNA intergenic spacer region polymorphism. *Appl. Environ. Microb.* **2002**, *68*, 358–366.

58. Katoh, K.; Misawa, K.; Kuma, K.; Miyata, T. MAFFT: A novel method for rapid multiple sequence alignment based on fast Fourier transform. *Nucleic Acids Res.* **2002**, *30*, 3059–3066.

59. Katoh, K.; Kuma, K.; Toh, H.; Miyata, T. MAFFT version 5: Improvement in accuracy of multiple sequence alignment. *Nucleic Acids Res.* **2005**, *33*, 511–518.

60. Chevreux, B.; Wetter, T.; Suhai, S. Genome Sequence Assembly Using Trace Signals and Additional Sequence Information. *Comput. Sci. Biol. Proc. German Conf. Bioinform.* **1999**, *99*, 45–56.

61. MAFFT Version 7. Multiple Alignment Program for Amino Acid or Nucleotide Sequences. http://mafft.cbrc.jp/alignment/software/ (accessed on 25 August 2011).

62. Gouy, M.; Guindon, S.; Gascuel, O. SeaView version 4: A multiplatform graphical user interface for sequence alignment and phylogenetic tree building. *Mol. Biol. Evol.* **2010**, *27*, 221–224.

63. SeaView, at the Université Lyon, Villeurbanne, France. http://pbil.univ-lyon1.fr/software/seaview.html/ (accessed 8 October 2011).

Novel Cold-Adapted Esterase MHlip from an Antarctic Soil Metagenome

Renaud Berlemont, Olivier Jacquin, Maud Delsaute, Marcello La Salla, Jacques Georis, Fabienne Verté, Moreno Galleni and Pablo Power

Abstract: An Antarctic soil metagenomic library was screened for lipolytic enzymes and allowed for the isolation of a new cytosolic esterase from the α/β hydrolase family 6, named MHlip. This enzyme is related to hypothetical genes coding esterases, aryl-esterases and peroxydases, among others. MHlip was produced, purified and its activity was determined. The substrate profile of MHlip reveals a high specificity for short *p*-nitrophenyl-esters. The apparent optimal activity of MHlip was measured for *p*-nitrophenyl-acetate, at 33 °C, in the pH range of 6–9. The MHlip thermal unfolding was investigated by spectrophotometric methods, highlighting a transition (Tm) at 50 °C. The biochemical characterization of this enzyme showed its adaptation to cold temperatures, even when it did not present evident signatures associated with cold-adapted proteins. Thus, MHlip adaptation to cold probably results from many discrete structural modifications, allowing the protein to remain active at low temperatures. Functional metagenomics is a powerful approach to isolate new enzymes with tailored biophysical properties (e.g., cold adaptation). In addition, beside the ever growing amount of sequenced DNA, the functional characterization of new catalysts derived from environment is still required, especially for poorly characterized protein families like α/β hydrolases.

Reprinted from *Biology*. Cite as: Berlemont, R.; Jacquin, O.; Delsaute, M.; La Salla, M.; Georis, J.; Verté, F.; Galleni, M.; Power, P. Novel Cold-Adapted Esterase MHlip from an Antarctic Soil Metagenome. *Biology* **2013**, *2*, 177-188.

1. Introduction

Lipases and esterases are enzymes active towards different substrates. Some of these enzymes are used to access carbon sources (e.g., cinnamoyl esterase) [1], some of them are regarded as pathogenic factors [2], others are involved in biocide degradation [3], but many of them have as yet uncharacterized physiological functions. Because of the increasing amount of available sequenced genomes and the possibility to construct DNA libraries from uncultured microbial consortia, many new hypothetical enzymes, including lipases and esterases, are now accessible. This finding provides the possibility to isolate new biotechnologically relevant catalysts from extreme environments [4] and to test the robustness of hypotheses derived from cultivable microorganisms [5].

Cold-adapted enzymes from psychrophilic organisms are supposed to display higher activity at low and moderate temperature when compared to their mesophilic homologs. Such a high activity at low temperature is associated with higher thermal instability [6]. In cold-adapted enzymes the active site is regarded as the most heat labile structural element whereas some other parts of the proteins can remain correctly folded over a wider range of temperatures [7]. Often, when increasing the temperature, the temperature of inactivation does not correspond to the apparent melting temperature of the protein [6,8].

Many cold-adapted enzymes have been characterized; xylanase [9–11], cellulase [12], amylase [13–15] and lipase-esterase [16,17], among others. Their analysis reveals the molecular basis of protein adaptation to cold. The adaptation to low temperatures is thought to proceed by minor structural changes of the protein through modification of non-covalent interactions leading to an increase of the flexibility of crucial parts of the protein rather than to a global increase of the flexibility [8]. A decrease of the amount of proline and arginine residues and an increase of the amount of glycine are also assumed to contribute to the increase of the flexibility, in some cases [12]. In addition, a more accessible active site is considered as an adaptation to low temperature [18].

Interestingly, the metagenomic approach offers a unique opportunity for isolating new enzymes from uncultivated psychrophilic microorganisms [19,20]. Although few cold-adapted enzymes have been isolated using the metagenomic approach [21–23], little is known about their thermal stability.

In this work a metagenomic library, constructed from an Antarctic soil sample, was partially screened for lipolytic activity. One clone displaying esterase activity was isolated. After production and purification, its adaptation to cold temperatures was analyzed regarding both the activity and the stability.

2. Results and Discussion

2.1. MHlip Isolation and Sequence Characterization

From the Antarctic soil metagenomic library [24], one clone producing a blue halo after one week incubation at 18 °C on SBA-tributyrin media was selected as a potential esterase/lipase producer.

Complete DNA sequencing of the insert (3 kb) revealed a unique putative gene, named *mhlip* (Genbank access. number. GU550075). The gene encodes a 262 amino acids hypothetical protein with a theoretical molecular mass of 28,075 Da and a pI of 5.2. The encoded protein, MHlip, displayed 83% amino acid identity to a putative cytosolic α/β hydrolase derived from *Acidovorax delafieldii* 2AN (Acc. Numb. ZP-04765427). The sequence also showed significant identity with many other mesophilic proteins including the arylesterase from *Pseudomonas fluorescens* (PFE-ESTE 26% id., acc. no. P22862.4), the non-heme bromoperoxidase (BPOA2, 30% id., acc. no. P29715.3) and the non-heme chloroperoxidase (PRXC-STRAU, 29% id., acc. no. O31168.1) from *Streptomyces aureofaciens*, and with the crystallized esterase YTXM from *Bacillus subtilis* (id. 26%, acc. no. P23974).

We constructed a phylogenetic tree using the MHlip sequence and some sequences derived from various esterase-related families [25]. Next, Pfam numbers were assigned to each sequence using Pfam_scan (http://pfam.sanger.ac.uk/) [26] (Figure 1). MHlip was associated with the α/β hydrolase family 6 (Pf12697). This fold is observed in many enzymes having multiple functions (*i.e.*, transacetylase, lipase, hydroxynitrile lyase, methylesterase, peptidase, haloperoxidase, carboxylesterase) [27].

Figure 1. Unrooted neighbour-joining tree built with enzymes belonging to different families of esterases. The Jones-Tailor-Thornton method was used to compute distances [28]. Higher bootstrap values are displayed at the nodes and expressed as percentages of 1000 replicates. The Pfam numbers (names) are displayed on the right side of the tree.

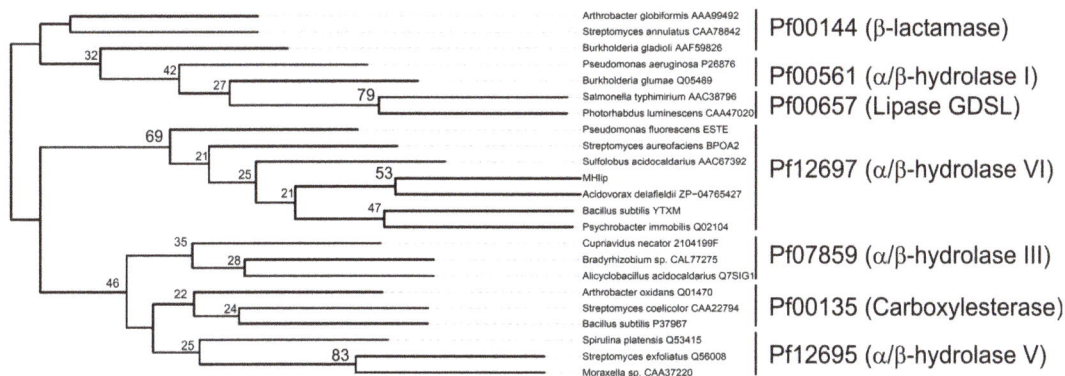

Sequences alignment highlighted all the residues considered as involved in substrate binding and catalysis (Figure 2) [29]. In addition, residues W_{29} and M_{95} assumed to limit the active site of PFE-ESTE are displayed. These two residues limit the diffusion of large substrates in the catalytic cleft, independently of any lid structure. Thus resulting in an enzyme having high affinity for short substrates (*i.e.*, pNPA) [29].

Figure 2. Sequences alignment of the MHlip enzyme and some related proteins. PFE-ESTE (P22862.4), aryl esterase from *P. fluorescens*; ZP-04765427, putative α/β hydrolase from *Acidovorax delafieldii*; BPO-A2 (P29715.3), bromoperoxidase from *Streptomyces aureofaciens* and YTXM (P23974.2), lipase from *Bacillus subtilis*. ★ residues forming the catalytic triad (Ser96, D213 and H241), ▲ residues forming the oxyanion hole involved in hydrogen bonding interactions, located upstream the active site (HG), ○ conserved residues involved in the formation of the GXSXG conserved pentapeptide containing the active-site serine, • residues involved in the active site occlusion in PFE-ESTE [29].

2.2. Biochemical Characterization

The MHlip-encoding gene was PCR-amplified and introduced in the pET22b vector in order to express a C-terminal His-tagged MHlip. The recombinant protein was purified from the cytoplasmic fraction of transformed *E. coli* BL21(DE3) using Ni-NTA columns.

Since MHlip was first isolated on tributyrin containing media, its activity was investigated on various p-nitrophenyl (pNP)-esters at pH 8.0 (Figure 3). MHlip is only active on small substrates such as pNPA; activity on pNPB is less than 5% of the pNPA hydrolysis and no activity was detected on substrates with longer acyl-chain lengths.

The pH-dependence of the MHlip activity on pNPA was investigated by monitoring the absorbance at 348 nm between pH 3.5 and 11.5, in a 20 mM pH-adjusted poly-buffer. MHlip was significantly active over a wide range of pH, from 4.5 to 9 (Figure 3).

Figure 3. MHlip substrate specificity measured at pH 8 and influence of the pH on the MHlip activity (the inset).

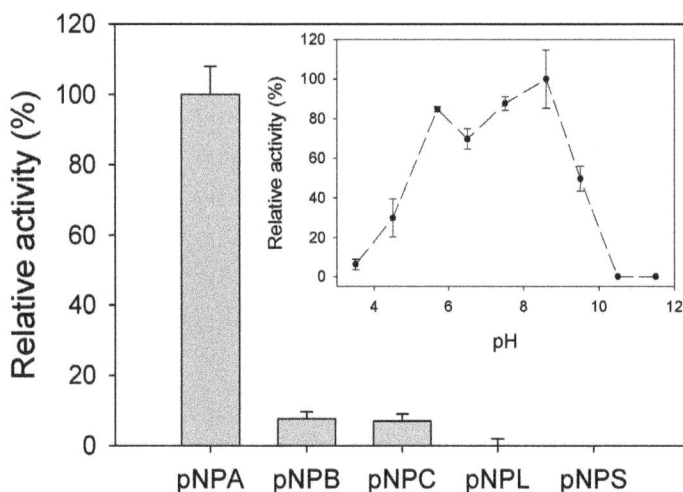

The thermal sensitivity of the MHlip activity towards pNPA was recorded at pH 8.0 in a thermal gradient ranging from 4 to 55 °C. The rate of pNPA hydrolysis rapidly increased from 4 to 35 °C and then quickly decreased to few percents of residual activity at 45 °C. Interestingly, MHlip remained significantly active when temperature decreased (20% at 4 °C) (Figure 4).

Kinetic parameters for pNPA hydrolysis were determined under initial rate conditions using a nonlinear regression analysis of the Michaelis–Menten equation. Hydrolysis was measured at 30 °C using pNPA as substrate at final concentrations ranging from 0 to 10 mM in 20 mM Tris-HCl (pH 8). For MHlip, the kinetic constants were: $K_m = 1.2 \sim 0.1$ mM, $k_{cat} = 3.13.10^{-2} \sim 0.2.10^{-2}$ sec^{-1} and $k_{cat}/K_m = 0.025$ mM^{-1} sec^{-1}.

2.3. Thermal Stability

MHlip fluorescence spectra recorded at 20 °C and 90 °C were compared. When MHlip was incubated at 80 °C its fluorescence emission spectrum, compared to the spectrum obtained at 20 °C, revealed its unfolding. A bathochromic effect was observed when the temperature increased as the maximum fluorescence was shifted to higher wavelengths (from 338 to 350 nm). In order to determine the MHlip melting temperature, the fluorescence emission at 338 nm was recorded in a thermal gradient, from 20 to 90 °C (data not shown).

This approach appeared to be inappropriate since the increase of temperature caused a decrease of the fluorescence emission at 338 nm, with no clear transition. The determination of the apparent melting temperature (Tm) was eventually achieved by recording the wavelength of the maximum fluorescence emission (λ_{max}) as a function of the temperature. At an excitation wavelength of 280 nm, a clear transition from a native to an unfolded state was observed at ~50 °C (Figure 4).

Figure 4. Influence of temperature on the MHlip activity (•). The enzyme was pre-incubated for 5 min at temperature ranging from 4 to 55 °C. Enzyme activity was determined under standard conditions on pNPA, at pH 8.0. The activity measured at 30 °C was taken as 100%. MHlip thermal denaturation following the maximum of fluorescence emission (λ_{max}) in a thermal gradient ranging from 20 to 80 °C (○).

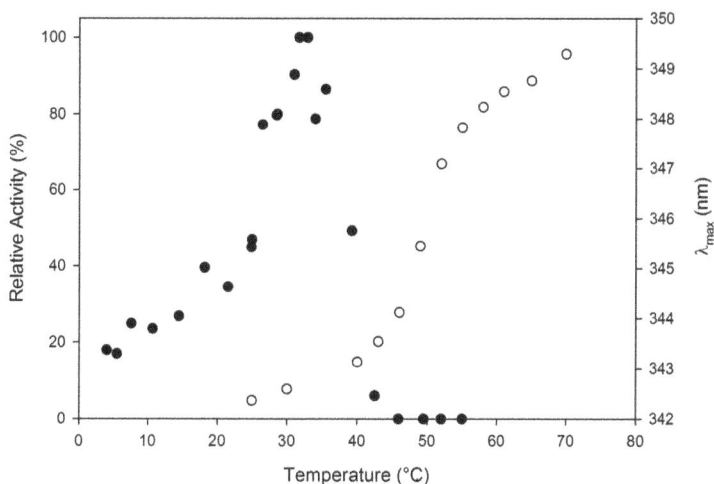

3. Experimental Section

3.1. Metagenomic Library Construction and Screening

A previously described short-insert metagenomic library (average insert size of 5.1 kb) was used for the present work [19]. Screening for lipolytic activity was carried out by plating 3,000 recombinant *E. coli* clones on Spirit-Blue Agar (SBA) Media (BD-Difco, MD) containing chloramphenicol (12.5 µg/mL) and 1% emulsified Tributyrin (Sigma-Aldrich). The plates were incubated at 37 °C for 24 hours and kept at 18 °C for one week.

3.2. Expression and Purification

Forward primer (5'-GAGCACATATGCCTTTCGCGCA-3') and the reverse primer (5'-GCTTCAGAGGCGCTCTCGAGCTTGTCGAGA-3'), containing *Nde*I and *Xho*I restriction sites, respectively (underlined), were used for PCR amplification of the full length MHlip-encoding gene. The obtained amplicon was ligated in an *NdeI/XhoI* double digested pET22b plasmid (Novagen, NJ) that allowed the production of a C_{term}-6×HisTag fused protein. After confirmation of the sequence, the pET22b:MHlip was introduced in competent *E. coli* BL21(DE3) cells (Novagen). Transformed cells were cultivated at 18 °C in LB media containing ampicillin (100 µg/mL). Heterologous protein expression was carried out for four hours by adding 0.4 mM isopropyl-D-1-thiogalactopyranoside (IPTG) when the OD_{600nm} reached ~0.5. After centrifugation, the cell pellet was resuspended in 20 mM Tris-HCl (pH 8.0) and the cells were disrupted by sonication (3 cycles of 30 sec at amplitude of 10–12 µm). Proteins from the cytoplasmic fractions were recovered by centrifugation at 20,000 × g for 40 min, and MHlip was purified by affinity chromatography on a 5 mL Ni-NTA columns (GE Healthcare) in a linear gradient of imidazole (from 0 to 250 mM). Protein concentration was determined by the Bicinchoninic Acid (BCA)-protein quantitation assay (Pierce) using serum albumin as standard. Finally, 30 mg of pure protein was obtained per litre of culture. The purified protein was subjected to automatic Edman degradation for the determination of the N-terminal amino acid sequence on an Applied Biosystems 492 Protein Sequencer (Perkin Elmer, Waltham, MA, USA). The resulting sequence (MPFAH) was found to match the predicted one. The purified protein was used for further characterization.

3.3. Activity Assays

Lipase/esterase activity assays were carried out on *p*-nitrophenyl-esters (pNP-Acetate, pNPA; pNP-Butyrate, pNPB; pNP-Caprilate, pNPC; and pNP-Laurate, pNPL; all purchased from Sigma) by spectrophotometric methods. The assay mixture was prepared as follow: 940 µL of 20 mM Tris-HCl (pH 8.0), 10 µL of 10 mM *p*-nitrophenyl-ester in acetonitrile, and 50 µL of enzyme solution. The release of *p*-nitrophenol was followed taking into account a molar extinction coefficient at 405 nm (ε_{405nm}) of 16,500 $M^{-1}.cm^{-1}$.

The pH-dependence of the enzyme's activity on pNPA, between pH 3.5 and 11.5, was investigated by monitoring the absorbance at 348 nm and replacing the Tris-HCl buffer by a 20 mM pH-adjusted polybuffer consisting of Tris, KCl, Na_2HPO_4, $CH_3COO.Na$ and dihydrogen citrate [30].

The thermal sensitivity of the MHlip activity toward pNPA was recorded as described above, over a temperature range between 4 and 55 °C.

3.4. Thermal Stability and Unfolding

In order to investigate the thermal stability of MHlip, the purified protein was submitted to increasing temperatures and the process of denaturation was followed by intrinsic fluorescence spectra. The spectrum of the native MHlip at 20 °C was recorded on a Perkin Elmer LS50B spectrofluorometer using a 1 cm cell path length. Maximum fluorescence emission was observed at a wavelength of 338 nm using an excitation wavelength of 280 nm. The Tm of the enzyme was determined by plotting the wavelength of the maximum emission (λ_{max}), measured at temperatures ranging from 20 to 80 °C.

4. Conclusions

The functional metagenomic is a powerful methodology to access the genetic material of environmental microorganisms. By the past, this approach has been used to isolated new enzymes from various environments [22–24,31–33]. Here, a new esterase form Antarctica is described.

MHlip belongs to the α/β-hydrolase family 6 and is part of a large group of uncharacterized proteins. The physiological function of these enzymes has not been clearly elucidated [34]. In addition, MHlip displays significant identity (~30%) to various haloperoxidases and proteases. Although similarity between some esterases, oxidases and proteases was reported [29,35,36], MHlip does not show any detectable activity when tested for the conversion of the substrate phenol red to bromophenol blue [37] or azo-casein hydrolysis [38] (data not shown).

Among the tested substrates, the MHlip activity is highly specific for short esters such as *p*-nitrophenyl-acetate. Similar specificity for short substrates (e.g., pNPA) has been observed for the aryl-esterase derived from the goat rumen metagenome [39] and for the aryl-esterase from *Pseudomonas fluorescens* (PFE) [29]. Concerning PFE, such a high specificity is the consequence of a sterically hidden active site. Indeed, the catalytic cleft of PFE is delimited by W28 and M96 [29,40] whereas cumbersome residues, L29 and Q97, are observed in the corresponding positions of MHlip, and could thus have a similar effect. The kinetic parameters for the pNPA hydrolysis suggest that the MHlip primary function is not yet well understood. However, the measured K_m for the pNPA hydrolysis (1.2 mM) is similar to some values obtained for previously characterized aryl-esterases, including the enzyme from *Lactobacillus casei* [41] and an aryl-esterase derived from the goat rumen metagenome [39].

The characterization of purified MHlip reveals its adaptation to cold temperatures. Indeed, MHlip retains 20% activity at 4 °C. This is consistent with the temperature of the Antarctic soil sample used for the metagenomic library construction [24]. The MHlip activity increases from 4 to ~30 °C and is lost at higher temperatures. This suggests that our screening method, including a long incubation at cold temperature, is suitable for allowing the isolation of this highly thermo-sensitive protein. The aryl-esterases derived form the goat rumen metagenome (estR5) [39] and from *Pseudomonas fluorescens* (PFE) [42] are regarded as mesophilic homologs of MHlip since their optimal activity is observed at 60 and 45 °C, respectively (Figure 5).

The MHlip unfolding begins at 30 °C. At this temperature the MHlip activity is optimal. When increasing the temperature above 30 °C, the protein undergoes a partial denaturation and quickly loses its activity. Complete denaturation is apparently achieved at a temperature higher than 65 °C. Based on the fluorescence transition curve, the apparent T_m is determined to be 50 °C.

The thorough characterization of cold active proteins unravels different ways for a protein to remain active at low temperatures [5–8,43]. Among these paths, increasing the length of the unstructured loop has been described in several cases [44,45]. However, MHlip sequence analysis does not reveal any extra stretches of amino acid.. Increasing the accessibility of the active site, as described for a cold adapted protease [18], was reported to increase the activity at low temperature. This is unlikely to occur here since MHlip may present cumbersome residues surrounding its active site. Moreover, the overall amino acid composition of MHlip is not significantly different from that of related proteins from mesophilic organisms. Nevertheless, this new enzyme, like most of the previously characterized cold adapted proteins, is an extremely heat labile enzyme.

Figure 5. Shift in the observed optimal temperature of activity (gray rectangle) and the melting temperature (Tm, white rectangle) of some characterized proteins. *, ** and *** stand for protein considered psychrophilic, mesophilic and thermophilic in the corresponding references, respectively.

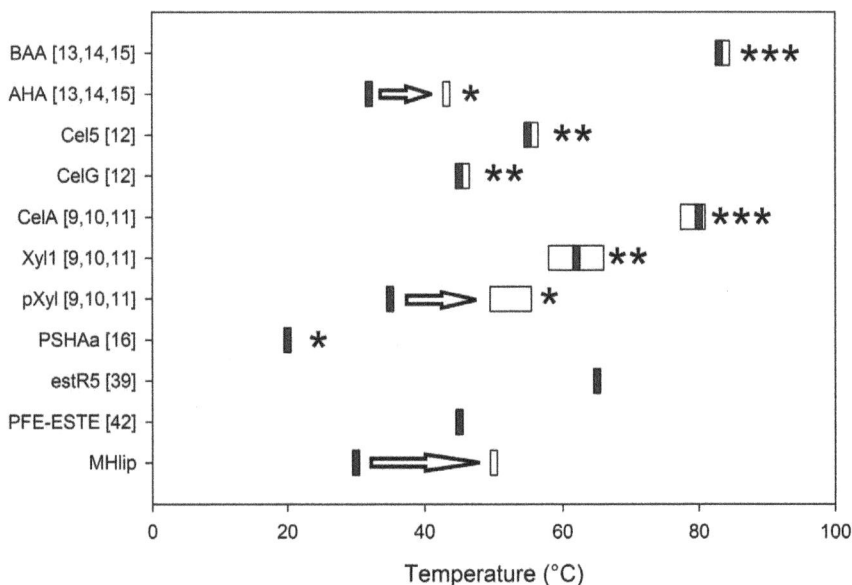

Indeed, the MHlip's loss of activity corresponds to the early stages of its unfolding. This suggests that the catalytic site is extremely thermo-sensitive compared to the overall protein structure. Shifts between the temperature at which proteins lose their activity and the unfolding temperature (Tm) have been recorded for many cold-adapted enzymes (Figure 5) [6,8,11–14]. Thus, even if no signature associated with cold-adapted protein can be observed, at the sequence level, we assume the MHlip adaptation to cold results from many discrete structural modifications that allow the protein to remain active when temperature is low.

Acknowledgments

This project was supported by the First-Postdoc Program, Grant no. 916984 from the Région Wallonne (Belgium) and by a grant from FNRS (FRFC, 2.4561.07) to MG.

P. Power is a researcher from the Consejo Nacional de Investigaciones Científicas y Técnicas (CONICET), Argentina, and was a recipient of a Postdoctoral fellowship from the Belgian Science Policy Office (BELSPO).

Authors thank Charles Gerday and Georges Feller for their skilful technical assistance and their discussion of the manuscript.

References and Notes

1. Dalrymple, B.P.; Swadling, Y.; Cybinski, D.H.; Xue, G.P. Cloning of a gene encoding cinnamoyl ester hydrolase from the ruminal bacterium butyrivibrio fibrisolvens e14 by a novel method. *FEMS Microbiol. Lett.* **1996**, *143*, 115–120.

2. McQueen, D.A.; Schottel, J.L. Purification and characterization of a novel extracellular esterase from pathogenic streptomyces scabies that is inducible by zinc. *J. Bacteriol.* **1987**, *169*, 1967–1971.

3. von der Haar, B.; Walter, S.; Schwapenheuer, S.; Schrempf, H. A novel fusidic acid resistance gene from streptomyces lividans 66 encodes a highly specific esterase. *Microbiology* **1997**, *143*, 867–874.

4. Schloss, P.D.; Handelsman, J. Metagenomics for studying unculturable microorganisms: Cutting the gordian knot. *Genome Biol.* **2005**, *6*, 229.

5. Casanueva, A.; Tuffin, M.; Cary, C.; Cowan, D.A. Molecular adaptations to psychrophily: The impact of 'omic' technologies. *Trends Microbiol.* **2010**, *18*, 374–381.

6. D'Amico, S.; Claverie, P.; Collins, T.; Georlette, D.; Gratia, E.; Hoyoux, A.; Meuwis, M.A.; Feller, G.; Gerday, C. Molecular basis of cold adaptation. *Philos Trans. R Soc. Lond. B Biol. Sci.* **2002**, *357*, 917–925.

7. Feller, G.; Gerday, C. Psychrophilic enzymes: Hot topics in cold adaptation. *Nat. Rev. Microbiol.* **2003**, *1*, 200–208.

8. Marx, J.C.; Collins, T.; D'Amico, S.; Feller, G.; Gerday, C. Cold-adapted enzymes from marine antarctic microorganisms. *Mar. Biotechnol.* **2007**, *9*, 293–304.

9. Collins, T.; De Vos, D.; Hoyoux, A.; Savvides, S.N.; Gerday, C.; Van Beeumen, J.; Feller, G. Study of the active site residues of a glycoside hydrolase family 8 xylanase. *J. Mol. Biol.* **2005**, *354*, 425–435.

10. Collins, T.; Hoyoux, A.; Dutron, A.; Georis, J.; Genot, B.; Dauvrin, T.; Arnaut, F.; Gerday, C.; Feller, G. Use of glycoside hydrolase family 8 xylanases in baking. *J. Cereal Sci.* **2006**, *43*, 79–84.

11. Collins, T.; Meuwis, M.A.; Gerday, C.; Feller, G. Activity, stability and flexibility in glycosidases adapted to extreme thermal environments. *J. Mol. Biol.* **2003**, *328*, 419–428.

12. Garsoux, G.; Lamotte, J.; Gerday, C.; Feller, G. Kinetic and structural optimization to catalysis at low temperatures in a psychrophilic cellulase from the antarctic bacterium pseudoalteromonas haloplanktis. *Biochem. J.* **2004**, *384*, 247–253.

13. Feller, G.; d'Amico, D.; Gerday, C. Thermodynamic stability of a cold-active alpha-amylase from the antarctic bacterium alteromonas haloplanctis. *Biochemistry* **1999**, *38*, 4613–4619.

14. Siddiqui, K.S.; Feller, G.; D'Amico, S.; Gerday, C.; Giaquinto, L.; Cavicchioli, R. The active site is the least stable structure in the unfolding pathway of a multidomain cold-adapted alpha-amylase. *J. Bacteriol.* **2005**, *187*, 6197–6205.

15. Siddiqui, K.S.; Poljak, A.; Guilhaus, M.; De Francisci, D.; Curmi, P.M.; Feller, G.; D'Amico, S.; Gerday, C.; Uversky, V.N.; Cavicchioli, R. Role of lysine *versus* arginine in enzyme cold-adaptation: Modifying lysine to homo-arginine stabilizes the cold-adapted alpha-amylase from pseudoalteramonas haloplanktis. *Proteins* **2006**, *64*, 486–501.

16. Aurilia, V.; Parracino, A.; Saviano, M.; Rossi, M.; D'Auria, S. The psychrophilic bacterium pseudoalteromonas halosplanktis tac125 possesses a gene coding for a cold-adapted feruloyl esterase activity that shares homology with esterase enzymes from gamma-proteobacteria and yeast. *Gene* **2007**, *397*, 51–57.

17. de Pascale, D.; Cusano, A.M.; Autore, F.; Parrilli, E.; di Prisco, G.; Marino, G.; Tutino, M.L. The cold-active lip1 lipase from the antarctic bacterium pseudoalteromonas haloplanktis tac125 is a member of a new bacterial lipolytic enzyme family. *Extremophiles* **2008**, *12*, 311–323.

18. Aghajari, N.; Van Petegem, F.; Villeret, V.; Chessa, J.P.; Gerday, C.; Haser, R.; Van Beeumen, J. Crystal structures of a psychrophilic metalloprotease reveal new insights into catalysis by cold-adapted proteases. *Proteins* **2003**, *50*, 636–647.

19. Handelsman, J.; Rondon, M.R.; Brady, S.F.; Clardy, J.; Goodman, R.M. Molecular biological access to the chemistry of unknown soil microbes: A new frontier for natural products. *Chem. Biol.* **1998**, *5*, R245–R249.

20. Rondon, M.R.; August, P.R.; Bettermann, A.D.; Brady, S.F.; Grossman, T.H.; Liles, M.R.; Loiacono, K.A.; Lynch, B.A.; MacNeil, I.A.; Minor, C.; *et al.* Cloning the soil metagenome: A strategy for accessing the genetic and functional diversity of uncultured microorganisms. *Appl. Environ. Microbiol.* **2000**, *66*, 2541–2547.

21. Heath, C.; Hu, X.P.; Cary, S.C.; Cowan, D. Identification of a novel alkaliphilic esterase active at low temperatures by screening a metagenomic library from antarctic desert soil. *Appl. Environ. Microbiol.* **2009**, *75*, 4657–4659.

22. Kim, E.Y.; Oh, K.H.; Lee, M.H.; Kang, C.H.; Oh, T.K.; Yoon, J.H. Novel cold-adapted alkaline lipase from an intertidal flat metagenome and proposal for a new family of bacterial lipases. *Appl. Environ. Microbiol.* **2009**, *75*, 257–260.

23. Berlemont, R.; Pipers, D.; Delsaute, M.; Angiono, F.; Feller, G.; Galleni, M.; Power, P. Exploring the antarctic soil metagenome as a source of novel cold-adapted enzymes and genetic mobile elements. *Revista Argentina de microbiologia* **2011**, *43*, 94–103.

24. Berlemont, R.; Delsaute, M.; Pipers, D.; D'Amico, S.; Feller, G.; Galleni, M.; Power, P. Insights into bacterial cellulose biosynthesis by functional metagenomics on antarctic soil samples. *ISME J.* **2009**, *3*, 1070–1081.

25. Arpigny, J.L.; Jaeger, K.E. Bacterial lipolytic enzymes: Classification and properties. *Biochem. J.* **1999**, *343*, 177–183.

26. Punta, M.; Coggill, P.C.; Eberhardt, R.Y.; Mistry, J.; Tate, J.; Boursnell, C.; Pang, N.; Forslund, K.; Ceric, G.; Clements, J.; *et al.* The pfam protein families database. *Nucleic Acids Res.* **2012**, *40*, D290–D301.

27. Marchot, P.; Chatonnet, A. Enzymatic activity and protein interactions in alpha/beta hydrolase fold proteins: Moonlighting *versus* promiscuity. *Protein Pept. Lett.* **2012**, *19*, 132–143.

28. Retief, J.D. Phylogenetic analysis using phylip. *Meth. in Mol. Biol.* **2000**, *132*, 243–258.

29. Cheeseman, J.D.; Tocilj, A.; Park, S.; Schrag, J.D.; Kazlauskas, R.J. Structure of an aryl esterase from pseudomonas fluorescens. *Acta. Crystallogr. D Biol. Crystallogr.* **2004**, *60*, 1237–1243.

30. Otero, C.; Fernández-Pérez, M.; Hermoso, J.A.; Ripoll, M.M. Activation in the family of candida rugosa isolipases by polyethylene glycol. *J. Mol. Catal. B Enzym.* **2005**, *32*, 225–229.

31. Chu, X.; He, H.; Guo, C.; Sun, B. Identification of two novel esterases from a marine metagenomic library derived from south china sea. *Appl. Microbiol. Biotechnol.* **2008**, *80*, 615–625.

32. Hong, K.S.; Lim, H.K.; Chung, E.J.; Park, E.J.; Lee, M.H.; Kim, J.C.; Choi, G.J.; Cho, K.Y.; Lee, S.W. Selection and characterization of forest soil metagenome genes encoding lipolytic enzymes. *J. Microbiol. Biotechnol.* **2007**, *17*, 1655–1660.

33. Wei, P.; Bai, L.; Song, W.; Hao, G. Characterization of two soil metagenome-derived lipases with high specificity for p-nitrophenyl palmitate. *Arch. Microbiol.* **2009**, *191*, 233–240.

34. Khalameyzer, V.; Fischer, I.; Bornscheuer, U.T.; Altenbuchner, J. Screening, nucleotide sequence, and biochemical characterization of an esterase from pseudomonas fluorescens with high activity towards lactones. *Appl. Environ. Microbiol.* **1999**, *65*, 477–482.

35. Itoh, N.; Kawanami, T.; Liu, J.Q.; Dairi, T.; Miyakoshi, M.; Nitta, C.; Kimoto, Y. Cloning and biochemical characterization of co(2+)-activated bromoperoxidase-esterase (perhydrolase) from pseudomonas putida if-3 strain. *Biochim. Biophys. Acta* **2001**, *1545*, 53–66.

36. Pelletier, I.; Altenbuchner, J. A bacterial esterase is homologous with non-haem haloperoxidases and displays brominating activity. *Microbiology* **1995**, *141*, 459–468.

37. Loo, T.L.; Burger, J.W.; Adamson, R.H. Bromination of phthalein dyes by the uterus of the dogfish, squalus acanthias. *Proc. Soc. Exp. Biol. Med.* **1963**, *114*, 60–63.

38. Polgar, L.; Szigetvari, A.; Low, M.; Korodi, I.; Balla, E. Metalloendopeptidase qg. Isolation from escherichia coli and characterization. *Biochem. J.* **1991**, *273*, 725–731.

39. Wang, G.; Meng, K.; Luo, H.; Wang, Y.; Huang, H.; Shi, P.; Pan, X.; Yang, P.; Yao, B. Molecular cloning and characterization of a novel sgnh arylesterase from the goat rumen contents. *Appl. Microbiol. Biotechnol.* **2011**, *91*, 1561–1570.

40. Park, S.; Morley, K.L.; Horsman, G.P.; Holmquist, M.; Hult, K.; Kazlauskas, R.J. Focusing mutations into the p. Fluorescens esterase binding site increases enantioselectivity more effectively than distant mutations. *Chem. Biol.* **2005**, *12*, 45–54.

41. Fenster, K.M.; Parkin, K.L.; Steele, J.L. Nucleotide sequencing, purification, and biochemical properties of an arylesterase from lactobacillus casei lila. *J. Dairy Sci.* **2003**, *86*, 2547–2557.

42. Liu, A.M.F.; Somers, N.A.; Kazlauskas, R.J.; Brush, T.S.; Zocher, F.; Enzelberger, M.M.; Bornscheuer, U.T.; Horsman, G.P.; Mezzetti, A.; Schmidt-Dannert, C.; *et al.* Mapping the substrate selectivity of new hydrolases using colorimetric screening: Lipases from bacillus thermocatenulatus and ophiostoma piliferum, esterases from pseudomonas fluorescens and streptomyces diastatochromogenes. *Tetrahedron Asymmetry* **2001**, *12*, 545–556.

43. Gerday, C.; Aittaleb, M.; Arpigny, J.L.; Baise, E.; Chessa, J.P.; Garsoux, G.; Petrescu, I.; Feller, G. Psychrophilic enzymes: A thermodynamic challenge. *Biochimica et biophysica acta* **1997**, *1342*, 119–131.

44. Sonan, G.K.; Receveur-Brechot, V.; Duez, C.; Aghajari, N.; Czjzek, M.; Haser, R.; Gerday, C. The linker region plays a key role in the adaptation to cold of the cellulase from an antarctic bacterium. *Biochem. J.* **2007**, *407*, 293–302.

45. Bauvois, C.; Jacquamet, L.; Huston, A.L.; Borel, F.; Feller, G.; Ferrer, J.L. Crystal structure of the cold-active aminopeptidase from colwellia psychrerythraea, a close structural homologue of the human bifunctional leukotriene a4 hydrolase. *J. Biol. Chem.* **2008**, *283*, 23315–23325.

Timescales of Growth Response of Microbial Mats to Environmental Change in an Ice-Covered Antarctic Lake

Ian Hawes, Dawn Y. Sumner, Dale T. Andersen, Anne D. Jungblut and Tyler J. Mackey

Abstract: Lake Vanda is a perennially ice-covered, closed-basin lake in the McMurdo Dry Valleys, Antarctica. Laminated photosynthetic microbial mats cover the floor of the lake from below the ice cover to >40 m depth. In recent decades, the water level of Lake Vanda has been rising, creating a "natural experiment" on development of mat communities on newly flooded substrates and the response of deeper mats to declining irradiance. Mats in recently flooded depths accumulate one lamina (~0.3 mm) per year and accrue ~0.18 μg chlorophyll-a cm^{-2} y^{-1}. As they increase in thickness, vertical zonation becomes evident, with the upper 2-4 laminae forming an orange-brown zone, rich in myxoxanthophyll and dominated by intertwined *Leptolyngbya* trichomes. Below this, up to six phycobilin-rich green/pink-pigmented laminae form a subsurface zone, inhabited by *Leptolyngbya, Oscillatoria* and *Phormidium* morphotypes. Laminae continued to increase in thickness for several years after burial, and PAM fluorometry indicated photosynthetic potential in all pigmented laminae. At depths that have been submerged for >40 years, mats showed similar internal zonation and formed complex pinnacle structures that were only beginning to appear in shallower mats. Chlorophyll-a did not change over time and these mats appear to represent resource-limited "climax" communities. Acclimation of microbial mats to changing environmental conditions is a slow process, and our data show how legacy effects of past change persist into the modern community structure.

Reprinted from *Biology*. Cite as: Hawes, I.; Sumner, D.Y.; Andersen, D.T.; Jungblut, A.D.; Mackey, T.J. Timescales of Growth Response of Microbial Mats to Environmental Change in an Ice-Covered Antarctic Lake. *Biology* **2013**, *2*, 151-176.

1. Introduction

The perennially ice covered lakes of the McMurdo Dry Valleys (MDVs), Antarctica, are amongst the most extreme lacustrine environments on Earth. Already set apart from most lakes by their endorheic and meromictic characters, the lakes also have extreme photoperiods, are persistently cold and are almost isolated from all physical forcing by perennial ice cover. In addition, they contain truncated, microbially-dominated food webs within which higher metazoans are largely absent. These conditions combine to result in minimal physical and biological perturbations of lake sediments, a situation that favors the formation of well-developed, complex microbial mat communities [1,2]. In response, microbial mats are both widespread and abundant in the lakes of the MDV [3,4], where they play a major role in carbon and nutrient cycling on a lake-wide basis [5–7].

While it is acknowledged that benthic microbial communities are significant components of the MDV lake systems, to date, quantitative information on their self-organization and growth is sparse, in part because accumulation rates tend to be slow and long duration observations would be required to assess how communities change. However, in recent decades, the MDV lakes have been undergoing gradual increases in lake level, in response to persistently warm summer temperatures [8,9]. This rising level provides a natural experiment into the rate and trajectory of microbial mat development on recently inundated soils and on the effects of changing environment on established microbial mat communities as levels increase. In Lake Joyce, for example, we have shown that reduced irradiant flux to deep mat communities has resulted in the extinction of distinctive components of the benthic biota, while inundation of substrates provide opportunities for colonization, albeit at a slow pace [10]. The relationship between the speed at which lake levels rise and how various communities respond has emerged as an important consideration in understanding ecosystem responses to change [7].

Lake Vanda, one of the MDV lakes, provides a particularly favorable opportunity to study the temporal characteristics of microbial mat development. In Lake Vanda, a punctuated rise in lake level has been well documented since 1973 [11], which allows us to construct a precise time-series of inundation. In addition, the upper waters of this lake are well mixed by convective processes [12] and the water column is very clear [13]. This means that conditions for microbial development in the zone inundated by lake level rise are not confounded by depth-related changes in water chemistry or by steep light gradients. The mats are internally laminated [14], and there is good evidence from other nearby lakes to suggest that these laminae record annual growth [5], allowing cross-validation of estimates of mat age. In this contribution, we present a detailed examination of the microbial mats in Lake Vanda, focusing on those that can be dated by recent changes in lake level. We use measurements made at similar depths and locations in 1998 and 2010 to describe temporal changes in mats and discuss these results in the context of the timescales of environmental change and ecosystem response.

2. Methods

2.1. Study Site

Lake Vanda lies in the Wright Valley, one of the MDVs in southern Victoria Land (77.52° S 161.67° E; Figure S1). It occupies a closed basin, and in 2010, it was a little over 75 m deep, with a perennial ice cover 3.5 to 4.0 m thick, though for several weeks each summer, the ice around parts of the lake shore melts to produce a discontinuous, open-water moat. The lake has an unusual physical structure, with temperature increasing with depth, from 4 °C just below the ice cover to >20 °C [15] (Figure S2). This inverse temperature gradient is due to solar heating and is stabilized by increasing solute content with depth [12]. Temperature and conductivity increase in two steps between 4 and 45 m, these being separated by a pycnocline at 20–25 m. These two upper steps are individually mixed by thermohaline convection [12], while below 45 m, there is a continuous gradual increase in temperature and conductivity.

The ice cover transmits 15%–20% of incident photosynthetically active radiation (PAR), heavily biased to wavelengths below 550 nm [14]. The water of the lake is also extraordinarily clear, with a vertical extinction coefficient for a downwelling PAR of 0.06 m^{-1} [13].

Several investigations of the plankton of Lake Vanda [16–18] report very sparse phytoplankton with a deep maximum of chlorophyll-a and photosynthesis at 55–65 m within the steep basal pycnocline [19]. Vincent and Vincent [16] provided strong evidence that phytoplankton abundance is limited by phosphorus supply. Zooplankton in the lake are restricted to five categories of ciliates and sparse numbers of the rotifer *Philodina gregaria* [20].

In contrast, benthic microbial mats are abundant and attain high biomass in Lake Vanda [4,14]. In 1980, these mats were reported to comprise mostly cyanobacteria, particularly species of *Phormidium* and *Lyngbya*, with pennate diatoms (species of *Navicula*, *Nitzschia*, *Caloneis* and *Stauroneis*) and occasional strands of moss [21]. In 1980, mats at a depth of 8.7 m (allowing for depth increase; see below; this would correspond to a depth ~18 m in 2010) were described as 2–5 mm thick and comprising a thin, orange-brown laminated mat overlying rock [21]. From 12–31 m (~21–40 m in 2010), Love *et al.* [21] described conical pinnacles with peaks 20–50 mm high and bases 30–50 cm in diameter, containing sub-mm thick laminations and calcite granules. Photographs included by Love *et al.* [21] suggest that these quoted sizes relate to the larger pinnacles and that smaller cones/pinnacles just a few mm in size were also present.

Photosynthetic activity has been demonstrated in benthic mats to at least 40 m [14], and modeling studies suggest that these benthic communities are responsible for the bulk of primary production in Lake Vanda [5]. Hawes and Schwarz [14,22] noted that the microbial mat communities growing in the seasonally melted "moat" regions around the edges of MDV lakes are quite distinct from those under perennial ice cover, lacking laminations or distinct vertical structure and adapted to high irradiance.

2.2. Lake Level Rise

Having no outflow, the depth of Lake Vanda is dependent on the balance between inflow via the Onyx River and ablative/evaporative losses from the lake surface [11]. A robust record of lake level for Lake Vanda has been maintained since 1969 ([11], http://www.mcmlter.org), and this, combined with comparison of depth *vs.* temperature profiles that date back to 1960 [15], allows a 50-year record of lake level change to be constructed (Figure 1). Over the past 40 years, the water balance has favored accrual, and Lake Vanda has risen in level, though there have also been several multi-year periods of stable lake levels. Comparison of temperature profiles from 1960 and recent years shows that the thickness of the lower convection cell and the deep density gradient have not changed and that water column discontinuities are at the same position relative to the bed of the lake. This suggests that level increase has been largely confined to an expansion of the upper convecting cell (Figure S2).

Figure 1. Change in Lake Vanda water level over time. Values on left scale are based on actual level measurements relative to fixed lakeside benchmarks from 1969 onwards and prior to this on reports of lake depths and water column structure. They are normalized to level in 2010 and show, for example, that in 1990, the lake was approximately 4 m shallower than in 2010. The right scale refers to the 2010 depth at which the underside of a 3.75 m thick ice cover was located in each year and shows, for example, that substrate at 10 m depth in 2010 had been below the lake ice cover since approximately 1987.

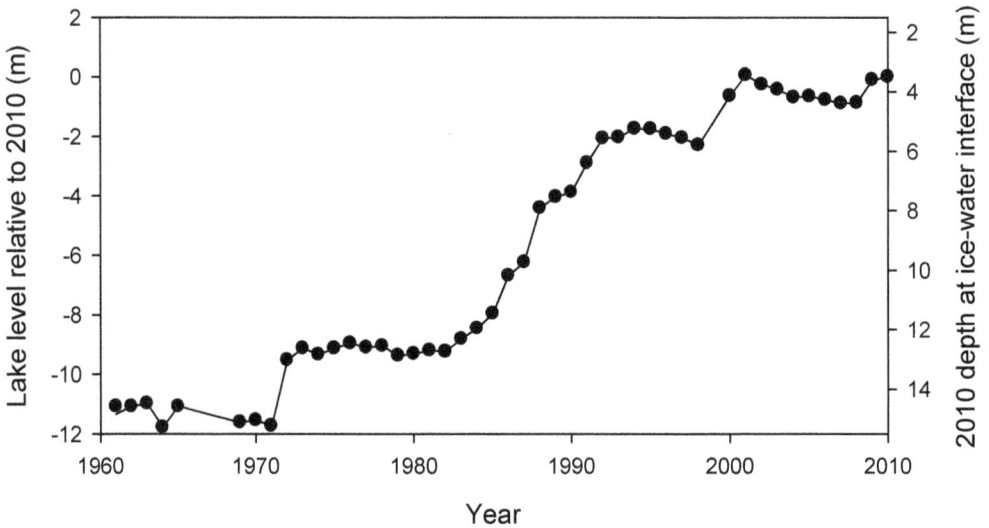

The increase in lake level constrains the time that specific depths have been below the 3.5–4 m thick perennial lake ice and, thus, the time available for growth of benthic microbial communities. For example, Figure 1 shows that the lake bed at 6 m in 2010 had been below the ice for ~18 years, at ~10 m for approximately 24 years and at ~14 m for 38 years.

2.3. Sampling

Water column temperature profile in November 2010 was obtained with a YSI 6600 Sonde. Sampling of microbial mats was by SCUBA divers operating through access holes made in the lake ice. The access holes were made at similar locations in the lake (77° 31.63′S, 161° 36.33′ E) in both 1998 and 2010. Principal microbial mat sampling depths were those that, in November–December 1998, were at 4, 6, 8, 10, 12, 16, 20 and 24 m. In November–December 2010, we resampled at 6, 10, 14, 18, 22 and 26 m depth, which, allowing for the 2 m lake level rise that had occurred between samplings, were approximately the same locations as the 1998 sampling. In addition, in 1998, samples were taken from the moat region of the lake at depths of 1 and 2 m and under ice at 6 and 10 m.

2.4. Macroscale Morphology

Benthic communities were comprised of undulating sheets of prostrate microbial mats from which pinnacles grew (Figure 2). The thickness of the prostrate mat to the underlying coarse sediments was measured using a sharpened, graduated rod. To obtain pinnacle height, a 0.4 m diameter circular quadrat was dropped haphazardly onto the mat surface, and pinnacles were measured as height above the prostrate mat surface systematically along a series of transects within the quadrat to ±1 mm by SCUBA divers using a graduated rod. The number of pinnacles measured in each quadrat varied with depth, with a minimum of 40 and a maximum of 120. Three quadrats were cast at each depth, and overall median and quartiles (25 and 75 percentiles) of pinnacle height were calculated after pooling all data from each depth.

Figure 2. Oblique image of a microbial mat community with well developed pinnacles rising from a prostrate background in Lake Vanda at 18 m depth in December 2010. The scale bar refers to the large central pinnacle.

2.5. Species Composition

Microbial mats in the upper part of Lake Vanda have an upper layer that is cohesive and contained all pigmented and some unpigmented laminae, as well as the vast majority of photosynthesis and aerobic respiration [14]. In 1998, samples of this cohesive layer were taken from the principal sample depths. Triplicate samples were cut from the mat at each sampling depth using a sharpened 60 mL syringe and placed in a darkened box for return to the ice surface. They were frozen and returned to New Zealand for microscope analysis. Examination was at magnifications of up to 1,000× and % abundance of recognizable morphotypes scored by counting subsamples from homogenized active layers, with a minimum count of 400 individuals. Cyanobacterial filaments and individual diatoms were considered a counting unit, this tending to overstate the absolute abundance of the latter. No coccoid cyanobacteria were encountered in these counts, and though we have identified these in other Antarctic microbial mats where they form colonies [23], we cannot exclude isolated coccoids in Lake Vanda mats.

In 2010, the differently pigmented zones within the mats were examined on site immediately after collection, again at magnifications of up to 1,000×. In addition, at depths of 10 and 18 m, mats were carefully dissected and the relative abundance of morphotypes within different zones evaluated. One sample was examined from each zone, and morphotypes were scored for abundance on a scale of 1 (present in some fields), 2 (present in all fields) and 3 (many trichomes in all fields). Morphotype identifications were made as described by Sutherland and Hawes [24], but cyanobacterial morphotypes from 1998 are reported only to a level where we can be confident that similar identities would have been ascribed in 2010. Evaluation of taxonomic status of cyanobacteria collected in 2010 by molecular approaches is ongoing and will be presented separately. For the current study, we consider only the consistency of morphotypes by depth, by location within the mat and over time.

2.6. Biomass and Pigments

In 1998, a microbial mat was cut out by divers using a knife, and the active layer was transferred to darkened boxes and returned to the surface (n = 5). Known-area samples were cut from these samples using a 25 mm diameter coring device. For depths of 4 to 12 m in 1998, the full mat thickness was taken as a single cohesive unit, whereas at greater depths, non-cohesive organic material was found below the upper cohesive and pigmented layer, but was not collected. Samples were collected haphazardly, though areas with unusually large or complex surface topography were avoided to optimize replication.

In 2010, divers cut through the mats to the underlying coarse sediment with a 38 mm diameter, sharpened plastic ring. The resulting core, which included the upper cohesive layer and any underlying organic material, was transferred to an individual container and returned to the surface. Five haphazardly selected replicates were taken at each depth, and once again, unusually large microbial features were avoided.

Prior to analysis, samples were stored for up to 2 months at −20 °C, then freeze-dried. Weighed aliquots were taken for analysis of loss of mass on acidification (LoAc—inferred as calcite content) and loss of mass on ignition (LoI), carbon and nitrogen content (also phosphorus in 1998) and chlorophyll-a. In 2010, aliquots were additionally taken for determination of other acetone-soluble pigments. Methods are fully described in Hawes et al., [10], but briefly, LoAc was determined by weight change before and after addition of 10% HCl to oven dried samples (60 °C), ash was determined after subsequent combustion at 450 °C for three hours and LoI calculated as oven-dry mass less LoAc and ash. In both years, LoAc and ash are expressed as a percentage of dry mass. Organic C and N contents were measured on weighed aliquots (after acidification) using a Carlo Erba Automated CHN analyzer. Organic P was estimated as reactive P after oxidation by acid persulfate. Because the sampling method used in 2010 included accumulated organic material to the underlying gravels, elemental contents are reported as mg cm^{-2}, while because only the cohesive pigmented upper layer was sampled in 1998, they are expressed as % dry weight.

For chlorophyll-a and other acetone-soluble pigment analysis, aliquots were extracted by ultrasonication (15–20 W for 30 seconds) in ice-cold 95% acetone and left in the dark for 24 h to complete extraction. After clarification by centrifugation, chlorophyll-a concentration was determined by spectrophotometry, without acidification [25]. In 2010, three extracts were randomly selected from the five replicates for each depth, transferred to a sealed glass vial and stored at −80 °C under nitrogen for up to 50 days before a more detailed analysis of pigments was undertaken by High Performance Liquid Chromatography (HPLC) [10]. A Dionex HPLC system, with PDA-100 diode array detector (300–800 nm), separated pigments according to the chromatographic method of Zapata *et al.* [26]. Pigments were quantified by absorption at 436 nm, with calibration by reference to commercially available standards.

Statistical treatment comprised calculation of means and standard deviations, and comparisons used targeted 1-way ANOVA, where appropriate. Sigmastat 3.5 (www.systat.com) was used for all statistical analyses.

2.7. Variable Chlorophyll-a Fluorescence

We examined vertical sections of mat samples returned to the surface using a pulse amplitude modulated (PAM) fluorescence analysis [27] to provide an estimate of the distribution of potential photosystem II (PSII) electron transport activity. Vertical sections, each approximately 5 mm thick, were cut with a scalpel from mat samples from 6, 10 and 18 m depth, laid on their sides and examined using a Walz Imaging-PAM fluorometer (Walz Mess- und Regeltechnik, Germany). Imaging PAM fluorometry allows resolution of the distribution of pigment systems in microbial mats and the activity of photosystem II (PSII) in two dimensions [10,27,28]). The instrument was fitted with a lens imaging an area of 30 × 23 mm and the saturation pulse method used to determine minimum (F_o) and maximum (F_m) fluorescence yields of dark-adapted samples to a pulsed blue light. From these values, we calculate variable fluorescence, F_v, as ($F_m - F_o$) and derived the maximum quantum yield of PSII as F_v/F_m. For a detailed description of the saturation pulse method, see Schreiber [27]. Care was taken to avoid exposure of samples to bright light after collection, and all measures were taken after at least 30 minutes of darkness. The measuring light of our instrument was provided by blue LEDs and peaked at 450 nm. While cyanobacteria often absorb this wavelength only weakly, we considered the wavelength to be acceptable given the spectral qualities of light under Lake Vanda ice, which contains almost no red and is dominated by blue and green wavelengths [14]. However, the method may be biased towards diatoms over cyanobacteria, since the former absorb blue light well, and results must be interpreted with care. Prior to use, the imaging system was adjusted using a fluorescent plastic target to ensure that equal fluorescent yield was returned from all parts of the image in the focal plane of the lens, according to manufacturer's recommendations.

The Imaging-PAM outputs fluorescence parameters as false-color images, the color scale ranging from red through violet indicating ranges of 0 to 1. Images of dark adapted fluorescence yield (F_o) were taken as a proxy of the distribution of chlorophyll-a. In this case, the images are returned in arbitrary units that can be quantitatively interpreted only within individual images. Images of the maximum photochemical yield (F_v/F_m) are dimensionless and can be compared between images. The Imaging-PAM instrument also records images of samples under red and near infra-red illumination and approximates chlorophyll-a absorption as A = 1 − (R/NIR). To ensure that absorption images were meaningful, care was taken to adjust the intensity of the R and NIR emitters such that a white calibration target returned an absorption of zero.

2.8. Microelectrodes

The distribution of net photosynthetic activity within microbial mats at 18 m was also inferred from the profiles of dissolved oxygen in mats and the immediate overlying water column. These profiles were measured by divers, *in situ*, using a Clark-type underwater O_2 microelectrode [29] (Unisense—www.unisense.com) attached to a manually operated micromanipulator mounted on an aluminum post [28]. Constraints on divers at 18 m depth under lake ice and instrument failures restricted the number of profiles that could be obtained. An underwater picoammeter (Unisense PA 3000U) provided the polarization voltage for the O_2 microelectrode, which had an outside tip diameter of 50 μm, a 90% response time of ~1 s and a stirring sensitivity of ~2%. The diver used the micromanipulator to move the electrode in 0.2-mm increments normal to the mat surface from a position above the diffusive boundary layer (DBL) to a depth of 6–7 mm into the mat. The position of the mat surface was estimated by the diver and confirmed by a break in the dissolved O_2 profile. The diffusive flux (J) of O_2 from the microbial mat into the overlying bottom water was calculated from the measured steady-state O_2 gradients (dC/dz) within the diffusive boundary layer (DBL), according to methods fully described in Vopel and Hawes [28].

3. Results

Throughout this section, all depths are referenced to water level in 2010. Our water column profile (Figure S2) identified upper and lower isohaline and isothermal convection cells overlying a continuous density gradient, and the locations of the discontinuities are consistent with other published profiles of the water column [12], confirming that lake level rise had been achieved by an increase in the thickness of the upper convection cell.

3.1. Macroscale Morphology

To the maximum depth sampled, the lake floor was covered by a flat to undulating mat from which pinnacles of varying height emerged (Figure 2). Mat thickness increased with depth on both sampling occasions and, over the 6–18 m depth interval, also increased between samplings (Table 1). The rate of increase in thickness increased with depth to 18 m, with 12 years of growth producing an increased thickness of 0.3–1.1 mm y^{-1} (Table 1). Pinnacle height also tended to increase with depth into the lake (Figure 3A), with a discontinuity in the rate of increase at 10–15 m and an apparent maximum height reached at ~22–26 m. Pinnacle height, even after log-transformation, was highly variable and not normally distributed, and a Kruskal-Wallis ANOVA by ranks yielded no significant differences (at $p = 0.05$) at specific depths between 1998 and 2010 samplings. In

2010, ANOVA by ranks clustered pinnacle height into three depth-groups, from 6 and 10 m (group median height 5 mm) and from 18, 22 and 26 m depth (group median height 24 mm), while those from 14 m (median height 12 mm) fell significantly between these groups. By plotting thickness and median pinnacle height against time under ice, where this last could be estimated, a consistent trend of a net increase over the early period of inundation was apparent (Figure 4). A highly significant linear increase in pinnacle height of ~0.3 mm y^{-1} (r^2 = 0.97, p < 0.01) was seen in the known-age part of the water column, that is, from 6–14 m. For thickness, a significant linear relationship was also apparent (r^2 = 0.80, n = 6, p < 0.05), suggesting a rate of growth of 0.14 mm y^{-1}. A curve fitting exercise comparing a range of models for these time-dependent data suggested that there was a tendency for rates of increase to decline after several decades (Table S1), though the estimates of initial rates of increase were very similar to those obtained from linear fits.

Table 1. Characteristics of microbial mats along a depth profile in Lake Vanda in 1998 and in 2010. Depths are referred to 2010 lake levels.

Depth in 2010 (m)	Approximate age in 1998 (y)	Total thickness of flat mat 1998 (number of laminae)	Total thickness of flat mat 2010 (number of laminae)	Annualized rate of increase in thickness (number of laminae)
6	6	1 mm (2–3)	4 mm (12)	0.3 (0.85)
10	10	1–2 mm (7)	8 mm (18)	0.5 (0.92)
14	28	4 mm (>12)	14 mm (>20)	0.8 (-)
18	>50	5 mm (>14)	18 mm (>20)	1.1 (-)
22	>50	>15 mm	25 mm	- (-)
26	>50	>50 mm	>45 mm	- (-)

3.2. Internal Morphology

In both 1998 and 2010, mats contained up to four zones distinguishable by pigmentation. An orange-brown zone overlay green and then purple zones, which, in turn overlay a non-pigmented zone. The shallowest mats did not contain green and purple zones in 1998, but at 6 m depth in 2010 and 10 m depth in 1998, all zones were present. On a finer scale, each zone contained multiple laminae, which varied in thickness from several hundred microns to more than a millimeter, with some undulations in thickness within individual laminae. Laminae were defined by bands of fine sediments separating otherwise organic-rich, mucilaginous material. Sediment bands varied from barely discernible to up to 1 mm thick, and a particularly dense band of sediment was seen across depths at the base of the 9th lamina from the surface.

In 1998, at the shallowest depth sampled (equivalent to 6 m in 2010), the mat was 1 mm thick and contained only an orange-brown zone, comprising 2-3 recognizable laminae. At the same location in 2010, the mat had increased in thickness to 4 mm and contained all color zones. Four orange-brown laminae, each approximately 0.2 mm thick, capped 6-7 green-purple laminae of ~0.5 mm thickness, giving a total of ~12 organo-sedimentary laminae of which ten were pigmented.

Figure 3. A. Pinnacle heights (median and quartiles, n > 100) at selected depths in Lake Vanda in 1998 and 2010. B. Concentrations of benthic chlorophyll-a (mean and s.d., n = 5) at various levels in Lake Vanda in 1998 and 2010. Both profiles are referred to depth in 2010.

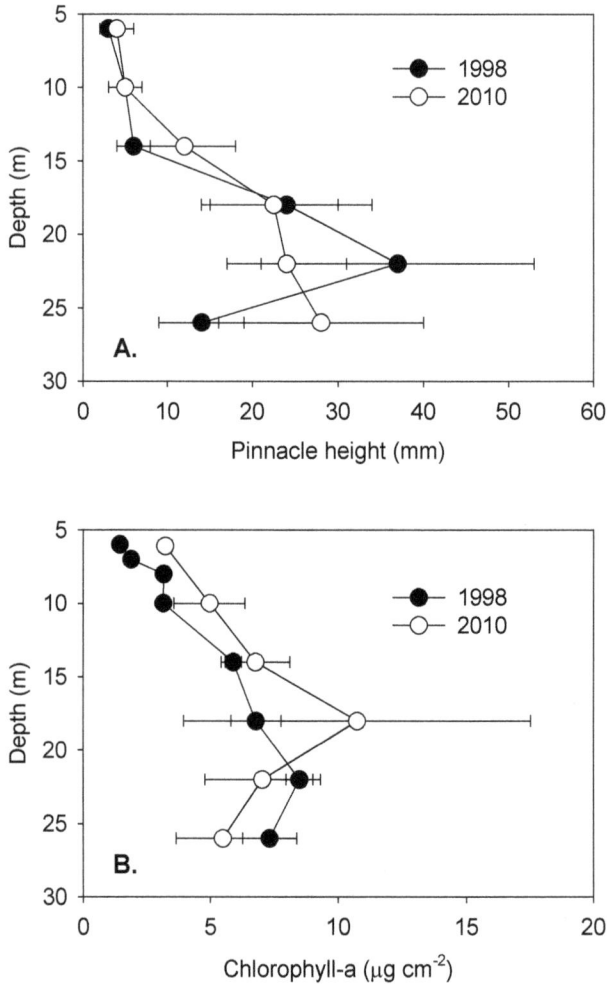

At a (2010) lake depth of 10 m, a 2 mm thick mat was present in 1998, with 7 recognizable laminae (3–4 orange, 3–4 green-purple). In 2010 the number of laminae at this location had increased to ~18, with four orange laminae overlying 6–8 green and purple laminae, fading to ~4 colorless laminae and becoming increasingly difficult to resolve. These observations showed that the subsurface laminae and zones were continuing to thicken after new laminae had formed.

At water depths exceeding 14 m, the clearly laminated and pigmented zones overlay an increasingly thick and less distinctly laminated mix of organic material and fine silts. Once the lamina count exceeded 20, counting became difficult. However, wherever counting was feasible, approximately 10–12 more laminae were present at any given depth horizon in 2010 than in 1998, consistent with the accumulation of one lamina per year (Table 1). The picture that emerges is of mats accumulating by annual lamina accrual, with the characteristics of laminae evolving from thin, orange-brown to thicker, purple or green and, ultimately, to colorless as they are progressively overgrown.

3.3. Species Composition

Cyanobacteria formed the matrix of mats, and the dominant morphotypes on both samplings, across all lake depths and locations within the mats, were identified as belonging to the genus *Leptolyngbya*. In 1998, *Leptolyngbya* comprised 40%–50% of the total counts at all depths, with the remainder ascribable to *Phormidium* (5%–20%) and *Oscillatoria* (2%–5%). In 1998, the 6 m mat also contained 5% *Nostoc*, a genus not seen in 2010, but which was also common (36 and 10% relative abundance, declining with depth) in the samples from within the moat of Lake Vanda.

In 2010, four morphotypes of *Leptolyngbya* were distinguished on the basis of cell dimensions. Morphotype diversity appeared to decline with depth in the lake, though our data do not allow robust statistical comparisons. At 10 m, three *Leptolyngbya* morphotypes were common across all color zones, and one other was present, whereas at 18 m depth, only one morphotype (1–2 μm wide, square cells with no constriction) was dominant in all color zones (Table S2).

Orange-brown surface zones consisted almost exclusively of *Leptolynbya*, usually oriented vertically and taking the form of ropes and columns of intertwined trichomes within laminae and tending to extend vertically out of the surface of the mat as tufts. Trichomes showed a tendency to orient horizontally at lamina interfaces. The green zone, if present, was immediately below the orange layer and comprised two or three laminations. *Leptolyngbya* was, again, overwhelmingly dominant. *Oscillatoria* cf. *sancta*, *Phormidium* cf. *autumnale*, *P.* cf. *murrayi* and a *Pseudanabaena* sp also occurred in the mats, but always at low abundance. Here, as in the orange-brown zone, *Leptolyngbya* trichomes tended to be oriented vertically and were intertwined. The transition from green to purple coincided with an increase in *Phormidium* and *Leptolyngbya* morphotypes that were visibly full of phycoerythrin. In the purple zone, trichomes were more widely spaced and less well-oriented than higher in the mat (Table S2).

Diatoms, notably Navicula muticopsis forma murrayi (14%–28% of total diatoms), Diadesmis contenta and D. contenta var. parallela (2%–9%), Hantzschia amphioxys var. maior (7%–34%) and Muelleria peraustralis (7%–28%) were common across depths, though less so close to the ice cover, and a coccoid eukaryote, tentatively identified as a species of Chrysosococcus, was at times frequent. No clear differences were evident in diatom relative abundances between depths, and methodological differences preclude a quantitative comparison between years. Similar species lists and degrees of dominance were, however, seen across years.

3.4. Biomass and Pigments

Quantitative comparisons between 1998 and 2010 are most robust for chlorophyll-a, where analytical methods were similar in the two years. In both years, chlorophyll-a increased with lake depth to a maximum at 18 m (Figure 3B), and in all but the deepest sample, chlorophyll-a concentrations were higher in 2010 than in 1998. However, only at 6 and 10 m were these increases statistically significant (ANOVA, $p < 0.05$). Within the 6–14 m depth band, chlorophyll-a increased steadily with age (Figure 4) equivalent to an annualized rate of 0.27 µg cm^{-2} y^{-1} (linear regression, $r^2 = 0.90$, n = 6, $p < 0.01$). The intercept of chlorophyll-a with age below ice is not zero, and this most likely reflects that the under-ice mats are developing not from barren ground, but from mats that were previously in the seasonally frozen marginal zone [14]. Comparison of other curve fits to this small data set suggested that the rate of increase in chlorophyll-a tended to decrease after some decades (Table S1), though similar rates of increase during the first 30 years after inundation are indicated. HPLC analysis of pigments in 2010 showed similar pigment profiles, with some pigments showing proportional changes between lake depths (Table 2). Two cyanobacterial pigments, myxoxanthophyll and nostoxanthin, showed a decline in relative abundance with depth to 14 m, whereas three pigments associated with diatoms, chlorophyll-c, diadinoxanthin and fucoxanthin, increased to 14 m. Two taxonomically widespread carotenoids, β-carotene and canthaxanthin, all tended to decline gradually with depth and reached minima at 26 m. Of the recognized chlorophyll-a derivatives, when expressed as ratio to chlorophyll-a, an allomer, phaeophytin and phaeophorbide increased with depth to 22 m, whereas chlorophyllides showed only a slight increase in relative abundance with depth. No bacteriochlorophylls were observed, though the technique used detects these when they are present.

Figure 4. Relationships between median pinnacle height, areal chlorophyll-a content and thickness of flat mat from a range of depths, expressed as time under ice. Solid symbols represent samples taken in 2010; open symbols in 1998. Linear regression models are fitted to each variable for all samples and are significant at $p < 0.01$. For the mat thickness, a sigmoid curve provided a better fit ($r^2 = 0.998$ cf. $r^2 = 0.959$ for the linear model). Pinn height—pinnacle height.

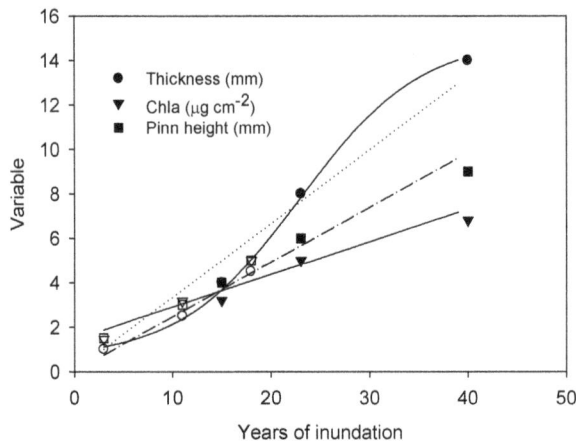

It is not possible to compare directly other biomass determinants between 1998 and 2010 due to slightly different sampling and analytical methods. However, in both years, the compositional changes with depth showed similar trends, and the differences between years were strongly associated with the ash content. In 1998, ash content was very high (90%) immediately below the ice, declining to ~65% from 8 to 18 m, then falling again to 30% at 22 m and below (Figure 5). In 2010, ash content was also very high immediately under the ice (99%), but on this sampling remained high to 14 m (Figure 5). However, as in 1998 ash content declined at greater depths, falling to 60% at 18 m and 20% at 26 m. The decline in % ash at 22–26 m depth was balanced by an increase in %LoAc, which in both years, reached 40%–50% here. Microscopy in 2010 supported these composition data, as shallow depth mats contained substantial amounts of both fines and sands, whereas at greater depths, sand was rare, and mineral inclusions were primarily small amounts of fine clay and precipitated calcite.

Figure 5. Biomass-related variables from 1998 and 2010. In 1998 (top) only % composition was measured, and all depths refer to the "active layer" of mats only. In 2010 (bottom), all depths, except 26 m, comprised cores to the underlying coarse sediments and were normalized to unit area. The 26 m sample represents only the active layer. In each case, the mean and standard deviation are plotted (n = 5).

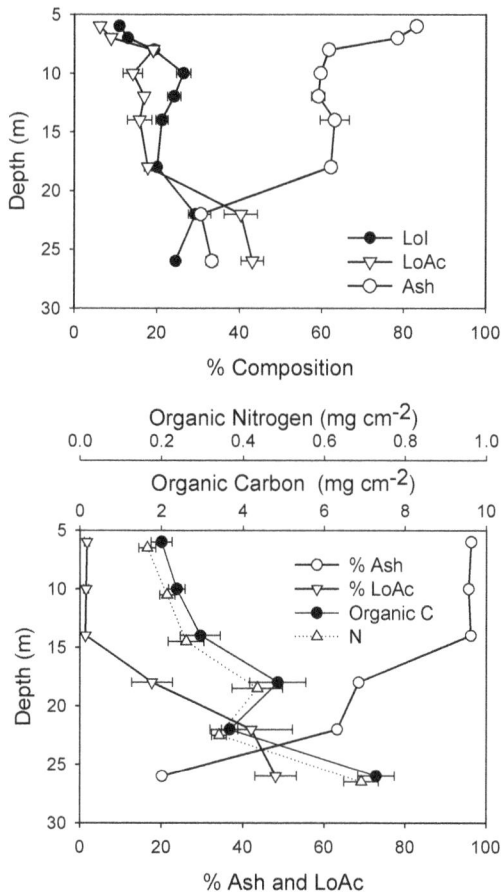

In 2010, when area-specific concentrations were measured, carbon and nitrogen both increased near exponentially to 18 m depth, closely paralleling the increase in chlorophyll-a (Figure 3), though at 22 m depth, both C and N were anomalously low. Atomic ratios of C:N in both 1998 and 2010 were consistent across depths at 12–16 (1998) and 12–14 (2010). In 1998, N:P atomic ratio averaged 40:1 (range 33–54:1) across the 6–26 m depth range, with no clear depth-related pattern.

Table 2. Relative abundance of selected pigments determined by high-performance liquid chromatography (HPLC) at the principal sampling depths in November 2010.

	6 m	10 m	14 m	18 m	22 m	26 m
Myxoxanthophyll	5.3	2.6	1.3	1.4	1.1	1.9
Nostoxanthin	1.8	1.0	0.5	0.6	0.7	0.9
Chlorophyll-c	0.4	2.1	3.7	3.9	4.1	3.9
Diadinoxanthin	0.2	1.2	2.0	2.2	2.4	1.0
Fucoxanthin	2.7	6.4	10.9	12.0	12.5	10.0
β-carotene	1.2	1.0	0.5	0.8	0.7	0.7
Canthaxanthin	1.6	1.2	0.9	0.9	0.9	0.6
Chl-a allomer	4.8	8.9	23.5	21.3	26.1	19.7
Phaeophorbide-a	3.0	11.0	25.4	29.3	30.6	23.3
Chlorophyllide-a	13.1	18.1	19.8	17.0	28.5	18.3
Phaeophytin-a	1.1	6.3	18.5	15.1	14.7	4.5

For the upper seven pigments, results are expressed as % total pigment, whereas the four chlorophyll-a derivatives at the bottom of the table as ratio to chlorophyll-a ($\times 100$).

3.5. Variable Chlorophyll-a Fluorescence

Imaging of variable chlorophyll-a fluorescence provided insights into both the location of areas where pigments were most abundant and of potential electron transport capacity. Four images of a vertical section of microbial mat from 10 m are shown in Figure 6. At the left, a photograph of the mat shows the orange, green and pink-dominated zones overlying a near-colorless underlayer; the adjacent image shows how F_m is distributed primarily into the laminae below the surface. Note that the cut surface of the section is indicated in the left hand image, mat components behind the cut surface are further from the camera and fluorescence yield cannot be directly compared. The third image from the left indicates that the yield of PSII under experimental conditions was rather evenly distributed, with F_v/F_m of 0.4–0.5, with the possible exception of the orange zone, where F_v/F_m values of ~0.25 were observed. The right hand image combines F_m and F_v/F_m, in that the intensity of the image is scaled to F_m, while the color is scaled to F_v/F_m. These images together show that potentially active photosystems are found in all of the pigmented layers and, indeed, that the highest values of F_v/F_m occurred within the purple and green zones, rather than close to the mat surfaces. Images also show bands of high pigment concentration at intervals below the mat surface, and comparison with the photograph shows how these are associated with lamina boundaries. Absorption images support the view that chlorophyll-a is distributed well down into the deeper laminae of the mats, indicated by absorption of red light, but not near infra-red light (Figure S3).

Images of example mat sections from 6 m (Figure S4) show similar organization to Figure 6, though with a markedly thinner cross section. The 6 m images are slightly oblique, but when the cut surface is examined once again the distribution of fluorescence and absorptions indicate a concentration of pigments and PSII yield in the sub-surface laminae. Finally, images from 14 m depth samples show further evolution of this arrangement in a thicker mat section (Figure S5). Images at the left show the arrangement of pigmented zones and the localization of maximal chlorophyll-a absorption in the deeper laminae, while F_m and F_v/F_m images indicate the maximum potential yield occurs in these deeper parts of the mats and that fluorescence is maximal at the lamina boundaries.

Figure 6. Distribution of fluorescence intensity (minimum fluorescence yield [F_o]—second image from left) and maximum yield of photosystem II (the maximum quantum yield of PSII [F_v/F_m]—third image) in part of a vertical section of a mat from 10 m depth. At far right is a maximum fluorescence yield (F_m)-weighted yield image, where the false color represents the yield and the intensity represents F_m. Color scales below fluorescence images represent imaging pulse amplitude modulated (PAM)-derived false-color values. At far left is a photographic image of the mat section at the same scale on which the annotated line shows the outline of the cut surface. Scale bars are 5 mm.

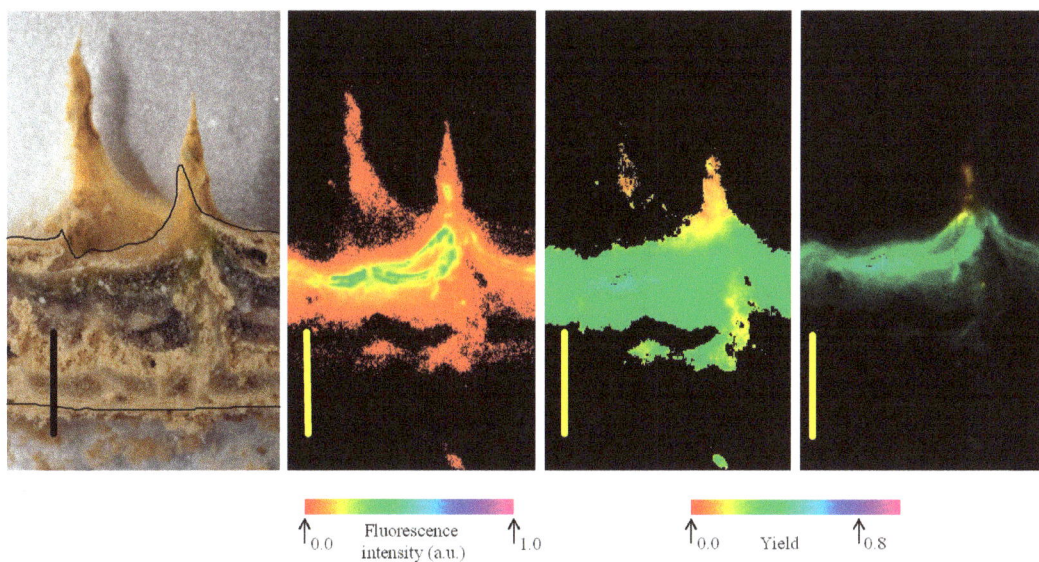

3.6. Oxygen Microprofiles

In situ dissolved oxygen microprofiles in prostrate mat at 18 m (Figure 7) were made at ambient temperature (4.2 °C, 602 µS cm^{-1} conductivity, pH 8.52 and a photon flux of 80 µmol m^{-2} s^{-1}). As is common in Antarctic lakes, dissolved oxygen concentration in the water column exceeded atmospheric saturation [28]). Profiles show a rapid rise in dissolved O$_2$ concentrations through the diffusive boundary layer and into the top part of the mat and a convex profile indicative of O_2 production via photosynthesis to at least 4 mm. This is consistent with fluorescence analyses that show photosynthetic potential extending into the older laminae. Penetration into the mat was insufficient to document the activity of the full thickness of the microbial mat. Estimation of areal oxygen evolution rate from the concentration gradient in the DBL of Figure 7 yields a rate of 0.3 to 0.4 µg oxygen cm^{-2} h^{-1}. This is of a similar magnitude to that reported earlier for this lake of -1 to $+3$ µg oxygen cm^{-2} h^{-1} from darkness to light saturation using *in vitro* incubation methods [14].

Figure 7. Two *in situ* profiles of dissolved oxygen within a prostrate mat at 18 m in Lake Vanda, obtained with microelectrodes. The vertical axis, in mm, locates the mat surface at 0 mm, with the water column as negative and the mat as positive displacements.

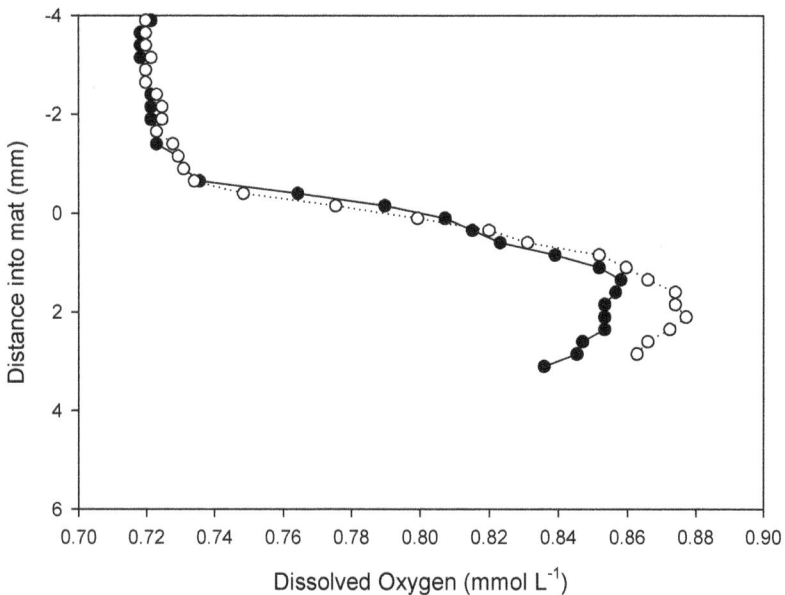

4. Discussion

Existing models of the growth and lamination of prostrate microbial mats in the MDV lakes are based on the accumulation of a new organic-rich layer during summer photosynthetic growth and the laying down of a sediment-rich layer during winter [5,6]. The increase in the number of laminae at specific locations in Lake Vanda between 1998 and 2010 was sufficiently close to 12 to support the view of these laminae as annual alternations between sediment deposition and mat growth. Our data suggest that mats tend to contain 2–3 laminae less than the number of years they have been within the main body of the lake, suggesting that annual banding does not occur within the seasonally frozen moat and becomes evident only 2–3 years after mats have emerged under the ice.

Vertical lamination is a characteristic of accreting microbial mats in clastic sedimentary environments. It is typically viewed as a consequence of filamentous cyanobacteria and diatoms moving through and overgrowing episodic sedimentary events and binding sediments in extracellular materials [2,30]. Lamination on an annual timescale, as seen in MDV lakes, is consistent with the seasonal rather than diel photoperiod at high latitude and the short pulse of flowing water and, hence, sediment influx during January [5,24]. Variability of the amount of sediment within laminae most likely reflects variations in summer sediment load, and unusually thick deposits are likely related to high sediment influx. In 2001–2002, a MDV-wide high flow event was recorded in response to unusually warm summer temperature [31], and we interpret the thick sediment layer seen across depths ~9 laminae down to be related to delivery of a large dose of sediment through this climate-driven event. It provides further mechanistic support for near annual nature of lamination, but caution is required in that sediment laminae can be expected to be less distinct during years with low sediment influx and are not completely reliable as annual phenomena. However, our observations show how changing climatic conditions can have complex effects on microbial communities, not just through rising lake levels and associated effects on irradiance, but also through changes to sediment flux.

Gradual accumulation of microbial mat biomass through this process is evident by the increase in thickness, areal C, N and chlorophyll-a content with depth and years since emergence below the ice. Where it is possible to date mat initiation, predictable rates of increase in thickness, chlorophyll-a and pinnacle size were observed, though it is evident that mat dynamics during this early phase of growth, which may last for 3–4 decades, changes once mats reach maturity. At maturity, ongoing growth at the surface is likely to be increasingly offset by decomposition and compression of deeper laminae, though gradual accumulation of organic material and thickness appears to continue. Pigment per unit area can also be expected to saturate, as the effects of increasing biomass on self-shading will be compounded by the effects of reduced irradiance. Even if pigment biomass accumulated to light-limited status, it will still be expected to decline with depth within the mature mat community due to declining irradiance [24]. Indeed, we would expect that the ultimate, light-limited chlorophyll-a biomass that could be achieved will be declining over time in the deep, mature mats, as irradiance is reduced by increasing water depth. Thus, the response of deep communities to lake level rise can be expected to be opposite to those growing in new habitat, with slower growth and declining photosynthetically active biomass, though our data are insufficient to quantify any such change.

Microbial mats are clearly a sink for C, N and P in Lake Vanda at all depths. The rates of accumulation of C in sediments at depth contours of known age, in the shallow part of the lake where mats are still accruing towards saturation biomass, can be estimated from Figure 5 as approximately 390 mg m^{-2} y^{-1}. This equates to an average of 104 µg oxygen cm^{-2} y^{-1} (using a photosynthetic quotient of 1:1) and extrapolating the very limited data on areal oxygen evolution presented here (0.3 µg oxygen cm^{-2} h^{-1} would generate 104 µg oxygen cm^{-2} in 14 days) suggests that such a rate of accrual is quite feasible given that sufficient irradiance is likely to be present for at least 100 days of photosynthesis. Using C:N and N:P mass ratios of approximately 14 and 90:1 measured here, this carbon accrual is equivalent to approximately 24 and 1.4 µg m^{-2} y^{-1} of N and P. Jones-Lee an Lee [32] provided data from which an annual external load of P to Lake Vanda can be estimated as 2.3 µg m^{-2} y^{-1}, a value similar to that calculated for mat sequestration and suggesting that mats may be an important sink for nutrients entering the upper part of the water column from outside of the lake.

While annual accretion of thin laminae of microbial mat at the surface is supported by our data, a growth model solely based on accrual of a new layer is only partially supported. Increases in the thickness of subsurface laminae relative to the surface, shifts in pigment contents and species composition and ongoing photosynthetic capacity with higher maximum yield of PSII than surface laminae suggest that the buried laminae remain significantly active for many years and contribute to mat thickening and pigment content.

In Lake Vanda, we found a shift in F_v/F_m, as well as pigmentation at the orange-pink transition. Shifts in pigment composition can result in alterations in apparent yields of photosystems to a given excitation wavelength [33], particularly in cyanobacteria, where phycobilins can be associated to PSII rather than PSI [34]. It is possible that the shifts in F_v/F_m at the pink-brown interface are thus methodological and may also involve the relative ability of different pigment complements to respond to the blue measuring light and to connect to the PSII centers.

Internal metabolic zonation is, however, a characteristic of microbial mats. Pearl and Pinckney [35] summarized how the consortium of microorganisms within a mat typically creates and exploits steep gradients of environmental conditions over mm length scales. Zonation of metabolism within typical temperate microbial mats sees oxygenic photosynthesis dominating surface zones on a diurnal basis, with anoxygenic photosynthesis and sulfide metabolism embedded in steep gradients of light, oxygen, pH, redox potential and sulfide concentrations [1,36], as also observed from mats in shallow Antarctic melt ponds [37,38]. However, such acute metabolic zonation does not appear to be common in the upper waters of MDV lakes during summer, where steep gradients of dissolved oxygen are not apparent [28]. The reasons for this are not clear, though Vopel and Hawes [28] did calculate low rates of oxygen consumption within mats and may also relate to perennially high ambient oxygen concentration in overlying waters that prevent hypoxia within surface sediments.

The shift towards dominance of pigment profiles by phycobilins over carotenoids with depth into mats and the corresponding change in appearance from orange (carotenoids) to green (phycocyanin) and pink (phycoerythrin) is consistent with acclimation to changing spectral characteristics inside the microbial mats. Irradiance becomes increasingly green and orange inside Lake Vanda mats [14], and such depth-in-mat related spectral adaptation is shown in other types of microbial mat [1,39], albeit reflecting a shift to sulfur bacteria rather than phycobilin-rich cyanobacteria. Internal zonation boundaries did not mark a shift in dominant cyanobacterial phototrophs, based on morphotype distribution or in photosynthetic process, since photosystem II was clearly active here. Indeed, profiles of dissolved oxygen in Lake Vanda mats show concentrations well above 100% saturation to extend through the orange layer and into the pink, and no bacteriochlorophylls were detected by HPLC pigment analysis. Chloroflexi 16S rRNA gene sequences have been detected in the Lake Vanda mats (Jungblut, personal observations). Ongoing research is using molecular techniques to determine whether taxonomic shifts undetectable by microscopy are occurring with these zones.

Lake Vanda is, however, highly oligotrophic, and zonation within mats may reflect differential supply of internally recycled nutrients. Quesada *et al.* [7] reported elevated nutrient concentrations within microbial mats relative to bulk lake water in nearby Lake Hoare, and other polar microbial mats from the Arctic can be internally nutrient replete [40], assisted by potential microbial recycling and scavenging processes [41]. The paradigm of microbial mats acquiring, then efficiently retaining and recycling, nutrients has been advanced to explain this disparity. A potential "crossed resource gradient" therefore exists in Lake Vanda mats, with irradiance flux decreasing and nutrient concentration increasing inside the mat. The role of nutrients in controlling benthic microbial growth in Lake Vanda has not been addressed, though the plankton in the upper waters are considered strongly P-limited [16,18]. Elemental ratios of N:P reported here for microbial mats are high relative to the median of freshwater habitats, while C:N ratio is only slightly above median [42], supporting the view that P may limit mat growth. P-limited growth may constrain the rate of biomass accrual in the recently flooded parts of the lake, and the availability of recycled nutrients within mats may support internal metabolism.

The picture that emerges from our data, consistent across depths and between sampling occasions, is that benthic biomass begins to accrue as discreet annual laminae within a few years of the emergence of benthic substrate under ice. New laminae are derived by the upwards growth of intertwined bundles of *Leptolyngbya* trichomes, rich in carotenoids (particularly myxoxanthophyll), giving rise to a thin new orange-brown surface layer, which, during the late summer and winter, receives a variable dose of fine sedimentary material. These tufts occasionally form pinnacles, initially a mm or less tall, which slowly extend over time. Slow, undisturbed accumulation of biomass, layer by layer, perhaps limited by rate of nutrient accrual, provides a simple explanation of the patterns of benthic biomass by depth (and hence by time) in the upper waters of Lake Vanda. It appears that this process may be a quantitatively significant sink for biologically active elements in that lake.

The pattern of accumulation in the part of the lake where inundation history is known is consistent with that history. The multi-variable discontinuity at 14–16 m in the lake (in 2010) relates to a zone that has been submerged by a surge in lake level rise since 1970–1980. The mat characteristics at this discontinuity, thickness of 25 mm accumulating at 0.3 mm y^{-1} and pinnacle height of 10 mm accumulating at 0.2 mm y^{-1}, are consistent with an age of ~50 years. At depths below the discontinuity, mats are substantially thicker with much larger pinnacles and a markedly higher organic carbon and carbonate content. Estimating the age of these mats requires extrapolation of relationships such as those in Figure 4. Thickness increase rates of 0.3 mm y^{-1} and pinnacle net extension of 0.2 mm y^{-1} suggest ages for 22 m mats of 60–100 years, though the questionable assumption that rates of change remain constant for such lengths of time renders estimates imprecise.

Rising levels in the MDV lakes are best related to summer degree days above freezing [8]. In recent decades, the region appears to be undergoing a period when summer climate is sufficiently warm for lake level to have been steadily rising. Microbial mats confirm that this increase has been ongoing since the mid 20th century and suggest that it may have begun perhaps 100 years ago. At the current rate of increase in lake level, microbial mats in Lake Vanda are able to keep pace with level rise in terms of initiating colonization. However, their slow rate of development means that they will take decades to reach their climax biomass and complexity. What is not addressed here is the fate of microbial mats in deeper parts of the lake where level change is affecting irradiance regime. There, a decline in microbial production can be expected as irradiance becomes increasingly limiting to sustainable biomass. Coring of the microbial mat at the junction between the two convection cells and below will allow the history of mat development to be better constrained, and sampling of mats from significantly below the chemocline will determine qualitative differences between the responses of mat communities there and those in the recently evolved upper cell.

5. Concluding Remarks

The level of Lake Vanda has been gradually rising for several decades, resulting in new habitat becoming available for colonization by microbial mats. This natural experiment allows temporal aspects of the development of these mats to be examined. Here, we show how mats are net sinks for nutrients and carbon in the lake, accumulating slowly through the elaboration of annual growth increments. Annual laminae retain photosynthetic capacity for many years after they have formed, acclimating to changing conditions associated with gradual burial under new layers. As mats grow, macroscopic structures emerge over decadal timescales from prostrate communities in the form of cm-scale, laminated pinnacles; at present, these are poorly developed in the upper 20 m of the lake compared to deeper waters. Rate of mat growth is such that, at the time of this study, areal biomass was not in equilibrium with ambient conditions; rather, the history of lake level change played a significant role in determining extant mat characteristics. The pace of change in Antarctica in relation to the rate at which ecosystems can respond will play a significant role in determining how ecosystems respond to climatic change.

Acknowledgments

This study was supported by funds from the NASA Exobiology Programme (**NNX08AO19G**) and the New Zealand Foundation for Research, Science and Technology (CO1X0306). We would like to acknowledge the logistic support of the US Antarctic Programme and Antarctica New Zealand and ongoing support and interaction with the Taylor Valley Long Term Ecological Research Programme (NSF grant 115245). The manuscript was much improved by comments from three anonymous reviewers.

References

1. Kühl, M.; Fenchel, T. Bio-optical Characteristics and the Vertical Distribution of Photosynthetic Pigments and Photosynthesis in an Artificial Cyanobacterial Mat. *Microbial. Ecol.* **2000**, *40*, 94–103.
2. Gerdes, G. What are microbial mats? In *Microbial Mats: Modern and Ancient Microorganisms in Stratified Systems*; Seckbach, J., Oren, A., Eds.; Springer Science: Dordrecht, Netherlands, 2010; pp. 3–25.

3. Wharton, R.A., Jr.; Parker, B.C.; Simmons, G.M., Jr. Distribution, species composition and morphology of algal mats in Antarctic dry valley lakes. *Phycologia* **1983**, *22*, 355–365.
4. Wharton, R.A., Jr. Stromatolitic mats in Antarctic lakes. In *Phanerozoic Stromatolites* II; Bertrand-Sarfati, J., Monty, C., Eds.; Kluwer: Dordrecht, Netherlands, 1994; pp. 53–70.
5. Hawes, I.; Moorhead, D.; Sutherland, D.; Schmeling, J.; Schwarz, A.-M. Benthic primary production in two perennially ice-covered Antarctic Lakes: Patterns of biomass accumulation with a model of community metabolism. *Ant. Sci.* **2001**, *13*, 18–27.
6. Moorhead, D.; Schmeling, J.; Hawes, I. Contributions of benthic microbial mats to net primary production in Lake Hoare, Antarctica. *Ant. Sci.* **2005**, *17*, 33–45.
7. Quesada, A.; Fernández-Valiente, E.; Hawes, I.; Howard-Williams, C. Benthic primary production in polar lakes and rivers. In *Polar Lakes and Rivers—Arctic and Antarctic Aquatic Ecosystems*; Vincent, W.F., Laybourn-Parry, J., Eds.; Oxford University Press: Oxford, UK, 2008; 179–196.
8. Wharton, R.A., Jr.; McKay, C.P.; Clow, G.D.; Andersen, D.T.; Simmons, G.M., Jr.; Love, F.G. Changes in ice cover thickness and lake level of Lake Hoare, Antarctica: Implications for local climate change. *J. Geophys. Res.* **1992**, *97*, 3503–3513.
9. Lyons, W.B.; Laybourn-Parry, J.; Welch, K.A.; Priscu, J.C. Antarctic lake systems and climate change. In *Trends in Antarctic Terrestrial and Limnetic Ecosystems*; Bergstrom, D.M., Ed.; Springer: Dordrecht, The Netherlands, 2006; pp. 273–295.
10. Chinn, T.J. Physical hydrology of the Dry Valley lakes. In *Antarctic Research Series: Physical and Biogeochemical Processes in Antarctic Lakes*; Green, W.J., Friedmann, E.I., Eds; American Geophysical Union: Washington, DC, USA, 1993; pp. 1–51.
11. Hawes, I.; Sumner, D.Y.; Andersen, D.T.; Mackey, T.J. Legacies of recent environmental change in benthic communities of Lake Joyce, a perennially ice covered Antarctic lake. *Geobioloy* **2011**, *9*, 397–410.
12. Spigel, R.H.; Priscu, J.C. Physical limnology of the McMurdo Dry Valley lakes. In *Ecosystem Dynamics in a Polar Desert: The McMurdo Dry Valleys, Antarctica*; Priscu, J.C., Ed.; American Geophysical Union: Washington, DC, USA, 1998; pp. 153–189.
13. Howard-Williams, C.; Schwarz, A.-M.; Hawes, I. Optical properties of the McMurdo Dry Valley Lakes, Antarctica. In *Ecosystem Dynamics in a Polar Desert: The McMurdo Dry Valleys, Antarctica*; Priscu, J.C., Ed.; American Geophysical Union: Washington, DC, USA, 1998; pp. 189–205.
14. Hawes, I.; Schwarz, A.-M. Absorption and utilization of low irradiance by cyanobacterial mats in two ice-covered Antarctic lakes. *J. Phycol.* **2001**, *37*, 5–15.
15. Armitage, K.B.; House, H.B. A limnological reconnaissance in the area of McMurdo Sound, Antarctica. *Limnol. Oceanogr.* **1962**, *7*, 36–41.
16. Vincent, W.F.; Vincent, C.L. Factors controlling phytoplankton production in Lake Vanda (728S). *Can. J. Fish. Aq. Sci.* **1982**, *39*, 1602–1609.
17. Parker, B.C.; Simmons, G.M.; Seaburg, K.G.; Cathey, D.D.; Allnutt, F.C.T. Comparative ecology of plankton communities in seven Antarctic oasis lakes. *J. Plankton Res.* **1982**, *4*, 271–286.

18. Priscu, J.C. Phytoplankton nutrient deficiency in lakes of the McMurdo dry valleys, Antarctica. *Freshwater Biol.* **1995**, *34*, 215–227.

19. Burnett, L.; Moorhead, D.; Hawes, I.; Howard-Williams, C. Environmental Factors Associated with Deep Chlorophyll Maxima in Dry Valley Lakes, South Victoria Land, Antarctica. *Arctic Ant. Alpine Res.* **2006**, *38*, 179–189.

20. Vincent, W.F.; James, M.R. Biodiversity in extreme environments: Lakes, pools and streams of the Ross Sea sector, Antarctica. *Biodiv. Cons.* **1996**, *5*, 1451–1471.

21. Love, F.G.; Simmons, G.M.; Parker, B.C.; Wharton, R.A.; Seaburg, K.G. Modern conophyton-like microbial mats discovered in Lake Vanda, Antarctica. *Geomicrobiol. J.* **1983**, *3*, 33–48.

22. Hawes, I.; Schwarz, A.-M. Photosynthesis in an extreme shade habitat: Benthic microbial mats from Lake Hoare, Antarctica. *J. Phycol.* **1999**, *35*, 448–459.

23. Jungblut, A.D.; Wood, S.A.; Hawes, I.; Webster-Brown, J.; Harris, C. The Pyramid Trough Wetland: Environmental and biological diversity in a newly created Antarctic protected area. *FEMS Microbial. Ecol.* **2012**, *82*, 356–366.

24. Sutherland, D.L.; Hawes, I. Annual growth layers as proxies of past growth conditions for benthic microbial mats in a perennially ice-covered Antarctic lake. *FEMS Microbial. Ecol.* **2009**, *67*, 279–292.

25. Marker, A.F.; Crowther, C.A.; Gunn, R.J.M. Methanol and acetone as solvents for estimation chlorophyll-a and phaeopigments by spectrophotomery. *Ergebn. Limnol.* **1980**, *14*, 52–69.

26. Zapata, M.; Rodriguez, F.; Garrido, J.L. Separation of chlorophylls and carotenoids from marine phytoplankton: A new HPLC method using reversed phase C8 column and pyridine-containing mobile phases. *Mar. Ecol. Prog. Ser.* **2000**, *195*, 29–45.

27. Schreiber, U. Pulse-amplitude (PAM) fluorometry and saturation pulse method. In *Chlorophyll Fluorescence: A Signature of Photosynthesis*; Papageorgiou, G., Govindjee, Eds.; Kluwer: Dordrecht, Netherlands, 2004; pp. 279–319.

28. Vopel, K.; Hawes, I. Photosynthetic performance of benthic microbial mats in Lake Hoare, Antarctica. *Limnol. Oceanogr.* **2006**, *51*, 1801–1812.

29. Revsbech, N.P. An oxygen microelectrode with a guard cathode. *Limnol. Oceanog.* **1989**, *34*, 474–478.

30. des Marais, D.J. Microbial mats and the early evolution of life. *Trends Ecol. Evol.* **1990**, *5*, 140–144.

31. Doran, P.T.; McKay, C.P.; Fountain, A.G.; Nylen, T.; McKnight, D.M.; Jaros, C.; Barrett, J.E. Hydrologic response to extreme warm and cold summers in McMurdo Dry Valleys, east Antarctica. *Ant. Sci.* **2008**, *20*, 499–509.

32. Jones-Lee, A.; Lee, G.F. The relationship between phosphorus load and eutrophication response in Lake Vanda. *Ant. Res. Ser.* **1993**, *59*, 197–214.

33. Schreiber, U.; Klughammer, C.; Kolbowski, J. Assessment of wavelength-dependent parameters of photosynthetic electron transport with a new type of multi-color PAM chlorophyll fluorometer. *Photosynth. Res.* **2012**, doi 10.1007/s11120-012-9758-1.

34. Bryant, D.A.; Frigaard, N.-U. Prokaryotic photosynthesis and phototrophy illuminated. *Trends Microbiol.* **2006**, *14*, 488–496.

35. Pearl, H.W.; Pinckney, J.L. A mini-review of microbial consortia: Their roles in aquatic production and biogeochemical cycling. *Microbial. Ecol.* **1996**, *31*, 225–247.

36. Revsbech, N.P.; Jørgensen, B.J.; Blackburn, T.H. Microelectrode studies of the photosynthesis and O_2, H_2S and pH profiles of a microbial mat. *Limnol. Oceanogr.* **1983**, *28*, 1062–1074.

37. Vincent, W.F.; Castenholz, R.W.; Downes, M.T.; Howard-Williams, C. Antarctic cyanobacteria: Light, nutrients and photosynthesis in the microbial mat environment. *J. Phycol.* **1993**, *29*, 745–755.

38. Jungblut, A.D.; Neilan, B.A. Cyanobacteria mats of the meltwater ponds on the McMurdo Ice Shelf (Antarctica). In *Microbial Mats: Modern and Ancient Microorganisms in Stratified Systems*; Seckbach, J., Oren, A., Eds.; Springer Science: Dordrecht, The Netherlands, 2010; pp. 499–514.

39. Jørgensen, B.; Cohen, Y.; Des Marais, D. Photosynthetic action spectra and adaptation to spectral light distribution in a benthic cyanobacterial mats. *Appl. Env. Microbiol.* **1987**, *53*, 879–886.

40. Bonilla, S.; Villeneuve, V.; Vincent, W.F. Benthic and planktonic algal communities in a high Arctic lake: Pigment structure and contrasting responses to nutrient enrichment. *J. Phycol.* **2005**, *41*, 1120–1130.

41. Varin, T.; Lovejoy, C.; Jungblut, A.D.; Vincent, W.F.; Corbeil, J. Metagenomic profiling of Arctic microbial mat communities as nutrient scavenging and recycling systems. *Limnol. Oceanogr.* **2010**, *55*, 1901–1911.

42. Elser, J.J.; Fagan, W.F.; Denno, R.F.; Dobberfuhl, D.R.; Folarin, A.; Huberty, A.; Interlandi, S.; Kilham, S.S.; McCauley, E.; Schulz, K.L.; Siemann, E.H.; Sterner, R.W. Nutritional constraints in terrestrial and freshwater food webs. *Nature* **2000**, *408*, 578–580.

Supplementary Material

Table S1. Results of regression model fitting to time series data. Three functions were fitted to each variable, linear (Chla = aT), rectangular hyperbolic (Chla = a(T)/(b+T)); and sigmoid (Chl-a = $a/(1+e^{(-(T-b)/c)})$), where T is time under perennial ice cover. In each case curves were fitted with and without intercepts. Cells indicate the adjusted r^2 of the model fit and the *p*-value for the regression ANOVA in parentheses. In all cases, n = 6. Models were fitted using SigmaPlot 10.0 (Systat Software). The best fit for each variable is indicated in italics.

Model	Mean Chlorophyll-a	Mean Thickness	Median Pinnacle Height
Linear + intercept	0.90 (0.002)	0.96 (0.0003)	0.99 (<0.0001)
Linear - intercept	0.73 (>0.05)	0.95 (0.005)	0.95 (<0.0001)
Hyperbola +	0.92 (0.011)	0.95 (0.005)	0.99 (0.0003)
Hyperbola -	*0.93 (0.001)*	0.94 (0.001)	*1.00 (<0.0001)*
Sigmoid +	0.82 (>0.05)	0.99 (0.006)	1.00 (0.002)
Sigmoid -	0.92 (0.01)	*0.99 (0.0003)*	*1.00 (<0.0001)*

Commentary

Interpretation of model fits is complicated, as accompanying time is a gradual deepening of the water and, thus, a change in growth conditions. In all three variables, rates of increase declined after several decades of growth, and whether this is due to deteriorating growth conditions (perhaps less irradiance as depth increased), limitation by self-shading or a combination of the two, it cannot be distinguished from our data.

Chlorophyll-a showed the best fit to the rectangular hyperbolic curves, with curve parameters suggesting that saturation would be achieved at 12.4 ± 3.2 µg cm^{-2}, with half saturation biomass reached after 33 ± 14 years. That translates to an initial (years 0–33) rate of accrual of approximately 0.18 µg cm^{-2} y^{-1}, slightly faster than that indicated by the linear + intercept model of 0.14 µg cm^{-2} y^{-1}. A saturating hyperbolic model is consistent with rapid initial accumulation of chlorophyll-a, followed by a declining rate of accrual as biomass begins self-shading or as ambient irradiance declined.

Table S2. Microscopic characterization of cyanobacterial morphotypes in flat microbial mats at 10 and 18 m in Lake Vanda. +: rare (present in some fields); ++: frequent (present in all fields); +++: common (many trichomes in all fields).

Morphotypes	Description	Cell width (µm)	10 m			18 m		
			Brown	Green	Purple	Brown	Green	Purple
Oscillatoria cf. *sancta*		~ 8	+	+		+	+	+
Phormidium cf. *autumnale*		6–7	+	+	+	+	+	+
Phormidium cf. *murrayi*		4–5	+			+		
Leptolyngbya sp. 1	pointy end cell, longer than wide	0.5–1	+++	+++	+++	+	+	+
Leptolyngbya sp. 2	square cells, longer than wide	1–2	+++	+++	+++	+		
Leptolygnbya sp. 3	square cells, longer than wide, no constriction at cross-wall	1–2	+++	+++	++	+++	+++	+++
Leptolyngbya sp. 4	rounded end cell, longer than wide, constriction at cross-wall	1–2		++		+	+	
Pseudanabaena sp.		~2		+	+	+		+
Unicellular cyanobacterium cf. *Aphanocapsa*						+		

Thickness was well described by all models, though the best fit was to a Sigmoid (logistic) function. Parameters suggest that inflection time from exponential early growth was at 25 years, when the mat would be 8 mm thick, with maximum thickness of 16 mm after ~50 years. Saturation of thickness would require annual growth to be balanced by annual loss, and this may not be reasonable for the Lake Vanda situation. Indeed, material thicker than 16 mm was found in the lake, and biomass appears to accumulate for much longer than 50 years. The accumulation rate of new mats may thus be governed by different processes than for older mature mats.

Pinnacle height was also well described by all models. The slightly better fit of the hyperbolic and sigmoid functions suggests that some degree of saturation of pinnacle size was developing as mats aged and as irradiance declined, as was suggested for thickness and chlorophyll-a. Hyperbolic and sigmoid models suggested that saturation would occur after 60 and 40 years, respectively, at a maximum median heights of approximately 20 mm. However, pinnacles of much taller than 20 mm were seen in the lake (Figure 3) and, as with thickness, pinnacle elongation during the development of new mats appears to be governed by other processes than those controlling development in older, mature mats.

Figure S1. Map showing the location of Lake Vanda in the Wright Valley, McMurdo Dry Valleys region of Antarctica.

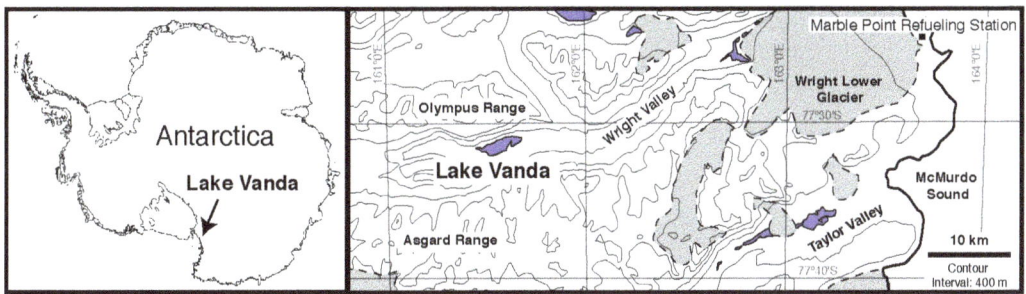

Figure S2. Water column structure in Lake Vanda in 1960–1961 and in 2010–2011. 1960–1961 data [15] were offset vertically by 11 m to allow for the increase in lake level. 2010 data were obtained with a YSI 6600 Sonde (http://www.ysi.com) in November 2010. The identification of upper and lower isohaline and isothermal convection cells overlying a continuous density gradient is consistent with other published profiles of the water column [12].

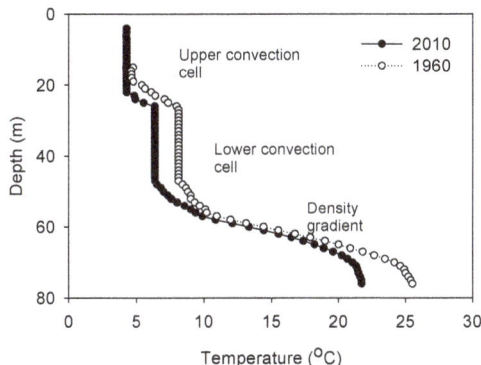

Figure S3. "Absorption" of red light, a proxy for the distribution of chlorophyll-a, by the microbial mat in Figure 9, collected at 10 m depth. Image A was taken in near infra-red light, image B in red light and image C is a false color image of chlorophyll-a absorption calculated as (1-R/NIR). Image D is a photograph of the mat cross section, and arrows indicate the locations of sediment-rich laminae. Scale bars are 5 mm, and color scale indicates 1-R/NIR.

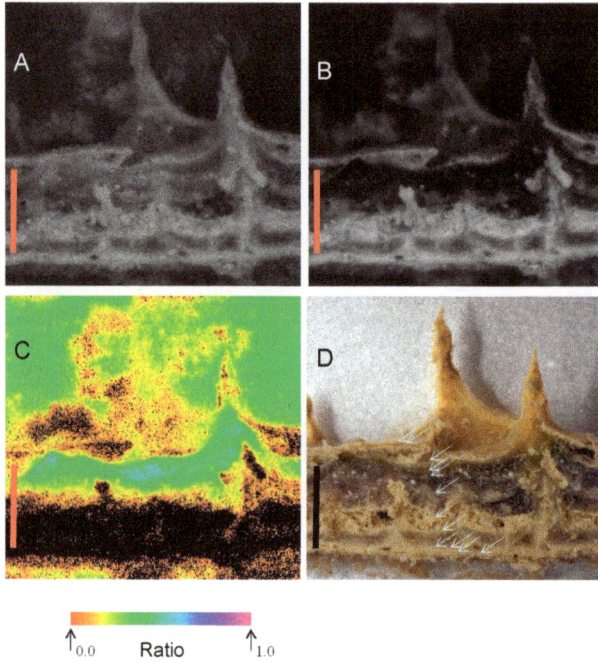

Figure S4. Vertical cross section of a microbial mat from 6 m depth. A and B are images obtained under infra-red and red light, and image C is the absorption, expressed as false color (as in Figure S3). Image D: distribution of maximal fluorescence intensity (F_m). Image E: maximum yield of photosystem II (F_v/F_m). Images D and E are combined in image F, which indicates F_v/F_m according to false color value and F_m as intensity. Color scales represent imaging PAM-derived false-color values. False color scales are as in Figure 9, and the horizontal bar indicates 5 mm.

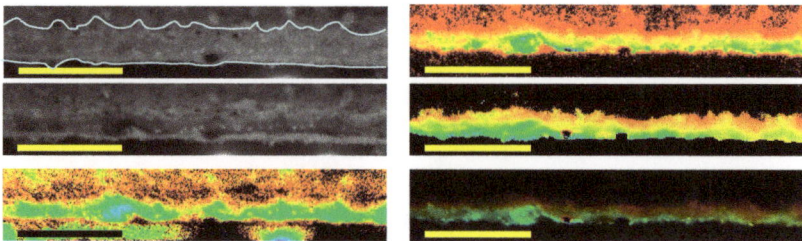

Figure S5. Vertical cross section of a microbial mat from 18 m depth. Image A is a photograph of the cross section, and the cut surface is outlined to indicate the part of each image that is in the focal plane; in all cases, the scale bar is 5 mm. Image B: distribution of maximal fluorescence intensity (F_m). Image C: maximum yield of photosystem II (F_v/F_m). Images B and C are combined in image E, which indicates F_v/F_m according to false color value and F_m as intensity. Color scales represent imaging PAM-derived false-color values. Image D is the "absorption" of the cross section, as defined in Figure S3, and indicates the distribution of chlorophyll-a.

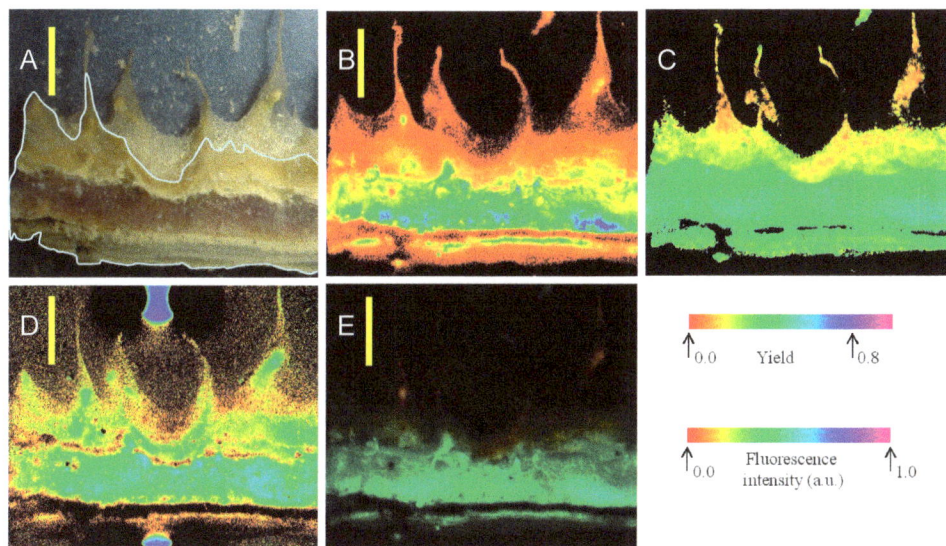

The Dynamic Arctic Snow Pack: An Unexplored Environment for Microbial Diversity and Activity

Catherine Larose, Aurélien Dommergue and Timothy M. Vogel

Abstract: The Arctic environment is undergoing changes due to climate shifts, receiving contaminants from distant sources and experiencing increased human activity. Climate change may alter microbial functioning by increasing growth rates and substrate use due to increased temperature. This may lead to changes of process rates and shifts in the structure of microbial communities. Biodiversity may increase as the Arctic warms and population shifts occur as psychrophilic/psychrotolerant species disappear in favor of more mesophylic ones. In order to predict how ecological processes will evolve as a function of global change, it is essential to identify which populations participate in each process, how they vary physiologically, and how the relative abundance, activity and community structure will change under altered environmental conditions. This review covers aspects of the importance and implication of snowpack in microbial ecology emphasizing the diversity and activity of these critical members of cold zone ecosystems.

Reprinted from *Biology*. Cite as: Larose, C.; Dommergue, A.; Vogel, T.M. The Dynamic Arctic Snow Pack: An Unexplored Environment for Microbial Diversity and Activity. *Biology* **2013**, *2*, 317-330.

1. Introduction

1.1. The Arctic, a Frozen Ecosystem

A large portion of the Earth is cold: about 14% of the biosphere is polar and 90% (by volume) is cold ocean (less than 5 °C). About two thirds of global freshwater is contained in ice and roughly 20% of the soil ecosystem exists as permafrost [1]. The Arctic, a vast circumpolar area consisting mainly of seasonally ice-covered ocean surrounded by continental land masses and islands, is an important part of the cryosphere, which can be defined as the portion of the Earth where water is in solid form [2]. The Arctic lies above 60°N and is characterized by a harsh climate, unique ecosystems and highly resilient biota [3]. Four million human residents of which approximately 10% are indigenous peoples inhabit many communities in eight countries: Canada, the Kingdom of Denmark (including Greenland and the Faroe Islands), Finland, Iceland, Norway, Russia, Sweden, and the United States of America (Alaska) [3].

Seasonal snow cover extends over a third of the Earth's land surface, covering up to 47 million km² [4] and is also an important feature of the Arctic. Snow cover can be considered as a dynamic habitat of limited duration [5] that acts as a medium and a mediator by transmitting and modifying interactions among microorganisms, plants, animals, nutrients, the atmosphere and soil [6]. Snow cover influences global energy and moisture budgets, thereby influencing climate [4]. The influence of seasonal snow cover on soil temperature, soil freeze-thaw processes, and permafrost has considerable impact on carbon exchange between the atmosphere and the ground and on the hydrological cycle in cold regions [7]. Snow cover acts as both an energy bank by storing and releasing energy and a radiation shield due to its high radiative properties that reflect as much as 80%–90% of the incoming radiation for fresh snow [4]. This high surface albedo reduces absorbed solar energy and lowers snow surface temperature [7].

Snow, a porous media with elevated air content [6], also has a high latent heat of fusion and acts as a heat sink as well as a ground insulator, since heat transfer is poor [4]. The extent and thickness of snow cover influences subsurface soil temperatures and soil metabolic activity [8] and its insulating properties protect soil surface organisms, such as vegetation, invertebrates and mammals against frost damage [4]. Furthermore, snow acts as a reservoir and as a transport medium for liquid water, moves as a particulate flux, and can be relocated by wind [6]. Physical metamorphism, phase changes and chemical transformations, which are modulated by interactions with the atmosphere and soil systems, control both the dynamics and the duration of the snow cover [9]. Thus snow cover is an important factor in the functioning of Arctic, and by extension, global ecosystems.

1.2. Snow Formation

Snow is formed in the atmosphere and consists of particles of ice that form in clouds. These crystals grow by vapor deposition and require atmospheric temperatures below 0 °C and the presence of supercooled water [10]. Ice formation is not spontaneous at temperatures above negative 40 °C (233 °K), so ice nucleation occurs mainly in the presence of substrates that act as catalysts. These substrates include dust, seasalt particles, sulfate, combustion products from industrial plants, volcanoes, forests and bacteria [6,11]. A recent report by Christner *et al.* [12] found that biological particles such as proteins or proteinaceous compounds play a significant role in the initiation of ice formation, especially when cloud temperatures are relatively warm. Once deposited, the snow cover forms as a result of snow crystal binding [13]. Snow crystals are subject to temperature gradients that generate water vapor fluxes between crystals. This results in the sublimation of parts of crystals and condensation on other parts, thus changing crystal size and shape and altering the physical properties of the snowpack. With each snowfall, the cover changes and the new layer may possess different properties than the preceding layer [14]. As snow ages, its physical properties, such as density, porosity, heat conductivity, hardness, specific surface area and albedo, evolve in response to thermodynamic stress and weather conditions [13]. Therefore, the composition of layered snow cover and ongoing changes in each of the layers are not only due to the circumstances of formation, but also to changing conditions over time.

1.3. Deposition and Incorporation of Impurities within Snowpacks

The snowpack is a receptor surface and storage compartment for nutrients, soluble inorganic, organic matter and contaminants that may or may not be attached to insoluble particles that are delivered by wet and dry deposition (reviewed by [11,15]). Their distribution within the snow is heterogeneous [16] and depends upon different physical processes such as atmospheric loading, wind speed, and snow metamorphism [11]. Nutrients exist in the atmosphere as trace gases such as

SO_2, CO_2, NO_X, N_2O or HNO_3 and as aerosols such as pollen, sea salt particles, mineral dust and sulfates [11]. Nutrients and contaminants can be delivered to the snowpack through wet and dry deposition. Wet deposition occurs when atmospheric components are scavenged and incorporated into growing or falling snow/rain as condensation or freezing nuclei by either particle impact, gas dissolution or by the collision of supercooled droplets with snow crystals [11]. Condensation and evaporation can alter the concentrations, resulting in the highly variable chemical composition of individual snow crystals.

Atmospheric scavenging and condensation largely condition the presence of major ions in the snowpack [11]. Dry deposition occurs when gases and particulates are transferred directly to the snow surface without the intermediate scavenging by precipitation. This pathway is dependent upon the atmospheric concentration of the chemical species, the stability or turbulence of the atmospheric boundary layer, as well as the capacity of the surface to retain the chemical species [11]. Once deposited, these species can be redistributed to the snowpack. Due to the permeability of the snowpack, gaseous diffusion occurs along a concentration gradient. Gases can also diffuse from the soil to the atmosphere [5]. Snow-air exchanges occur when the vapor diffuses through the air-filled pore space to the top of the snowpack and from there through a boundary layer to the atmosphere [15]. The penetration of gases and particles through the snowpack is dependent upon its physical-chemical properties, the geometry of the pore space, vapor pressure gradients and wind pressure [11]. Wind advection can accelerate solute transport within snow pores, even when the resistance to molecular diffusive transport is too great to allow gas exchange [15].

1.4. Snow Metamorphism and Impurity Cycling

The snowpack evolves chemically over time [17]. Physical processes of snow metamorphism also lead to the redistribution of chemical species. On a crystal, molecules diffuse from convex to concave sites, thus transforming crystals to small round snow grains that evaporate and distill onto larger grains once in close proximity. The grains grow rapidly by diffusion, which is initiated by temperature gradients within the snowpack and facilitated by the quasi-liquid surface layer of snow crystals that gives molecules high mobility. During this process, impurities are excluded from the crystals and concentrate at the grain boundaries and pore spaces of the snow [11]. The layered nature of the snowpack, which is composed of a heterogeneous mixture of grains of various sizes, water saturation levels, densities, and ice layers that reduce the permeability to air and water [14], is also important in the redistribution of solutes. Chemicals can be lost from the snow through degradation, volatilization and runoff with meltwater [15]. Impurities can be transformed within the snowpack and also returned to the atmosphere. Snow also transmits atmospherically derived impurities such as nutrients, microorganisms, particles and contaminants to meltwater-fed systems. Snow is, thus, a mediator favoring exchanges among different environmental compartments.

1.5. Snow Melt and Ecosystem Transfer

Melting can occur at air temperatures below 0 °C when solar radiation is intense enough and penetrates into the snowpack [18]. The top snow layers melt first and meltwater percolates downward towards the base of the snowpack. Initially, meltwater is retained in the capillaries and pore walls where it fills 5%–10% of the pore space before becoming more mobile [19,20]. As melting progresses, the water mobilizes solutes and contaminants from the pore walls. Preferential flow may develop due to the non-homogenous nature of the snow and lead to accelerated percolation and solvent concentration of meltwater in certain areas [11]. If the weather conditions prevent further melt, the highly concentrated meltwater may refreeze as a layer within the snowpack and becomes stationary. If multiple freeze/thaw cycles occur, each cycle will increase the solute concentrations of the meltwater, which becomes highly concentrated as it advances deeper into the snowpack [21]. Usually, meltwater reaches the ground, refreezes and develops into a solid layer. The first flush is highly concentrated, with preferential elution of certain solutes [22]. The most soluble ions are removed first and ionic concentrations taper off as melting proceeds [23,24]. Based on laboratory and field studies, fractionation of solutes into meltwater has been shown to occur in all snowpacks, with variable concentration factors [23,25,26]. Roughly 80% of solutes are removed from the snowpack by the first 30% of meltwater [11,27] and fractionation increases with the age of snow, by repeated melting and freezing and slow meltwater flow [23,28,29]. Soluble ions are removed first ([30,31]), followed by, in some cases, the preferential elution of some ions (e.g. SO_4^{2-}, Ca^{2+}, Mg^{2+}, K^+, Na^+) over others (NO_3^-, NH_4^+, Cl^-, F^-) [32]. Species such as non-polar organic molecules are also found in meltwater, but are less easily mobilized by percolating water due to their weak water solubility [33]. Particulate material can also be removed during percolation, but usually remains in the snow until the final stages of melting [33–35]. Rain events during the snowmelt period may lead to increases in solute and contaminant load [15].

During snowmelt, snow impurities are released to meltwater-fed catchments, soil and aquatic systems, potentially delivering a pulse of highly concentrated solutes, contaminants and microorganisms. Longer melt periods have been shown to lead to increased evaporation of chemicals to the atmosphere, thus reducing contaminant loading to terrestrial and aquatic ecosystems, while short melt periods deliver greater proportions of stored contaminants [15]. The snow cover lasts several months, thereby leading to longer solute accumulation periods, and arctic ecosystems are especially at risk for pulse exposure since the melt period is short [15].

2. Biology of the Cryosphere

2.1. Colonization and Activity in Cold Environments

Microorganisms exist in several extreme cold environments such as glacial ice [36–38], sea ice [39], Arctic biofilms [40], Arctic snow [41,42], supercooled clouds [43] and Antarctic permafrost [44]. Although both poles are different, it is likely that some of the microbial colonization pathways described for Antarctica, such as atmospheric circulation, ocean currents, birds, fishes, marine mammals and human vectors apply to the Arctic as well [45]. Aerial transport has long been viewed as a major transport route given that spore formers, such as Gram-positive bacteria and fungi, are able to survive long-range transport [46]. Due to the cold conditions and the limited supply of liquid water, snow and ice have long been only considered as entrapment and storage systems for microorganisms that were thought to enter as vegetative and resting cells, transported by wind-blown particles, aerosols and ice crystals. These cells would then be buried by subsequent snowfall events before being transferred to other systems upon snowmelt [47]. However, this view started to change with a number of studies that examined microbial diversity, ecology and function in the cryosphere. Whether the microorganisms found in cold environments are metabolically active and reproducing remains unclear, but it is assumed that certain microbial species are at least able to survive [2].

The occurrence of related phylotypes from geographically-diverse cold environments has been reported [39], suggesting that adaptation for survival, persistence and activity at low temperatures might be a common feature of these species and that they might possess common adaptive strategies [1]. The bacterial classes most frequently reported are *Proteobacteria* (*Alphaproteobacteria*, *Betaproteobacteria* and *Gammaproteobacteria*), *Bacteroidetes* group, low and high G+C Gram-positive genera, and *Cyanobacteria* [1,37,48,49]. Moreover, microorganisms might be metabolically active at low temperatures down to −20 °C [50,51] and very low rates of metabolic activity might be sustained for up to 10^4 to 10^6 years and at temperatures as low as −40 °C [52].

However, these studies focused on characterizing bacteria in ice or permafrost and relatively little is known about life in snow, despite the extent and importance of seasonal snow. The snow cover might support a microbial community composed of snow algae, bacteria, yeasts and snow fungi [5,53,54]. Snow algae represent an ecologically and physiologically specialized group that can form visible blooms, however their development is dependent on the availability of liquid water and are only active during the spring and summer, when air temperatures are above zero (reviewed in [54]). Snow algae have been studied relatively extensively [54–57], but data on bacteria inhabiting seasonal snow cover are sparse, especially for polar snowpacks. Carpenter (2000) reported low rates of DNA synthesis and the presence of Thermus-Deinococcus-like organisms in Antarctic snow [58], while Amato *et al.* (2007) used culture-based methods to isolate 10 bacterial strains belonging to *Proteobacteria*, *Firmicutes* and *Actinobacteria* from a snowpit dug on a polythermal glacier in Svalbard (Norway) [41]. Both studies focused on bacterial density and activity, but important questions about diversity, community structure, population dynamics and function remain unanswered. Using a 16S rRNA gene (rrs) clone library approach on snow and meltwater from Svalbard, Larose *et al.* (2010) observed high levels of diversity, similar to those of Arctic pack ice and Arctic microbial mats [39,42]. Significant differences in diversity among sample types were also reported and these may be related to seasonal changes in the snow environment (*i.e.*, pH, water content and temperature). Segawa *et al.*, 2005, also observed seasonal changes in bacterial flora and biomass in mountain snow in Japan, with increases in biomass during the melting season (March to October) that were attributed to nutrient and/or environmental conditions in the snow [59]. These results highlight the links between environmental conditions and changes in community structure.

The snowpack appears to be a diverse habitat and many studies suggest the occurrence of related phylotypes from geographically diverse, but predominantly cold environments [42,46,59]. However, the seasonal evolution of the microbial community and the physiological state of the organisms within the snowpack are topics that remain to be addressed.

2.2. Life in the Cold Lane

In order to colonize and survive in cold environments such as snowpacks, microorganisms must overcome a number of physiological stress parameters such as cold temperatures (less than 5 °C), high levels of solar radiation, desiccation and freeze/thaw cycles [1]. These harsh environmental conditions vary temporally as well as spatially and necessitate physiological acclimation. In the Arctic, because of the high latitudes, a pronounced seasonality causes gradual, yet extreme, changes in the photoperiod, irradiance, and temperature. During the springtime melt period, snow undergoes temperature shifts across the freezing point of water, leading to a more dynamic environment, but also to an increase in freeze/thaw cycles [60].

Different survival strategies at low temperatures have been observed in bacteria: reduction of cell size and capsular polysaccharide coat thickness, changes in fatty acid and phospholipid membrane composition, decrease of the fractional volume of cellular water, increase of the fraction of ordered cellular water, energy synthesis by catalysis of redox reactions of ions in aqueous veins in ice or in thin aqueous films in permafrost [52]. Moreover, many species that have been isolated form spores that provide high resistance levels, while others have thick cell walls or polysaccharide capsules that resist freeze/thaw cycles [1]. Cold tolerance has been shown to involve down-regulation of enzymes involved in major metabolic processes such as glycolysis, anaerobic respiration, ATP synthesis, fermentation, electron transport, sugar metabolism as well as the metabolism of lipids, amino acids, nucleotides and nucleic acids [61]. However, up-regulation and overexpression of several enzymes and proteins (cold shock proteins, *etc.*) may enhance survivability during freeze-thaw cycles [61]. Other adaptive strategies include the production of pigments such as oligosaccharide mycosporine-like amino acids, scytonemins, carotenoids, phycobiliproteins and chlorophylls that offer a broad strategy to cope with high irradiance [60].

Ability to attach to surfaces also provides bacteria with adaptive strategies. Junge *et al.* (2004) reported that particle-associated bacteria were more active than free-living cells as temperatures dropped and that they also produced exopolysaccharides (EPS). Bacteria growing in microbial mats were also shown to form EPS [60]. The EPS production favors attachment [51] and protects against freezing, dessication, viral and bacterial attacks [60]. Moreover, it is likely that the presence of species with specialized mechanisms of stress resistance may provide a protective effect on other members of the community. For example, certain pigments such as oligosaccharide mycosporine-like amino acids and scytonemin are located outside the cells and may benefit non-producing microorganisms against radiation damage [60].

2.3. Microorganisms—Active Members of the Cryosphere?

Recent reports suggest that microorganisms impact nutrient dynamics, composition and abundance [16], that they may shift surface albedo of snow and ice [62,63] and that they impact hydrochemistry [64]. Critical processes controlling biogenic trace gas (e.g., CO_2, CH_4, N_2O, and NO) fluxes are carried out by microorganisms [65]. Within the snowpack, microbiological activities such as carbon fixation by algal communities may modify the nutrient cycle [5]. The importance of bacteria in governing redox conditions and their role in Fe, S, N and P cycling is now acknowledged [16]. The role of bacteria in carbon cycling in Antarctic surface snow has recently been highlighted. Antony *et al.* (2012) reported that snow bacteria were able to use a wide range of low and high molecular weight carbon substrates and suggested that these organisms could potentially govern snow chemistry [66]. Microorganisms may also be responsible for the metabolism and transformation of contaminants such as pesticides [67] and mercury [40,68], both pollutants of arctic ecosystems. Mercury is an excellent example of the possible importance of microorganisms in pollutant fate in snowpacks.

The Arctic is experiencing mercury (Hg) toxicity [68] and Hg concentrations are increasing. Mercury exists in several forms in the environment: elemental (Hg°), divalent form (Hg^{2+}) and an organo-metallic form of which methylmercury (MeHg) is the most important. The MeHg organic form is the most toxic of the three forms, even at very low exposure doses [69]. Mercury is mainly emitted to the atmosphere in its gaseous form (Hg°), but also in the oxidized form (reactive gaseous mercury, RGM) or in the particle-bound form (particulate mercury, PM). Hg° has a relatively long atmospheric residence time (between 0.5 and 1.5 years) and average atmospheric concentrations have been estimated to 1.7 g/m^3 for the Northern Hemisphere [70]. RGM and PM have shorter lifetimes and tend to be deposited near their sources [70]. Mercury reaches polar ecosystems mainly as Hg°; however due to the cyclical nature of Hg transformations (transport-deposition-re-emission), even mercury originally emitted as RGM and PM can be transported to the Arctic [71]. Similar to other contaminants, Hg can be deposited after atmospheric scavenging by precipitation and dry deposition. Once RGM is formed in the atmosphere, snow can act as an efficient surface for its sorption. In addition, active growth of snow and ice crystals from the vapor phase readily scavenges available RGM [72].

In 1995 at Alert, Canada, Schroeder *et al.* (1998) measured the episodic near-total depletion of Hg° from the atmosphere during the spring [73]. These events, termed Atmospheric Mercury Depletion Events (AMDEs), were observed in parallel to the depletion of ozone [74] and led to intense field, laboratory and theoretical studies to determine which reactions were involved. In particular, mercury was shown to undergo rapid oxidation and deposition via photochemically-initiated reactions believed to involve reactive marine halogens, mainly Br and BrO [75–77]. These reactions transform Hg° to PM and RGM species that can then be deposited onto the snow. It has been estimated that AMDEs increase polar mercury deposition by 100 tonnes a year [71], yet the post-depositional fate of this Hg remains uncertain, although it could undergo a series of possible transformations once deposited [78]. Using a bacterial *mer-lux* biosensor, Larose *et al.*, 2011 detected Hg in Arctic snow in a bioavailable chemical form, *i.e.* able to interact with microorganisms, and showed that fresh snowfall events contributed to higher proportions of BioHg than mercury depletion events [79]. Therefore, Hg is deposited in a chemical state that allows for biological uptake and transformation. Different simultaneous biotic and abiotic processes alter the chemical state of mercury and thereby its toxicity in the environment. Four different reactions control mercury speciation: methylation, demethylation, reduction and oxidation [80] and microorganisms can carry out each of these transformations. Microorganisms are able to methylate mercury. Bacteria have been isolated from Arctic snowpacks [41] and microbial activity has been measured at temperatures down to −20 °C [50]. Constant *et al.* 2007, reported increases in the

MeHg:THg ratio and positive correlations with bacterial colony counts and particles. These results led to the hypothesis that MeHg was being formed within the snowpack, despite the absence of correlation with sulfate-reducing bacteria (SRB), the principal methylators in anoxic environments [81]. Recently, Larose *et al.*, 2010 proposed a mechanism by which bioavailable Hg may undergo methylation by microorganisms in an aerobic process involving biogenic sulfur molecules [17].

In order to cope with the toxicity of Hg and MeHg, bacteria have developed specialized resistance mechanisms. For example, bacteria possessing the *mer* operon are able to detoxify Hg via MerA [80]. The genes that encode MerA have been isolated both from a variety of environments including soil [82] (150), Siberian permafrost [83] and Arctic biofilms [40] and from bacteria [80] and archaea [84]. Some bacteria are able to detoxify both BioHg and MeHg, while others are only able to transform inorganic mercury via the *mer* operon resistance pathway [80]. Based on results from a cultivation study of Arctic snow bacteria, Moller *et al*, 2011 were able to demonstrate that mercury-resistant bacteria accounted for almost a third of cultivatable organisms and that 25% of these were able to completely reduce mercury, thus limiting the supply of Hg available for methylation [85]. These mercury resistant bacteria may therefore help lower the risk of methylmercury entering Arctic food chains.

3. General Conclusion and Perspectives

The Arctic climate is changing. A 20th century warming trend has been documented in the Arctic, with air temperatures over land areas increasing by as much as 5 °C and increased temperatures over sea ice [3]. In addition, precipitation has also increased. Other changes include a 2.9% per decade decrease in Arctic sea-ice extent (1978–1996), the thinning of sea-ice, an increase in melt days par summer, the warming of Atlantic water flowing into the Arctic Ocean, the thinning of the oceanic surface layer and the increase in ground temperatures and resulting permafrost melt [86]. Decreases in snow and ice cover, increased plant growth, increased primary production of terrestrial algae in freshwater lakes, and the northward movement of the tree line in the most-warmed Arctic regions have also been reported [3]. These changes are probably linked to human activities that are clearly influencing the climate, with arctic (and polar) environments subjected to substantial warming and increases in precipitation over the 21st century.

Climate change is also expected to alter contaminant loading and transformations in the Arctic. The predicted warming of air temperatures at lower latitudes will have direct effects on contaminants through increased volatility, more rapid degradation and altered partitioning between phases [87], while increased precipitation could lead to more scavenging of contaminants by rain and snow, thereby augmenting inputs to aquatic and terrestrial ecosystems [3]. Extended ice-free areas in the Arctic Ocean may favor both atmospheric scavenging by precipitation in addition to seawater partitioning [3] and certain contaminants might evade from surface seawater more rapidly [87].

While it appears that microbial life is well adapted to cold ecosystems, the response of these populations to changing environments is mostly unknown. Climate change may alter microbial functioning by increasing growth rates and substrate use due to increased temperature. This may lead to changes in process rates. Another impact could be the restructuring of microbial communities [65]. Biodiversity may increase as the Arctic warms and population shifts occur as non-heat tolerant species disappear in favour of more heat tolerant ones. For example, a circumpolar shift was seen in the fossil remains of algae and invertebrates in the mid to late 19th century probably due to climate change. In order to predict how ecological processes will evolve as a function of global change, it is essential to identify which populations participate in each process, how they vary physiologically, and how the relative abundance, activity and community structure will change under altered environmental conditions [65].

Acknowledgments

The authors would like to acknowledge the contribution of the entire AWIPEV staff. This research was supported by grants from EC2CO/CYTRIX (Programme National INSU), LEFE, IPEV CHIMERPOL program (399) and CL would like to acknowledge the FQRNT (le Fonds Québécois de la Recherche sur la Nature et les Technologies) for a PhD research fellowship. AD would like to thank the IUF, the Fond France Canada pour la Recherche and la Région Rhône-Alpes for supporting this research.

References and Notes

1. Priscu, J.C.; Christner, B.C. Earth's icy biosphere. In *Microbial Diversity and Bioprospecting*; Bull, A.T., Ed.; American Society for Microbiology: Washington, DC, USA, 2004; pp. 130–145.
2. Miteva, V. Bacteria in snow and glacier ice. In *Psychrophiles: From Biodiversity to Biotechnology*; Margesin, R.E.A., Ed.; Springer-Verlag: Berlin, Heidelberg, Germay, 2008; pp. 31–47.
3. AMAP. *Amap Assessment 2009: Human Health in the Arctic. Arctic Monitoring and Assessment Programme*; AMAP: Oslo, Norway, 2009.
4. Hinkler, J.; Hansen, B.U.; Tamstorf, M.P.; Sigsgaard, C.; Petersen, D. Snow and snow-cover in central northeast greenland. *Adv. Ecol. Res.* **2008**, *40*, 175–195.
5. Jones, H.G. The ecology of snow-covered systems: A brief overview of nutrient cycling and life in the cold. *Hydrol. Process* **1999**, *13*, 2135–2147.
6. Pomeroy, J.W.; Brun, E. Physical properties of snow. In *Snow Ecology. An Interdisciplinary Examination of Snow-Covered Ecosystems*; Jones, H.G., Pomeroy, J.W., Walker, D.A., Hoham, R.W., Eds.; Cambridge University Press: Cambridge, UK, 2001; pp. 45–126.
7. Zhang, T. Influence of the seasonal snow cover on the ground thermal regime: An overview. *Rev. Geophys.* **2005**, *43*, RG4002.
8. Larsen, K.S.; Grogan, P.; Jonasson, S.; Michelsen, A. Respiration and microbial dynamics in two subarctic ecosystems during winter and spring thaw: Effects of increased snow depth. *Arct. Antarct. Alp. Res.* **2007**, *39*, 268–276.
9. Jones, H.G.; Pomeroy, J.W.; Walker, D.A.; Hoham, R.W. *Snow Ecolog*; Cambridge University Press: Cambridge, UK, 2001; p. 398.
10. Libbrecht, K.G. The physics of snow crystals. *Rep. Prog. Phys.* **2005**, *68*, 855–895.

11. Kuhn, M. The nutrient cycle through snow and ice, a review. *Aquat. Sci.* **2001**, *63*, 150–167.

12. Christner, B.C.; Morris, C.E.; Foreman, C.M.; Cai, R.; Sands, D.C. Ubiquity of biological ice nucleators in snowfall. *Science* **2008**, *319*, 1214.

13. Jordan, R.E.; Albert, M.R.; Brun, E. Physical processes within snow and their parameterization. In *Snow and Climate*; Armstrong, R.L., Brun, E., Eds.; Cambridge University Press: Cambridge, UK, 2008; pp. 12–69.

14. Colbeck, S.C. The layered character of snow covers. *Rev. Geophys.* **1991**, *29*, 81–96.

15. Daly, G.L.; Wania, F. Simulating the influence of snow on the fate of organic compounds. *Environ. Sci. Technol.* **2004**, *38*, 4176–4186.

16. Hodson, A.; Anesio, A.M.; Tranter, M.; Fountain, A.; Osborn, M.; Priscu, J.; Laybourn-Parry, J.; Sattler, B. Glacial ecosystems. *Ecol. Monogr.* **2008**, *78*, 41–67.

17. Larose, C.; Dommergue, A.; De Angelis, M.; Cossa, D.; Averty, B.; Marusczak, N.; Soumis, N.; Schneider, D.; Ferrari, C. Springtime changes in snow chemistry lead to new insights into mercury methylation in the arctic. *Geochimica Et Cosmochimica Acta* **2010b**, *74*, 6263–6275.

18. Kuhn, M. Micro-meteorological conditions for snow melt. *J. Glaciol.* **1987**, *33*, 24–26.

19. Colbeck, S.C. The physical aspects of water flow through snow. *Adv. Hydrosciences* **1978**, *11*, 165–206.

20. Davis, R.E. Links between snowpack physics and snowpack chemistry. In *Seasonal snowpacks nato asi series g*; Davies, T.D., Martyn, T., Jones, H.G., Eds.; Springer-Verlag: Berlin, Germany, 1991; Volume 28; pp. 115–138.

21. Meyer, T.; Wania, F. Organic contaminant amplification during snowmelt. *Water Res.* **2008**, *42*, 1847–1865.

22. Hodson, A. Biogeochemistry of snowmelt in an antarctic glacial ecosystem. *Water Resour. Res.* **2006**, *42*, W11406.

23. Colbeck, S.C. A simulation of the enrichment of atmospheric pollutants in snow cover runoff. *Water Resour. Res.* **1981**, *17*, 1383–1388.

24. Goto-Azuma, K.; Nakawo, M.; Han, J.; Watanabe, O.; Azuma, N. Melt-induced relocation of ions in glaciers and in a seasonal snowpack. *IAHS Publ.* **1994**, *223*, 287–298.

25. Johannessen, M.; Henriksen, A. Chemistry of snow melt water: Changes in concentration during melting. *Water Resour. Res.* **1978**, *14*, 615–619.

26. Davies, T.D.; Vincent, C.E.; Brimblecombe, P. Preferential elution of strong acids from a norwegian ice cap. *Nature* **1982**, *300*, 161–163.

27. Brimblecombe, P.; Tranter, M.; Tsiouris, S.; Davies, T.D.; Vincent, C.E. The chemical evolution of snow and meltwater. *IAHS Publ.* **1986**, *155*, 283–295.

28. Johannessen, M.; Dale, T.; Gjessing, E.T.; Henriksen, A.; Wright, R.F. Acid precipitation in norway: The regional distribution of contaminants in snow and the chemical concentration processes during snow melt. *IAHS Publ.* **1977**, *118*, 116–120.

29. Davis, R.E.; Petersen, C.E.; Bales, R.C. Ion flux through a shallow snowpack: Effects of initial conditions and melt sequences. *IAHS Publ.* **1995**, *228*, 115–126.

30. Tranter, M.; Brimblecombe, P.; Davies, T.D.; Vincent, C.E.; Abrahams, P.W.; Blackwood, I. A composition of snowfall, snowpack and meltwater in the scottish highlands—Evidence for preferential elution. *Atmos. Environ.* **1986**, *20*, 517–525.

31. Meyer, T.; Lei, Y.D.; Muradi, I.; Wania, F. Organic contaminant release from melting snow. 2. Influence of snow pack and melt characteristics. *Environ. Sci. Technol.* **2008**, *43*, 663–668.

32. Eichler, A.; Schwikowski, M.; Gäggeler, H.W. Meltwater induced relocation of chemical species in alpine firn. *Tellus* **2001**, *53B*, 192–203.

33. Meyer, T.; Lei, Y.D.; Wania, F. Measuring the release of organic contaminants from melting snow under controlled conditions. *Environ. Sci. Technol.* **2006**, *40*, 3320–3326.

34. Hodgkins, R.; Tranter, M.; Dowdeswell, J.A. The hydrochemistry of runoff from a 'coldbased' glacier in the high arctic (scott turnerbreen, svalbard). *Hydrol. Process.* **1998**, *12*, 87–103.

35. Lyons, W.B.; Welch, K.A.; Fountain, A.G.; Dana, G.L.; Vaughn, B.H.; McKnight, D.M. Surface glaciochemistry of taylor valley, southern victoria land, antarctica, and its relation to stream chemistry. *Hydrol. Processes* **2003**, *17*, 115–130.

36. Christner, B.C.; Mosley-Thompson, E.; Thompson, L.G.; Zagorodnov, V.S.; Sandman, K.; Reeve, J.N. Recovery and identification of viable bacteria immured in glacial ice. *Icarus* **2000**, *144*, 479–485.

37. Christner, B.C.; Mosley-Thompson, E.; Thompson, L.G.; Reeve, J.N. Isolation of bacteria and 16s rdnas from lake vostok accretion ice. *Environ. Microbiol.* **2001**, *3*, 570–577.

38. Skidmore, M.L.; Foght, J.M.; Sharp, M.J. Microbial life beneath a high arctic glacier. *Appl. Environ. Microbiol.* **2000**, *66*, 3214–3220.

39. Brinkmeyer, R.; Knittel, K.; Jurgens, J.; Weyland, H.; Amann, R.; Helmke, E. Diversity and structure of bacterial communities in arctic *versus* antarctic pack ice. *Appl. Environ. Microbiol.* **2003**, *69*, 6610–6619.

40. Poulain, A.J.; Ni Chadhain, S.M.; Ariya, P.A.; Amyot, M.; Garcia, E.; Campbell, P.G.C.; Zylstra, G.J.; Barkay, T. Potential for mercury reduction by microbes in the high arctic. *Appl. Environ. Microbiol.* **2007**, *73*, 2230–2238.

41. Amato, P.; Hennebelle, R.; Magand, O.; Sancelme, M.; Delort, A.M.; Barbante, C.; Boutron, C.; Ferrari, C. Bacterial characterization of the snow cover at spitzberg, svalbard. *FEMS Microbiol. Ecol.* **2007**, *59*, 255–264.

42. Larose, C.; Berger, S.; Ferrari, C.; Navarro, E.; Dommergue, A.; Schneider, D.; Vogel, T.M. Microbial sequences retrieved from environmental samples from seasonal arctic snow and meltwater from svalbard, norway. *Extremophiles* **2010**, *14*, 205–212.

43. Sattler, B.; Puxbaum, H.; Psenner, R. Bacterial growth in super cooled cloud droplets. *Geophys. Res. Lett.* **2001**, *28*, 239–242.

44. Yergeau, E.; Newsham, K.K.; Pearce, D.A.; Kowalchuk, G.A. Patterns of bacterial diversity across a range of antarctic terrestrial habitats. *Environ. Microbiol.* **2007**, *9*, 2670–2682.

45. Vincent, W.F. Evolutionary origins of antarctic microbiota: Invasion, selection and endemism. *Antarct. Sci.* **2000**, *12*, 374–385.

46. Harding, T.; Jungblut, A.D.; Lovejoy, C.; Vincent, W.F. Microbes in high arctic snow and implications for the cold biosphere. *Appl. Environ. Microbiol.* **2011**, *77*, 3234–3243.

47. Cowan, D.A. Tow, L.A. Endangered antacrctic environments. *Annu. Rev. Microbiol.* **2004**, *58*, 649–690.

48. Liu, Y.; Yao, T.; Jiao, N.; Kang, S.; Zeng, Y.; Huang, S. Microbial community structure in moraine lakes and glacial meltwaters, mount everest. *FEMS Microbiol. Lett.* **2006**, *265*, 98–105.

49. Liu, Y.; Yao, T.; Jiao, N.; Kang, S.; Xu, B.; Zeng, Y.; Huang, S.; Liu, X. Bacterial diversity in the snow over tibetan plateau glaciers. *Extremophiles* **2009**, *13*, 89–99.

50. Christner, B.C. Incorporation of DNA and protein precursors into macromolecules by bacteria at −15 degrees c. *Appl. Environ. Microbiol.* **2002**, *68*, 6435–6438.

51. Junge, K.; Eicken, H.; Jody, W. Bacterial activity at −2 to −20 °C in arctic wintertime sea ice. *Appl. Environ. Microbiol.* **2004**, *70*, 550–557.

52. Price, P.B.; Sowers, T. Temperature dependence of metabolic rates for microbial growth, survival and maintenance. *Proc. Natl. Acad. Sci. USA* **2004**, *101*, 4631–4636.

53. Bachy, C.; Lopez-Garcia, P.; Vereshchaka, A.; Moreira, D. Diversity and vertical distribution of microbial eukaryotes in the snow, sea ice and seawater near the north pole at the end of the polar night. *Front. Microbiol.* **2011**, *2*, 106.

54. Komárek, J.; Nedbalová, L. Green cryosestic algae. In *Algae and Cyanobacteria in Extreme Environments*; Seckbach, J., Ed.; Springer: Amsterdam, Netherlands, 2007; Volume 11; pp. 321–342.

55. Hoham, R.W. Optimal temperatures and temperature ranges for growth of snow algae. *Arct. Alp. Res.* **1975**, *7*, 13–24.

56. Hoham, R.W.; Duval, B. Microbial ecology of snow and freshwater ice with emphasis on snow algae. In *Snow Ecology: An Interdisciplinary Examination of Snow-covered*; Jones, H.G., Pomeroy, J.W., Walker, D.A., Hoham, R.W., Eds.; Cambridge University Press: Cambridge, UK, 2001; pp. 168–228.

57. Stibal, M.; Elster, J.; Sabacka, M.; Kastovska, K. Seasonal and diel changes in photosynthetic activity of the snow alga *chlamydomonas nivalis* (*chlorophyceae*) from svalbard determined by pulse amplitude modulation fluorometry. *FEMS Microbiol. Ecol.* **2007**, *59*, 265–273.

58. Carpenter, E.J.; Lin, S.; Capone, D.G. Bacterial activity in south pole snow. *Appl. Environ. Microbiol.* **2000**, *66*, 4514–4517.

59. Segawa, T.; Miyamoto, K.; Ushida, K.; Agata, K.; Okada, N.; Kohshima, S. Seasonal change in bacterial flora and biomass in mountain snow from the tateyama mountains, japan, analyzed by 16s rrna gene sequencing and real-time pcr. *Appl. Environ. Microbiol.* **2005**, *71*, 123–130.

60. Mueller, D.R.; Vincent, W.F.; Bonilla, S.; Laurion, I. Extremotrophs, extremophiles and broadband pigmentation strategies in a high arctic ice shelf ecosystem. *FEMS Microbiol. Ecol.* **2005**, *53*, 73–87.

61. Qiu, Y.; Vishnivetskaya, T.A.; Lubman, D.M. Proteomic insights: Cryoadaptation of permafrost bacteria. In *Permafrost Soils*; Springer: New York, NY, USA, 2009; pp. 169–181.

62. Thomas, W.H.; Duval, B. Sierra nevada, california, USA, snow algae: Snow albedo changes, algal-bacterial interrelationships, and ultraviolet radiation effects. *Arct. Antarct. Alp. Res.* **1995**, *27*, 389–399.

63. Yallop, M.L.; Anesio, A.M.; Perkins, R.G.; Cook, J.; Telling, J.; Fagan, D.; Macfarlane, J.; Stibal, M.; Barker, G.; Bellas, C.; *et al.* Photophysiology and albedo-changing potential of the ice algal community on the surface of the greenland ice sheet. *ISME J.* **2012**, *6*, 2302–2313.

64. Tranter, M.; Sharp, M.J.; Lamb, H.R.; Brown, G.H.; Hubbard, B.P.; Willis, I.C. Geochemical weathering at the bed of haut glacier d'arolla, switzerland—a new model. *Hydrol. Processes* **2002**, *16*, 959–993.

65. Schimel, J.P.; Gulledge, J. Microbial community structure and global trace gases. *Global Change Biol.* **1998**, *4*, 745–758.

66. Antony, R.; Mahalinganathan, K.; Krishnan, K.P.; Thamban, M. Microbial preference for different size classes of organic carbon: A study from antarctic snow. *Environ. Monit. Assess.* **2012**, *184*, 5929–5943.

67. Stibal, M.; Telling, J.; Cook, J.; Mak, K.M.; Hodson, A.; Anesio, A.M. Environmental controls on microbial abundance and activity on the greenland ice sheet: A multivariate analysis approach. *Microb. Ecol.* **2012**, *63*, 74–84.

68. Barkay, T.; Poulain, A.J. Mercury (micro)biogeochemistry in polar environments. *FEMS Microbiol. Ecol.* **2007**, *59*, 232–241.

69. Ullrich, S.M.; Tanton, T.W.; Abdrashitova, S.A. Mercury in the aquatic environment: A review of factors affecting methylation. *Crit. Rev. Env. Sci. Technol.* **2001**, *31*, 241–293.

70. Lindberg, S.; Bullock, R.; Ebinghaus, R.; Engstrom, D.; Feng, X.B.; Fitzgerald, W.; Pirrone, N.; Prestbo, E.; Seigneur, C. A synthesis of progress and uncertainties in attributing the sources of mercury in deposition. *Ambio* **2007**, *36*, 19–32.

71. Ariya, P.A.; Dastoor, A.P.; Amyot, M.; Schroeder, W.H.; Barrie, L.; Anlauf, K.; Raofie, F.; Ryzhkov, A.; Davignon, D.; Lalonde, J.; *et al.* The arctic: A sink for mercury. Tellus B. *Chem. Phys. Meteorol.* **2004**, *56*, 397–403.

72. Douglas, T.A.; Sturm, M.; Simpson, W.R.; Brooks, S.; Lindberg, S.E.; Perovich, D.K. Elevated mercury measured in snow and frost flowers near arctic sea ice leads. *Geophys. Res. Lett.* **2005**, *32*, 4.

73. Schroeder, W.H.; Anlauf, K.G.; Barrie, L.A.; Lu, J.Y.; Steffen, A.; Schneeberger, D.R.; Berg, T. Arctic springtime depletion of mercury. *Nature* **1998**, *394*, 331–332.

74. Barrie, L.A.; Bottenheim, J.W.; Schnell, R.C.; Crutzen, P.J.; Rasmussen, R.A. Ozone destruction and photochemical reactions at polar sunrise in the lower arctic atmosphere. *Nature* **1988**, *334*, 138–141.

75. Skov, H.; Christensen, J.H.; Goodsite, M.E.; Heidam, N.Z.; Jensen, B.; Wahlin, P.; Geernaert, G. Fate of elemental mercury in the arctic during atmospheric mercury depletion episodes and the load of atmospheric mercury to the arctic. *Environ. Sci. Technol.* **2004**, *38*, 2373–2382.

76. Lu, J.Y.; Schroeder, W.H.; Barrie, L.A.; Steffen, A.; Welch, H.E.; Martin, K.; Lockhart, L.; Hunt, R.V.; Boila, G.; Richter, A. Magnification of atmospheric mercury deposition to polar regions in springtime: The link to tropospheric ozone depletion chemistry. *Geophys. Res. Lett.* **2001**, *28*, 3219–3222.

77. Lindberg, S.E.; Brooks, S.; Lin, C.J.; Scott, K.J.; Landis, M.S.; Stevens, R.K.; Goodsite, M.; Richter, A. Dynamic oxidation of gaseous mercury in the arctic troposphere at polar sunrise. *Environ. Sci. Technol.* **2002**, *36*, 1245–1256.

78. Steffen, A.; Douglas, T.; Amyot, M.; Ariya, P.; Aspmo, K.; Berg, T.; Bottenheim, J.; Brooks, S.; Cobbett, F.; Dastoor, A.; *et al.* A synthesis of atmospheric mercury depletion event chemistry in the atmosphere and snow. *Atmos. Chem. Phys.* **2008**, *8*, 1445–1482.

79. Larose, C.; Dommergue, A.; Maruszczak, N.; Coves, J.; Ferrari, C.P.; Schneider, D. Bioavailable mercury cycling in polar snowpacks. *Environ. Sci. Technol.* **2011**, *45*, 2150–2156.

80. Barkay, T.; Miller, S.M.; Summers, A.O. Bacterial mercury resistance from atoms to ecosystems. *FEMS Microbiol. Rev.* **2003**, *27*, 355–384.

81. Constant, P.; Poissant, L.; Villemur, R.; Yumvihoze, E.; Lean, D. Fate of inorganic mercury and methyl mercury within the snow cover in the low arctic tundra on the shore of hudson bay (Québec, Canada). *J. Geophys. Res.* **2007**, *112*, D21311.

82. Oregaard, G.; Sorensen, S.J. High diversity of bacterial mercuric reductase genes from surface and sub-surface floodplain soil (Oak Ridge, USA). *ISME J.* **2007**, *1*, 453–467.

83. Mindlin, S.; Minakhin, L.; Petrova, M.; Kholodii, G.; Minakhina, S.; Gorlenko, Z.; Nikiforov, V. Present-day mercury resistance transposons are common in bacteria preserved in permafrost grounds since the upper pleistocene. *Res. Microbiol.* **2005**, *156*, 994–1004.

84. Schelert, J.; Dixit, V.; Hoang, V.; Simbahan, J.; Drozda, M.; Blum, P. Occurrence and characterization of mercury resistance in the hyperthermophilic archaeon sulfolobus solfataricus by use of gene disruption. *J. Bacteriol.* **2004**, *186*, 427–437.

85. Moller, A.K.; Barkay, T.; Abu Al-Soud, W.; Sorensen, S.J.; Skov, H.; Kroer, N. Diversity and characterization of mercury-resistant bacteria in snow, freshwater and sea-ice brine from the high arctic. *FEMS Microbiol. Ecol.* **2011**, *75*, 390–401.

86. Anisimov, O.; Fitzharris, B. Polar regions (arctic and antarctic). In *Intergovernmental Panel on Climate Change 2001: Impacts, Adaptation, and Vulnerability*; McCarthy, J.J., Canziani, O.F., Leary, N.A., Dokken, D.J., White, K.S., Eds.; Cambridge University Press: Cambridge, UK, 2001; pp. 801–841.

87. Macdonald, R.W.; Harner, T.; Fyfe, J. Recent climate change in the arctic and its impact on contaminant pathways and interpretation of temporal trend data. *Sci. Total Environ.* **2005**, *342*, 5–86.

Micro-Eukaryotic Diversity in Hypolithons from Miers Valley, Antarctica

Jarishma K. Gokul, Angel Valverde, Marla Tuffin, Stephen Craig Cary and Don A. Cowan

Abstract: The discovery of extensive and complex hypolithic communities in both cold and hot deserts has raised many questions regarding their ecology, biodiversity and relevance in terms of regional productivity. However, most hypolithic research has focused on the bacterial elements of the community. This study represents the first investigation of micro-eukaryotic communities in all three hypolith types. Here we show that Antarctic hypoliths support extensive populations of novel uncharacterized bryophyta, fungi and protists and suggest that well known producer-decomposer-predator interactions may create the necessary conditions for hypolithic productivity in Antarctic deserts.

Reprinted from *Biology*. Cite as: Gokul, J.K.; Valverde, A.; Tuffin, M.; Cary, S.C.; Cowan, D.A. Micro-Eukaryotic Diversity in Hypolithons from Miers Valley, Antarctica. *Biology* **2013**, *2*, 331-340.

1. Introduction

Microbial life in terrestrial Antarctica soils is subjected to extreme low temperatures, low water availability, high salinity, high UV radiation, and low nutrient availability [1]. However, despite the many adverse environmental constraints this extreme ecosystem has been shown to support extensive microbial biomass [2].

Much of the microbial research in Antarctic terrestrial and aquatic ecosystems has focused on the bacterial populations, and to a lesser extent on the archaea [3,4] and viruses [5,6]. In contrast, eukaryotic microorganisms have received much less attention [7].

Hypolithic communities in the Dry Valleys region of eastern Antarctica colonize the ventral surface of quartz rocks at the rock-soil interface [8–11]. Hypoliths can be envisioned as a stress-avoidance strategy, where the overlying rock creates a favorable sub-lithic microhabitat with greater physical stability, increased water availability, desiccation buffering, and UV protection [9,12]. As they are typically dominated by cyanobacteria [8,13] or bryophytes [9], hypolithons represent an important contribution to regional productivity [14,15]. Fungal dominated hypoliths have also been described in Antarctica [9].

In an earlier study [11] we characterized the bacterial and eukaryotic phylogenetic diversity of two hypolith types: Type I (cyanobacteria-dominated) and Type III (moss-dominated). Here, we extend this research to Type II hypolithons (fungal-dominated) with a focus on the micro-eukaryotic communities. We also compare all three types of eukaryal communities in terms of habitat preferences.

2. Results and Discussion

Environmental DNA was used as template for construction of three separate clone libraries using universal 18S rRNA, 18S-28S rRNA (ITS), and microalgal 18S rRNA-specific PCR primers

(Table 1). A total of 31 unique phylotypes was found (Table 2). Most of the sequences showed low identity values, indicating that the majority of sequences might represent novel taxa. Rarefaction curves (not shown) showed that more extensive sequencing would be required to capture the complete diversity within micro-eukaryotic communities in hypoliths. Incomplete sampling might be aggravated by the inherent limitations of the PCR approach, since several groups of micro-eukaryotes (e.g., multinucleated fungi) have multiple rRNA gene copy numbers that would be preferentially amplified because of primer competition [16].

Phylotypic analyses demonstrate that diverse communities of micro-eukaryotes inhabit Antarctic hypoliths (Table 2), showing both a broad range of taxa and a large functional diversity, including phototrophs (bryophyta) and a variety of heterotrophic organisms (fungi and protists). However, most of the clones showed low identity values, indicating that the majority of sequences might represent novel taxa. Further studies, using a polyphasic approach (*i.e.*, including a combination of genotypic and phenotypic approaches) will be necessary to confirm this hypothesis.

The phylotypic abundance data indicates that ascomycetes were present in all three hypolith types, but also possible habitat preferences for certain groups of eukaryotes. For example, amoebozoa were only found in Type I hypolithons (cyanobacteria dominated) whereas cercozoa were present only in Type III hypolithons (moss dominated) (Figure 1). Cyanobacteria can modify the surrounding environment [17], and play critical roles in the structuring of hypolithic communities [12,18]. For example, cyanobacteria produce UV-screening pigments, enzymes, and carotenoids that quench reactive oxygen species, solute-binding materials, water absorbing gels, antifreeze compounds, and ice-nucleating substances [19], which will reduce oxidative, osmotic, freeze-thaw, and dehydration stresses for all organisms embedded within the matrix. In contrast to open soil, hypoliths are also rich in inorganic nutrients, organic carbon and bacteria [18] that may provide substrates for eukaryotic heterotrophs such as protists and the metazoan microfauna. The presence of saprophytic, phagotrophic, parasitic and predatory eukaryotes would increase the inherent capacity for nutrient and energy transfer, thereby increasing trophic complexity and potential resilience to environmental change [12].

Fungal sequences were classified into 13 ascomycete phylotypes (Table 2). Most (86%) were related to the genus Acremonium, while some sequences were affiliated to Stromatonectria and Verrucaria (7% each); although it is worth noting that sequence comparisons of the ITS region gave low similarity values (76%–92%). There have been a limited number of molecular diversity studies of hypolithic [11] and soil [7,24] fungi in the Antarctica, some of which show contrasting results. For example, Fell *et al.* [7] found that both ascomycetes (lichen-forming and decomposers) and basidiomycetes (decomposers and nematode pathogens) were widely distributed in soils, whereas Khan *et al.* [11] reported ascomycetes as the only members in hypolithic fungal communities. This apparent dichotomy in fungal distribution between open soils and hypolithic communities offers a potential line for future research.

Figure 1. Relative distribution of phylotypes.

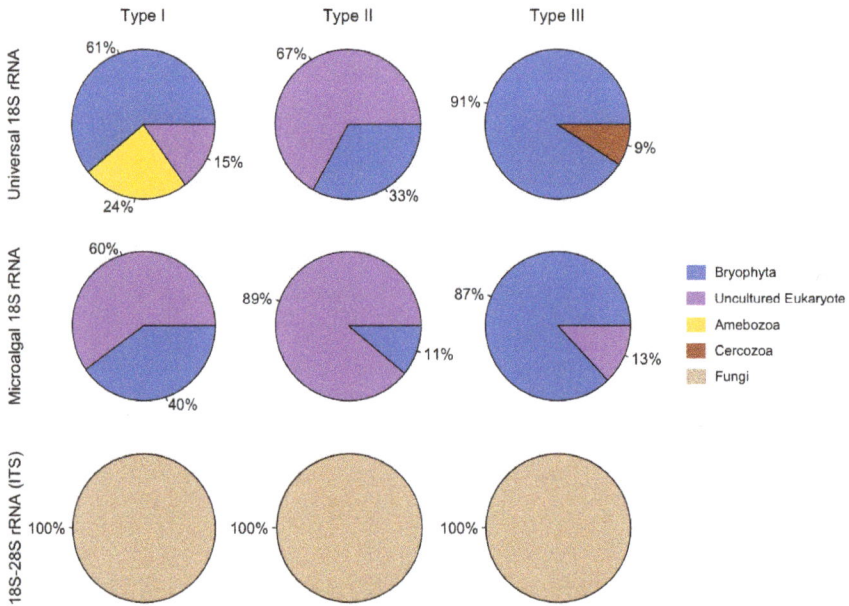

Table 1. PCR primers used to amplify universal 18S rRNA, Internal Transcribed Spacer (ITS) and microalgal 18S regions of ribosomal RNA genes from eukaryotic microorganisms, and their respective PCR cycling conditions.

Primer Set	Sequence (5'- 3')	Region of Amplification	PCR Parameters	Reference
EukA EukB	AACCTGGTTGATCCTGCCAGT TGATCCTTCTGCAGGTTCACCTAC	18S rRNA gene	94 °C for 3 min; 30 cycles: 94 °C for 45 s, 50 °C for 1 min, 72 °C for 3 min; 72 °C for 20 min	[20]
ITS1F ITS4	CTTGGTCATTTAGAGGAAGTAATC CTCCGCTTATTGATATGC	ITS1-ITS2	94 °C for 5 min; 35 cycles: 94 °C for 1 min, 50 °C for 1 min, 72 °C for 1 min; 72 °C for 20 min	[21,22]
P45 P47	ACCTGGTTGATCCTGCCAGT TCTCAGGCTCCCTCTCCGGA	Microalgal 18S rRNA gene	94 °C for 1 min; 37 cycles: 92 °C for 50 s, 57 °C for 50 s, 72°C for 50 s; 72°C for 10 min	[23]

Much of the knowledge of fungal communities in Antarctica is based on culture-dependent techniques. To date, over 1,000 non-lichenized fungal species have been recorded [25], including numerous representatives of all of the major fungal groups (ascomycetes, basidiomycetes, zygomycetes and chytrids) but only a single member of the Glomeromycota. The most complete list of Antarctic fungal species known from culturing and collection comprises approximately 68% ascomycetes, 23% basidiomycetes and 5% zygomycetes, with the remaining 4% consisting of oomycetes, chytrids and myxomycetes [26].

Basidiomycetes are commonly associated with old or decaying wood. The complete absence of higher plants in east Antarctica, in particular those with woody components, may represent a significant constraint on the diversity of Antarctic basidiomycetes. Indeed, a recent study has shown that when exotic organic substrates were buried, there was a significant increase in fungal colony-forming units (CFU) in soils in direct contact with the introduced, sterile cellulosic substrates compared to background soil levels [27]. Fungi are often also found in association with bryophyte communities and are thought to exploit the release of dissolved organic C from moss structures due to damage caused by freeze-thaw cycles [28].

The vast majority of Antarctic fungi are mesophiles capable of growing at low temperatures [29]. This, together with their widespread occurrence, could suggest that many Antarctic fungi are particularly resilient cosmopolitan species and therefore likely to be relatively recent colonists [25].

Three OTUs belonging to protists were found (Table 2). Sequences related to amoebozoa showed relatively high homology, 98% and 96%, to members of the genera *Saccamoeba* and *Platyamoeba*, respectively. A cercozoa sequence was closely related to the genus *Cercomonas* (99%). Members of the genera *Sacchamoeba*, *Platyamoeba* and *Cercomonas* have been found previously in Antarctica [30,31] and their abundance and species diversity is greater than that of the nematodes. The protists have generally received much less attention than the bacteria and much of the research on protists is focused on aquatic ecosystems, where these organisms predate bacteria and contribute to the remineralization of major-, minor- and micro-nutrients [32]. In soils, cercozoa and amoeba are known to colonize the pore spaces of soils; however, the smaller pore spaces might provide a more protected or favorable environment with increasing moisture, which might sustain a higher bacterial population and in general, have higher organic matter than a coarser soil fraction with large pore spaces [33]. Amoebae, which are thought to be more resilient than nematodes because of rapid encysting abilities and short life cycle, are thought to be the major predators of bacteria and to make a substantial contribution to carbon and nutrient cycling [34].

Sequences belonging to bryophytes were classified into nine phylotypes (Table 2). In the Antarctic Dry Valleys, bryophytes represent the most "advanced" terrestrial photoautotrophs and are all primary producers providing an additional pathway other than vascular plants for C to enter the soil [35]. However, to our knowledge, there is little information available on hypolithic bryophyte traits. Temperate bryophytes are rich in secondary metabolites such as terpenes and phenolics, which are likely to be allelopathic, affecting decomposition and therefore nutrient cycling [36]. Moreover, mosses can enhance the amount of water infiltrating the soil [37]. Bryophytes do not have stomata and lose water readily from their tissues [38]. They also lack roots, and therefore, they are unable to extract water from depth in a drying soil. Mosses instead depend on the availability of water in the environment, from either humid air, the surface substrate or precipitation. As many species are adapted to survive long periods of desiccation, we suggest that bryophytes may contribute significantly to primary production in periods of moisture sufficiency.

Table 2. Affiliation of the clones sampled from hypolithic communities.

Representative Clone	Accession No.	Closest Sequence Match	Accession No.	Identity	Type[a]
Eukaryote 18S rRNA					
Euk75-A1	KC352912	Uncultured *Eucalypta* (Bryophyta)	Y17871	81%	2/I
Euk75-A7	KC352913	Uncultured *Tortula ruralis* (Bryophyta)	AF023682	78%	2/I
Euk75-A8	KC352914	Uncultured *Tortula ruralis* (Bryophyta)	AF023682	86%	2/I
Euk75-A12	KC352915	Uncultured *Tortula ruralis* (Bryophyta)	AF023682	86%	2/I
Euk75-B2	KC352916	Uncultured *Tortula ruralis* (Bryophyta)	AF023682	92%	2/I
Euk75-B9	KC352917	*Saccamoeba limax* (Amoebozoa)	AF293902	98%	2/I
Euk75-C4	KC352918	*Platyamoeba contorta* (Amoebozoa)	DQ229954	96%	2/I
Euk134-C6	KC352919	*Pottia truncata* (Bryophyta)	X95935	99%	4/II
Euk134-D11	KC352920	Uncultured eukaryote	HM490274	100%	4/II
Euk50-B10	KC352921	Uncultured eukaryote	EF024087	91%	5/III
Euk50-D10	KC352922	*Cercomonas plasmodialis* (Cercozoa)	AF411268	99%	5/III
Microalgal 18S RNA					
P50-A4	KC352936	*Mnium hornum* (Bryophyta)	X80985	95%	5/III
P50-B3	KC352937	Uncultured eukaryote	EF024845	99%	5/III
P50-B6	KC352938	*Mnium hornum* (Bryophyta)	X80985	91%	5/III
P50-E7	KC352939	Uncultured eukaryote	EF526889	99%	5/III
P75-E	KC352940	*Bryoxiphium norvegicum* (Bryophyta)	AF223008	85%	2/I
P134-A2	KC352941	Uncultured eukaryote	FN394778	100%	4/II
P134-A11	KC352942	Uncultured eukaryote	HM490274	100%	4/II
IGS 28S-18S rRNA					
ITS65-A1	KC352923	Uncultured *Acremonium* (Ascomycota)	HE977538	82%	1/I
ITS65-A2	KC352924	Uncultured *Acremonium* (Ascomycota)	HE977544	85%	1/I
ITS65-A5	KC352925	Uncultured *Acremonium* (Ascomycota)	HE977544	85%	1/I
ITS65-A8	KC352926	Uncultured *Acremonium* (Ascomycota)	HE977544	78%	1/I
ITS65-B6	KC352927	Uncultured *Acremonium* (Ascomycota)	HE977544	88%	1/I
ITS65-C12	KC352928	Uncultured *Acremonium* (Ascomycota)	HE977544	88%	1/I

Table 2. *Cont.*

Representative Clone	Accession No.	Closest Sequence Match	Accession No.	Identity	Type[a]
IGS 28S-18S rRNA					
ITS65-D2	KC352929	Uncultured *Acremonium* (Ascomycota)	HE977544	87%	1/I
ITS65-D10	KC352930	*Stromatonectria caraganae* (Ascomycota)	HQ112288	80%	1/I
ITS134-A1	KC352931	*Verrucaria* sp. (Ascomycota)	FJ664858	92%	2/II
ITS50-B11	KC352932	Uncultured *Acremonium* (Ascomycota)	HE977544	84%	5/III
ITS50-D2	KC352933	Uncultured *Acremonium* (Ascomycota)	HE977538	80%	5/III
ITS50-E3	KC352934	Uncultured *Acremonium* (Ascomycota)	HE977538	86%	5/III
ITS50-H7	KC352935	Uncultured *Acremonium* (Ascomycota)	HE977538	91%	5/III

[a]Sample no./hypolithon type.

3. Experimental Section

3.1. Samples Collection

Six hypolith samples from all three Types (2 samples per Type) were obtained from Miers Valley (S78°05.01'–S78°05.921', E163°49.496'–E163°48.149'), Antarctica, in January 2011. The hypolithons were selected and classified as Type I, II or III during sampling. Classification was based on gross morphology of the biomass present on the ventral surfaces of the rocks [8,9]. Samples were recovered aseptically and stored in WhirlPak® bags at 4 °C in the field and during transport. Long term storage was at −80 °C in the laboratory, prior to further analysis.

3.2. DNA Extractions and PCR Amplifications

Total soil DNA was extracted using the method described by Von Sigler [39]. Briefly, 1mL of extraction buffer (50 mM NaCl; 50mM Tris-HCl at pH 7.6; 50 mM EDTA; 5% SDS) was added to 1 g of each soil sample in 2 mL vials containing 0.5 g mesh sea-sand. Then, 1 μL of 1M dithiothreitol (DTT) was added and mixed. Samples were shaken for 15 minutes at maximum speed (Vortex Genie 2; Scientific Industries Inc., USA) followed by 3 minutes of centrifugation at 14,000 × g. The supernatant was carefully decanted and 0.5× volumes of chloroform/isoamyl alcohol (24:1) was added to the tubes and mixed. This was followed by centrifugation at 14,000 × g for 3 minutes. The aqueous phase was transferred to a 2 mL sterile tube, and an equal volume of chloroform was added, vortexed and centrifuged as before. The aqueous phase was transferred to a sterile 1.5 mL and precipitated using sodium acetate and isopropanol. After centrifugation the pellet was washed by the addition of 70% ethanol, dry and resuspended in 25 μL of sterile distilled water.

The presence of DNA was confirmed by gel electrophoresis on 1% agarose gels, viewed using the AlphaImager 3400 imaging system (Alpha Innotech Co., USA) and quantified using a NanoDrop® ND-1000 UV/Vis Spectrophotometer (NanoDrop Technologies, USA).

The primers and parameters used for the PCR amplifications are described in Table 1. Reactions (25 μL) consisted of ~20 ng metagenomic DNA, 1x DreamTaq™ buffer, 0.2 mM of each dNTP, 0.5 μM of each primer and 0.2 U DreamTaq™ DNA polymerase (Fermentas, USA). PCR products were verified on 1% agarose gels and purified with the GFX™ PCR DNA and Band Purification Kit (GE healthcare, USA) and quantified using the NanoDrop® ND-1000 (NanoDrop Technologies, USA).

3.3. Clone Library Construction and Phylogenetic Analysis

Clone libraries were constructed after pooling equal amounts of amplicons from the individual samples for each hypolith type. Aliquots of the pooled products were cloned into *Escherichia coli* GeneHogs™ (Invitrogen) using pGEM-T cloning kit (Promega, USA) and transformants were selected by blue-white screening. The presence of the correctly sized insert was verified by colony PCR using the M13F and M13R vector primers (Fermentas, USA). ARDRA analysis (using *Alu*I and *Hae*III) was used to de-replicate clones. Restriction patterns were visualized on 2% agarose gels and analyzed using Gel-compare II (Applied Maths, Keistraat, Belgium). Plasmid DNA, from a representative of each unique restriction pattern, was extracted with QIAprep Spin Miniprep kit (Qiagen GmbH, Germany) and sequenced using the vector primer M13F with an ABI 3130 DNA Sequencer (Applied Biosystems).

Putative chimeric sequences were filtered using Bellerophon [40]. Sequences of >97% identity (for 18S rRNA amplicons) and >95% (for ITS amplicons) were grouped into OTUs using CD-HIT suite [41]. Taxonomic assignments of representative OTUs were determined by BLAST searches of the NCBI GenBank database (http://www.ncbi.nlm.nih.gov/). Sequences obtained in this study were deposited in the NCBI GenBank database under accession numbers KC352912-KC352942.

4. Conclusions

Hypoliths were examined at a single time point using only one molecular approach (*i.e.*, clone libraries). Thus, an in-depth analysis is necessary to elucidate the "true" diversity of the micro-eukaryotes existing in Antarctic hypolithons. However, in spite of its limitations, this baseline study gives insight to the existing micro-eukaryotic community supported by hypoliths in the Miers Dry Valley, Antarctica. We show that these communities are represented by a wide diversity of lower eukaryotes (bryophyta, fungi and protists). The presence of these organisms supports the concept that hypolithic communities constitute complex multi-domain food webs in an environment which is generally considered to be characterised by low diversity and complexity.

The phototrophic bryophyte component is thought to provide a significant contribution to primary productivity in periods of moisture sufficiency. Protists feed on bacterial populations and contribute to standing biomass. Fungi participate in decomposition and recycling, maintaining the balance of nutrients in the discrete and "self-contained" hypolithic microhabitats. The hypothesis that the partitioned activity of co-colonizers may create the necessary conditions for sustained hypolithic productivity is currently being tested in our research group.

Acknowledgments

The authors gratefully acknowledge the National Research Foundation (South Africa) and Antarctica New Zealand for support of this research.

References

1. Cary, S.C.; McDonald, I.R.; Barrett, J.E.; Cowan, D.A. On the rocks: The microbiology of Antarctic Dry Valley soils. *Nat. Rev. Microbiol.* **2010**, *8*, 129–138.
2. Franzmann, P.D. Examination of Antarctic prokaryotic diversity through molecular comparisons. *Biodivers. Conser.* **1996**, *5*, 1295–1305.
3. Campanaro, S.; Williams, T.J.; Burg, D.W.; De Francisci, D.; Treu, L.; Lauro, F.M.; Cavicchioli, R. Temperature-dependent global gene expression in the Antarctic archaeon Methanococcoides burtonii. *Environ. Microbiol.* **2011**, *13*, 2018–2038.
4. Massana, R.; Taylor, L.J.; Murray, A.E.; Wu, K.Y.; Jeffrey, W.H.; DeLong, E.F. Vertical distribution and temporal variation of marine planktonic archaea in the Gerlache Strait, Antarctica, during early spring. *Limnol. Oceanogr.* **1998**, *43*, 607–617.
5. Gardner, H.; Kerry, K.; Riddle, M.; Brouwer, S.; Gleeson, L. Poultry virus infection in Antarctic penguins. *Nature* **1997**, *387*, 245–245.
6. Yau, S.; Lauro, F.M.; DeMaere, M.Z.; Brown, M.V.; Thomas, T.; Raftery, M.J.; Andrews-Pfannkoch, C.; Lewis, M.; Hoffman, J.M.; Gibson, J.A.; Cavicchioli, R. Virophage control of antarctic algal host-virus dynamics. *Proc. Natl. Acad. Sci. USA* **2011**, *108*, 6163–6168.
7. Fell, J.W.; Scorzetti, G.; Connell, L.; Craig, S. Biodiversity of micro-eukaryotes in Antarctic Dry Valley soils with <5% soil moisture. *Soil Biol. Biochem.* **2006**, *38*, 3107–3119.
8. Cowan, D.A. Cryptic microbial communities in Antarctic deserts. *Proc. Natl. Acad. Sci. USA* **2009**, *106*, 19749–19750.
9. Cowan, D.A.; Khan, N.; Pointing, S.B.; Cary, S.C. Diverse hypolithic refuge communities in the McMurdo Dry Valleys. *Antarc. Sci.* **2010**, *22*, 714–720.
10. Cowan, D.A.; Pointing, S.B.; Stevens, M.I.; Cary, S.C.; Stomeo, F.; Tuffin, I.M. Distribution and abiotic influences on hypolithic microbial communities in an Antarctic Dry Valley. *Polar Biol.* **2011**, *34*, 307–311.
11. Khan, N.; Tuffin, M.; Stafford, W.; Cary, C.; Lacap, D.C.; Pointing, S.B.; Cowan, D. Hypolithic microbial communities of quartz rocks from Miers Valley, McMurdo Dry Valleys, Antarctica. *Polar Biol.* **2011**, *34*, 1657–1668.
12. Chan, Y.; Lacap, D.C.; Lau, M.C.Y.; Ha, K.Y.; Warren-Rhodes, K.A.; Cockell, C.S.; Cowan, D.A.; McKay, C.P.; Pointing, S.B. Hypolithic microbial communities: Between a rock and a hard place. *Environ. Microbiol.* **2012**.

13. Pointing, S.B.; Chan, Y.; Lacap, D.C.; Lau, M.C.Y.; Jurgens, J.A.; Farrell, R.L. Highly specialized microbial diversity in hyper-arid polar desert. *Proc. Natl. Acad. Sci. USA* **2009**, *106*, 19964–19969.

14. Cockell, C.S.; Stokes, M.D. Widespread colonization by polar hypoliths. *Nature* **2004**, *431*, 414.

15. Cowan, D.A.; Sohm, J.A.; Makhalanyane, T.P.; Capone, D.G.; Green, T.G.A.; Cary, S.C.; Tuffin, I.M. Hypolithic communities: important nitrogen sources in Antarctic desert soils. *Environ. Microbiol. Rep.* **2011**, *3*, 581–586.

16. Potvin, M.; Lovejoy, C. PCR-Based Diversity Estimates of Artificial and Environmental 18S rRNA Gene Libraries. *J. Eukaryot. Microbiol.* **2009**, *56*, 174–181.

17. Jones, C.G.; Lawton, J.H.; Shachak, M. Organisms as ecosystem engineers. *Oikos* **1994**, *69*, 373–386.

18. Pointing, S.B.; Belnap, J. Microbial colonization and controls in dryland systems. *Nat. Rev. Microbiol.* **2012**, *10*, 551–562.

19. Zakhia, F.; Jungblut, A.; Taton, A.; Vincent, W.; Wilmotte, A. Cyanobacteria in cold environments. In *Psychrophiles: From Biodiversity to Biotechnology*; Margesin, R.S.F., Marx, J.C., Gerday, C., Ed.; Springer-Verlag: Berlin, 2007; pp. 121–135.

20. Diez, B.; Pedros-Alio, C.; Marsh, T.L.; Massana, R. Application of denaturing gradient gel electrophoresis (DGGE) to study the diversity of marine picoeukaryotic assemblages and comparison of DGGE with other molecular techniques. *Appl. Environ. Microbiol.* **2001**, *67*, 2942–2951.

21. Gardes, M.; Bruns, T.D. Its primers with enhanced specificity for basidiomycetes—application to the identification of mycorrhizae and rusts. *Mol. Ecol.* **1993**, *2*, 113–118.

22. White, T.J.; Bruns, T.; Lee, S.; Taylor, J. Amplification and direct sequencing of fungal ribosomal RNA genes for phylogenetics. In *PCR Protocols: A Guide to Methods and Applications*; Innis, M.A., Gelfand, D.H., Sninsky, J.J., White, T.J., Eds.; Academic Press: San Diego, CA, USA, 1990.

23. Dorigo, U.; Berard, A.; Humbert, J.F. Comparison of eukaryotic phytobenthic community composition in a polluted river by partial 18S rRNA gene cloning and sequencing. *Microb. Ecol.* **2002**, *44*, 372–380.

24. Arenz, B.E.; Held, B.W.; Jurgens, J.A.; Farrell, R.L.; Blanchette, R.A. Fungal diversity in soils and historic wood from the Ross Sea Region of Antarctica. *Soil Biol. Biochem.* **2006**, *38*, 3057–3064.

25. Bridge, P.D.; Spooner, B.M. Non-lichenized Antarctic fungi: transient visitors or members of a cryptic ecosystem? *Fungal Ecol.* **2012**, *5*, 381–394.

26. Bridge, P.; Spooner, B.; Roberts, P. List of Non-lichenized Fungi from the Antarctic Region. Available online: http://www.antarctica.ac.uk/bas_research/data/access/fungi/ (Accessed on 10 December 2012).

27. Arenz, B.E.; Held, B.W.; Jurgens, J.A.; Blanchette, R.A. Fungal colonization of exotic substrates in Antarctica. *Fungal Divers.* **2011**, *49*, 13–22.

28. Wynn-Wmilliams, D.D. Seasonal fluctuations in microbial activity in antarctic moss peat. *Biol. J. Linn. Soc.* **1980**, *14*, 11–28.

29. Onofri, S.; Zucconi, L.; Tosi, S. *Continental Antarctic Fungi*; IHW-Verlag: Munich, Germany, 2007; p. 247.

30. Bamforth, S.S.; Wall, D.H.; Virginia, R.A. Distribution and diversity of soil protozoa in the McMurdo Dry Valleys of Antarctica. *Polar Biol.* **2005**, *28*, 756–762.

31. Nakai, R.; Abe, T.; Baba, T.; Imura, S.; Kagoshima, H.; Kanda, H.; Kohara, Y.; Koi, A.; Niki, H.; Yanagihara, K.; Naganuma, T. Eukaryotic phylotypes in aquatic moss pillars inhabiting a freshwater lake in East Antarctica, based on 18S rRNA gene analysis. *Polar Biol.* **2012**, *35*, 1495–1504.

32. Calbet, A.; Landry, M.R. Phytoplankton growth, microzooplankton grazing, and carbon cycling in marine systems. *Limnol. Oceanogr.* **2004**, *49*, 51–57.

33. Elliott, E.T.; Anderson, R.V.; Coleman, D.C.; Cole, C.V. Habitable pore-space and microbial trophic interactions. *Oikos* **1980**, *35*, 327–335.

34. Wardle, D. *Communities and Ecosystems: Linking the Aboveground and Belowground Components*; Princeton University Press: Princeton, USA, 2002; p. 400.

35. Lange, O.L.; Kidron, G.J.; Budel, B.; Meyer, A.; Kilian, E.; Abeliovich, A. Taxonomic composition and photosynthetic characteristics of the biological soil crusts covering sand dunes in the western negev desert. *Funct. Ecol.* **1992**, *6*, 519–527.

36. Mues, R. Chemical constituents and biochemistry. In *Bryophyte Biology*; Shaw, A.J., Goffinet, B., Eds.; Cambridge University Press: Cambridge, UK, 2000; pp. 150–181.

37. Liu, L.-C.; Li, S.-Z.; Duan, Z.-H.; Wang, T.; Zhang, Z.-S.; Li, X.-R. Effects of microbiotic crusts on dew deposition in the restored vegetation area at Shapotou, northwest China. *J. Hydrol.* **2006**, *328*, 331–337.

38. Proctor, M.C.F. The bryophyte paradox: Tolerance of desiccation, evasion of drought. *Plant Ecol.* **2000**, *151*, 41–49.

39. Von Sigler, W. DNA Extraction from soil, sediment and plant tissue: 50–50–50 buffer-chloroform/phenol method. Available online: http://www.eeescience.utoledo.edu/Faculty/ Sigler/-Von_Sigler/LEPR_Protocols_files/DNA%20extraction%20-%20soil.pdf (accessed on DD/MM/YY).

40. Huber, T.; Faulkner, G.; Hugenholtz, P. Bellerophon: A program to detect chimeric sequences in multiple sequence alignments. *Bioinformatics* **2004**, *20*, 2317–2319.

41. Huang, Y.; Niu, B.; Gao, Y.; Fu, L.; Li, W. CD-HIT Suite: A web server for clustering and comparing biological sequences. *Bioinformatics* **2010**, *26*, 680–682.

The Effect of Freeze-Thaw Conditions on Arctic Soil Bacterial Communities

Niraj Kumar, Paul Grogan, Haiyan Chu, Casper T. Christiansen and Virginia K. Walker

Abstract: Climate change is already altering the landscape at high latitudes. Permafrost is thawing, the growing season is starting earlier, and, as a result, certain regions in the Arctic may be subjected to an increased incidence of freeze-thaw events. The potential release of carbon and nutrients from soil microbial cells that have been lysed by freeze-thaw transitions could have significant impacts on the overall carbon balance of arctic ecosystems, and therefore on atmospheric CO_2 concentrations. However, the impact of repeated freezing and thawing with the consequent growth and recrystallization of ice on microbial communities is still not well understood. Soil samples from three distinct sites, representing Canadian geographical low arctic, mid-arctic and high arctic soils were collected from Daring Lake, Alexandra Fjord and Cambridge Bay sampling sites, respectively. Laboratory-based experiments subjected the soils to multiple freeze-thaw cycles for 14 days based on field observations (0 °C to −10 °C for 12 h and −10 °C to 0 °C for 12 h) and the impact on the communities was assessed by phospholipid fatty acid (PLFA) methyl ester analysis and 16S ribosomal RNA gene sequencing. Both data sets indicated differences in composition and relative abundance between the three sites, as expected. However, there was also a strong variation within the two high latitude sites in the effects of the freeze-thaw treatment on individual PLFA and 16S-based phylotypes. These site-based heterogeneities suggest that the impact of climate change on soil microbial communities may not be predictable *a priori*; minor differential susceptibilities to freeze-thaw stress could lead to a "butterfly effect" as described by chaos theory, resulting in subsequent substantive differences in microbial assemblages. This perspectives article suggests that this is an unwelcome finding since it will make future predictions for the impact of on-going climate change on soil microbial communities in arctic regions all but impossible.

Reprinted from *Biology*. Cite as: Kumar, N.; Grogan, P.; Chu, H.; Christiansen, C.T.; Walker, V.K. The Effect of Freeze-Thaw Conditions on Arctic Soil Bacterial Communities. *Biology* **2013**, *2*, 356-377.

1. Introduction

1.1. Will Climate Change Stress Arctic Soil Communities, and What Are the Likely Ecological Impacts?

Although low temperatures in the Arctic result in vast tracts of frozen ground or permafrost, the temperature of the soil is ameliorated by an insulating snow pack. As a result, snow depth and timing of first snow accumulation are important for the survival of subnivean life [1,2]. Prior to snow accumulation in autumn, and during the spring melt, dynamically fluctuating air temperatures are common and can result in freeze-thaw cycle (FTC) events in surface soils. Such freeze-thaw fluctuations are of ecological interest because of their possible impacts on soil microbial communities, soil carbon and nutrient transformations, as well as plant productivity [3–8]. In a changing climate, the Arctic is expected to undergo substantial warming with a projected increase in average air temperature of 4–8 °C during this century [9,10]. Although this may impact all seasons, our particular interest is in earlier spring warming, as well as the potential decrease in snow cover that together may result in more FTC incidents [11]. Climate change scenarios also predict increased variability in climate, with greater amplitude fluctuations in air temperature and precipitation, which may further enhance the frequency of soil FTCs [10,12].

Freeze-thaw events have been linked to declines in soil microbial biomass carbon [13–15], a proxy for microbial community size. In extreme cases, FTCs have been associated with microbial dieback of 40–60% [5,16,17]. Even a single FTC can cause the death of up to 50% of microbes [18]. Nevertheless, there is conspicuous lack of consensus among studies on the effects of FTCs on soil microbial biomass and activities [8] with other reports showing a subtle or insignificant impact (e.g., [4,6,19,20]). Some apparent inconsistencies between experiments could be attributed to the methods or the analysis, but others may depend on soil type and the severity of the experimental FTC regimes compared to naturally occurring FTCs. Both regional and landscape topographic location may be critical since they determine local climate. As a result, soils derived from sites subjected to "harsh and changing environmental conditions" would be expected to contain a relatively high abundance of indigenous FTC-resistant species [21]. In consequence, perhaps arctic soils from such locations will show little impact from any additional freeze-thaw stresses related to climate change.

Soils subjected to freeze-thaw regimes may release labile carbon and nutrients from lysed microbial cells and this has been associated with short-term peak respiratory pulses of N_2O and CO_2 [6,14,22,23]. A single FTC resulted in respiratory losses accounting for up to 15% of microbial biomass carbon [22]. Potentially, then, FTCs could have a significant impact on tundra carbon balance. Indeed, Schimel and colleagues [7] calculated that a freeze-thaw event could release carbon to the atmosphere corresponding to as much as 25% of the net annual primary production in an Alaskan tussock tundra region. Although release of more carbon will further exacerbate climate change, freeze-thaw induced microbial loss is crucial for arctic nutrient dynamics. Arctic vegetation growth is strongly limited by nutrient availability [24] in part due to strong microbial immobilization of soil nutrients [25]. Therefore, the release of microbial nitrogen and phosphorous from microbial cells that were lysed by FTCs could stimulate plant production and, hence, carbon uptake from the atmosphere. This would help to counteract the CO_2 release associated with lysis-enhanced respiration. Overall, since the carbon to nutrient ratio of plants is higher than that of microbes or soil [26], FTCs could eventually contribute to a net decrease in atmospheric CO_2 concentration. However, this prediction is critically dependent on plants being able to acquire the released nutrients at the time of microbial lysis [27]. To date, there is little evidence for this except for evergreen shrubs or perennial sedges in the arctic spring freeze-thaw period, and for graminoids in the early autumn [27–29]. If valid, however, FTC-mediated nutrient release could then ultimately shift plant community structure in favor of functional groups that can best capitalize on pulses of these liberated molecules. Nutrients released from FTC-lysed microbial cells that are not taken up by plants, may be acquired by surviving soil microbes, leached downslope, or lost to the atmosphere via dentrification (for nitrogen only).

Similar to FTCs, the drying and subsequent rewetting of soils may strongly affect microbes due to the rapidly changing osmotic potentials [7]. During a rewetting event, microorganisms release cytoplasmic nutrients, constituting up to 60% of the microbial biomass carbon [30,31], resulting in short-term pulses of enhanced CO_2 release and nutrient availability (e.g., [31–33]). CO_2 release and changes in microbial community composition following rewetting are usually less pronounced in soils frequently exposed to fluctuations in soil water potential *in situ* (e.g., [31,33,34]). Again, this suggests that community adaptations for stress resistance are shaped by local climate history. While drying/rewetting events have been principally addressed in relation to episodic rainfall, arctic soils are often subjected to the combination of freeze-thaw and drying-rewetting stresses in late winter [35]. Added to these stresses, in late winter, arctic soils are dried by sublimation due to the increase in sunlight, particularly in soils without much snow cover and adjacent to darker vegetation and roots, which can adsorb solar radiation [35]. As warmer air temperatures initiate above ground, snow and ice melt, with water percolating down into the frozen soil through these sublimed crevices, soil pores, frost-induced cracks, and dendritic channels [36–38].

1.2. Freeze-Thaw: Survival of the Fittest, or an Assemblage of Defenses?

Temperature changes can be challenging to microbial communities. Low temperatures and FTCs can affect protein structure and function, membrane fluidity and be associated with cellular damage due to the impact of oxidative and osmotic stresses [39,40]. Internal ice formation is largely avoided *in situ*, but the protective effect of an increase in cellular solute concentration [39] can itself result in damage, as can thawing leading to rapid changes in osmotic potential. External ice formed at low rates of cooling consists of large ice crystals [41], which are potentially harmful. During prolonged periods near 0 °C, or during freeze-thaw, ice recrystallization can result in still larger ice crystals that may contribute to further damage. Despite these many challenges, psychrophiles and psychrotolerant microbes have developed a range of physiological adaptations to survive freeze stress, allowing them to remain active at, and below, freezing conditions [42–45]. At subzero temperatures, microbial activity is controlled by the availability of unfrozen water films on soil particles [46,47], and by substrate limitations [48]. Adaptations include metabolic adjustments [49], and may involve a switch from the utilization of carbon-rich litter during thaw periods to the recycling of nitrogen-rich internal products as well as dead microbes during freeze intervals [48,50,51].

Different isolates show strikingly different susceptibilities to FTCs. For example, *Chryseobacterium* sp. C14 showed no loss of viability after 48 FTCs, resulting in a level of recovery that was three orders of magnitude higher than more vulnerable strains [52]. This species conferred some benefit to other isolates, demonstrating that experiments investigating the effect of FTCs and spring runoff should utilize assemblages, rather than individual isolates. Consortia containing cooperative species could be relatively resilient when faced with the multiple stresses associated with seasonal changes. This could partially explain the little impact seen in response to freeze-thaw stress in several studies, as well as a more marked effect in others (e.g., [6] *vs.* [15]).

Whether FTCs are the cause or not, it is now fairly well established that the active microbial soil community changes seasonally, resulting in distinct summer and winter arctic [53], subarctic [54], and alpine [55,56] ecosystems. Generally, fungi dominate the tundra in winter and to a lesser degree in summer when bacterial abundance rises in the relatively warm soils [55,57]. Such seasonal assemblage shifts could reflect differential stress susceptibility or the capacity to have a vulnerability complemented by other members of the consortium. If the enhanced resilience of soil microbial communities to FTCs can indeed be attributed to adaptation to a particular local climate associated with a geographic region [21], this prompts us to consider that arctic soils from climatically distinct locations could then show substantial variation in their responses to FTCs related to climate change. It was this speculation that prompted us to undertake a small, but multi-spatial scale analysis; we report our results as part of this perspectives article in order to underscore the need for further investigation.

2. Experimental Section: The Effect of Simulated Freeze-Thaw Cycles on Latitudinally Distinct Soils

We hypothesized that rapid temperature changes that result in soil freeze-thaw fluctuations could alter soil microbial diversity. Evidence for multiple FTCs was apparent at a low arctic site (Figure 1) and we speculated that the FTCs seen at this geographic location could serve as a proxy for the impact of more extreme future climate change at higher latitudes. A recent analysis of climatic trends over the past ~50 years across Canada (albeit largely but not entirely based on data from relatively southerly weather stations) indicates that the frequency of soil FTCs is generally higher at sites with relatively warm mean annual air temperatures (*i.e.*, at lower latitudes), and especially in relatively warm and dry winters [58]. Furthermore, these data suggest that climate change will increase the frequency of soil FTCs at most sites over the next 50 years [58]. Accordingly, we sampled replicate soils from three distinct sites in the Canadian low, mid- and high Arctic and used spring air and soil temperature data collected *in situ* at the low arctic location as the basis for FTC treatment of soils from all three sites. As indicated, we present our perspective on the effect of freeze-thaw events on soil microbes, and show the results of our biochemical analyses on the impact of freeze-thaw events on these three geographically distant assemblages.

2.1. Soil Collection Sites, Freeze-Thaw Regime and Respiration Monitoring

Soil samples were collected in triplicate from three different sites in the Canadian Arctic: Daring Lake, Cambridge Bay, and Alexandra Fjord representing the low (64°52' N 111°35' W), mid- (69°11' N 104°45' W), and high Arctic (78°53' N 75°47' W), respectively. Some soil biochemical and climatic variables for these sites are listed in Table 1. At each site, soils were obtained from three separate but similar locations (20–100 m apart) close to the top of exposed ridges where soil freeze-thaw fluctuations are most likely because of thin or absent snowcover. Samples of the top 2–3 cm of the soil organic layer were taken in spring (Daring Lake) and summer (the higher latitude sites) and shipped to our lab within several days and stored at −20 °C until processed [59]. Ideally, soils would be subjected to FTCs immediately, but climate-prescribed seasonal differences in the collection dates, transport availability to remote sites, and the requirement to randomize the microcosms in the experimental apparatus dictated otherwise. The Daring Lake site was on top of a wind-exposed upland esker consisting of dry heath soils with a dominant vegetation of lichens *Cladina* sp., and dwarf shrubs, *Ledum decumbens*, *Betula glandulosa*, and *Empetrum nigrum*. Shrub cover at the Cambridge Bay site was dominated by *Dryas integrifolia*, with some sedges *Carex* sp., willows *Salix* sp. and mosses. The Alexandra Fjord soils were collected from a high arctic oasis with vegetation mainly consisting of sedges *Eriophorum* sp., *Carex* sp. and arctic willow, *Salix arctica*.

Figure 1. Temperatures in the air (40 cm above the ground surface) and in the soil (2 cm depth) at two locations at least 2 m apart beneath birch hummock tundra vegetation at the Daring Lake site. Temperatures were measured in the spring (2005) with dates as indicated on the X-axis. Data were collected every 30 min using copper-constantan thermocouples and a datalogger.

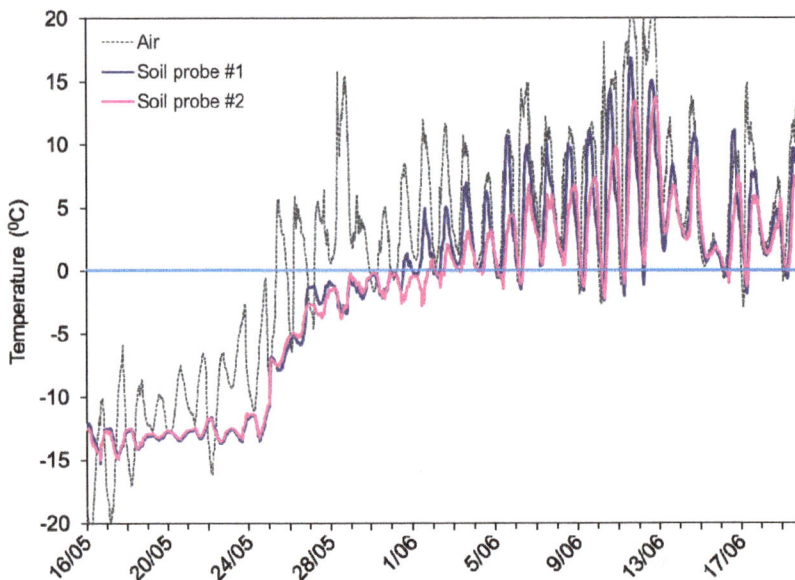

After thawing, obvious roots and stones were quickly removed by hand prior to biochemical analysis. Triplicates were composited, packed in sealed polypropylene containers (100 g soil in each of three 218 mL containers for each site), and subjected to multiple freeze-thaw temperature fluctuations. In order to use temperatures that were consistent with the low arctic site, we examined temperature records of the soil (2 cm depth into the soil organic layer) and air temperatures (40 cm above the ground surface) from the Daring Lake weather station (Figure 1). Since the exposed sites did not have the benefit of insulating snow cover, we made the assumption that the temperature of the top centimeters of soil would be the same as the air temperature. Indeed, records show that the likelihood of freeze-thaw diminishes greatly with depth from the soil surface down into the deeper soil horizons, and that soil FTCs may occur in either spring or fall. Climate change may eventually shift climates northwards so that the mid- and high arctic sites experience conditions more similar to current temperature patterns at the low arctic site [55]. Therefore, we exposed soil from all three sites to a 14-day treatment period of the same temperature regime. This approximated the air temperatures just above the snow surface observed at the low-arctic site in the 2005 field season (*i.e.*, oscillating from -10 °C to 0 °C and back once per day; Figure 1). We connected the soil containers to a CO_2 detector, a gas switcher and a computer for data acquisition (Qubit Systems, Kingston, Ontario, Canada) in order to monitor microcosm respiration throughout the incubation period.

Table 1. Comparison of some soil biochemical and climatic variables including pH, organic layer (org layer) depth, soil carbon and nitrogen, mean annual temperature (AT), mean growing season temperature (GST), average annual precipitation (AP), snow depth, and average number of days above 0 °C for the three sites, Daring Lake (DL), Cambridge Bay (CB) and Alexandra Fjord (AF).

Site	pH	Org layer (cm)	Soil C (%)	Soil N (%)	Soil P (ppm)	AT (°C)	GST (°C)	AP (mm)	Snow depth (cm)	Days above 0 °C
DL [1]	4.3	2–5	20	0.7	21	−9	3	150	29	127
CB [2]	6.6	5–6	24	1.4	7	−14	6	140	31	79
AF [3]	5.6	NA	12	0.8	4	−15	10	250	NA	~65

[1] data from this study and Chu *et al.*, 2010 [60], with climatic data from Bob Reid, Department of Indian and Northern Affairs, Water Resources Division, NWT; note that the days above 0 °C is reported as the diel average temperature above 0 °C. [2] data from Chu *et al.*, 2010 [60], with climatic data from the National Climate Data and Information Archive, Environment Canada (1971–2000); note that the days above 0 °C is reported as the minimum diel temperature above 0 °C. [3] data from Chu *et al.*, 2010 [60], with climatic data averaged from Labine, 1994 and Rayback and Henry, 2006 [61,62], with days above 0 °C reported as "frost-free days" for Quttinirpaaq National Park, Ellesmere Island (Environment Canada); NA, indicates that the information is not currently available.

2.2. Soil Phospholipid Fatty Acid and DNA Analyses

Each of the triplicate soil samples from each site was subjected to phospholipid fatty acid (PLFA) analysis. Fatty acid methyl esters were extracted as described using the Microbial Identification System (Microbial ID Inc. [MIDI], Newark, DE, USA) as previously cited [63,64]. Briefly, soil samples (3 g) were saponified (100 °C in 3 mL 3.75 M NaOH in 50% methanol, 30 min), methylated (80 8C in 6 mL of 6 M HCl in 54% for 10 min), extracted (in 3 mL of a mix of equal volumes of methyl-tert-butyl ether/hexane for 10 min), and washed (1.2% NaOH for 5 min). This procedure and the gas chromatography of the resulting esters were conducted by Keystone Labs, Edmonton, Alberta. Only those fatty acids that were the most abundant (>1% of chromatographic peak areas for either control or treated samples) were considered for analysis. The total average peak area for the triplicate samples and controls were converted to ratios of PFLA peak areas in the experimental series over their corresponding controls to facilitate comparisons between untreated and treated soils. This was done because the focus of these experiments was not to describe the fatty acid composition at each site in detail but determine if FTCs would perturb PLFA profiles.

DNA was extracted using a soil isolation kit (PowerSoil™ DNA Isolation Kit, MO BIO Laboratories, Inc., Carlsbad, CA, USA) as per the manufacturer's instructions. Polymerase chain reaction denaturing gradient gel electrophoresis (PCR-DGGE) was conducted as previously described [65] except that the DNA was not treated with ethidium monoazide. PCR-DGGE was performed three or more times on each sample to ensure reproducibility of the gel patterns.

For pyrosequencing, DNA was extracted from all 18 soil samples (controls and FTC-treated for each of the three geographic regions). The DNA samples were quantified using a Nanodrop spectrophotometer (NanoDrop-1000; Ver. 3.7.1). After multiple initial PCR and agarose gel analyses [65], all subsequent procedures were performed at the Research and Testing Laboratory (RTL: Lubbock, TX, USA) with tag-encoded FLX amplicon pyrosequencing (TEFAP) performed in accordance with established protocols [66,67]. Bacterial primers Gray28F (5'-TTTGATCNTGGCTCAG-3') and Gray519r (5'-GTNTTACNGCGGCKGCTG-3') were used to amplify ~500 bp fragments spanning the V1 to V3 hypervariable regions of the bacterial 16S ribosomal RNA (rRNA) genes. Initial generation of the sequencing library used a one-step PCR with a total of 30 cycles, a mixture of Hot Start and HotStar high fidelity Taq polymerases, and amplicons originating and extending from the 28F primer. Analysis utilized a Roche 454 FLX instrument with Titanium reagents based on RTL protocols [68]. After sequencing, all failed sequence reads, low quality sequence ends and tags and primers were removed, with non-bacterial rRNA gene sequences and chimeras removed using B2C2 [69] as previously described [70]. To identify the bacteria in the remaining sequences, sequences were denoised, assembled into clusters and compared with 16S bacterial sequences curated at the National Center for Biotechnology Information (NCBI) using a distributed MegaBLAST .NET algorithm [71]. Using RDP ver 9 [72] to determine quality and the .NET and C# analysis pipeline, the MegaBLAST outputs were compiled, validated and further analyzed as previously described [70]. These were subsequently used for sequence identity (percent of total length query sequence aligned with a given database sequence) and validated using taxonomic distance methods, and classified at the appropriate taxonomic levels based upon standard conventions. Specifically, 16S rRNA gene sequences with 90% or more base pair identity with existing sequences in the database were resolved at the family level. Similarly, those sequences with scores between 85–90% were resolved at the order level, 80–85% at the class level and 77%–80% at the phylum level [70].

2.3. Community Responses of Freeze-Thaw Stresses in Microcosms: Results

When subjected to the alternating temperature cycles, probes inserted into the microcosm cores showed that temperatures of −10 °C and 0 °C were achieved, and ice crystals or water vapor were visible on the outer surface of the microcosms in the appropriate distinct cycling periods. There were no overall differences between the CO_2 levels derived from the soil microcosms originating from the different geographic regions and therefore these nine profiles are not shown here. Nevertheless, respiration monitoring was consistent with FTCs, showing alternating periods of no detectable CO_2 discharge followed by a modest burst of CO_2 as the soil thawed (not shown). This could reflect either physical release of trapped gas, carbon mineralization of solutes from microbes that were lysed by the freeze-thaw treatment, or simply more favorable temperatures (at thaw) for microbial activity. However, these alternatives were not investigated for this study. After the two-week incubation period, there were again no significant differences in the total cumulative respiration between microcosms or in respiration rates compared to the beginning of the experiment.

The impact of FTCs on the soil community profiles was further examined using three additional culture-independent methods: fatty acid analysis, PCR-DGGE and 16S rRNA gene sequencing. Although downstream PLFA analysis may not be recommended in experiments that subject soils to temperature variations that can result in changes to membrane lipids [73], each of the experimental microcosms was treated in the same way. Using relative FA abundance gives a measure of the FTC-mediated impact rather than an assessment of the community profile *per se*. As expected, fatty acid profiles in untreated soils differed in richness and composition among sites. For example, the Daring Lake soils contained 17 different signatures with relative abundance $\geq 1\%$, while the Cambridge Bay and Alexandra Fjord soils had 25 and 24 signatures, respectively (Figure 2). FTC-treatment had a differential effect on the soils, depending upon their origin and the particular fatty acid. Overall, fungal-associated fatty acids (e.g., 18:2 ω6,9c and 18:1 ω9c; but note that 18:1 ω9c may be an unreliable fungal signature since it is found in certain Solirubrobacterales [74,75]), did not appear to show dramatic changes in response to FTCs, as suggested by observations in previous studies indicating that fungi tend to dominate winter soils [55,57]. Overall, however, since PLFA composition varied among the sites it was difficult to determine additional general trends. In Alexandra Fjord soil samples, for example, a fatty acid indicative of the Gram-positive Actinobacteria of the Order Actinomycetales (18:0 10-methyl, tuberculostearic acid) increased 15.8-fold after freeze-thaw treatment. Indeed, the ratio of the abundance of individual fatty acids before and after the freeze-thaw treatment most readily showed the overall response patterns. The Daring Lake soils appeared relatively resilient to FTCs since only 18% (3/17) of the most abundant fatty acids showed >10% increase or decrease in the abundance ratio. In contrast, the FTC-mediated impact on PLFAs was much greater for the more northerly sites with substantive changes in 56% (14/25) and 33% (8/24) of the fatty acids derived from Cambridge Bay and Alexandra Fjord soils, respectively. Furthermore, the range of ratio changes (*i.e.*, the magnitude of increase or decrease in response to the FTC treatment) varied least among the Daring Lake soils (0.8–1.4) and most among the more northerly sites (Cambridge Bay: 0.5–1.4 and Alexandra Fjord: 0.8–1.8, excepting one signature that increased almost 16-fold).

Although each of the microcosms containing soil derived from the same site had identical banding patterns using PCR-DGGE analysis of the 16S ribosomal RNA genes, and were distinct from the banding patterns obtained from different sites, there were no clear, regular differences in band patterns after FTCs (not shown). Unlike the dramatic changes in DGGE community profiles that are evident after more stressful treatments (e.g., nanoparticle exposure, [65] and ultraviolet radiation [76]), our results initially suggested, similar to others [6], that FTCs did not appear to radically shift bacterial community structure in a predictable, consistent way.

Figure 2. Mean ratios of the most abundant fatty acids (>1% of chromatographic peak areas) from low arctic (Daring Lake), mid arctic (Cambridge Bay) and high arctic (Alexandra Fjord) soils, both before and after daily freeze-thaw treatments for 14 days. Freeze-thaw-treated (n = 3)/untreated (n = 3) are represented along with standard errors (lines on the bars), with no change in mean abundance after treatment indicated by 1. Increases and decreases relative to mean control values are shown as bars with mean values >1 or <1, respectively. Fatty acids are named according to standard nomenclature but abbreviated where appropriate and represent from bottom to top as: 9:00, 10:00, 11:0 iso, 10:0 3OH, 12:00, 13:0 iso, 13:0 anteiso, 12:1 3OH, 14:0 iso, 14:00, 15:0 iso, 15:0 anteiso, 14:0 3OH/16:1 iso I, 16:0 N alcohol, 16:1 w7c/16:1 w6c, 16:1 w6c/16:1 w7c, 16:00, 16:0 2OH, unknown 16.586, 17:0 10-methyl, 17:1 anteiso A, 17:1 w7c, 18:3 w6c (6,9,12), 18:2 w6 9c/18:0 ante, 18:1 w9c, 18:1 w7c, 18:1 w5c, 18:00, 19:1 w11c/19:1 w9c, 19:0 cyclo w10c/19w6, 17:0 2OH, 18:0 10-methyl TBSA, 20:0 iso, 20:00 (the value for 18:0 10-methyl TBSA in Alexandra Fjord soil was 15.8 with a standard error of 1.62).

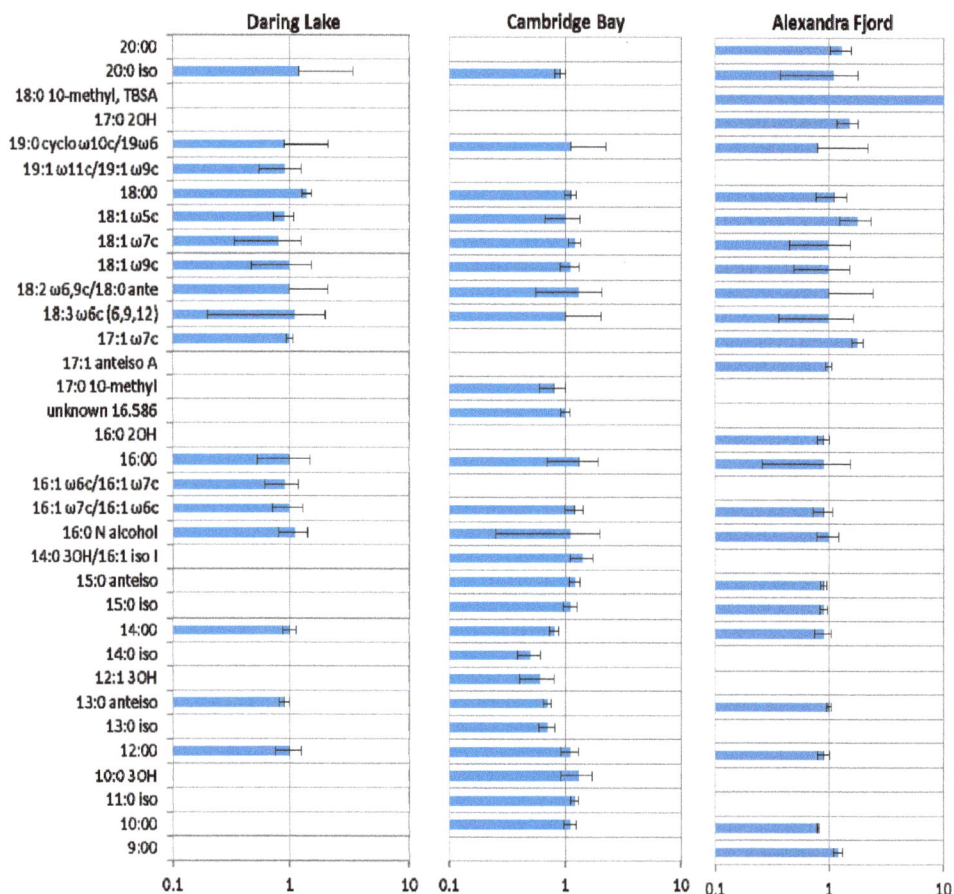

Due to the challenge in interpreting the modest and seemingly inconsistent changes we observed in the electropherograms, 16S rRNA gene sequencing was undertaken so that any differences in the relative abundance of specific community members could be better quantified. As has been documented in other studies, DNA sequence analysis is a sensitive technique (e.g., [59,77]). Because each sample was limited to a survey of 3,000 bacterial sequence reads, we focused on bacterial phylogenetic community structure at the Order taxonomic level, and found clear differences among sites both in richness (number of Orders) and evenness (relative abundances of the Orders). For example, the Daring Lake soils contained bacteria that were classified using a cut-off level of an abundance ≥1%, into 16 different Orders, while the Cambridge Bay and Alexandra Fjord bacteria were grouped into 26 and 20 Orders, respectively (Figure 3). Solirubrobacterales, Rhizobiales, Nitrosomonadales and Acidobacteriales dominated the Daring Lake community. At the higher latitudes, Rhizobiales also dominated along with Rhodospirilalles (Cambridge Bay) and Actinomycetales (Alexandra Fjord). Soil bacterial community structure at sites similar to ours across the Arctic appears to be strongly influenced by soil pH [59]. Since soil pH varied across the sites investigated here (4.3, 5.6 and 6.6), our results suggest that even at the Order taxonomic level, pH may have a strong influence on tundra soil bacterial community structure.

Figure 3. Bacterial phylogenetic composition within the soil assemblages of Daring Lake, Cambridge Bay and Alexandra Fjord, both before (control; C), and after multiple freeze-thaw cycles (FTC). Sequence identity was established after pyrosequencing of the 16S rRNA genes, classified into Orders, and the means of those with ≥1% abundance (for either treatment or control groups) presented as discrete categories, with groupings of less abundant Orders shown as "Others."

Figure 3. *Cont.*

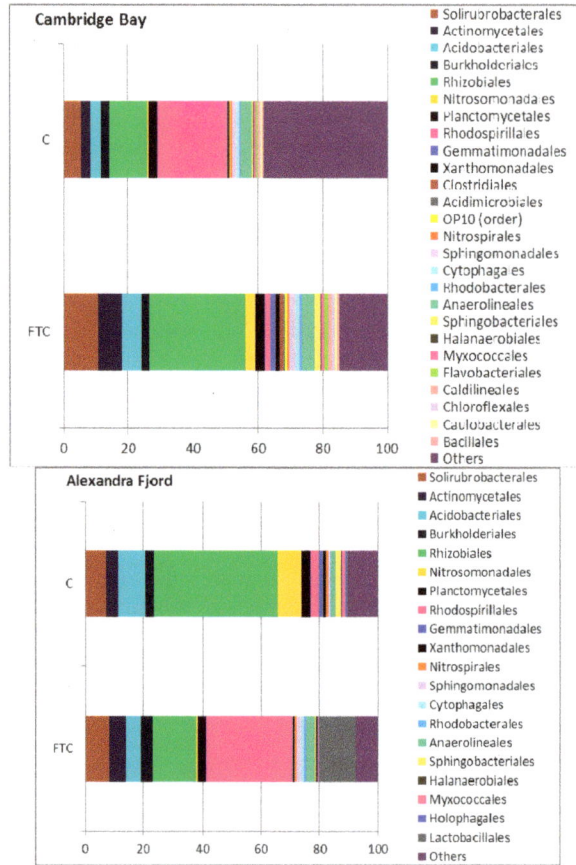

Perhaps most surprisingly, the effect of FTCs on the phylogenetic composition differed depending on the originating site (Figures 3 and 4). After freeze-thaw treatment, the overall abundance of the major Orders from the Daring Lake site did not appear to change. This was apparent even when the data was mined for abundance at the Family level (not shown). Similarly, the same proportion of "minor" Orders (<1%) was found both prior to and after FTCs in the low arctic microcosms. In striking contrast, profiles from the higher latitude soils appeared to be more perturbed by FTCs. In particular, Cambridge Bay soils initially showed a very high level of diversity with "minor" Orders making up over 38% of the mean phylogenetic profile (Figure 3). However, this level of diversity was reduced to less than 15% after freeze-thaw treatments. Alexandra Fjord soils were not as diverse, with minor groups making up 10% of the profile, which was reduced to 7% after FTCs. In the higher latitude soils the abundant Orders also showed an impact by FTCs, with contrasting patterns in the direction and magnitude of abundance changes in response to the treatment (Figures 3 and 4). For example, in the Cambridge Bay soils the Nitrosomonadales, Acidomicrobiales, Actinomycetales, Gemmatimonadales, Sphingobacteriales, Rhizobiales, and Solirubrobacterales were all increased at least twofold as a result of the freeze-thaw treatment, while the Rhodospirilalles and the Halanaerobiales were reduced by at least a factor of two. By contrast, in the Alexandra Fjord freeze-thaw-treated microcosms, the mean relative abundance of Rhodospirilalles increased approximately 10-fold. The Gram-positive Lactobacillales appeared only after FTCs in the Alexandra Fjord soil (Figures 3 and 4). Thus, changes in the relative proportions of Orders were clearly seen in the higher latitude samples, but perhaps just as significant, the sample variation, as shown by the error bars was striking (Figure 4).Therefore, there seems to be a clear distinction on the impact of FTCs in the higher latitude soils compared to the Daring Lake site.

Figure 4. The effect of the freeze-thaw treatment on the relative abundance of each of the bacterial Orders (with those with ≥1% abundance for either treatment or control groups shown) in the soils from Daring Lake, Cambridge Bay and Alexandra Fjord. Data are ratios of the mean abundance before and after freeze-thaw treatment, with no change in the mean abundance in the Order after treatment indicated by 1. Increases and decreases relative to control values are shown as bars with mean values >1 or <1, respectively. Standard errors of the means are shown as lines on the bars. Sequences were obtained in triplicate for all controls and treatment samples. However, caution must be used in examining the FTC-treated Alexandra Fjord microcosms since one of the isolated DNA samples could not be optimally pyrosequenced; only operational taxonomic units representing the most abundant orders were reported in one of the three replicates and, therefore, this particular file set was not considered in the analysis.

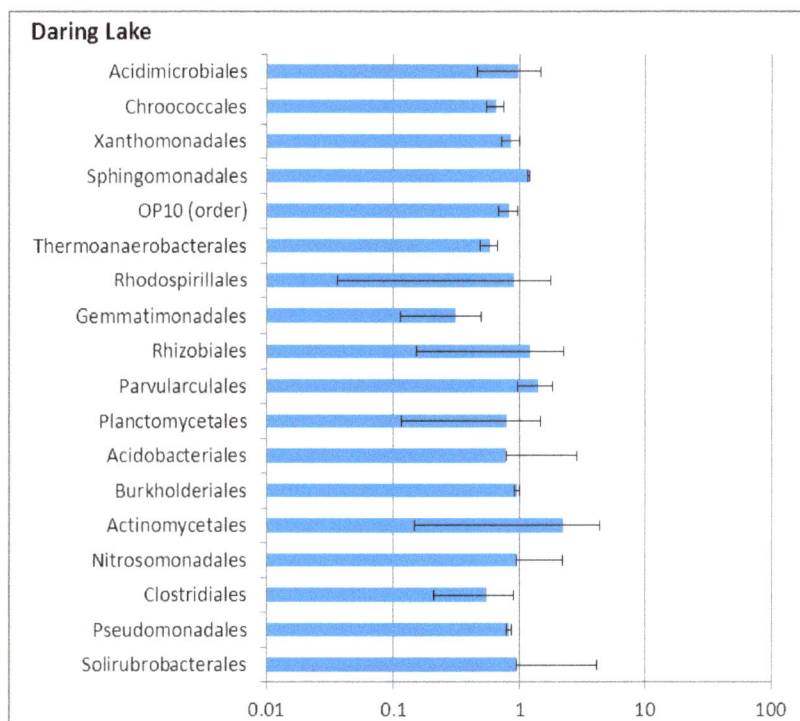

Figure 4. *Cont.*

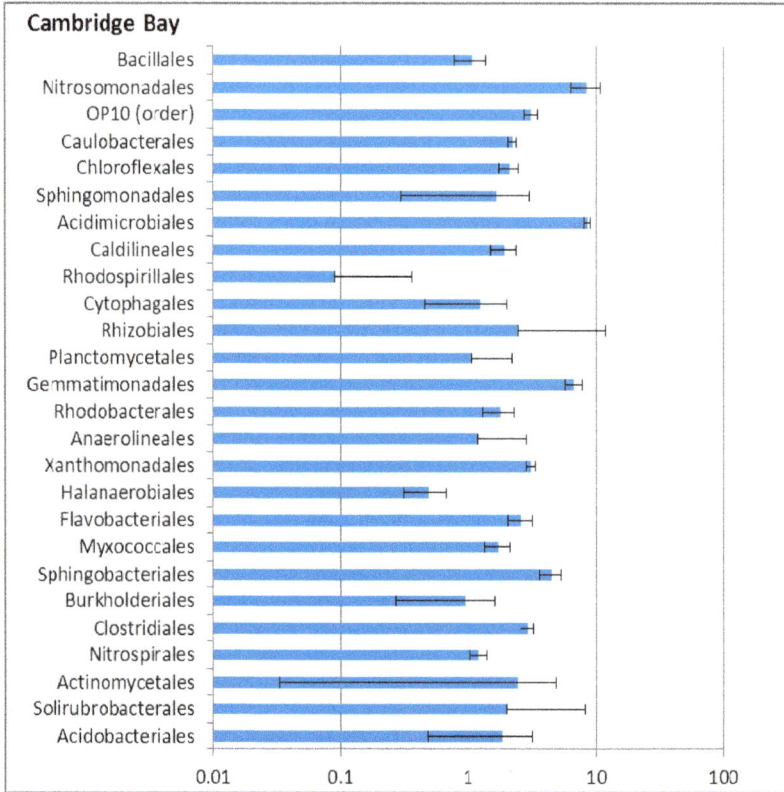

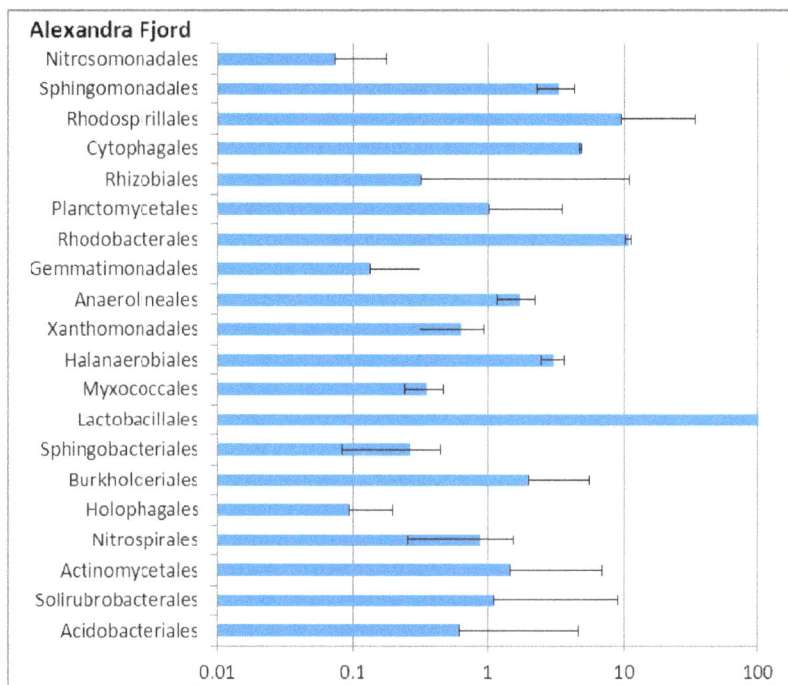

Alexandra Fjord

(Chart showing relative abundance of bacterial Orders: Nitrosomonadales, Sphingomonadales, Rhodospirillales, Cytophagales, Rhizobiales, Planctomycetales, Rhodobacterales, Gemmatimonadales, Anaerolineales, Xanthomonadales, Halanaerobiales, Myxococcales, Lactobacillales, Sphingobacteriales, Burkholceriales, Holophagales, Nitrospirales, Actinomycetales, Solirubrobacterales, Acidobacteriales; x-axis 0.01, 0.1, 1, 10, 100)

3. Discussion and Conclusions

3.1. Implications of the Microcosm Investigations

The frequency of FTCs in arctic environments is predicted to increase as a consequence of climate change [10,12]. Recordings of soil temperature fluctuations in late winter-early spring and in the autumn at our low arctic site showed multiple freeze-thaw events (Figure 1), which may well foreshadow future conditions at higher latitudes. Thus, we used these temperature fluctuations to design freeze-thaw stress conditions for soils collected at three latitudinally distinct sites, ranging from the Canadian low to high arctic. Under FTC conditions and at temperatures fluctuating around 0 °C, ice recrystallization can be very damaging to cell membranes [44], and this, coupled with the osmotic stress from solute concentration changes due to ice and snowmelt, may ultimately result in cell death [34]. Although, as previously indicated, some psychrophiles and psychrotolerant species contain antifreeze proteins and osmoprotectants that shield both themselves and others in the consortium from such stresses [39,52,78–82], many microorganisms die despite these adaptations [35,56].

PCR-DGGE profiles showed some modest changes in banding pattern, but it was not until the effects of FTC were examined in the communities by pyrosequence and fatty acid analysis that the differential impact on soils from the different sites was apparent. The bacterial assemblage from the low arctic Daring Lake site was relatively unperturbed by FTCs, as clearly evident from the phylogenetic composition. Here, we observed no change in the relative abundance of any of the major soil bacterial Orders or Families after FTC treatment and no obvious loss of diversity. This suggests a strong microbial community resilience (at these taxonomic levels) to the imposed environmental stress. Soils were deliberately sampled on relatively exposed ridges where FTCs are likely because of absent or very limited snowcover and full exposure to dynamic air temperature fluctuations. As well, because the Daring Lake site is situated at a lower latitude, it could be expected to undergo more FTCs than the more northerly sites [55]. Since the freeze-thaw experimental regime was based on temperature fluctuations measured in late winter and spring at this general location (Figure 1), it is perhaps not surprising [21] that indigenous soil microbes were relatively well adapted to survive these conditions. Similarly, hardy Antarctic bacterial communities have been reported to be relatively unresponsive across a range of FTC treatments varying in intensity [83]. In effect, the Daring Lake site, like other arctic and Antarctic environments that frequently experience freeze-thaw challenges, would be expected to be inhabited by stress-tolerant bacteria [21,83].

Contrary to the FTC-tolerant microbial assemblage from the low arctic site, mid- and high arctic soil communities appeared to be more affected by FTCs. In these higher latitude soils many of the major fatty acids showed a marked change; 56% and 33% of the fatty acids from the Cambridge Bay and Alexandra Fjord sites, respectively, showed more than a 10% change in abundance after freeze-thaw treatment, compared to 18% for the Daring Lake community. Pyrosequencing analysis revealed a similar trend with changes in the overall abundance of the bacterial Orders present prior to FTC treatment. The Alexandra Fjord soils were impacted by FTCs as evidenced by a decline in overall bacterial diversity in the minor Orders (10% to 7%), coupled with notable shifts in phylogenetic diversity. For example, after freeze-thaw treatments, there was an increased relative abundance of Lactobaccillales, Rhodospiralles, and Rhodobacterales, all of which have been reported from the Antarctic, or high arctic ice shelves and soils [84–86], suggesting that species within these groups may have low temperature adaptations. Nevertheless, the range of abundance values varied between microcosm replicates, thereby indicating that there was a lack of consistency in the magnitude of these increases. It is possible, then, that not only methods and analysis, but also a variable biological response, may underlie the apparent discrepancies between previously published studies [8,13–21] on the impact of FTCs.

For the mid-arctic site, the most notable shift in community structure was the dramatic decrease in the diversity contributed by minor Orders; this dropped more than half (38% to 15%) after FTCs (Figure 3), with some major Orders then appearing to make up more of the phylogenic composition. This striking reduction in diversity is similar to the results of FTC selection on cultured enrichments, which ultimately resulted in the recovery of microbes that were characterized by the production of osmolytes and inhibitors of ice recrystallization [52,82]. As significant as the reduction of diversity, however, was the high variation seen in the three replicate microcosms for this site compared to the relatively small variation in the Daring Lake soils (Figure 4). To illustrate, the FTC-mediated decrease in abundance of Rhodospiralles and Halanaerobiales in Cambridge Bay soils was affiliated with error bars spanning almost an order of magnitude. Thus, species within these Orders are likely susceptible to FTC-mediated conditions, but the magnitude of this vulnerability could have been mitigated in a particular microcosm by the concurrent freeze-thaw selection of a chance group of beneficial, commensal species. These latter species may be part of the Orders present in low abundance. In this regard, it would be of interest to increase the number of sequence reads so that minor groups could be more accurately enumerated, as well as to replicate each microcosm a dozen or more times to determine if any patterns in this apparent unpredictability emerge. Overall, however, the observed variation between microcosms derived from the high latitude soils strongly suggests that FTC effects on microbial community composition and relative abundance may not be absolutely predictable and that stochastic influences may play a significant role in the outcome.

Taken together, we speculate that soil microbes collected at the mid and high arctic sites were not strongly selected *in situ* for community adaptations to FTCs, compared to the Daring Lake site. Perhaps because of differences in local climate, whereby the more northerly sites typically experience fewer FTCs [55], both the higher latitude sites were noticeably impacted by freeze-thaw treatments. In particular, Cambridge Bay microcosms showed an overall reduction in diversity. The question then becomes: to what extent do disturbed consortia function differently compared to the original community? Bacteria are some of the most abundant and diverse organisms on earth [87], and our knowledge of community structure and its linkages to microbial biogeochemical activity is severely limited. For instance, studies on the temperature response of arctic and Antarctic microbial communities report not only variations in bacterial species, but highly divergent changes in the gene expression of the major functional genes such as nifH, nosZ and amoA involved in the soil nitrogen cycle [83,86,88,89]. In this case then, it is very difficult to predict directional effects of climate warming on nitrogen cycling by the community as a whole, and likewise, more generally, for all other critical biogeochemical processes. Furthermore, changes in microbial composition may not affect ecosystem process rates if the post-disturbed community (with a different composition/structure) contains taxa that are functionally equivalent to those that were previously abundant in the pre-disturbed community [90]. Likewise, new taxa in the post-disturbed community may function differently, but still maintain pre-disturbance community process rates. In summary, the inherent genetic variability and potential for rapid acclimation and adaptation in soil bacterial communities undoubtedly plays an important role in retaining a full range of ecosystem functions even after such environmental stress.

3.2. Implications for Predictions on the Effect of Climate Change

It is undoubtedly a challenge to predict directional shifts in arctic bacterial communities in response to climate change. However, what makes this difficult task virtually impossible is the evidence of stochastic variation that we have reported here. Since different microbes can employ very different tactics to evade or mitigate the impact of FTCs, including the synthesis of specialized proteins, the maintenance of a fluid membrane, the production of osmolytes and even commensal or mutualistic relationships, to mention just a few, it is perhaps no wonder that every microcosm which was exposed to novel or at least unusual freeze-thaw treatments resulted in a different community structure. We initially hypothesized that soils from higher latitudinal sites might be especially vulnerable to soil temperature changes recorded at the low arctic site, as a proxy to predict the effect of future climate change. Even if one argues that the climatic histories with respect to FTCs for each site were not that different, the fundamental point is that our data clearly show strong differences in the variability of freeze-thaw responses among sites. We now speculate that we might not have needed to collect samples at such vast distances. A similar variable response may well have been obtained using soils collected from different topographical locations at a single site if those locations varied in FTC history as a selective force on the bacterial community. We further hypothesize that if indeed the soil communities at the Daring Lake site had been "pre-selected" for relatively high numbers and amplitudes of FTCs over countless seasons, then our results together provide an explanatory mechanism for the pattern of responses.

We propose that freeze-thaw fluctuations are not near-catastrophic stresses, but can exert a sufficiently strong selective pressure to allow resource distribution and community adaptation. Adaptation in our microcosms, which harbored communities approximating 10^{10} individuals [91] and numbering more than 10^4 different species is perhaps, in retrospect, not surprisingly a product of stochastic, chaotic mechanisms. A "butterfly effect," as defined by chaos theory [92], results when slight initial disturbances can give rise to larger differences in a subsequent state. In our case, we posit that minor perturbations in our microcosm communities gave rise to unique species assemblages in each "non-adapted" microcosm derived from the higher latitude soils. The consequence of these observations is rather sobering; regretfully, we must submit that the effect of climate change on arctic soils may be inherently unpredictable.

Acknowledgments

NSERC (Canada) grants and a Queen's University Research Chair to VKW financially supported this research. G. Palmer is thanked for his help with the equipment for respiration analysis, our colleagues are appreciated for their assistance with the collection of soils, and K. Moniz is acknowledged for her help with the DNA purification.

References and Notes

1. Olsson, P.Q.; Sturm, M.; Racine C.H.; Romanovsky, V.; Liston G.E. Five Stages of the Alaskan Arctic Cold Season with Ecosystem Implications. *Arctic Antarct. Alp. Res.* **2003**, *35*, 74–81.
2. Brooks, P.D.; Grogan, P.; Templer, P.H.; Groffman, P.; Öquist, M.G.; Schimel, J.P. Carbon and Nitrogen Cycling in Snow-Covered Environments. *Geogr. Compass.* **2011**, *5*, 682–699.
3. Sulkava, P.; Huhta, V. Effects of Hard Frost and Freeze-Thaw Cycles on Decomposer Communities and N Mineralisation in Boreal Forest Soil. *Appl. Soil Ecol.* **2003**, *22*, 225–239.
4. Grogan, P.; Michelsen, A.; Ambus, P.; Jonasson, S. Freeze-Thaw Regime Effects on Carbon and Nitrogen Dynamics in Sub-Arctic Heath Tundra Mesocosms. *Soil Biol. Biochem.* **2004**, *36*, 641–654.
5. Yanai, Y.; Toyota, K.; Okazaki, M. Effects of Successive Soil Freeze-Thaw Cycles on Soil Microbial Biomass and Organic Matter Decomposition Potential of Soils. *Soil Sci. Plant Nutr.* **2004**, *50*, 821–829.
6. Sharma, S.; Szele, Z.; Schilling, R.; Munch, J.C.; Schloter, M. Influence of Freeze-Thaw Stress on the Structure and Function of Microbial Communities and Denitrifying Populations in Soil. *Appl. Environ. Microbiol.* **2006**, *72*, 2148–2154.

7. Schimel, J.; Balser, T.C.; Wallenstein, M. Microbial Stress-Response Physiology and its Implications for Ecosystem Function. *Ecology* **2007**, *88*, 1386–1394.

8. Henry, H.A.L. Soil Freeze-Thaw Cycle Experiments: Trends, Methodological Weaknesses and Suggested Improvements. *Soil Biol. Biochem.* **2007**, *39*, 977–986.

9. *ACIA: Arctic Climate Impact Assessment*; Cambridge University Press: Cambridge, UK, 2005; pp. 1042.

10. *IPCC, Climate Change 2007: The Physical Science Basis. Contribution of Working Group I to the Fourth Assessment Report of the Intergovernmental Panel on Climate Change*; Solomon, S., Qin, D., Manning, M., Chen, Z., Marquis, M., Averyt, K.B., Tignor, M., Miller, H.L., Eds.; Cambridge University Press: Cambridge, UK; New York, NY, USA, 2007.

11. Groffman, P.M.; Driscoll, C.T.; Fahey, T.J.; Hardy, J.P.; Fitzhugh, R.D.; Tierney, G.L. Colder Soils in a Warmer World: A Snow Manipulation Study in a Northern Hardwood Forest Ecosystem. *Biogeochemistry* **2001**, *56*, 135–150.

12. Kattsov, V.M.; Källén, E.; Cattle, H.; Christensen, J.; Drange, H.; Hanssen-Bauer, I.; Jóhannesen, T.; Karol, I.; Räisänen, J.; Svensson, G.; *et al.* Future Climate Change: Modeling and Scenarios for the Arctic. In *Arctic Climate Impact Assessment*; Cambridge University Press: New York, NY, USA, **2005**; pp. 99–150.

13. Brooks, P.D.; Williams, M.W.; Schmidt, S.K. Inorganic Nitrogen and Microbial Biomass Dynamics Before and During Spring Snowmelt. *Biogeochemistry* **1998**, *43*, 1–15.

14. Schimel, J.P.; Clein, J.S. Microbial Response to Freeze-Thaw Cycles in Tundra and Taiga Soils. *Soil Biol. Biochem.* **1996**, *28*, 1061–1066.

15. Larsen, K.S.; Jonasson, S.; Michelsen, A. Repeated Freeze-Thaw Cycles and Their Effects on Biological Processes in Two Arctic Ecosystem Types. *Appl. Soil Ecol.* **2002**, *21*, 187–195.

16. Larsen, K.S.; Grogan, P.; Jonasson, S.; Michelsen, A. Respiration and Microbial Dynamics in Two Subarctic Ecosystems During Winter and Spring Thaw: Effects of Increased Snow Depth. *Arctic Antarct. Alp. Res.* **2007**, *39*, 268–276.

17. Christiansen, C.T.; Svendsen, S.H.; Schmidt, N.M.; Michelsen, A. High Arctic Heath Soil Respiration and Biogeochemical Dynamics During Summer and Autumn Freeze-in—Effects of Long-term Enhanced Water and Nutrient Supply. *Glob. Change Biol.* **2012**, *18*, 3224–3236.

18. Soulides, D.A.; Allison, F.E. Effect of Drying and Freezing Soils on Carbon Dioxide Production, Available Mineral Nutrients, Aggregation, and Bacterial Population. *Soil Sci.* **1961**, *91*, 291–298.

19. Lipson, D.A.; Monson, R.K. Plant-Microbe Competition for Soil Amino Acids in the Alpine Tundra: Effects of Freeze-Thaw and Dry-Rewet Events. *Oecologia* **1998**, *113*, 406–414.

20. Buckeridge, K.M.; Cen, Y.-P.; Layzell, D.B.; Grogan, P. Soil Biogeochemistry During the Early Spring in Low Arctic Mesic Tundra and the Impacts of Deepened Snow and Enhanced Nitrogen Availability. *Biogeochemistry* **2010**, *99*, 127–141.

21. Männistö, M.K.; Tiirola, M.; Haggblom, M.M. Effect of Freeze-Thaw Cycles on Bacterial Communities of Arctic Tundra Soil. *Microb. Ecol.* **2009**, *58*, 621–631.

22. Skogland, T.; Lomeland, S.; Goksoyr, J. Respiratory Burst after Freezing and Thawing of Soil—Experiments with Soil Bacteria. *Soil Biol. Biochem.* **1988**, *20*, 851–856.

23. Herrmann, A.; Witter, E. Sources of C and N Contributing to the Flush in Mineralization upon Freeze-Thaw Cycles in Soils. *Soil Biol. Biochem.* **2002**, *34*, 1495–1505.

24. Shaver, G.R.; Chapin, F.S. Response to Fertilization by Various Plant-Growth Forms in an Alaskan Tundra—Nutrient Accumulation and Growth. *Ecology* **1980**, *61*, 662–675.

25. Jonasson, S.; Michelsen, A.; Schmidt, I.K. Coupling of Nutrient Cycling and Carbon Dynamics in the Arctic, Integration of Soil Microbial and Plant Processes. *Appl. Soil Ecol.* **1999**, *11*, 135–146.

26. Shaver, G.R.; Billings, W.D.; Chapin, F.S., III; Giblin, A.E.; Nadelhoffer, K.J.; Oechel, W.C.; Rastetter, E.B. Global Change and the Carbon Balance of Arctic Ecosystems. *BioScience* **1992**, *42*, 433–441.

27. Edwards, K.A.; Jefferies, R.L. Nitrogen uptake by *Carex aquatilis* during the winter-spring transition in a low Arctic wet meadow. *J. Ecol.* **2010**, *98*, 737–744.

28. Grogan, P.; Jonasson, S. Controls on Annual Nitrogen Cycling in the Understorey of a Sub-Arctic Birch Forest. *Ecology* **2003**, *84*, 202–218.

29. Larsen, K.S.; Michelsen, A.; Jonasson, S.; Beier, C.; Grogan, P. Nitrogen Uptake during Fall, Winter and Spring Differs among Plant Functional Groups in a Subarctic Heath Ecosystem. *Ecosystems* **2012**, *15*, 927–939.

30. Bottner, P. Response of Microbial Biomass to Alternate Moist and Dry Conditions in a Soil Incubated with C-14-labeled and N-15-labelled Plant-Material. *Soil Biol. Biochem.* **1985**, *17*, 329–337.

31. Kieft, T.L.; Soroker, E.; Firestone, M.K. Microbial Biomass Response to a Rapid Increase in Water Potential When Dry soil is Wetted. *Soil Biol. Biochem.* **1987**, *19*, 119–126.

32. Clein, J.S.; Schimel, J.P. Reduction in Microbial Activity in Birch Litter due to Drying and Rewetting Events. *Soil Biol. Biochem.* **1994**, *26*, 403–406.

33. Fierer, N.; Schimel, J.P. A Proposed Mechanism for the Pulse in Carbon Dioxide Production Commonly Observed Following the Rapid Rewetting of a Dry Soil. *Soil Sci. Soc. Am. J.* **2003**, *67*, 798–805.

34. Franzluebbers, A.J.; Haney, R.L.; Honeycutt, C.W.; Schomberg, H.H.; Hons, F.M. Flush of Carbon Dioxide Following Rewetting of Dried Soil Relates to Active Organic Pools. *Soil Sci. Soc. Am. J.* **2000**, *64*, 613–623.

35. Jefferies, R.L.; Walker, N.A.; Edwards, K.A.; Dainty, J. Is the Decline of Soil Microbial Biomass in Late Winter Coupled to Changes in the Physical State of Cold Soils? *Soil Biol. Biochem.* **2010**, *42*, 129–135.

36. Kane, D.L.; Stein, J. Water-Movement into Seasonally Frozen Soils. *Water Resour. Res.* **1983**, *19*, 1547–1557.

37. Marsh, P.; Woo, M.K. Wetting Front Advance and Freezing of Meltwater within a Snow Cover 1. Observations in the Canadian Arctic. *Water Resour. Res.* **1984**, *20*, 1853–1864.

38. Kane, D.L.; Hinkel, K.M.; Goering, D.J.; Hinzman, L.D.; Outcalt, S.I. Non-Conductive Heat Transfer Associated with Frozen Soils. *Glob. Planet. Change* **2001**, *29*, 275–292.

39. Mazur, P. Theoretical and Experimental Effects of Cooling and Warming Velocity on Survival of Frozen and Thawed Cells. *Cryobiology* **1966**, *2*, 181–192.

40. Mazur, P. Freezing of Living Cells: Mechanisms and Implications. *Am. J. Physiol.* **1984**, *247*, C125–C142.
41. Deal, P.H. Freeze-Thaw Behaviour of a Moderately Halophilic Bacterium as a Function of Salt Concentration. *Cryobiology* **1970**, *7*, 107–112.
42. Mikan, C.J.; Schimel, J.P.; Doyle, A.P. Temperature Controls of Microbial Respiration in Arctic Tundra Soils Above and Below Freezing. *Soil Biol. Biochem.* **2002**, *34*, 1785–1795.
43. Panikov, N.S.; Flanagan, P.W.; Oechel, W.C.; Mastepanov, M.A.; Christensen, T.R. Microbial Activity in Soils Frozen to Below −39 °C. *Soil Biol. Biochem.* **2006**, *38*, 785–794.
44. Casanueva, A.; Tuffin, M.; Cary, C.; Cowan, D.A. Molecular Adaptations to Psychrophily: The Impact of "Omic" Technologies. *Trends Microbiol.* **2010**, *18*, 374–381.
45. Wilson, S.L.; Walker, V.K. Selection of Low-Temperature Resistance in Bacteria and Potential Applications. *Environ. Technol.* **2010**, *31*, 943–956.
46. Oquist, M.G.; Sparrman, T.; Klemedtsson, L.; Drotz, S.H.; Grip, H.; Schleucher, J.; Nilsson, M. Water Availability Controls Microbial Temperature Responses in Frozen Soil CO_2 Production. *Glob. Change Biol.* **2009**, *15*, 2715–2722.
47. Tilston, E.L.; Sparrman, T.; Oquist, M.G. Unfrozen Water Content Moderates Temperature Dependence of Sub-Zero Microbial Respiration. *Soil Biol. Biochem.* **2010**, *42*, 1396–1407.
48. Clein, J.S.; Schimel, J.P. Microbial Activity of Tundra and Taiga Soils at Subzero Temperatures. *Soil Biol. Biochem.* **1995**, *27*, 1231–1234.
49. Schimel, J.P.; Mikan, C. Changing Microbial Substrate Use in Arctic Tundra Soils through a Freeze-Thaw Cycle. *Soil Biol. Biochem.* **2005**, *37*, 1411–1418.
50. Michaelson, G.J.; Ping, C.L. Soil Organic Carbon and CO_2 Respiration at Subzero Temperature in Soils of Arctic Alaska. *J. Geophys. Res. Atmos.* **2003**, *108*, D2.
51. Schimel, J.P.; Bilbrough, C.; Welker, J.A. Increased Snow Depth Affects Microbial Activity and Nitrogen Mineralization in Two Arctic Tundra Communities. *Soil Biol. Biochem.* **2004**, *36*, 217–227.
52. Walker, V.K.; Palmer, G.R.; Voordouw, G. Freeze-Thaw Tolerance and Clues to the Winter Survival of a Soil Community. *Appl. Environ. Microb.* **2006**, *72*, 1784–1792.
53. McMahon, S.K.; Wallenstein, M.D.; Schimel, J.P. A Cross-Seasonal Comparison of Active and Total Bacterial Community Composition in Arctic Tundra Soil Using Bromodeoxyuridine Labeling. *Soil Biol. Biochem.* **2011**, *43*, 287–295.
54. Monson, R.K.; Lipson, D.L.; Burns, S.P.; Turnipseed, A.A.; Delany, A.C.; Williams, M.W.; Schmidt, S.K. Winter Forest Soil Respiration Controlled by Climate and Microbial Community Composition. *Nature* **2006**, *439*, 711–714.
55. Schadt, C.W.; Martin, A.P.; Lipson, D.A.; Schmidt, S.K. Seasonal Dynamics of Previously Unknown Fungal Lineages in Tundra Soils. *Science* **2003**, *301*, 1359–1361.
56. Lipson, D.A.; Schmidt, S.K. Seasonal Changes in an Alpine Soil Bacterial Community in the Colorado Rocky Mountains. *Appl. Environ. Microb.* **2004**, *70*, 2867–2879.
57. Lipson, D.A.; Schadt, C.W.; Schmidt, S.K. Changes in Soil Microbial Community Structure and Function in an Alpine Dry Meadow Following Spring Snow Melt. *Microb. Ecol.* **2002**, *43*, 307–314.

58. Henry, G. Climate Change and Soil Freezing Dynamics: Historical Trends and Projected Changes. *Clim. Change* **2008**, *87*, 421–434.

59. Chu, H.Y.; Neufeld, J.D.; Walker, V.K.; Grogan, P. The Influence of Vegetation Type on the Dominant Soil Bacteria, Archaea, and Fungi in a Low Arctic Tundra Landscape. *Soil Sci. Soc. Am. J.* **2011**, *75*, 1756–1765.

60. Chu, H.; Fierer, N.; Lauber, C.L.; Caporaso, J.G.; Knight, R.; Grogan, P. Soil Bacterial Diversity in the Arctic is not Fundamentally Different from that Found in Other Biomes. *Environ. Microb.* **2010**, *12*, 2998–3006.

61. Labine, C. Meteorology and Climatology of the Alexandra Fjord Lowland. In *Ecology of a Polar Oasis, Alexandra Fiord, Ellesmere Island, Canada*; Svoboda, J., Freedman, B., Eds.; Captus University Publications: Toronto, Canada, 1994; pp. 23–39.

62. Rayback, S.A.; Henry, G.H.R. Reconstruction of Summer Temperature for a Canadian High Arctic Site from Retrospective Analysis of the Dwarf Shrub, *Cassiope tetragona*. *Arctic Antarct. Alp. Res.* **2006**, *38*, 228–238.

63. Sasser, M. *Identification of Bacteria by Gas Chromatography of Cellular Fatty Acids*; MIDI Technical note #101: Newark, DE, USA, 1990, revised 2001.

64. Klose, S.; Acosta-Martinez, V.; Ajwa, H.A. Microbial Community Composition and Enzyme Activities in a Sandy Loam Soil after Fumigation with Methyl Bromide or Alternative Biocides. *Soil Biol. Biochem.* **2006**, *38*, 1243–1254.

65. Kumar, N.; Shah, V.; Walker, V.K. Perturbation of an Arctic Soil Microbial Community by Metal Nanoparticles. *J. Hazard. Mater.* **2011**, *190*, 816–822.

66. Dowd, S.E.; Callaway, T.R.; Wolcott, R.D.; Sun, Y.; McKeehan, T.; Hagevoort, R.G.; Edrington, T.S. Evaluation of the Bacterial Diversity in the Feces of Cattle Using 16S rDNA Bacterial Tag-Encoded FLX Amplicon Pyrosequencing (bTEFAP). *BMC Microbiol.* **2008**, *8*, 125.

67. Ishak, H.D.; Plowes, R.; Sen, R.; Kellner, K.; Meyer, E.; Estrada, D.A.; Dowd, S.E.; Mueller, U.G. Bacterial Diversity in *Solenopsis invicta* and *Solenopsis geminata* Ant Colonies Characterized by 16S amplicon 454 Pyrosequencing. *Microb. Ecol.* **2011**, *61*, 821–831.

68. Research and Testing Laboratory. Available online: http://www.researchandtesting.com/ (accessed on 17 December 2012).

69. Gontcharova, V.; Youn, E.; Wolcott, R.D.; Hollister, E.B.; Gentry, T.J.; Dowd, S.E. Black Box Chimera Check (B2C2): A Windows-Based Software for Batch Depletion of Chimeras from Bacterial 16S rRNA Gene Datasets. *Open Microbiol. J.* **2010**, *4*, 47–52.

70. Bailey, M.T.; Dowd, S.E.; Parry, N.M.; Galley, J.D.; Schauer, D.B.; Lyte, M. Stressor Exposure Disrupts Commensal Microbial Populations in the Intestines and Leads to Increased Colonization by *Citrobacter rodentium*. *Infect. Immun.* **2010**, *78*, 1509–1519.

71. Dowd, S.E.; Zaragoza, J.; Rodriguez, J.R.; Oliver, M.J.; Payton, P.R. Windows.NET Network Distributed Basic local Alignment Search Toolkit (W.ND-BLAST). *BMC Bioinformatics* **2005**, *6*, 93.

72. Cole, J.R.; Wang, Q.; Cardenas, E.; Fish, J.; Chai, B.; Farris, R.J.; Kulam-Syed-Mohideen, A.S.; McGarrell, D.M.; Marsh, T.; Garrity, G.M.; *et al*. The Ribosomal Database Project: Improved Alignments and New Tools for rRNA Analysis. *Nucleic Acids Res.* **2009**, *37*, D141–D145.

73. Suutari, M.; Laakso, S. Microbial Fatty-Acids and Thermal Adaptation. *Crit. Rev. Microbiol.* **1994**, *20*, 129–137.

74. Singleton, D.R.; Furlong, M.A.; Peacock, A.D.; White, D.C.; Coleman, D.C.; Whitman, W.B. *Solirubrobacter pauli* gen. nov., sp. nov., A Mesophilic Bacterium within the *Rubrobacteridae* Related to Common Soil Clones. *Int. J. Syst. Evol. Microbiol.* **2003**, *53*, 485–490.

75. Kim, M.K.; Na, J.R.; Lee, T.H.; Im, W.T.; Soung, N.K.; Yang, D.C. *Solirubrobacter soli* sp. nov., Isolated from Soil of a Ginseng Field. *Int. J. Syst. Evol. Microbiol.* **2007**, *57*, 1453–1455.

76. Paulino-Lima, I.G.; Azua-Bustos, A.; Vicuña, R.; González-Silva, C.; Salas, L.; Teixeira, L.; Rosado, A.; da Costa Leitao, A.A.; Lage, C. Isolation of UVC-tolerant Bacteria from the Hyperarid Atacama Desert, Chile. *Microb. Ecol.* **2012**, doi:10.1007/s00248-012-0121-z.

77. Collins, D.; Luxton, T.; Kumar, N.; Shah, S.; Walker, V.K.; Shah, V. Assessing the Impact of Copper and Zinc Oxide Nanoparticles on Soil: A Field Study. *PLoS One* **2012**, *7*, e42663.

78. Mackey, B.M. Lethal and Sublethal Effects of Refrigeration, Freezing and Freeze-Drying on Micro-Organisms. *Soc. Appl. Bacteriol. Symp. Ser.* **1984**, *12*, 45–75.

79. Xu, H.; Griffith, M.; Patten, C.L.; Glick, B.R. Isolation and Characterization of an Antifreeze Protein with Ice Nucleation Activity from the Plant Growth Promoting Rhizobacterium *Pseudomonas putida* GR12-2. *Can. J. Microbiol.* **1998**, *44*, 64–73.

80. Raymond, J.A.; Fritsen, C.H. Semipurification and Ice Recrystallization Inhibition Activity of Ice-active Substances Associated with Antarctic Photosynthetic Organisms. *Cryobiology* **2001**, *43*, 63–70.

81. Gilbert, J.A.; Hill, P.J.; Dodd, C.E.R.; Laybourn-Parry, J. Demonstration of Antifreeze Protein Activity in Antarctic Lake Bacteria. *Microbiology* **2004**, *150*, 171–180.

82. Wilson, S.L.; Frazer, C.; Cumming, B.F.; Nuin, P.A.S.; Walker, V.K. Cross-tolerance between Osmotic and Freeze-Thaw Stress in Microbial Assemblages from Temperate Lakes. *FEMS Microbiol. Ecol.* **2012**, *82*, 405–415.

83. Yergeau, E.; Kowalchuk, G.A. Responses of Antarctic Soil Microbial Communities and Associated Functions to Temperature and Freeze-Thaw Cycle Frequency. *Environ. Microbiol.* **2008**, *10*, 2223–2235.

84. Teixeira, L.; Peixoto, R.S.; Cury, J.C.; Sul, W.J.; Pellizari, V.H.; Tiedje, J.; Rosado, A.S. Bacterial Diversity in Rhizosphere Soil from Antarctic Vascular Plants of Admiralty Bay, Maritime Antarctica. *ISME J.* **2010**, *4*, 989–1001.

85. Bottos, E.M.; Vincent, W.F.; Greer, C.W.; Whyte, L.G. Prokaryotic Diversity of Arctic Ice Shelf Microbial Mats. *Environ. Microbiol.* **2008**, *10*, 950–966.

86. Deslippe, J.R.; Egger, K.N.; Henry, G.H.R. Impacts of Warming and Fertilization on Nitrogen-Fixing Microbial Communities in the Canadian High Arctic. *FEMS Microbiol. Ecol.* **2005**, *53*, 41–50.

87. Whitman, W.B.; Coleman, D.C.; Wiebe, W.J. Prokaryotes: The Unseen Majority. *Proc. Natl. Acad. Sci. USA* **1998**, *95*, 6578–6583.

88. Walker, J.K.M.; Egger, K.N.; Henry, G.H.R. Long-Term Experimental Warming Alters Nitrogen-Cycling Communities but Site factors Remain the Primary Drivers of Community Structure in High Arctic Tundra Soils. *ISME J.* **2008**, *2*, 982–995.

89. Lamb, E.G.; Han, S.; Lanoil, B.D.; Henry, G.H.R.; Brummell, M.E.; Banerjee, S.; Siciliano, S.D. A High Arctic Soil Ecosystem Resists Long-Term Environmental Manipulations. *Glob. Change Biol.* **2011**, *17*, 3187–3194.

90. Schimel, J.P.; Gulledge, J. Microbial Community Structure and Global Trace Gases. *Glob. Change Biol.* **1998**, *4*, 745–758.

91. Banerjee, S.; Siciliano, S.D. Evidence of High Microbial Abundance and Spatial Dependency in Three Arctic Soil Ecosystems. *Soil Sci. Soc. Am. J.* **2011**, *75*, 2227–2232.

92. Lorenz, E.N. Deterministic Nonperiodic Flow. *J. Atmos. Sci.* **1963**, *20*, 130–141.

Microbial Competition in Polar Soils: A Review of an Understudied but Potentially Important Control on Productivity

Terrence H. Bell, Katrina L. Callender, Lyle G. Whyte and Charles W. Greer

Abstract: Intermicrobial competition is known to occur in many natural environments, and can result from direct conflict between organisms, or from differential rates of growth, colonization, and/or nutrient acquisition. It has been difficult to extensively examine intermicrobial competition *in situ*, but these interactions may play an important role in the regulation of the many biogeochemical processes that are tied to microbial communities in polar soils. A greater understanding of how competition influences productivity will improve projections of gas and nutrient flux as the poles warm, may provide biotechnological opportunities for increasing the degradation of contaminants in polar soil, and will help to predict changes in communities of higher organisms, such as plants.

Reprinted from *Biology*. Cite as: Bell, T.H.; Callender, K.L.; Whyte, L.G.; Greer, C.W. Microbial Competition in Polar Soils: A Review of an Understudied but Potentially Important Control on Productivity. *Biology* **2013**, *2*, 533-554.

1. Introduction

Although many ecosystem processes are dependent on the growth and activity of multiple species, the productivity of particular individuals can often be limited by the presence of competitors. The classic ecological example of Connell's barnacles [1] demonstrates that the area potentially occupied by a particular species (*Chthamalus stellatus*), can be greater than its true distribution in the presence of a competitor (*Balanus balanoides*). Such relationships promote biodiversity in many environments, as they prevent complete dominance by the small number of organisms that are best adapted to quickly processing limiting nutrients [2–4]. In such cases, competition can constrain specific functions of a community, as the survival and activity of certain organisms limits the resources and habitat available to the most productive species.

When considering the microbial world, productivity can be defined as the rate and efficiency with which any target metabolic function occurs. Growth and accumulation of biomass are easy to picture as productive processes, but the degradation of substrates or the cycling of nutrients can also be considered productive from a microbial perspective (e.g., allowing increased activity or growth), and sometimes from a human perspective (e.g., reduction of environmental contaminants). Although microbial productivity is a universally important component of biogeochemical cycling across environments, the factors that control productivity are especially interesting in polar soils.

Firstly, climate warming and other human disturbances are exposing formerly frozen landscapes to increased temperatures, which will likely lead to more rapid cycling of stored organic material and nutrients. Even small amounts of warming can have large effects on microbial community structure and function in polar soils [5,6], which will inevitably shift the competitive dynamic between taxa. On the other hand, the short Arctic summer limits the highly active period for many microorganisms. For human applications, such as the use of native microbial populations in bioremediation, this means maximizing microbial activity over a short period of time. The exploitation of intermicrobial competition has previously been explored for applied purposes such as the treatment of pathogens (e.g., [7,8]), optimization of agriculture (e.g., [9,10]), and food preservation [11], and has recently been investigated as a means to optimize bioremediation in the Arctic [12].

This review will highlight the factors that are known to influence microbial abundance and community structure in polar soils, and how these shifts affect important functions that are mediated by microbial communities. Very few studies have explicitly shown how microbial competition affects function in soils (fewer still in polar regions), but we will attempt to point out areas in which competition may play an important role in limiting or promoting the activity of specific microbial functions. While future warming will likely lead to more active microbial populations, it may also shift the competitive dynamic between microorganisms (Figure 1). Understanding how competition affects key microbial processes will improve predictions of future gas and nutrient fluxes, and may open important biotechnological opportunities.

1.1. Microbial Diversity and Productivity

Studies on the relationship between biodiversity and productivity have been performed in many areas of ecology, but have yielded inconsistent results (e.g., [13–15]). The addition of species should be expected to increase the productivity of specific functions when species niches are complementary. This may not be the case when the activity of certain key organisms is limited by a lack of resources and space, or by direct inhibition from competitors. An analysis of 180 two-species bacterial cultures showed that almost all pairings resulted in competitive relationships that reduced CO_2 production relative to monocultures of each species [16]. In multi-species communities, the presence or absence of specific key phylotypes appears to be more important than the overall number of microbial strains in determining productivity in some cases [17–19]. Reducing diversity and/or microbial biomass has even been shown to lead to higher productivity with respect to certain functions such as decomposition, nutrient uptake, and bioremediation [12,20–22].

1.2. Microbial Competition in Polar Soils

Several reviews have highlighted the extent and importance of intermicrobial competition in natural environments [23,24], but few studies have characterized competition in polar environments, with only a handful examining competition among polar soil microorganisms (Table 1). A single gram of soil may contain thousands of microbial species [25] as well as a complex network of interactions. Despite the fact that many polar soils frequently experience extreme cold temperatures, low water content, and intermittently available nutrients, recent molecular studies have shown that the microbial diversity and community composition in these regions resembles what has been observed at lower latitudes [26,27]. Interspecies relationships will also be dynamic, as the growing portion of an Arctic soil community has been shown to vary substantially throughout the year [28].

Figure 1. Large environmental shifts such as climate change will alter many aspects of polar soil environments that will shift the growth and activity of microbial species. Although some changes may benefit multiple species in isolation, changes in competitive interactions may determine the ultimate productivity of the whole community. In this scenario, climate change causes changes in both temperature and plant communities. Species A is promoted disproportionately by temperature and suppresses species B, leading to higher productivity (purple circle) by species A, and thus by the overall community. Species B gains a competitive advantage in the new plant community, and suppresses species A, but is not as productive as species A, leading to a decline in overall productivity. It is mostly unknown which factors will be the most important in determining competitive outcomes following climate change, and thus changes in productivity are difficult to predict.

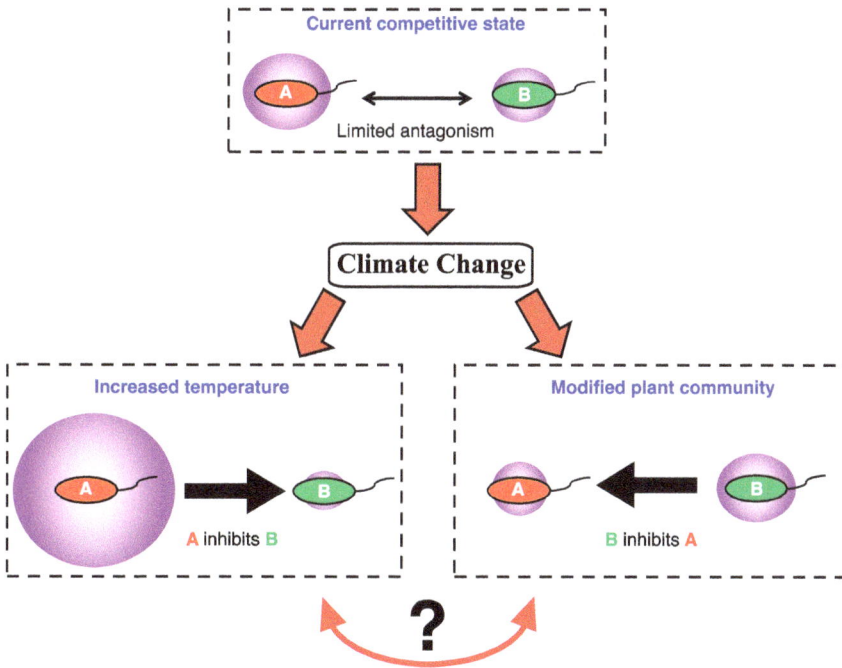

Table 1. Studies that have examined intermicrobial competition in polar soils.

Habitat	Antagonists	Function(s) affected	Proposed mechanism(s) of competition	* Special notes	Reference
In vitro					
Moss-covered and barren soil in Svalbard, Norway	*Actinobacteria (Arthrobacter), Gammaproteobacteria (Pseudomonas), Firmicutes (Paenibacillus), Bacteroidetes (Flavobacterium)*	Growth of individual strains	Antimicrobial production; differential growth rates	Competition varied at different incubation temperatures	[29]
Various Antarctic soils	**Antimicrobial producers:** *Actinobacteria (Arthrobacter), Firmicutes (Planococcus), Gammaproteobacteria (Pseudomonas);* **Affected:** *Firmicutes (Listeria, Staphylococcus, Brocothrix), Gammaproteobacteria (Salmonella, Escherichia, Pseudomonas)*	Growth of individual strains	Antimicrobial production	Producers were Antarctic bacteria, while affected bacteria were food-borne pathogens	[11]
King George Island, Antarctica	**Antimicrobial producers:** *Bacteroidetes (Pedobacter), Gammaproteobacteria (Pseudomonas);* **Affected:** *Gammaproteobacteria (Salmonella, Escherichia, Klebsiella, Enterobacter, Vibrio), Firmicutes (Bacillus)*	Growth of individual strains	Antimicrobial production	Producers were Antarctic bacteria, while affected bacteria were food-borne pathogens	[30]
Tundra wetland soil, Ural, Russia	Methanogens and homoacetogenic *Firmicutes (Acetobacterium)*	H_2 consumption	Differential H_2 affinity	Competition was modeled based on changing H_2 affinities at various temperatures; some strains isolated from pond and fen sediments	[31]
In situ					
Unvegetated contaminated soil in Alert, Nunavut, Canada	*Alpha-, Beta-, Gammaproteobacteria, Actinobacteria*	Assimilation of added monoammonium phosphate	Differential nutrient uptake	Alphaproteobacteria most effectively assimilated added nutrients	[32]
Soil microcosms					
Lowland soil, Devon Island, Nunavut, Canada	Archaeal and bacterial nitrifiers, fungal and bacterial denitrifiers	N_2O production, nitrate availability, biomass of microbial domains	Differential nutrient uptake	Effects varied with temperature	[33]

Microbial activity has been demonstrated at temperatures as low as −15 °C [34], but the effects of microbial competition on biogeochemical flux are likely to be most substantial over the summer, as warmer temperatures lead to higher overall activity. Nevertheless, extreme cold can restrict polar microorganisms to small brine pockets at subzero temperatures [35], which may lead to enhanced competition between microorganisms that remain active over winter, as they have reduced opportunities to separate spatially. Competition in polar soils will occur passively due to differential adaptations to soil and environmental conditions, but also actively, as a number of polar soil microorganisms are known to produce inhibitory concentrations of antimicrobial compounds [11,29,30]. The outcome of a change in the abundance of specific groups may substantially affect biogeochemical processes when scaled to entire polar landscapes.

2. Factors Influencing the Relative Success of Polar Microorganisms

Adaptations to certain environmental factors, such as extreme cold, will be widespread in polar microbial communities. As in lower latitude soils, taxa will vary in their competitiveness under different environmental conditions, and in the presence of specific co-occurring taxa. This variation will play a large role in determining microbial community composition in polar soils, and will ultimately influence the functional potential of these communities. Assuming that species are not equally efficient at performing a given function (e.g., substrate degradation), small shifts in environmental factors may have substantial effects on the growth and productivity or key microorganisms, as they are limited by competitors that are better adapted to the environment. Below we discuss some of the factors that are known to affect microbial community composition in polar soils.

2.1. Environmental Factors

At least at a coarse taxonomic scale, the soil environment appears to be more influential than geography in determining the relative abundance of microorganisms. Recent studies have shown that a main determinant of bacterial composition in polar soils is pH [26,36–39]. Among bacteria, the major shift due to pH is the increasing abundance of *Acidobacteria* below pH 6 [26,39]. Some studies have observed no effect of pH on bacterial communities in polar soils [40,41], but most of the soils examined had a pH of ~6 or higher. Soil pH has also been shown to correlate somewhat with fungal community composition in polar soils [42,43], while an extensive study of culturable fungal abundance across Antarctic soils showed that fungal abundance declines significantly with increasing pH [44], suggesting an increased importance of bacterial communities.

Other main determinants of community composition include organic matter [41,45], and nitrogen concentration [45–47]. Arctic soils with low organic matter content (<10% dry weight of soil) have been shown to favor *Actinobacteria*, while soils with higher organic matter (>10% dry weight of soil) favored an abundance of *Proteobacteria* [41]. High concentrations of nitrogen have generally promoted *Actinobacteria* and *Firmicutes* across biomes [46], as well as *Alpha* and *Gammaproteobacteria* in some Arctic tundra soils [47], although the effect of nitrogen on community composition may largely depend on existing soil organic matter [41]. Fungi in both the Arctic and Antarctic appear to be influenced by C:N ratios [42–44], although nutrient additions have sometimes failed to impact certain fungal groups [48,49].

Water content has also been correlated with the bacterial and archaeal community structure of polar soils [41,50,51], although it may have a greater impact on fungi and other microeukaryotes [44,52,53]. Oxygen has also been suggested as an important influence on community structure [54], although this has not been thoroughly tested independent of other factors. Oxygen concentrations will be closely related to soil water saturation, and will determine the dominant forms of metabolism that can occur in soil. Competition may play an important role at anoxic interfaces, when both aerobic and anaerobic forms of metabolism can occur. Other influences on community composition that have been identified from polar environments include phosphorous [43], micronutrients such as potassium and calcium [50], salinity [55], UV radiation [56], and soil particle size [40].

Seasonally changing temperatures will also affect the relative abundance of microorganisms. Two main types of microorganisms remain active in cold environments, and these are the stenopsychrophiles (those that do not grow well or at all at high temperatures (>20 °C)) and eurypsychrophiles (those that have wide temperature growth ranges and may grow optimally at high temperatures) [57,58]. Shifting incubation temperatures from 4 °C to 18 °C was shown to affect the growth rate of different Arctic bacterial isolates differently and ultimately influenced the outcome of competition between them [29]. Similarly, growth temperature has been shown to affect the outcome of competition between cold-adapted marine microbial strains [59,60]. Potential biomass and growth rate can also be decoupled in cold-adapted microbes [61,62]. For instance, psychrophilic bacteria and yeast developed a higher overall biomass at 1 °C than at 20 °C, even though growth rates were highest at 20 °C incubation, while the biomass of mesophiles was highest at 20 °C [62].

2.2. Biotic Interactions

The abundance of higher organisms tends to decrease with increasing latitude [63], and this may alter the biotic relationships in polar soils. It has been suggested that the simplified trophic structures of Antarctic soils may lead to an increased importance of abiotic factors in determining community composition and biomass [64], yet reduced complexity at higher trophic levels may lead to communities that are dominated more strongly by microbial processes. Although decreased microbial functional and taxonomic diversity has been observed in higher latitude Antarctic soils [37], it is known that highly diverse microbial communities exist at lower latitudes of the Antarctic [37,65], and throughout the Arctic [26,27]. The best-studied interactions are those that occur between co-occurring microorganisms, and between microorganisms and plants, although other polar soil inhabitants such as viruses and bacterivores are known to exert important top-down controls on the biomass and composition of microbial populations [66,67].

Mechanisms that are involved in intermicrobial cooperation and antagonism at lower latitudes have also been identified in polar and/or subpolar soils. For instance, active quorum sensing genes have been identified in a soil from subarctic Alaska [68]. Chemotaxis is an important strategy to competitively position consumers near nutrients, carbon or to evade toxic chemicals, and while little is known about its importance in cold regions [69], it has been identified in an Arctic *Pseudomonas* isolate [70]. As mentioned earlier, various polar microorganisms are known to produce antimicrobial compounds [11,29,30], while antibiotic resistance genes have even been identified from Arctic permafrost cores [71]. It is unknown how frequently horizontal gene transfer occurs, but a number of mobile elements have been identified from Antarctic soils, with evidence of past transfer events [72,73].

Plant and microbial communities also interact in a variety of ways, where mycorrhizal fungi are the most directly influenced due to their symbiotic relationships with plant root systems. The composition of root-associated fungal communities in the high Arctic has been shown to vary by plant species [43] and successional stage [74], while interactions between plant species and nutrient availability also influence fungal abundance [75]. Interactions between plant species are important as well, as the removal of one shrub species led to decreased ectomycorrhizal colonization of another [48]. Mycorrhizae have even been shown to facilitate carbon transfer between individual *Betula nana* plants in the Arctic tundra, increasing the ability of this plant to compete with neighboring species [76], but also presumably increasing the suitable habitat for its fungal symbionts. Bacterial and archaeal communities have also been influenced by the composition of plant communities in the Arctic [77], although sequencing of various plant assemblages in the Antarctic showed little influence of plant type on bacterial composition [65].

3. Important Microbial Functions Potentially Affected by Competition in Polar Soils

Functional redundancy is no longer assumed to be widespread in microbial communities and increasing the relative or absolute abundance of specific taxa is likely required to optimize productivity [19,78]. In mixed communities, it is often not the most productive members that dominate, as relative abundance is determined by adaptations to the abiotic and biotic components of the environment. A number of important biogeochemical processes are microbially-mediated in polar regions, and there is evidence that these processes are limited by constraints on key microbial taxa.

3.1. Greenhouse Gas Flux

One of the greatest concerns associated with the warming of polar regions is a potential increase in greenhouse gas production by soil microorganisms, which will further accelerate climate change [79]. The main reasons for this projection are that previously frozen organic matter will become available for degradation, and that microbial activity, previously restricted by low temperatures, is expected to increase. The production and mitigation of gases such as methane and nitrous oxide is restricted to specific microbial groups, so inevitably the factors that control the abundance and activity of these groups will have a major impact on future gas fluxes. While many active microorganisms release CO_2, the rate and extent of this process will also vary with the abundance and activity of specific key groups.

The abundance and composition of methanogenic and methanotrophic microbial communities have received substantial research attention, particularly in Arctic soils. Huge methane deposits exist in permafrost [80], and even warming to -3 °C and -6 °C has led to methane emissions from permafrost cores [81]. The influence of competition on methanotrophic communities has not been specifically investigated in polar regions, but a simulated disturbance in rice paddy soil showed that as methanotrophic communities reinhabited the underpopulated soil environment, type II methanotrophs dominated due to their more rapid growth rates, thus reducing methanotrophic diversity and evenness [82]. In response to this shift, methane uptake rates more than doubled, and the authors suggest that under natural conditions, methanotroph activity is constrained by competition. Stable isotope probing of high Arctic methanotrophs showed that type I methanotrophs represented the main active community, and methane oxidation was enhanced by amendment with nitrate mineral salts [83], suggesting that this group may be limited by nutrient competition under natural conditions.

In contrast, methanogens appear to be limited mainly be competition for H_2. Incubations of methanogenic and homoacetogenic strains isolated from Arctic soil and sediment were conducted at varying temperatures and concentrations of H_2, and modeling of these relationships demonstrated that methanogens would sometimes be outcompeted by homoacetogens at low temperatures and high partial pressures of H_2 [31]. Depending on the composition of nutrients present in soil, methanogens may have difficulty gaining access to H_2. By manipulating nutrient concentrations, it was observed in an anoxic rice paddy soil that nitrate, iron, and sulfate reducers were all more successful in H_2 acquisition than methanogens when H_2 was limiting [84]. The amount of methane production per methanogenic cell was shown to vary by several orders of magnitude in different subglacial Arctic and Antarctic environments [85], which demonstrates that reducing constraints on these populations could lead to large increases in methane production.

Nitrous oxide (N_2O) is another important greenhouse gas, with a warming potential 300 times that of CO_2 [86]. Although N_2O is frequently the result of incomplete denitrification, nitrifiers can also release N_2O as a byproduct of nitrification and/or incomplete nitrifier denitrification [87]. Interestingly, nitrifier release of N_2O has been shown to be the primary source of N_2O emitted from soils of Devon Island in the high Canadian Arctic [33]. This process appears to be mainly regulated by intermicrobial competition. Denitrifier activity was not enhanced, even following nitrate addition in water-saturated soils, but the inhibition of fungi led to large N_2O release by denitrifiers, without a subsequent decrease in nitrifier N_2O production [33]. This suggests that fungi and denitrifiers compete for nitrate, and that this competition mitigates N_2O release in the Arctic.

Although many organisms produce CO_2 as a byproduct of activity, competition between microorganisms can limit the amount that is produced by each, relative to the same organisms in isolation [16]. CO_2 output is also closely linked with the breakdown of soil organic matter, which is discussed in the following section.

3.2. Biodegradation

The decomposition of carbon compounds in soil is a key component of the carbon cycle, and is a precursor to the release of carbon-based greenhouse gases. The decomposition of soil organic matter occurs primarily as a result of microbial activity, and catabolic pathways for extracting energy and carbon from complex hydrocarbon substrates are widespread across microbial taxa. Although all soil microbial groups require some form of carbon substrate, they vary in their rate of carbon substrate use, meaning that the promotion or suppression of specific groups will affect rates of organic matter degradation in polar soils. This may apply equally to the degradation of naturally occurring organic matter, and of contaminating hydrocarbons. For instance, across 71 soils from various ecosystems, *Acidobacteria* were negatively correlated with carbon mineralization, while *Bacteroidetes* and *Betaproteobacteria* were positively correlated with this process [88]. *Betaproteobacteria* were also positively correlated with the degradation of diesel across Arctic soils, but were not always promoted following its addition [41].

The potential for decomposition of natural carbon stores is especially large in the Arctic, where nearly half of the world's below ground carbon may be contained [89]. Similar genes involved in transforming complex organic matter were identified from various microbial groups in metagenomes and metatranscriptomes from high Arctic peat [90], suggesting that competition for substrates is likely to occur. Although certain microorganisms may specialize in the use of different carbon compounds, competition may still occur for other limiting nutrients and space, resulting in the reduced growth of at least one population. This has been shown with bacterial and fungal populations from lower latitude soils [91]. Without explicit microbial competition studies for polar soils, it is often difficult to separate environmental constraints on activity from effects of community structure and activity. The uptake of added carbon in soils from three representative tundra environments was essentially equal, while subsequent release of methane and CO_2 varied substantially [92]. It is unclear whether other metabolic routes would be available, as these soils varied widely in water content and likely in oxygen availability.

Competition should similarly be expected to influence the degradation of certain contaminants in polar soils, especially compounds that resemble soil organic matter such as petroleum hydrocarbons. Many microorganisms in polar soils have evolved metabolic pathways to exploit petroleum hydrocarbons as sources of carbon and energy [93,94]. Despite a widespread ability to catabolize these molecules, petroleum-metabolizing bacteria differ in both rate and extent of hydrocarbon degradation [95–97]. This suggests that the most efficient hydrocarbon degraders may not be promoted naturally, which does appear to be the case, as soil parameters such as organic matter determine which bacteria dominated diesel-contaminated Arctic soils [41]. Nutrient amendments that are applied generally to soil to stimulate the activity of hydrocarbon degraders may actually promote suboptimal hydrocarbon-degrading communities if specific taxa make better use of these nutrients. Following the addition of monoammonium phosphate to contaminated high Arctic soils, the *Alphaproteobacteria* more efficiently assimilated added nitrogen than did the other major active groups [32], although other groups such as the *Gammaproteobacteria* have been associated with efficient remediation at this site [37,98].

The reduction or modification of microbial competition may also represent a biotechnological opportunity for the treatment of contaminated polar soils. In macroecological systems, the loss of key predators has led to reduced constraints on herbivore populations, which have subsequently depleted available vegetation [99]. In the context of bioremediation, this is a desirable outcome, and in fact the fumigation of soils contaminated with 2,4-dichlorophenoxyacetic acid to reduce native microbial populations led to much higher contaminant reduction by introduced strains [22]. Similarly, the inhibition of certain portions of a microbial community in a diesel-contaminated high Arctic soil led to increased degradation [12], suggesting that natural competitive networks may limit bioremediation efficiency.

3.3. Plant Productivity

It is not only competition within the microbial community that can affect ecosystem productivity. Plants are the main source of primary biosynthetic material in terrestrial ecosystems, and are a major global carbon pool [100]. Many microorganisms form symbiotic relationships with plants, and it is thought that over 85% of plant nitrogen may be supplied by fungi in Arctic tundra [101]. Nevertheless, antagonistic relationships between microorganisms and plants are known to occur in polar soils. Of these, the best studied involve competition for limiting nutrients such as nitrogen and phosphorus. Reduced nutrient uptake by plants inevitably limits potential biomass, and may seriously impact primary productivity in polar soils.

In Arctic terrestrial environments, the microbial biomass holds a disproportionate amount of the available nutrients when compared with lower latitude ecosystems [102]. In high latitude soils, plant biomass is also likely to be limited by extreme environmental factors such as freezing temperatures, and long-term snow cover. Nutrient additions can promote plant growth several-fold and this effect is more pronounced in the absence of soil microorganisms [102]. When nutrients do become available, they are often quickly assimilated by microorganisms. Irrespective of the form of nitrogen added, 40–50 times more nitrogen ended up in microbial biomass than in plants, in highly acidic (pH 4.6) and mildly acidic (pH 6.4) Arctic tundra soil [103]. This indicates that effective competition for nitrogen may be widespread across microbial taxa, as distinct microbial communities should be expected to exist in these soils [26]. This competitive relationship has also been shown explicitly, as soil sterilization led to increased nitrogen and phosphorus uptake by an Arctic graminoid (*Festuca vivipara*), and increased plant growth, while glucose addition stimulated microbial nutrient uptake, leading to lower plant nutrient acquisition [104].

Plants appear to be more competitive in nutrient acquisition over time, as the ultimate distribution of nutrient pools depends on temporal trends such as the turnover of microbial biomass and plant roots [105]. Although microbial biomass declined in the absence of plants in an Arctic salt marsh, added nitrogen was retained for longer than it was when plants were present [106], suggesting that plants retain nitrogen following microbial turnover. Clemmensen *et al.* [107] also demonstrated that in Arctic soils dominated by *Betula nana*, microbial communities were initially far more efficient at acquiring added nitrogen, but that plants obtained a larger share after less than a month of incubation. Changing seasonal conditions are also likely to affect competitive relationships. In the Arctic, microorganisms appear to accumulate nutrients over the winter [106], but may lose nutrients to plant roots each spring [106,108]. Although plant competition for nutrients is generally considered only for inorganic nutrient sources, plants in polar soils have also been shown to use amino acids and peptides [109–111]. In fact, the Antarctic hair grass (*Deschampsia antarctica*) competes successfully with microbial populations for amino acids and peptides, and assimilates peptides much more efficiently than other nitrogen sources [109].

3.4. Nutrient Cycling

Although plant-microbe competition for nutrients has been better studied in polar regions, intermicrobial competition may also play an important role in determining the size and composition of nutrient pools. The ability to efficiently acquire limiting nutrients is essential to microbial growth and activity. In addition, certain nutrients will be oxidized or reduced as by-products or end products of metabolic pathways. The combination of nutrient-acquiring and -transforming activities by polar soil microorganisms will determine the size of nutrient pools that are maintained in soils, and that are available to higher trophic levels.

Nitrogen is especially likely to be the subject of widespread competition, as nitrogen availability often limits biomass growth in terrestrial environments [112,113]. The relative abundance of different nitrogen forms will determine which microorganisms will be involved in this competition, as many microorganisms are known to preferentially assimilate NH_4^+ over NO_3^-, while some are entirely unable to assimilate NO_3^- [114–116]. In Arctic tundra soils, ectomycorrhizal fungi were shown to select nitrogen sources other than NO_3^- while effectively sequestering other nitrogen in their mycelia, which may have affected nitrogen selection and use by co-occurring microbes [107]. Similarly, L-alanine and its peptides were equally mineralized by three distinct Antarctic soil microbial communities, while D-alanine was mineralized to different extents and at different rates by each [117], showing that the form of available nitrogen will likely impact which microorganisms are able to remain active in specific soils.

Certain microbial groups are known to be important in nitrogen uptake, and may limit the activity of competitors. Inhibition of fungi in an Arctic tundra soil led to large increases in available NO_3^- [33], while *Alphaproteobacteria* assimilated between 2 and 10 times more added nitrogen than other major active groups in a hydrocarbon-contaminated Arctic soil [32]. While it has been previously suggested that the addition of nitrogen will favour the growth of specific copiotrophic organisms [46,118], it appears that at least in hydrocarbon-contaminated Arctic soils, nitrogen-based fertilizer enhances the competitive advantage of different taxa, depending on soil properties [41]. Competition for nitrogen as both an energy and biosynthetic source may also limit the activity of nitrogen-limited microorganisms. It has been suggested that transformations such as denitrification, which has been observed in hydrocarbon-contaminated Antarctic soils, may limit the nitrogen available to hydrocarbon-degrading taxa [119] as has been observed at lower latitudes [120].

Polar soil microorganisms are also likely to compete for other macronutrients such as phosphorus and sulfur, as well as a variety of micronutrients. A better understanding of the active and potential metabolic routes in polar soils is required in order to speculate on what role such competition might play in affecting important biogeochemical processes.

4. The Effects of Environmental Change on Competition

Human activities are causing unprecedented change in the previously isolated polar regions, and a large part of this change is due to rapid climate warming. Much research has been devoted to the effects of warming on polar terrestrial ecosystems, but potential shifts in biogeochemistry are difficult to predict since so many factors are likely to be affected. Although microorganisms are projected to better adapt to this change than other organisms due to their wide physiological range and rapid turnover rate [121], the resulting communities may be substantially changed. The physiology of individual microorganisms will be directly affected by warming, while changes in plant communities and/or soil parameters will likely favor different microbial communities. How these factors will combine to alter competitive relationships between microorganisms in polar soils is unknown, but this will affect the productivity of functions ranging from methane emission to nutrient cycling (Figure 1).

Although some functional redundancy probably exists within natural soil microbial communities, previous disturbances that have altered community composition have frequently shifted microbially-mediated ecosystem processes [122].

Since many of the microorganisms inhabiting seasonally-thawed polar soils are psychrotolerant rather than psychrophilic, increasing temperature should be expected to increase the potential metabolism of many microbial taxa. How this translates into community productivity will depend greatly on competitive interactions. Increased temperature was shown to substantially increase antagonism between many bacterial isolates from Arctic soils, possibly due to increased production of antimicrobials, or shifts in relative growth rates [29]. Long-term warming manipulations in the Arctic led to changes in both bacterial and fungal populations, with increased species evenness among fungi, and decreased evenness among bacteria [5]. Warming manipulations in Antarctic soils led to a more generalist microbial population, as a large decrease in functional richness did not coincide with a decrease in taxonomic richness, suggesting that more species may have been competing to process the same substrates [6]. Such changes may be short-lived, as communities should eventually adapt to new ecological equilibria. Specialization can also rapidly evolve in mixed communities [123], and this divergence may lessen competitive constraints, leading to more rapid resource use.

A major indirect effect of climate change on microbial communities will arise from changes in plant communities. In the Arctic, the abundance of mycorrhizal plants declines towards the north [124], but climate warming will increase the northward expansion of these plants, increasing bacterial-fungal interactions. Following glacier retreat in the high Arctic, the diversity of ectomycorrhizal fungi increased with increasing plant succession [74]. Warming has also resulted in increased plant success in competing for nutrients with microorganisms in both the Arctic [125] and Antarctic [109]. Interestingly, microorganisms may also better compete with each other by shaping these changing plant communities, and promoting species that favor their growth. Belowground transfer of carbon between *Betula nana* plants in the Arctic was increasingly mediated by fungi with increasing temperature, and helped to establish the dominance of this species [76]. How such changes will affect microbial community productivity in the long-term remains to be seen. Following a 16-year warming experiment in the high Arctic, many changes were observed in the plant communities, while few changes were noted in microbial community structure, or the release of greenhouse gases [126]. This points to a need to understand whether changes in microbial interactions and function following environmental change are transient, or a component of a new community dynamic.

5. Studying Competition in Natural Communities

To date, most studies that have examined microbial competition have involved combining a few target species in culture. A key challenge in determining competition in natural communities is that it is difficult to isolate the interactions of specific taxonomic groups. Broad-scale analyses of microbial co-occurrence patterns can establish which taxa are likely to interact frequently, as well as those that are negatively correlated [127]. Many extensive microbial community datasets are now available from polar soils (e.g., [6,26,41]), and meta-analyses may enable prediction of which taxa interact antagonistically. In addition, future studies combining metatranscriptomics and metagenomics will be able to determine whether gene:transcript ratios are equivalent across taxa capable of performing the same function. The advent of high-throughput SIP-proteomic technologies will allow comparisons between transcript and protein abundance [128]. Such studies will help in determining whether the most productive taxa are dominant in particular soils.

Finally, direct manipulation of the abundance of specific taxa within soil may lead to a better understanding of the interactions between key microbial groups. Chloroform fumigation and antibiotic addition can alter microbial diversity and composition in soils, and the resulting effect on activity can then be measured [12,22,129]. In addition, the suppression of specific activities may help in quantifying the contributions of metabolic pathways to bioremediation. This has been used previously to determine the effects of nitrification [130–132], nitrogen assimilation [133], denitrification [132,134,135], and sulfate reduction [136] on the nutrient dynamics in soils and sediments. In the future, more specific gene inactivation may also be possible, as RNA external guided sequences have been used in culture to inhibit the expression of targeted mRNA sequences [137], and may eventually be adapted for use in natural environments. Such innovative approaches will be necessary to enhance our understanding of competition in natural microbial communities to include the complex network of interactions that undoubtedly occur.

6. Conclusions

Although the importance of microbial interspecies interactions is well recognized, such dynamics have been difficult to assess on a wide scale in natural communities. Certain processes depend upon synergistic interactions, but the niches occupied by particular taxa are often reduced by the growth and activities of co-occurring species that require the same resources and/or space. A characterization of microbial competition in polar soils is desirable for several reasons:

1. Polar soils contain large stores of organic material and nutrients. The extent to which microbial competition can limit rates of decomposition and nutrient cycling will affect climate change predictions and future management plans.
2. By purposefully altering the soil environment, microbial competition may be either increased or reduced, possibly opening biotechnological opportunities such as enhanced bioremediation.
3. Microbial composition and activity also affect the activity and growth of other organisms such as plants, and vice versa. Competition between these groups is also likely to affect the composition and functioning of each.

Future studies that correlate genomic and functional information will help to identify microbial groups that are key to high productivity across polar soils, while manipulation of these communities may reveal some of the constraints that are placed on function due to the coexistence of antagonistic species.

192

Acknowledgements

This work was supported by an NSERC postgraduate scholarship to TH Bell.

References

1. Connell, J.H. The influence of interspecific competition and other factors on the distribution of the barnacle *Chthamalus stellatus*. *Ecology* **1961**, *42*, 710–723.
2. Huisman, J.; Weissing, F.J. Biodiversity of plankton by species oscillations and chaos. *Nature* **1999**, *402*, 407–410.
3. Van Nes, E.H.; Scheffer, M. Large species shifts triggered by small forces. *Am. Nat.* **2004**, *164*, 255–266.
4. Beninca, E.; Huisman, J.; Heerkloss, R.; Johnk, K.D.; Branco, P.; van Nes, E.H.; Scheffer, M.; Ellner, S.P. Chaos in a long-term experiment with a plankton community. *Nature* **2008**, *451*, 822–825.
5. Deslippe, J.R.; Hartmann, M.; Simard, S.W.; Mohn, W.W. Long-term warming alters the composition of arctic soil microbial communities. *FEMS Microbiol. Ecol.* **2012**, *82*, 303–315.
6. Yergeau, E.; Bokhorst, S.; Kang, S.; Zhou, J.Z.; Greer, C.W.; Aerts, R.; Kowalchuk, G.A. Shifts in soil microorganisms in response to warming are consistent across a range of antarctic environments. *ISME J.* **2012**, *6*, 692–702.
7. Barrett, L.G.; Bell, T.; Dwyer, G.; Bergelson, J. Cheating, trade-offs and the evolution of aggressiveness in a natural pathogen population. *Ecol. Lett.* **2011**, *14*, 1149–1157.
8. Kreth, J.; Merritt, J.; Shi, W.Y.; Qi, F.X. Competition and coexistence between *Streptococcus mutans* and *Streptococcus sanguinis* in the dental biofilm. *J. Bacteriol.* **2005**, *187*, 7193–7203.
9. Lopez-Garcia, S.L.; Vazquez, T.E.E.; Favelukes, G.; Lodeiro, A.R. Rhizobial position as a main determinant in the problem of competition for nodulation in soybean. *Environ. Microbiol.* **2002**, *4*, 216–224.
10. van Elsas, J.D.; Chiurazzi, M.; Mallon, C.A.; Elhottova, D.; Kristufek, V.; Salles, J.F. Microbial diversity determines the invasion of soil by a bacterial pathogen. *Proc. Natl. Acad. Sci. USA* **2012**, *109*, 1159–1164.
11. O'Brien, A.; Sharp, R.; Russell, N.J.; Roller, S. Antarctic bacteria inhibit growth of food-borne microorganisms at low temperatures. *FEMS Microbiol. Ecol.* **2004**, *48*, 157–167.
12. Bell, T.H.; Yergeau, E.; Juck, D.; Whyte, L.G.; Greer, C.W. Alteration of microbial community structure affects diesel degradation in an arctic soil. *FEMS Microbiol. Ecol.* **2013**, in press.
13. Bullock, J.M.; Pywell, R.F.; Burke, M.J.W.; Walker, K.J. Restoration of biodiversity enhances agricultural production. *Ecol. Lett.* **2001**, *4*, 185–189.
14. Doherty, J.M.; Callaway, J.C.; Zedler, J.B. Diversity-function relationships changed in a long-term restoration experiment. *Ecol. Appl.* **2011**, *21*, 2143–2155.

15. Fargione, J.; Tilman, D.; Dybzinski, R.; Lambers, J.H.; Clark, C.; Harpole, W.S.; Knops, J.M.H.; Reich, P.B.; Loreau, M. From selection to complementarity: Shifts in the causes of biodiversity-productivity relationships in a long-term biodiversity experiment. *Proc. Roy. Soc. B* **2007**, *274*, 871–876.

16. Foster, K.R.; Bell, T. Competition, not cooperation, dominates interactions among culturable microbial species. *Curr. Biol.* **2012**, *22*, 1845–1850.

17. Peter, H.; Beier, S.; Bertilsson, S.; Lindström, E.S.; Langenheder, S.; Tranvik, L.J. Function-specific response to depletion of microbial diversity. *ISME J.* **2011**, *5*, 351–361.

18. Salles, J.F.; Poly, F.; Schmid, B.; Le Roux, X. Community niche predicts the functioning of denitrifying bacterial assemblages. *Ecology* **2009**, *90*, 3324–3332.

19. Strickland, M.S.; Lauber, C.; Fierer, N.; Bradford, M.A. Testing the functional significance of microbial community composition. *Ecology* **2009**, *90*, 441–451.

20. Degens, B.P. Decreases in microbial functional diversity do not result in corresponding changes in decomposition under different moisture conditions. *Soil Biol. Biochem.* **1998**, *30*, 1989–2000.

21. Griffiths, B.S.; Ritz, K.; Bardgett, R.D.; Cook, R.; Christensen, S.; Ekelund, F.; Sørensen, S.J.; Bååth, E.; Bloem, J.; de Ruiter, P.C.; *et al.* Ecosystem response of pasture soil communities to fumigation-induced microbial diversity reductions: An examination of the biodiversity-ecosystem function relationship. *Oikos* **2000**, *90*, 279–294.

22. Fournier, G.; Fournier, J.C. Effect of microbial competition on the survival and activity of 2,4-d-degrading *Alcaligenes xylosoxidans* subsp. *Denitrificans* added to soil. *Lett. Appl. Microbiol.* **1993**, *16*, 178–181.

23. Hibbing, M.E.; Fuqua, C.; Parsek, M.R.; Peterson, S.B. Bacterial competition: Surviving and thriving in the microbial jungle. *Nat. Rev. Microbiol.* **2010**, *8*, 15–25.

24. Little, A.E.F.; Robinson, C.J.; Peterson, S.B.; Raffa, K.E.; Handelsman, J. Rules of engagement: Interspecies interactions that regulate microbial communities. *Annu. Rev. Microbiol.* **2008**, *62*, 375–401.

25. Roesch, L.F.; Fulthorpe, R.R.; Riva, A.; Casella, G.; Hadwin, A.K.M.; Kent, A.D.; Daroub, S.H.; Camargo, F.A.O.; Farmerie, W.G.; Triplett, E.W. Pyrosequencing enumerates and contrasts soil microbial diversity. *ISME J.* **2007**, *1*, 283–290.

26. Chu, H.Y.; Fierer, N.; Lauber, C.L.; Caporaso, J.G.; Knight, R.; Grogan, P. Soil bacterial diversity in the arctic is not fundamentally different from that found in other biomes. *Environ. Microbiol.* **2010**, *12*, 2998–3006.

27. Neufeld, J.D.; Mohn, W.W. Unexpectedly high bacterial diversity in arctic tundra relative to boreal forest soils, revealed by serial analysis of ribosomal sequence tags. *Appl. Environ. Microb.* **2005**, *71*, 5710–5718.

28. McMahon, S.K.; Wallenstein, M.D.; Schimel, J.P. A cross-seasonal comparison of active and total bacterial community composition in arctic tundra soil using bromodeoxyuridine labeling. *Soil Biol. Biochem.* **2011**, *43*, 287–295.

29. Prasad, S.; Manasa, P.; Buddhi, S.; Singh, S.M.; Shivaji, S. Antagonistic interaction networks among bacteria from a cold soil environment. *FEMS Microbiol. Ecol.* **2011**, *78*, 376–385.

30. Wong, C.M.V.L.; Tam, H.K.; Alias, S.A.; Gonzalez, M.; Gonzalez-Rocha, G.; Dominguez-Yevenes, M. *Pseudomonas* and *pedobacter* isolates from king george island inhibited the growth of foodborne pathogens. *Pol. Polar Res.* **2011**, *32*, 3–14.

31. Kotsyurbenko, O.R.; Glagolev, M.V.; Nozhevnikova, A.N.; Conrad, R. Competition between homoacetogenic bacteria and methanogenic archaea for hydrogen at low temperature. *FEMS Microbiol. Ecol.* **2001**, *38*, 153–159.

32. Bell, T.H.; Yergeau, E.; Martineau, C.; Juck, D.; Whyte, L.G.; Greer, C.W. Identification of nitrogen-incorporating bacteria in petroleum-contaminated arctic soils by using [^{15}n]DNA-based stable isotope probing and pyrosequencing. *Appl. Environ. Microb.* **2011**, *77*, 4163–4171.

33. Siciliano, S.D.; Ma, W.K.; Ferguson, S.; Farrell, R.E. Nitrifier dominance of arctic soil nitrous oxide emissions arises due to fungal competition with denitrifiers for nitrate. *Soil Biol. Biochem.* **2009**, *41*, 1104–1110.

34. Steven, B.; Niederberger, T.D.; Bottos, E.M.; Dyen, M.R.; Whyte, L.G. Development of a sensitive radiorespiration method for detecting microbial activity at subzero temperatures. *J. Microbiol. Methods* **2007**, *71*, 275–280.

35. D'Amico, S.; Collins, T.; Marx, J.C.; Feller, G.; Gerday, C. Psychrophilic microorganisms: Challenges for life. *EMBO Rep.* **2006**, *7*, 385–389.

36. Fierer, N.; Jackson, R.B. The diversity and biogeography of soil bacterial communities. *Proc. Natl. Acad. Sci. USA* **2006**, *103*, 626–631.

37. Yergeau, E.; Schoondermark-Stolk, S.A.; Brodie, E.L.; Dejean, S.; DeSantis, T.Z.; Goncalves, O.; Piceno, Y.M.; Andersen, G.L.; Kowalchuk, G.A. Environmental microarray analyses of antarctic soil microbial communities. *ISME J.* **2009**, *3*, 340–351.

38. Chong, C.W.; Pearce, D.A.; Convey, P.; Tan, I.K.P. The identification of environmental parameters which could influence soil bacterial community composition on the antarctic peninsula: A statistical approach. *Antarct Sci.* **2012**, *24*, 249–258.

39. Mannisto, M.K.; Tiirola, M.; Haggblom, M.M. Bacterial communities in arctic fjelds of finnish lapland are stable but highly ph-dependent. *FEMS Microbiol. Ecol.* **2007**, *59*, 452–465.

40. Ganzert, L.; Lipski, A.; Hubberten, H.W.; Wagner, D. The impact of different soil parameters on the community structure of dominant bacteria from nine different soils located on livingston island, south shetland archipelago, antarctica. *FEMS Microbiol. Ecol.* **2011**, *76*, 476–491.

41. Bell, T.H.; Yergeau, E.; Maynard, C.; Juck, D.; Whyte, L.G.; Greer, C.W. Predictable bacterial composition and hydrocarbon degradation in arctic soils following diesel and nutrient disturbance. *ISME J.* **2013**, doi:10.1038/ismej.2013.1031.

42. Dennis, P.G.; Rushton, S.P.; Newsham, K.K.; Lauducina, V.A.; Ord, V.J.; Daniell, T.J.; O'Donnell, A.G.; Hopkins, D.W. Soil fungal community composition does not alter along a latitudinal gradient through the maritime and sub-antarctic. *Fungal Ecol.* **2012**, *5*, 403–408.

43. Fujimura, K.E.; Egger, K.N. Host plant and environment influence community assembly of high arctic root-associated fungal communities. *Fungal Ecol.* **2012**, *5*, 409–418.

44. Arenz, B.E.; Blanchette, R.A. Distribution and abundance of soil fungi in antarctica at sites on the peninsula, ross sea region and mcmurdo dry valleys. *Soil Biol. Biochem.* **2011**, *43*, 308–315.

45. Powell, S.M.; Bowman, J.P.; Ferguson, S.H.; Snape, I. The importance of soil characteristics to the structure of alkane-degrading bacterial communities on sub-antarctic macquarie island. *Soil Biol. Biochem.* **2010**, *42*, 2012–2021.

46. Ramirez, K.S.; Craine, J.M.; Fierer, N. Consistent effects of nitrogen amendments on soil microbial communities and processes across biomes. *Glob. Change Biol.* **2012**, *18*, 1918–1927.

47. Campbell, B.J.; Polson, S.W.; Hanson, T.E.; Mack, M.C.; Schuur, E.A.G. The effect of nutrient deposition on bacterial communities in arctic tundra soil. *Environ. Microbiol.* **2010**, *12*, 1842–1854.

48. Urcelay, C.; Bret-Harte, M.S.; Diaz, S.; Chapin, F.S. Mycorrhizal colonization mediated by species interactions in arctic tundra. *Oecologia* **2003**, *137*, 399–404.

49. Robinson, C.H.; Saunders, P.W.; Madan, N.J.; Pryce-Miller, E.J.; Pentecost, A. Does nitrogen deposition affect soil microfungal diversity and soil n and p dynamics in a high arctic ecosystem? *Glob. Change Biol.* **2004**, *10*, 1065–1079.

50. Stomeo, F.; Makhalanyane, T.P.; Valverde, A.; Pointing, S.B.; Stevens, M.I.; Cary, C.S.; Tuffin, M.I.; Cowan, D.A. Abiotic factors influence microbial diversity in permanently cold soil horizons of a maritime-associated antarctic dry valley. *FEMS Microbiol. Ecol.* **2012**, *82*, 326–340.

51. Hoj, L.; Rusten, M.; Haugen, L.E.; Olsen, R.A.; Torsvik, V.L. Effects of water regime on archaeal community composition in arctic soils. *Environ. Microbiol.* **2006**, *8*, 984–996.

52. Fell, J.W.; Scorzetti, G.; Connell, L.; Craig, S. Biodiversity of micro-eukaryotes in antarctic dry valley soils with <5% soil moisture. *Soil Biol. Biochem.* **2006**, *38*, 3107–3119.

53. Bridge, P.D.; Newsham, K.K. Soil fungal community composition at mars oasis, a southern maritime antarctic site, assessed by pcr amplification and cloning. *Fungal Ecol.* **2009**, *2*, 66–74.

54. Liebner, S.; Harder, J.; Wagner, D. Bacterial diversity and community structure in polygonal tundra soils from samoylov island, lena delta, siberia. *Int. Microbiol.* **2008**, *11*, 195–202.

55. Aislabie, J.M.; Jordan, S.; Barker, G.M. Relation between soil classification and bacterial diversity in soils of the ross sea region, antarctica. *Geoderma* **2008**, *144*, 9–20.

56. Tosi, S.; Onofri, S.; Brusoni, M.; Zucconi, L.; Vishniac, H. Response of antarctic soil fungal assemblages to experimental warming and reduction of uv radiation. *Polar Biol.* **2005**, *28*, 470–482.

57. Feller, G.; Gerday, C. Psychrophilic enzymes: Hot topics in cold adaptation. *Nat. Rev. Microbiol.* **2003**, *1*, 200–208.

58. Cavicchioli, R. Cold-adapted archaea. *Nat. Rev. Microbiol.* **2006**, *4*, 331–343.

59. Harder, W.; Veldkamp, H. Competition of marine psychrophilic bacteria at low temperatures. *Antonie Van Leeuwenhoek* **1971**, *37*, 51–63.

60. Nedwell, D.B.; Rutter, M. Influence of temperature on growth rate and competition between two psychrotolerant antarctic bacteria: Low temperature diminishes affinity for substrate uptake. *Appl. Environ. Microb.* **1994**, *60*, 1984–1992.

61. Knoblauch, C.; Jorgensen, B.B. Effect of temperature on sulphate reduction, growth rate and growth yield in five psychrophilic sulphate-reducing bacteria from arctic sediments. *Environ. Microbiol.* **1999**, *1*, 457–467.

62. Margesin, R. Effect of temperature on growth parameters of psychrophilic bacteria and yeasts. *Extremophiles* **2009**, *13*, 257–262.

63. Hillebrand, H. On the generality of the latitudinal diversity gradient. *Am. Nat.* **2004**, *163*, 192–211.

64. Hogg, I.D.; Cary, S.C.; Convey, P.; Newsham, K.K.; O'Donnell, A.G.; Adams, B.J.; Aislabie, J.; Frati, F.; Stevens, M.I.; Wall, D.H. Biotic interactions in antarctic terrestrial ecosystems: Are they a factor? *Soil Biol. Biochem.* **2006**, *38*, 3035–3040.

65. Teixeira, L.C.R.S.; Peixoto, R.S.; Cury, J.C.; Sul, W.J.; Pellizari, V.H.; Tiedje, J.; Rosado, A.S. Bacterial diversity in rhizosphere soil from antarctic vascular plants of admiralty bay, maritime antarctica. *ISME J.* **2010**, *4*, 989–1001.

66. Allen, B.; Willner, D.; Oechel, W.C.; Lipson, D. Top-down control of microbial activity and biomass in an arctic soil ecosystem. *Environ. Microbiol.* **2010**, *12*, 642–648.

67. Newsham, K.K.; Rolf, J.; Pearce, D.A.; Strachan, R.J. Differing preferences of antarctic soil nematodes for microbial prey. *Eur. J. Soil Biol.* **2004**, *40*, 1–8.

68. Williamson, L.L.; Borlee, B.R.; Schloss, P.D.; Guan, C.H.; Allen, H.K.; Handelsman, J. Intracellular screen to identify metagenomic clones that induce or inhibit a quorum-sensing biosensor. *Appl. Environ. Microb.* **2005**, *71*, 6335–6344.

69. Deming, J.W. Psychrophiles and polar regions. *Curr. Opin. Microbiol.* **2002**, *5*, 301–309.

70. Lifshitz, R.; Kloepper, J.W.; Scher, F.M.; Tipping, E.M.; Laliberte, M. Nitrogen-fixing pseudomonads isolated from roots of plants grown in the canadian high arctic. *Appl. Environ. Microb.* **1986**, *51*, 251–255.

71. D'Costa, V.M.; King, C.E.; Kalan, L.; Morar, M.; Sung, W.W.L.; Schwarz, C.; Froese, D.; Zazula, G.; Calmels, F.; Debruyne, R.; *et al.* Antibiotic resistance is ancient. *Nature* **2011**, *477*, 457–461.

72. Ma, Y.F.; Wang, L.; Shao, Z.Z. *Pseudomonas*, the dominant polycyclic aromatic hydrocarbon-degrading bacteria isolated from antarctic soils and the role of large plasmids in horizontal gene transfer. *Environ. Microbiol.* **2006**, *8*, 455–465.

73. Martinez-Rosales, C.; Fullana, N.; Musto, H.; Castro-Sowinski, S. Antarctic DNA moving forward: Genomic plasticity and biotechnological potential. *FEMS Microbiol. Lett.* **2012**, *331*, 1–9.

74. Fujiyoshi, M.; Yoshitake, S.; Watanabe, K.; Murota, K.; Tsuchiya, Y.; Uchida, M.; Nakatsubo, T. Successional changes in ectomycorrhizal fungi associated with the polar willow *salix polaris* in a deglaciated area in the high arctic, svalbard. *Polar Biol.* **2011**, *34*, 667–673.

75. Sundqvist, M.K.; Giesler, R.; Graae, B.J.; Wallander, H.; Fogelberg, E.; Wardle, D.A. Interactive effects of vegetation type and elevation on aboveground and belowground properties in a subarctic tundra. *Oikos* **2011**, *120*, 128–142.

76. Deslippe, J.R.; Simard, S.W. Below-ground carbon transfer among *betula nana* may increase with warming in arctic tundra. *New Phytol.* **2011**, *192*, 689–698.

77. Chu, H.Y.; Neufeld, J.D.; Walker, V.K.; Grogan, P. The influence of vegetation type on the dominant soil bacteria, archaea, and fungi in a low arctic tundra landscape. *Soil Sci. Soc. Am. J.* **2011**, *75*, 1756–1765.

78. Reed, H.E.; Martiny, J.B.H. Testing the functional significance of microbial composition in natural communities. *FEMS Microbiol. Ecol.* **2007**, *62*, 161–170.

79. Singh, B.K.; Bardgett, R.D.; Smith, P.; Reay, D.S. Microorganisms and climate change: Terrestrial feedbacks and mitigation options. *Nat. Rev. Microbiol.* **2010**, *8*, 779–790.

80. Wagner, D.; Liebner, S. Global warming and carbon dynamics in permafrost soils: Methane production and oxidation. In *Permafrost Soils*; Margesin, R., Ed.; Springer Berlin Heidelberg: Berlin, Heidelberg, Germany, 2009; pp. 219–236.

81. Wagner, D.; Gattinger, A.; Embacher, A.; Pfeiffer, E.M.; Schloter, M.; Lipski, A. Methanogenic activity and biomass in holocene permafrost deposits of the lena delta, siberian arctic and its implication for the global methane budge. *Glob. Change Biol.* **2007**, *13*, 1089–1099.

82. Ho, A.; Luke, C.; Frenzel, P. Recovery of methanotrophs from disturbance: Population dynamics, evenness and functioning. *ISME J.* **2011**, *5*, 750–758.

83. Martineau, C.; Whyte, L.G.; Greer, C.W. Stable isotope probing analysis of the diversity and activity of methanotrophic bacteria in soils from the canadian high arctic. *Appl. Environ. Microb.* **2010**, *76*, 5773–5784.

84. Achtnich, C.; Bak, F.; Conrad, R. Competition for electron donors among nitrate reducers, ferric iron reducers, sulfate reducers, and methanogens in anoxic paddy soil. *Biol. Fertil. Soils* **1995**, *19*, 65–72.

85. Stibal, M.; Wadham, J.L.; Lis, G.P.; Telling, J.; Pancost, R.D.; Dubnick, A.; Sharp, M.J.; Lawson, E.C.; Butler, C.E.H.; Hasan, F.; *et al.* Methanogenic potential of arctic and antarctic subglacial environments with contrasting organic carbon sources. *Glob. Change Biol.* **2012**, *18*, 3332–3345.

86. IPCC. *Climate Change 2007: The Physical Science Basis*; Cambridge University Press: Cambridge, UK, 2007.

87. Wrage, N.; Velthof, G.L.; van Beusichem, M.L.; Oenema, O. Role of nitrifier denitrification in the production of nitrous oxide. *Soil Biol. Biochem.* **2001**, *33*, 1723–1732.

88. Fierer, N.; Bradford, M.A.; Jackson, R.B. Toward an ecological classification of soil bacteria. *Ecology* **2007**, *88*, 1354–1364.

89. Tarnocai, C.; Canadell, J.G.; Schuur, E.A.G.; Kuhry, P.; Mazhitova, G.; Zimov, S. Soil organic carbon pools in the northern circumpolar permafrost region. *Glob. Biogeochem. Cycles* **2009**, *23*, doi: 10.1029/2008GB003327.

90. Tveit, A.; Schwacke, R.; Svenning, M.M.; Urich, T. Organic carbon transformations in high-arctic peat soil: Key functions and microorganisms. *ISME J.* **2013**, *7*, 299–311.

91. Meidute, S.; Demoling, F.; Bååth, E. Antagonistic and synergistic effects of fungal and bacterial growth in soil after adding different carbon and nitrogen sources. *Soil Biol. Biochem.* **2008**, *40*, 2334–2343.

92. Zak, D.R.; Kling, G.W. Microbial community composition and function across an arctic tundra landscape. *Ecology* **2006**, *87*, 1659–1670.

93. Greer, C.W.; Whyte, L.G.; Niederberger, T.D. Microbial communities in hydrocarbon-contaminated temperate, tropical, alpine, and polar soils. In *Handbook of Hydrocarbon and Lipid Microbiology*; Timmis, K.N., Ed.; Springer Berlin Heidelberg: Berlin, Germany; Heidelberg, Germany, 2010; pp. 2313–2328.

94. Aislabie, J.; Saul, D.J.; Foght, J.M. Bioremediation of hydrocarbon-contaminated polar soils. *Extremophiles* **2006**, *10*, 171–179.

95. Ciric, L.; Philp, J.C.; Whiteley, A.S. Hydrocarbon utilization within a diesel-degrading bacterial consortium. *FEMS Microbiol. Lett.* **2010**, *303*, 116–122.

96. Sorkhoh, N.A.; Ghannoum, M.A.; Ibrahim, A.S.; Stretton, R.J.; Radwan, S.S. Crude-oil and hydrocarbon-degrading strains of *rhodococcus-rhodochrous* isolated from soil and marine environments in kuwait. *Environ. Pollut.* **1990**, *65*, 1–17.

97. Whyte, L.G.; Hawari, J.; Zhou, E.; Bourbonnière, L.; Inniss, W.E.; Greer, C.W. Biodegradation of variable-chain-length alkanes at low temperatures by a psychrotrophic *rhodococcus* sp. *Appl. Environ. Microb.* **1998**, *64*, 2578–2584.

98. Yergeau, E.; Sanschagrin, S.; Beaumier, D.; Greer, C.W. Metagenomic analysis of the bioremediation of diesel-contaminated canadian high arctic soils. *PLoS One* **2012**, *7*, e30058.

99. Beschta, R.L.; Ripple, W.J. Large predators and trophic cascades in terrestrial ecosystems of the western united states. *Biol. Conserv.* **2009**, *142*, 2401–2414.

100. Falkowski, P.; Scholes, R.J.; Boyle, E.; Canadell, J.; Canfield, D.; Elser, J.; Gruber, N.; Hibbard, K.; Hogberg, P.; Linder, S.; *et al.* The global carbon cycle: A test of our knowledge of earth as a system. *Science* **2000**, *290*, 291–296.

101. Hobbie, J.E.; Hobbie, E.A. [15]n in symbiotic fungi and plants estimates nitrogen and carbon flux rates in arctic tundra. *Ecology* **2006**, *87*, 816–822.

102. Jonasson, S.; Michelsen, A.; Schmidt, I.K. Coupling of nutrient cycling and carbon dynamics in the arctic, integration of soil microbial and plant processes. *Appl. Soil Ecol.* **1999**, *11*, 135–146.

103. Nordin, A.; Schmidt, I.K.; Shaver, G.R. Nitrogen uptake by arctic soil microbes and plants in relation to soil nitrogen supply. *Ecology* **2004**, *85*, 955–962.

104. Schmidt, I.K.; Michelsen, A.; Jonasson, S. Effects of labile soil carbon on nutrient partitioning between an arctic graminoid and microbes. *Oecologia* **1997**, *112*, 557–565.

105. Hodge, A.; Robinson, D.; Fitter, A. Are microorganisms more effective than plants at competing for nitrogen? *Trends Plant Sci.* **2000**, *5*, 304–308.

106. Buckeridge, K.M.; Jefferies, R.L. Vegetation loss alters soil nitrogen dynamics in an arctic salt marsh. *J. Ecol.* **2007**, *95*, 283–293.

107. Clemmensen, K.E.; Sorensen, P.L.; Michelsen, A.; Jonasson, S.; Strom, L. Site-dependent n uptake from n-form mixtures by arctic plants, soil microbes and ectomycorrhizal fungi. *Oecologia* **2008**, *155*, 771–783.

108. Edwards, K.A.; McCulloch, J.; Kershaw, G.P.; Jefferies, R.L. Soil microbial and nutrient dynamics in a wet arctic sedge meadow in late winter and early spring. *Soil Biol. Biochem.* **2006**, *38*, 2843–2851.

109. Hill, P.W.; Farrar, J.; Roberts, P.; Farrell, M.; Grant, H.; Newsham, K.K.; Hopkins, D.W.; Bardgett, R.D.; Jones, D.L. Vascular plant success in a warming antarctic may be due to efficient nitrogen acquisition. *Nat. Clim. Change* **2011**, *1*, 50–53.

110. Henry, H.A.L.; Jefferies, R.L. Plant amino acid uptake, soluble n turnover and microbial n capture in soils of a grazed arctic salt marsh. *J. Ecol.* **2003**, *91*, 627–636.

111. Chapin, F.S.; Moilanen, L.; Kielland, K. Preferential use of organic nitrogen for growth by a nonmycorrhizal arctic sedge. *Nature* **1993**, *361*, 150–153.

112. Vitousek, P.M.; Howarth, R.W. Nitrogen limitation on land and in the sea: How can it occur. *Biogeochemistry* **1991**, *13*, 87–115.

113. Vitousek, P.M.; Aber, J.D.; Howarth, R.W.; Likens, G.E.; Matson, P.A.; Schindler, D.W.; Schlesinger, W.H.; Tilman, D. Human alteration of the global nitrogen cycle: Sources and consequences. *Ecol. Appl.* **1997**, *7*, 737–750.

114. Imsenecki, A.A.; Popova, L.S.; Kirillova, N.F. Effect of nitrogen source on growth of *arthrobacter simplex* and its biosynthesis of cholinesterase. *Mikrobiologiâ* **1976**, *45*, 614–619.

115. Rice, C.W.; Tiedje, J.M. Regulation of nitrate assimilation by ammonium in soils and in isolated soil microorganisms. *Soil Biol. Biochem.* **1989**, *21*, 597–602.

116. Recous, S.; Mary, B.; Faurie, G. Microbial immobilization of ammonium and nitrate in cultivated soils. *Soil Biol. Biochem.* **1990**, *22*, 913–922.

117. Hill, P.W.; Farrell, M.; Roberts, P.; Farrar, J.; Grant, H.; Newsham, K.K.; Hopkins, D.W.; Bardgett, R.D.; Jones, D.L. Soil- and enantiomer-specific metabolism of amino acids and their peptides by antarctic soil microorganisms. *Soil Biol. Biochem.* **2011**, *43*, 2410–2416.

118. Fierer, N.; Lauber, C.L.; Ramirez, K.S.; Zaneveld, J.; Bradford, M.A.; Knight, R. Comparative metagenomic, phylogenetic and physiological analyses of soil microbial communities across nitrogen gradients. *ISME J.* **2012**, *6*, 1007–1017.

119. Powell, S.M.; Ferguson, S.H.; Snape, I.; Siciliano, S.D. Fertilization stimulates anaerobic fuel degradation of antarctic soils by denitrifying microorganisms. *Environ. Sci. Technol.* **2006**, *40*, 2011–2017.

120. Roy, R.; Greer, C.W. Hexadecane mineralization and denitrification in two diesel fuel-contaminated soils. *FEMS Microbiol. Ecol.* **2000**, *32*, 17–23.

121. Callaghan, T.V.; Bjorn, L.O.; Chernov, Y.; Chapin, T.; Christensen, T.R.; Huntley, B.; Ims, R.A.; Johansson, M.; Jolly, D.; Jonasson, S.; *et al.* Biodiversity, distributions and adaptations of arctic species in the context of environmental change. *AMBIO* **2004**, *33*, 404–417.

122. Allison, S.D.; Martiny, J.B.H. Resistance, resilience, and redundancy in microbial communities. *Proc. Natl. Acad. Sci. USA* **2008**, *105*, 11512–11519.

123. Lawrence, D.; Fiegna, F.; Behrends, V.; Bundy, J.G.; Phillimore, A.B.; Bell, T.; Barraclough, T.G. Species interactions alter evolutionary responses to a novel environment. *PLoS Biol.* **2012**, *10*, e1001330.

124. Olsson, P.A.; Eriksen, B.E.; Dahlberg, A. Colonization by arbuscular mycorrhizal and fine endophytic fungi in herbaceous vegetation in the canadian high arctic. *Can. J. Bot.* **2004**, *82*, 1547–1556.

125. Schmidt, I.K.; Jonasson, S.; Shaver, G.R.; Michelsen, A.; Nordin, A. Mineralization and distribution of nutrients in plants and microbes in four arctic ecosystems: Responses to warming. *Plant Soil* **2002**, *242*, 93–106.

126. Lamb, E.G.; Han, S.; Lanoil, B.D.; Henry, G.H.R.; Brummell, M.E.; Banerjee, S.; Siciliano, S.D. A high arctic soil ecosystem resists long-term environmental manipulations. *Glob. Change Biol.* **2011**, *17*, 3187–3194.

127. Barbéran, A.; Bates, S.T.; Casamayor, E.O.; Fierer, N. Using network analysis to explore co-occurrence patterns in soil microbial communities. *ISME J.* **2012**, *6*, 343–351.

128. Pan, C.L.; Fischer, C.R.; Hyatt, D.; Bowen, B.P.; Hettich, R.L.; Banfield, J.F. Quantitative tracking of isotope flows in proteomes of microbial communities. *Mol. Cell. Proteomics* **2011**, *10*, M110.006049.

129. Griffiths, B.S.; Kuan, H.L.; Ritz, K.; Glover, L.A.; McCaig, A.E.; Fenwick, C. The relationship between microbial community structure and functional stability, tested experimentally in an upland pasture soil. *Microb. Ecol.* **2004**, *47*, 104–113.

130. Deni, J.; Penninckx, M.J. Nitrification and autotrophic nitrifying bacteria in a hydrocarbon-polluted soil. *Appl. Environ. Microb.* **1999**, *65*, 4008–4013.

131. Powell, S.J.; Prosser, J.I. Inhibition of ammonium oxidation by nitrapyrin in soil and liquid culture. *Appl. Environ. Microb.* **1986**, *52*, 782–787.

132. Bremner, J.M.; McCarty, G.W.; Yeomans, J.C.; Chai, H.S. Effects of phosphoroamides on nitrification, denitrification, and mineralization of organic nitrogen in soil. *Commun. Soil Sci. Plant* **1986**, *17*, 369–384.

133. Myrold, D.D.; Posavatz, N.R. Potential importance of bacteria and fungi in nitrate assimilation in soil. *Soil Biol. Biochem.* **2007**, *39*, 1737–1743.

134. Bremner, J.M.; Yeomans, J.C. Effects of nitrification inhibitors on denitrification of nitrate in soil. *Biol. Fertil. Soils* **1986**, *2*, 173–179.

135. Yeomans, J.C.; Bremner, J.M. Effects of urease inhibitors on denitrification in soil. *Commun. Soil Sci. Plant* **1986**, *17*, 63–73.

136. Winfrey, M.R.; Ward, D.M. Substrates for sulfate reduction and methane production in intertidal sediments. *Appl. Environ. Microb.* **1983**, *45*, 193–199.

137. Shen, N.; Ko, J.H.; Xiao, G.P.; Wesolowski, D.; Shan, G.; Geller, B.; Izadjoo, M.; Altman, S. Inactivation of expression of several genes in a variety of bacterial species by egs technology. *Proc. Natl. Acad. Sci. USA* **2009**, *106*, 8163–8168.

Composition, Diversity, and Stability of Microbial Assemblages in Seasonal Lake Ice, Miquelon Lake, Central Alberta

Anna Bramucci, Sukkyun Han, Justin Beckers, Christian Haas and Brian Lanoil

Abstract: The most familiar icy environments, seasonal lake and stream ice, have received little microbiological study. Bacteria and Eukarya dominated the microbial assemblage within the seasonal ice of Miquelon Lake, a shallow saline lake in Alberta, Canada. The bacterial assemblages were moderately diverse and did not vary with either ice depth or time. The closest relatives of the bacterial sequences from the ice included Actinobacteria, Bacteroidetes, Proteobacteria, Verrucomicrobia, and Cyanobacteria. The eukaryotic assemblages were less conserved and had very low diversity. Green algae relatives dominated the eukaryotic gene sequences; however, a copepod and cercozoan were also identified, possibly indicating the presence of complete microbial loop. The persistence of a chlorophyll *a* peak at 25–30 cm below the ice surface, despite ice migration and brine flushing, indicated possible biological activity within the ice. This is the first study of the composition, diversity, and stability of seasonal lake ice.

Reprinted from *Biology*. Cite as: Bramucci, A.; Han, S.; Beckers, J.; Haas, C.; Lanoil, B. Composition, Diversity, and Stability of Microbial Assemblages in Seasonal Lake Ice, Miquelon Lake, Central Alberta. *Biology* **2013**, *2*, 514-532.

1. Introduction

Remote, polar floating ice systems, such as sea ice [1] and perennial lake ice [2–4], harbor dynamic and diverse microbial ecosystems that play important roles in the biogeochemistry, biology, and functioning of the underlying waters and surrounding environments. However, the most familiar icy environments, including the ice that forms on lakes and streams each winter in many temperate environments, have not been studied microbiologically. While there are accounts of the phytoplankton and zooplankton winter dynamics in some northern lakes [5,6], there are none describing bacterial dynamics.

Sea ice harbors algal communities that have high rates of primary productivity, with global totals estimated to be as high as 63 to 70 Tg C year^{-1} [7]. Bacterial production in sea ice is coupled to microalgae growth [8]. Bacteria might provide algae with inorganic nutrients for prolonged sympagic survival [9]. Furthermore, diverse populations of microheterotrophs (e.g., protozoans, dinoflagellates, ciliates, and amoebae) are present and active in sea ice; thus, these systems include a complete microbial loop [10,11].

Saline lakes are found on every continent on earth, with total volumes roughly equaling the volume of terrestrial freshwater lakes [12]. The hundreds of brackish to saline lakes of the Canadian Great Plains are economically, agriculturally, and ecologically important for the region [13,14]. The lakes support numerous algal species that have been extensively documented [15–18], as well as complete microbial food webs [19–21]. However, little is known about how the lake water organisms are influenced by the annual freeze-thaw cycles of the upper waters or the progression from being an open-water to becoming an ice-covered lake.

Here we report the first characterization of microbial diversity in seasonal lake ice. We explored the inter- and intra-seasonal shifts in microbial assemblage composition both within the lake ice and underlying lake water of Miquelon Lake, Alberta CA. This study indicates the likely degree of microbial activity occurring throughout the winter across the frozen Albertan plains, a value which has likely been sorely underappreciated. The central hypotheses of this study are as follows: (1) the seasonal lake ice has communities similar in composition to other floating ice systems (e.g., polar sea ice or perennial lake ice); (2) the distribution of some microbial populations is limited to specific depths in the ice and/or points in the season; (3) seasonally frozen lakes maintain an actively functioning ecosystem and microbial food web throughout the winter.

2. Methods

2.1. Study Site

Miquelon Lake, located in Miquelon Lake Provincial Park Edmonton, Alberta at 53.25° N, 112.90° W [14] is small (surface area: 8.72 km²), shallow (mean depth: 2.7 m), secluded (residence time of water: >100 years), and brackish (6–9 ppt) [14]. Miquelon Lake waters are dominated by microbial life; higher trophic levels are absent. However, algae and cyanobacteria are abundant in this mesotrophic system [14]. The lake is fully mixed until the freeze-in (Jan-April), which leads to weak stratification in the underlying waters during this time.

2.2. Sample Collection and Processing

Ice cores and underlying lake water samples were collected every two weeks throughout the 4-month 2009/2010 winter season. Two 9-cm-diameter ice cores were collected with a Kovacs Mark II corer (Kovacs Enterprises Inc.; Lebanon, NH): one for biological sampling and one for bulk salinity measurements. Ice thickness measurements were taken at the time of sampling by measuring the length of the ice core. Surface water samples were collected in 1 L sterile acid washed Nalgene® bottles (VWR). Two Ice Mass Balance Buoys (IMB) (MetOcean/CRREL, Darmouth, Nova Scotia, and SAMS IMB, Oban, Scotland) obtained *in situ* measurements of air and ice temperatures throughout the winter. Temperature sensors are accurate to 0.1 °C [22].

The bulk salinity core was sectioned on site into 3–4 cm pieces and placed into sterile plastic tubs to melt. Measurements of water temperature (*in situ*), lake water salinity (*in situ*), and bulk ice core melt salinity were acquired using a MultiLine® IDS WTW Cond 330i conductivity meter (Wissenschaftlich-Technische Werkstätten (WTW) Inc./Xylem Inc., Weilheim, Germany), which was calibrated prior to use according to manufacturer's specifications.

The biology core was kept frozen at −20 °C in the dark until processing. This core was aseptically sectioned into ~5 cm sections (varying from 3–8 cm, depending on natural fractures in the ice) using a flame sterilized 15 cm drywall saw. To ensure aseptic sampling procedures one test core was sectioned and melted in sheaths and four aliquots were taken from the outer ice, middle outer, middle inner, and inner most ice respectively. These sheaths were tested on DGGE and were found to be identical based on DGGE analysis, indicating no contamination; thus, no further decontamination efforts were performed. The sections were melted in the dark at 4 °C. Subsamples for cell enumeration (25 mL) and chlorophyll *a* concentrations (60 mL) were removed and prepared as described below. The remaining water (100 to 300 mL) was filtered through 0.22 μm pore size, 47 mm diameter polysulfone filters (Pall Corporation; East Hills, NY, USA). Filters were stored frozen at −80 °C in sterile sealed Seal-a-Meal® bags (Sunbeam® Products Inc.; Neosho, MO, USA).

Surface water samples were kept in the dark at 4 °C and processed within 24 hours of sampling following the same procedure as the melted ice core segments. Approximately 900 mL of water was filtered for subsequent DNA processing. Brine salinity and brine volume were calculated from measured ice temperature and salinity as described previously [23].

2.3. Cell Enumeration

Formalin-fixed (3.7% v/v) subsamples of the ice cores and lake water were filtered on polycarbonate black membrane filters (pore size: 0.22 μm; diameter: 25 mm; Whatman; VWR) and stained with 4',6-diamidino-2-phenylindole (DAPI) (Sigma-Aldrich) for 15 minutes in the dark, and bacterial abundances were determined by fluorescence microscopy as previously described [24]. The analyses were limited to non-filamentous and non-autofluorescing bacterial morphotypes (<5 μm cell length). The volume of water examined varied from 1 to 5 mL of water, depending on the cell concentration. A procedural blank of 5 mL of sterile Nanopure water was examined to ensure sterile technique. The cell counts were run in triplicate and standard deviations ranged from 0.2 to 1×10^6 cells/mL.

2.4. Chlorophyll a Measurements

To measure Chlorophyll *a* (Chl-*a*) concentrations, 60 mL sample aliquots were filtered though a 25 mm precombusted (500 °C for 12 h) Whatman GF/F glass fiber filter in the dark. Filters were stored frozen at −20 °C in the dark until processing. Duplicate samples were taken randomly and used as quality controls throughout the extraction and measurement process: the quality controls totaled 10% of the total number of samples. The precision of the extraction method was assessed using percent relative standard deviation (%RSD) of the duplicates, the %RSD was always below 5% for all quality control samples.

Chl-*a* was extracted by overnight incubation in 95% ethanol in the dark using a standard spectrofluorometric approach [25]. The minimum detection limit of this protocol was ~3.3 μg/L. Concentrations were determined based on a daily standard curve of Chl-*a* from *Anacystis nidulans* (Sigma).

2.5. Nucleic Acid Extraction

DNA was extracted using the FastDNA® extraction kit according to the manufacturer's protocol (MP Biomedicals, Solon, OH, USA). DNA was eluted in 200 μL of warm DNAse-free commercial water (Life Technologies, Grand Island, NY, USA). The solution was buffered to $1\times$ TE concentration (10 mM Tris, pH 8.0, 1 mM NaEDTA) and stored at -20 °C.

2.6. Denaturing Gradient Gel Electrophoresis (DGGE)

Partial bacterial 16S rRNA genes were amplified as previously described [26] with primers 341F and 518R, with a 40-mer GC clamp on the 341F primer (GC-341f; Table 1; [27]). Eukaryotic-specific primers GC-Euk1a and Euk516r [28] were used for amplification of 18S rRNA gene (Table 1, [29,30], and references therein). All PCRs were preformed in triplicate and pooled [31].

Table 1. PCR primers used.

	Primer set	Target	Sequence (5'-3')	Reference
General Primers	341F	Bact 16S rRNA	CCTACGGGAGGCAGCAG	[32]
	518R	Bact 16S rRNA	ATTACCGCGGCTGCTGG	[32]
	Euk1A	Euk 18S rRNA	CTGGTTGATCCTGCCAG	[28]
	Euk516R	Euk 18S rRNA	ACCAGACTTGCCCTCC	[28]
	A21F external	Arc 16S rRNA	TTCCGGTTGATCCYGCCGGA	[33]
	344F internal	Arc 16S rRNA	ACGGGGCGCAGCAGGCGCGA	[30]
	519R internal	Arc 16S rRNA	GGTDTTACCGCGGCKGCTG	[20]
DGGE Primers	341F *	Bact 16S rRNA	CCTACGGGAGGCAGCAG	[32]
	518R	Bact 16S rRNA	ATTACCGCGGCTGCTGG	[32]
	Euk1A	Euk 18S rRNA	CTGGTTGATCCTGCCAG	[28]
	Euk516R *	Euk 18S rRNA	ACCAGACTTGCCCTCC	[28]

* GC clamp (40 bp) added for DGGE-PCR [27]. 5'-CGCCCGCCGCGCCCCGCGCCCGTCCCGCCGCCCCCGCCCC-3'.

DGGE was performed using a D-CODE system (BioRad, Hercules, CA, USA) as previously described [26]. For each sample, 400 ng of DNA were loaded. Bands were visualized after staining the gel for 15–30 minutes in SYBR Green stain (Molecular Probes, Eugene, OR, USA), according to the manufacturer's instructions.

DGGE banding patterns were analyzed with the program GelCompar II (version 4.0; Applied Maths, Austin, TX, USA) using a 2% band position tolerance to determine band locations. The cladograms were generated using an Unweighted Pair Group Method (UPGMA) based on Dice correlation coefficients, which are based on the presence/absence of a band regardless of absolute band intensity, as previously described [26].

2.7. qPCR Analysis

To assess variation in relative abundance of domain-level gene copy number with time, DNA from lake ice and water samples was homogenized, resulting in one bulk sample for ice (all ice core depths and sampling dates) and one bulk sample for the underlying lake water (all sampling dates). The relative abundance of Bacteria, Eukarya, and Archaea small subunit (SSU) rRNA genes in the lake ice and lake water samples was determined using general bacterial, eukaryal, and two sets of archaeal primers (Table 1).

qPCR was performed in triplicate 10 μL reactions containing 5 μL Rotor-Gene SYBR green PCR kit (Qiagen, Inc.), 1 μM concentration of primers, 2 μL template and 1 μL Qiagen RNase-Free water. Reactions were performed in a Rotor-Gene Q (Qiagen, CA, USA) qPCR machine. PCR conditions were 40 cycles at 95 °C for 10 s and 60 °C for 15 s. Gene copy number was calculated relative to an *E. coli* genomic DNA standard for Bacterial DNA and 16S rRNA environmental gene clones for Archaea and Eukarya. Two experimental replicates were performed and data combined for analysis. Primers were tested for cross-reactivity to the standards—no cross reactivity was observed.

2.8. Clone Library Construction

Two lake-ice and two lake-water clone libraries were constructed: one bacterial and one for eukaryal for each. SSU rRNA genes were PCR amplified using the general primers (without GC clamps) for Bacteria and Eukarya (Table 1) as previously described [26]. PCR products were cloned using the TOPO® TA Cloning® Kit (Invitrogen) according to the manufacturer's instructions. Libraries of clones were randomly selected from the Bacteria, lake ice (n = 123), Bacteria, lake water (n = 191), Eukarya, lake ice (n = 123), and Eukarya, lake water (n = 39) samples.

2.9. Restriction Fragment Length Polymorphism (RFLP)

Preliminary grouping of clones was performed by RFLP analysis using *HhaI* and *MspI*, as previously described [34]. Clone insert orientation was determined by unidirectional PCR with only the M13F primer in the master mix. The 5' end sequence of one representative clone for each 10 members of an operational taxonomic unit (OTU) was determined with M13F or M13R. All clones chosen for sequencing were reanalyzed via DGGE prior to sequencing to confirm band position in reference to the original samples [26]. Good's coverage [35] was determined manually.

2.10. Phylogenetic Analysis

Sequences were trimmed, sections of ambiguous base pair matching were removed, and gaps were eliminated using standard methods [36]. Chimeric sequences as determined by DECIPHER [37] were excluded from analysis. Sequences were aligned in Genious 5.5.8 (Biomatters Ltd.; New Zealand) using 25 alignment iterations and the FastAligner function. All alignments were refined manually and shared gaps were eliminated. Maximum likelihood-based phylogenetic analysis was conducted with the PHYML module in Genious [38] using sequences with length ranging from 300–600 bp for the final analysis. Bootstrap support (100 iterations) is shown at the nodes.

2.11. Nucleotide Sequence Accession Numbers

Sequences are deposited in Genbank with the accession numbers KC592375-KC592385.

3. Results

Throughout the 2009–2010 winter season (November-April), air temperatures at Miquelon Lake ranged from a low of −40 °C (9 December 2013) to a high of +10 °C (29 March 2010), with an average winter air temperature of −10.6 °C. During that time period, Miquelon lake ice grew from 0 to 0.4 m in total thickness, had internal ice temperatures ranging from approximately −1 to −4 °C (Figure 1). Seasonal average ice temperatures and underlying water temperatures were very stable. Miquelon Lake water salinity ranged from 10.2 to 12.5 ppt. Brine salinity, which is directly determined by ice temperature, varied with depth, ranging from a high of 60 ppt (hypersaline, at ~1.7× higher than that of standard seawater) to a low of ~10 ppt (brackish, at ~3.5× lower that of standard seawater) (Figure 1). The brine volumes average 10% of the total ice volume throughout this season, with the lowest brine volume occurring at the same depth as the highest brine salinities (Figure 1).

Figure 1. Environmental variables for representative dates during the 2009–2010 winter season at Miquelon Lake, Alberta, Canada. Solid lines are for measured parameters in ice; dashed lines are for measured parameters in the underlying water. Ice temperature and bulk salinity were measured directly. Brine salinity and brine volume were calculated from measured ice temperature and bulk ice salinity according to Cox and Weeks [23]. (**a**) 3 December 2009, (**b**) 29 December 2009, (**c**) 14 January 2010, (**d**) 11 February 2010.

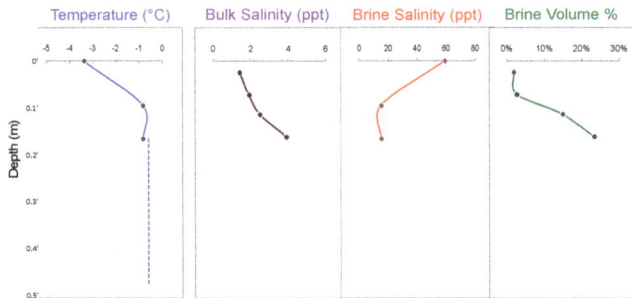

(**a**)

Figure 1. *Cont.*

(b)

(c)

(d)

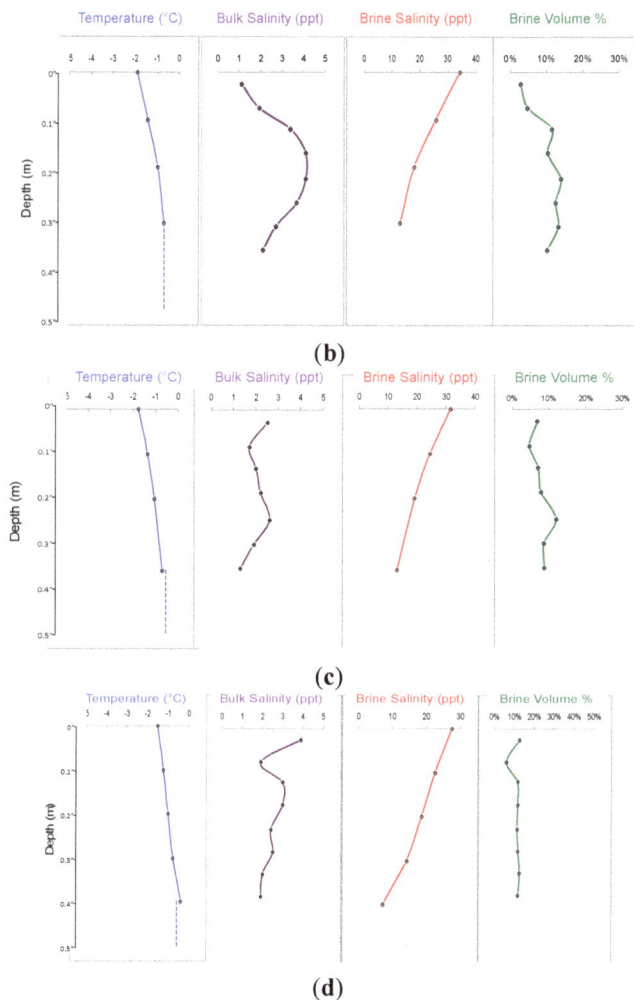

Microscopic enumeration showed 3.7 × 10⁶ (±0.30 × 10⁶) cells mL⁻¹ in the ice and 8.8 × 10⁶ (±10.4 × 10⁶) cells mL⁻¹ in the lake water, ca. 30% of which are auto-fluorescing (and therefore likely photosynthetic) cells. With the exception of the water on 14 January 2010, which was 5× higher than other dates, the variance was <10% for all dates and depths; thus the overall average values are given. qPCR showed SSU rRNA relative gene copy number of 50% Bacteria, 50% Eukarya, and <1% for Archaea (data not shown). Note that due to differences in genome size and SSU rRNA gene copy number, this result does not indicate equal abundance or biomass of Bacteria or Eukarya; only that both Bacteria and Eukarya are abundant while Archaea are exceedingly rare.

There was a sustained Chl-*a* peak at a depth of ~0.25 to 0.3 m throughout the season (Figure 2). The peak, which was 2 to 2.5 times higher at this depth than at any other depth in the core, was sustained for the months where ice was thick enough to reach this depth despite flushing of the brine and ice growth.

Figure 2. Chl-*a* distribution in Miquelon Lake ice on six sampling dates throughout the 2009–2010 winter season: 3 December 2009, 10 December 2009, 29 December 2009, 14 January 2010, 31 January 2010, and 11 February 2010. Depths indicated are the midpoint of ice core segment processed for biological sampling.

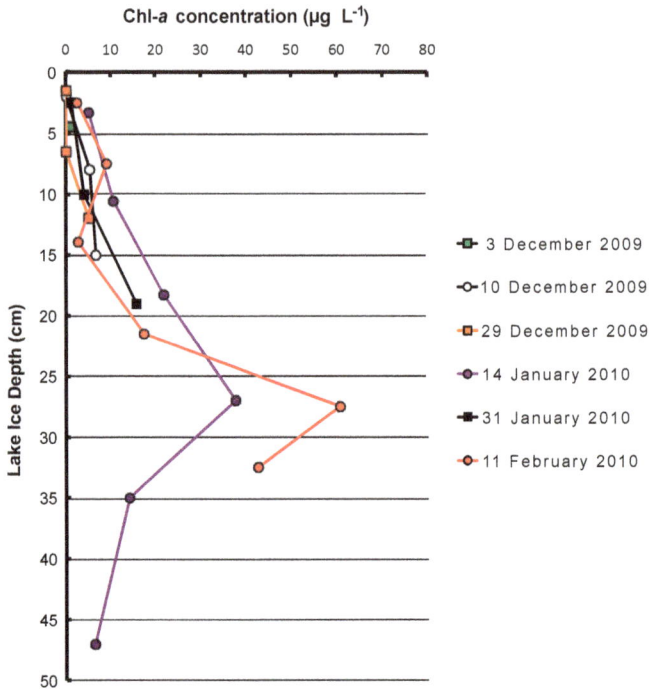

DGGE cluster analysis of Bacteria and Eukarya SSU rRNA genes shows a conserved microbial assemblage within the ice (Figure 3). Bacterial assemblages were nearly invariant with depth, both early in the season with thin ice (29 December 2009) and late in the season with thicker ice (11 February 2010) (Figure 3a). The bacterial assemblage was highly similar between the two dates as well, sharing >94% similarity. Ice bacterial assemblages were ~66% similar to those in the underlying water and both were nearly invariant throughout the season (Figure 3c).

Although still highly similar, eukaryotic assemblages were more variable with depth, showing >70% similarity; however, this decreased similarity may be an artifact of the small numbers of bands (*i.e.*, changes in a single band could have an outsized impact on similarity) (Figure 3b). Both ice and water eukaryotic assemblages were essentially invariant throughout the season, showing >90% similarity between dates (Figure 3d). The ice eukaryotic assemblages were highly similar to those in the underlying water, showing >80% similarity (Figure 3d). Variability with depth on a single date exceeded that between dates for the eukaryotes, indicating there was minimal change in the eukaryotic assemblage over time (Figure 3b,d). Overall, the eukaryotic assemblage had lower band richness than the bacterial assemblage.

Figure 3. Cluster analysis of DGGE profiles from Miquelon Lake. Cladogram generated by Unweighted Pair Group Method with Arithmetic Mean (UPGMA) of Dice correlation coefficients (which reflect only band presence/absence, not band intensity). (**a**) Similarity of depth profile of Bacteria 16S rRNA genes in lake ice from two representative dates: 29 December 2009 and 11 February 2010. (**b**) Similarity of depth profile of Eukarya 18S rRNA genes in lake ice from two representative dates: 29 December 2009 and 11 February 2010. (**c**) Similarity of Bacteria 16S rRNA genes from homogenized ice cores or water samples from six representative dates: 3 December 2009, 10 December 2009, 29 December 2009, 14 January 2010, 31 January 2010, and 11 February 2010. (**d**) Similarity of Eukarya 18S rRNA genes from homogenized ice cores or water samples from six representative dates: 3 December 2009, 10 December 2009, 29 December 2009, 14 January 2010, 31 January 2010, and 11 February 2010.

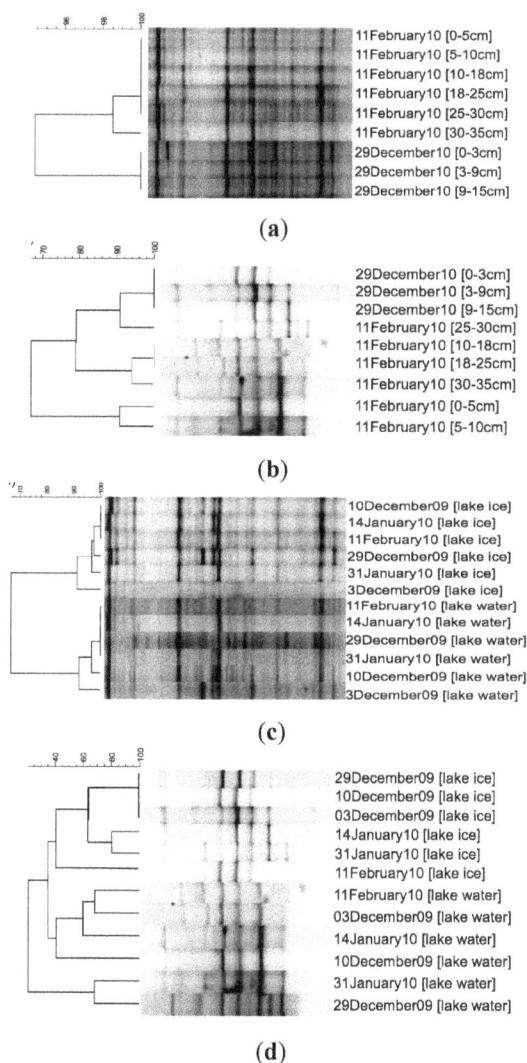

(a)

(b)

(c)

(d)

In order to identify the origin of the dominant bands and elucidate the differences between the ice and water consortia, we constructed a bacterial and eukaryal clone library for the bulked ice (homogenized all ice depths and all sample dates) and bulked water (homogenized all sample dates). The dominant band in the bacterial DGGE (Figure 3a) (initially making up 71% of the bacterial ice clone library and 41% of the bacterial water clone library) was identified as a chloroplast rRNA gene sequence closely related to those from *Nannochloropsis oceanica* and *Chlorella minutissima* (Figure 4). These clones were excluded from further analysis. The Bacteria clone library had a total of 314 clones (with 213 clones remaining after removal of the chloroplast rRNA gene sequences). These clones were grouped into 19 unique operational taxonomic units (OTU) by RFLP analysis. Eleven of these OTU were found in both ice and the underlying waters, with the remainder only present in one of the clone libraries (Figure 5). Miquelon Lake ice and water show a surprising rank abundance curve (Figure 5), with half of the OTUs being represented by roughly equivalent numbers of clones and the remainder of the OTUs comprising a short tail of singletons. A more standard rank-abundance curve, where a few OTUs dominate the clone library and the remaining OTUs are a long tail of singletons, is the pattern seen in the Eukarya rank abundance curve [Eukaryal Lake Ice (n = 123), and Eukaryal Lake Water (n = 39) after removal of the chloroplast rRNA gene sequences] (Figure 5).

Figure 4. Phylum-level distribution of: (**a**) bacteria ice; and (**b**) bacteria water clone libraries. The "other" category includes all phyla that were represented by <10 clones in each library (see main text for more details). Chloroplast rRNA gene sequences are separated because they were not included in subsequent analyses.

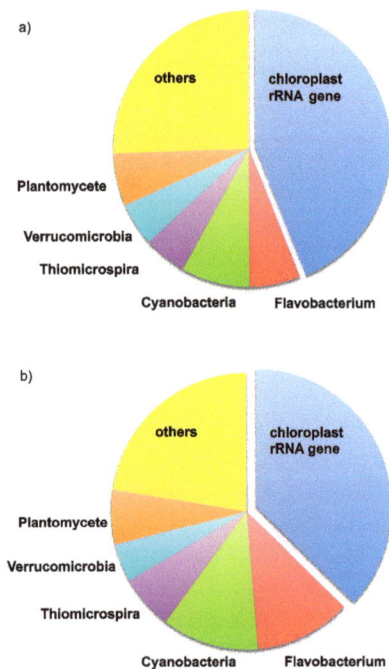

Figure 5. Rank-abundance curves for: (**a**) bacteria; and (**b**) Eukarya clone libraries. Operational taxonomic units (out) were numbered in order of total number of clones in both libraries. Hatched bars show the number of clones in the ice clone library; solid bars show the number of clones in the water clone library.

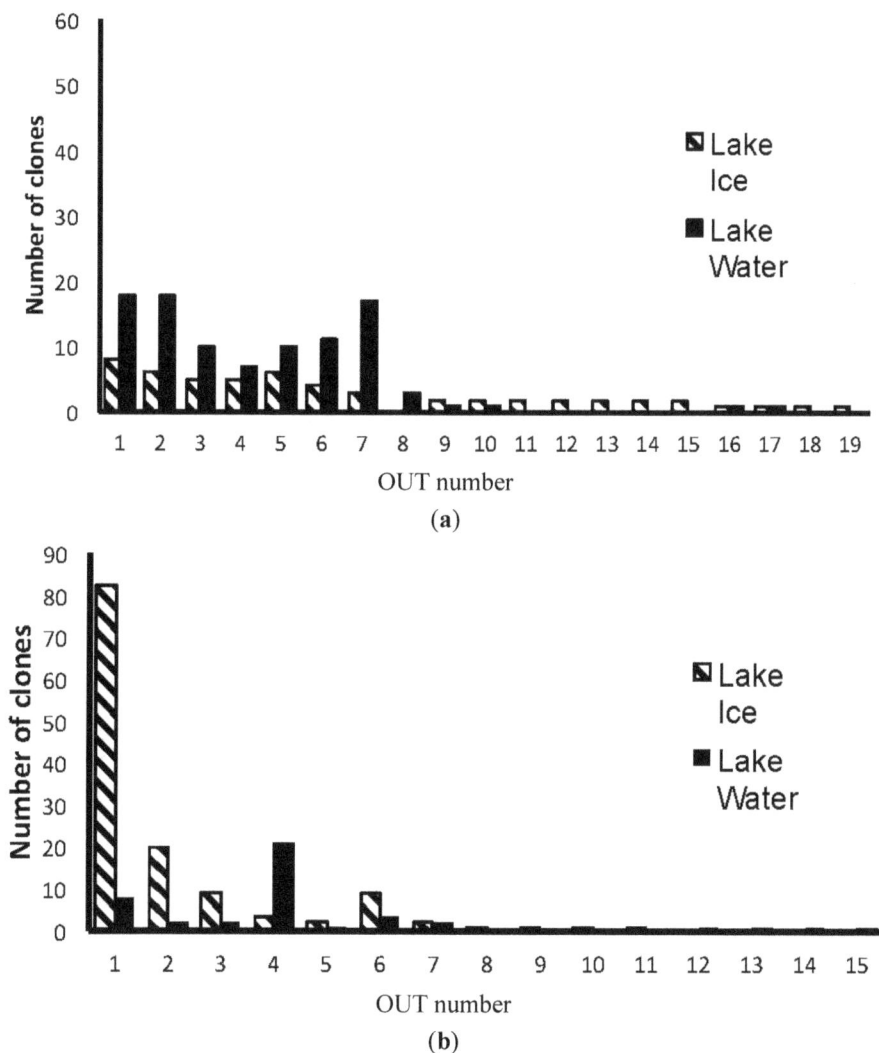

(a)

(b)

Based on rarefaction curves, both the ice and water bacterial diversity was sampled to near completion (Figure 6). Separately, these libraries accounted for 50% and 72% of the overall predicted diversity, based on Good's coverage estimate for the ice cover and water, respectively; however, the combined estimate for bacterial coverage is 92%. The ice Eukarya diversity was sampled to completion, but the coverage of the water clone library was lower (Figure 6). Good's coverage estimation agreed with lower coverage for the Eukaryal water library (ice: 95% and water: 37%), but the combined estimate for ice and water had 97.8% coverage.

Figure 6. Collector's curves for: (**a**) Eukarya ice; (**b**) Eukarya water; (**c**) bacteria ice; and (**d**) bacteria water clone libraries. The top line in each graph shows the hypothetical line if each clone belonged to a novel OTU.

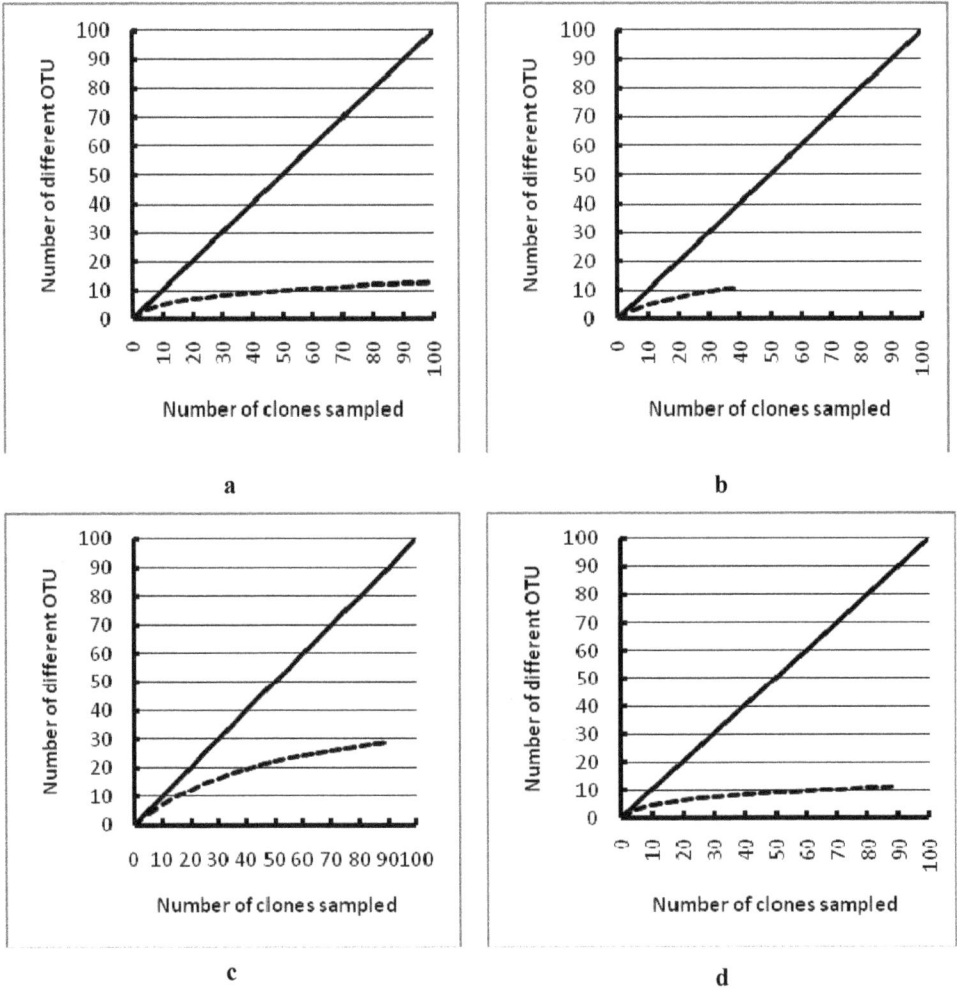

a

b

c

d

The closest relatives of the OTUs obtained in the bacterial clone library include the phyla Actinobacteria, Bacteroidetes, Proteobacteria, Verrucomicrobia, and Cyanobacteria (Figure 7). The eukaryotic clone library was dominated almost entirely by OTU with nearest neighbors from green algae, including *Chlamydomonas*, *Chlorella*, and other chlorophytes (Figure 8a). However, a copepod and cercozoan were also identified in both the ice and underlying waters (Figure 8b), indicating the possible presence of a complete microbial food web within the ice-cover of this seasonally frozen lake.

Figure 7. Maximum likelihood phylogenetic tree of Bacteria 16S rRNA genes from Miquelon Lake water and ice and relatives from the Genbank database. Scale bar represents 1 nucleotide change for each 10 nucleotides of sequence. Bootstrap support greater than 50 (of 100 replicates) is shown at nodes. Accession numbers for publically available sequences are given in parentheses. Miquelon Lake OTU are shown in bold; the relative abundance in the ice and water clone libraries is shown in brackets.

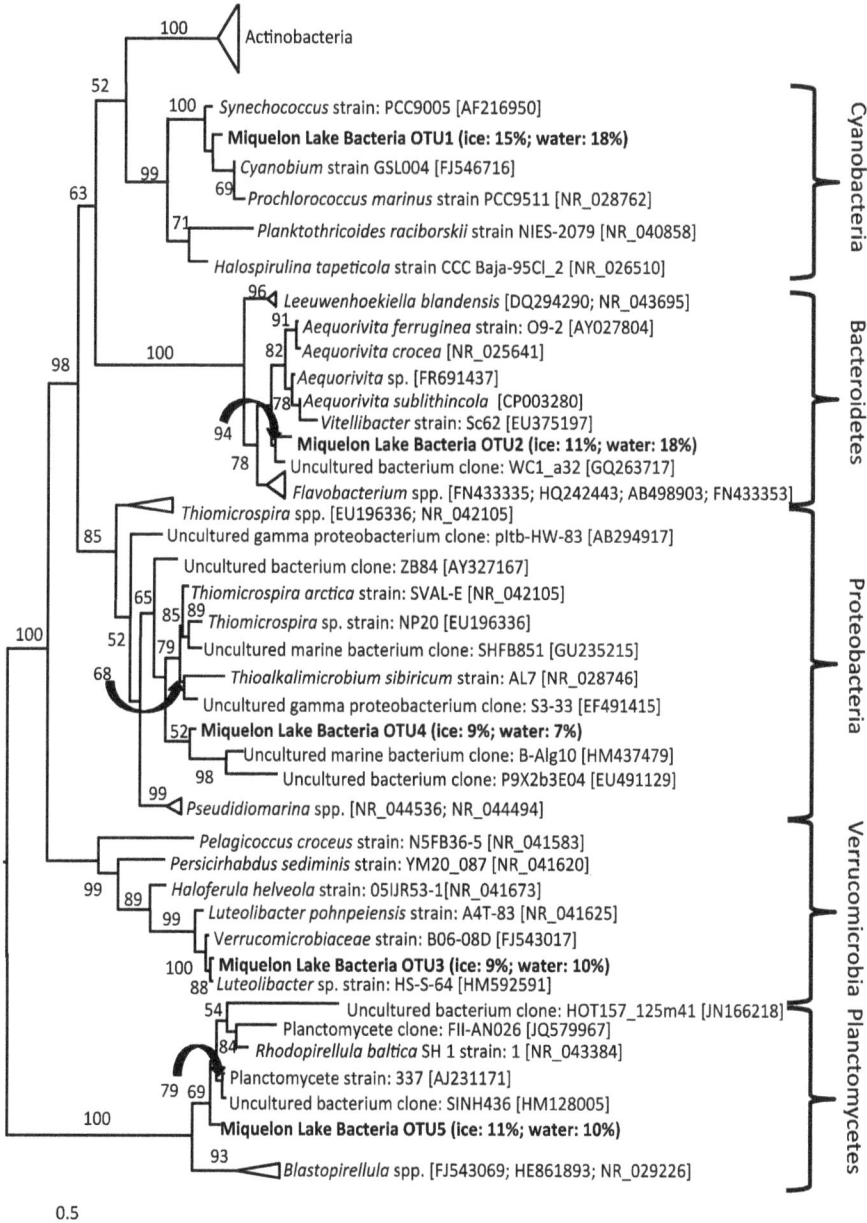

Figure 8. Maximum likelihood phylogenetic tree of Eukarya 18S rRNA genes from Miquelon Lake water and ice and relatives from the Genbank database. Scale bar represents 1 nucleotide change for each 10 nucleotides of sequence. Bootstrap support greater than 50 (of 100 replicates) is shown at nodes. Accession numbers for publically available sequences are given in parentheses. Miquelon Lake OTU are shown in bold; the relative abundance in the ice and water clone libraries is shown in brackets. (**a**) Detail showing relationships of Miquelon Eukarya OTU 1-4 within the Chlorophyta (green algae). (**b**) Phylogenetic position of Miquelon Eukarya OTU 5 and 6 within the Eukarya as a whole.

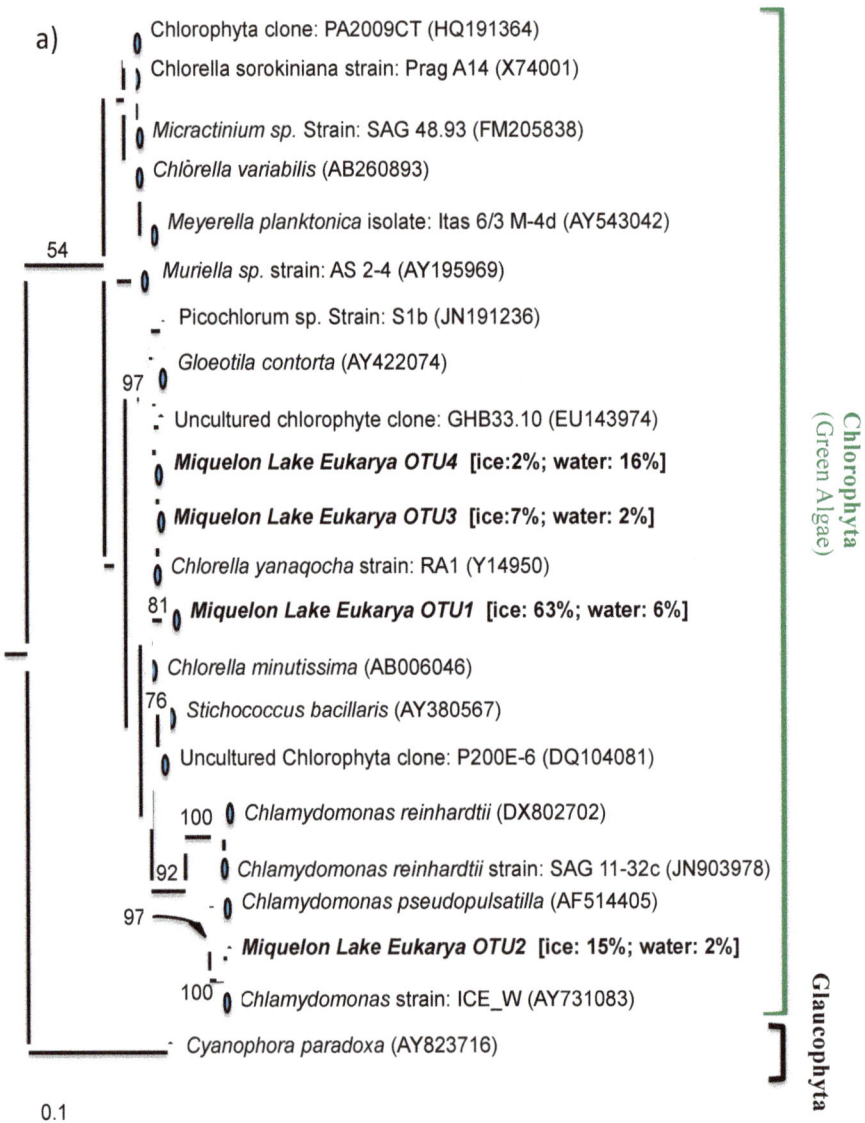

(a)

Figure 8. *Cont.*

b)

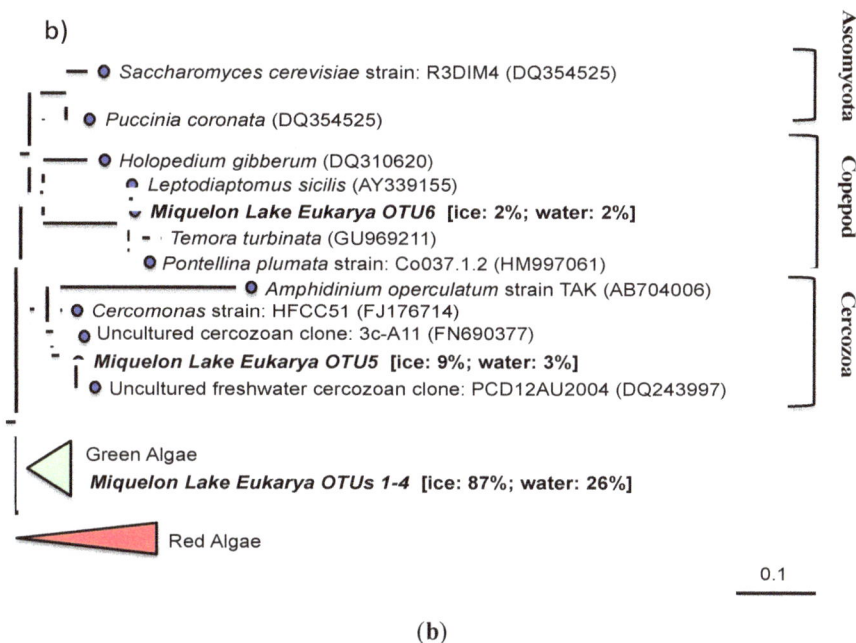

(b)

0.1

4. Discussion

Our findings support four main conclusions: (1) the ice assemblage composition is essentially invariant with depth in the ice; (2) the ice assemblage composition is essentially invariant throughout the season; (3) the ice and lake assemblages might be active throughout the winter; and (4) there may be a complete microbial loop within the ice-cover of this seasonally frozen briny lake in central Alberta.

The invariance in ice assemblage composition with depth might be an indication that the observed variations in temperature, brine volume, and brine salinity (Figure 1) did not have a significant impact on the composition of the microbial ice assemblages. Although Miquelon Lake ice does not reach the extremes sometimes seen in sea ice [1], it is surprising that the variation between brackish (~10 ppt) to hypersaline (>60 ppt) and subzero temperatures did not lead to more obvious changes in the assemblage composition. This lack of variation may be due to a high degree of microbial mixing within the brine channels; a premise supported by the observation that the majority (83%) of the bacterial clones were representatives of OTU that were present in both the lake ice and waters. Mixing events that force underlying water to flush through the brine channels and onto the top of the overlying ice have been observed at Miquelon Lake. The temporal stability of the lake ice and water microbial assemblages demonstrate that time in ice does not affect the microbial composition.

The presence of a seasonally stable Chl-*a* peak indicates some biological activity and growth because this Chl-*a* peak remains at the same ice depth regardless of brine movement and flushing and ice growth and loss, indicating that they must grow in order to maintain the same position.

Biological activity may therefore be sustained within the ice throughout the winter, similar to activity that has been observed in sea ice and perennial lake ice [8,24,39]. Over-winter microbial biological activity could theoretically have significant, previously unrecognized implications for nutrient mass balance and productivity estimates in saline lake systems.

It was somewhat surprising that Miquelon Lake clones were not similar to clones and isolates from sea ice or the ice covered lakes of Antarctica. Miquelon Lake ice bacteria were predominantly related to sequences originally found in cold-water lakes, springs, and saline environments. This finding implies that the freezing process does not strongly influence the microbes in the perennial lake ice. Two clones in the lake ice and waters were over 97% similar to human skin biota and wastewater treatment waters, likely representing human contamination. Some contamination might be expected as the source waters for the lake incorporate significant agricultural drainage.

We were unable to consistently detect Archaea in Miquelon lake ice or water by PCR and qPCR measurements indicated that they comprise <1% of the total assemblage. Archaea, which are active in seawater during winter [40], have not been identified consistently within sea ice and represented very low percentages of the total ice-population when found [39]. It is not clear why Archaea appear to be rare in icy environments.

5. Conclusions

Briny lakes play important roles in global ecosystem processes. It has been assumed previously that these lakes become essentially biologically inactive in winter and that their ice is biologically inert. Here we have demonstrated that microbes are entrained in the seasonal ice and may maintain biological activity throughout the winter. SSU rRNA genes for primary producers, bacteria, bacteriovores, and bacteriovore predators were observed in Miquelon Lake ice and the underlying water, raising the possibility of a complete microbial loop within the lake ice. These organisms may be actively cycling organic carbon and nutrients throughout the winter. Further studies will help clarify the role of winter microbial ecosystem dynamics in overall ecosystem function and structure.

Acknowledgments

We thank the NSERC Discovery Grant to Brian Lanoil and the Alberta Innovates—Technology Futures Grant to Christian Haas for funding this research. We thank Arleen Oatway and the University of Alberta Advanced Microscopy Facility for assistance with microscopy. We thank Cheryl Nargang and the University of Alberta Molecular Biology Service Unit (MBSU) for assistance with DNA sequencing.

References

1.   Mock, T.; Thomas, D.N. Recent advances in sea-ice microbiology. *Environ. Microbiol.* **2005**, *7*, 605–619.

2. Dieser, M.; Nocker, A.; Priscu, J.C.; Foreman, C.M. Viable microbes in ice: Application of molecular assays to McMurdo Dry Valley lake ice communities. *Antarct. Sci.* **2010**, *22*, 470–476.

3. Foreman, C.M.; Dieser, M.; Greenwood, M.; Cory, R.M.; Laybourn-Parry, J.; Lisle, J.T.; Jaros, C.; Miller, P.L.; Chin, Y.P.; Mcknight, D.M. When a habitat freezes solid: Microorganisms over-winter within the ice column of a coastal Antarctic lake. *FEMS Microbiol. Ecol.* **2011**, *76*, 401–412.

4. Priscu, J.C.; Fritsen, C.H.; Adams, E.E.; Giovannoni, S.J.; Paerl, H.W.; McKay, C.P.; Doran, P.T.; Gordon, D.A.; Lanoil, B.D.; Pinckney, J.L. Perennial antarctic lake ice: An oasis for life in a polar desert. *Science* **1998**, *280*, 2095–2098.

5. Bowers, J.A.; Cooper, W.E.; Hall, D.J. Midwater and epibenthic behaviors of Mysis relicta Loven: Observations from the Johnson-Sea-Link II submersible in Lake Superior and from a remotely operated vehicle in northern Lake Michigan. *J. Plankton Res.* **1990**, *12*, 1279–1286.

6. Vanderploeg, H.A.; Bolsenga, S.J.; Fahnenstiel, G.L.; Liebig, J.R.; Gardner, W.S. Plankton ecology in an ice-covered bay of Lake Michigan: Utilization of a winter phytoplankton bloom by reproducing copepods. *Hydrobiologia* **1992**, *243/244*, 175–183.

7. Lizotte, M.P. The contributions of sea ice algae to Antarctic marine primary production. *Am. Zool.* **2001**, *41*, 57–73.

8. Kottmeier, S.T.; Sullivan, C.W. Sea ice microbial communities (SIMCO)—9. Effects of temperature and salinity on rates of metabolism and growth of autotrophs and heterotrophs. *Polar Biol.* **1988**, *8*, 293–304.

9. McGrath Grossi, S.; Kottmeier, S.T.; Sullivan, C.W. Sea ice microbial communities. III. Seasonal abundance of microalgae and associated bacteria, McMurdo Sound, Antarctica. *Microb. Ecol.* **1984**, *10*, 231–242.

10. Garrison, D.L.; Buck, K.R.; Silver, M.W. Microheterotrophs in the ice-edge zone. *Antarct. J. US* **1984**, *19*, 109–111.

11. Kottmeier, S.T.; Sullivan, C.W. Bacterial biomass and production in pack ice of Antarctic marginal ice edge zones. *Deep Sea Res. Oceanogr. Res. Pap.* **1990**, *37*, 1311–1330.

12. Last, W.M. Geolimnology of salt lakes. *Geosci. J.* **2002**, *6*, 347–369.

13. Bowman, J.S.; Sachs, J.P. Chemical and physical properties of some saline lakes in Alberta and Saskatchewan. *Saline Syst.* **2008**, *4*, 1–17.

14. Swanson, H.; Zurawell, R. Miquelon Lake water quality monitoring report. Provincial Park Lakes Monitoring Program. Technical Report for Monitoring and Evaluation Branch, Environmental Assurance Division, Alberta Environment: Edmonton, Canada, February 2006. Available online: http://environment.gov.ab.ca/info/library/7730.pdf (accessed on 12 December 2011).

15. Bierhuizen, J.F.H.; Prepas, E.E. Relationships between nutrients, dominant ions, and phytoplankton standing crop in prairie saline lakes. *Can. J. Fish. Aquat. Sci.* **1985**, *42*, 1588.

16. Evans, J.C.; Prepas, E.E. Potential effects of climate change on ion chemistry and phytoplankton communities in prairie saline lakes. *Limnol. Oceanogr.* **1996**, *41*, 1063.

17. Hammer, U.T. The Saline Lakes of Saskatchewan I. Background and Rationale for Saline Lakes Research. *Int. Rev. Gesamten Hydrobiol. Hydrog.* **1978**, *63*, 173.

18. Haynes, R.C.; Hammer, U.T. The saline lakes of Saskatchewan. IV Primary production of phytoplankton in selected saline ecosystems. *Int. Rev. Gesamten Hydrobiol.* **1978**, *63*, 337.

19. Grasby, S.E.; Londry, K.L. Supporting a Mid-Continent Marine Ecosystem: An Analogue for Martian Springs? *Astrobiology* **2007**, *7*, 662.

20. Sørensen, K.B.; Teske, A. Stratified communities of active archaea in deep marine subsurface sediments. *Appl. Environ. Microbiol.* **2006**, *72*, 4596–4603.

21. Sorokin, D.Y.; Tourova, T.P.; Lysenko, A.M.; Muyzer, G. Diversity of culturable halophilic sulfur-oxidizing bacteria in hypersaline habitats. *Microbiology* **2006**, *152*, 3013–3023.

22. Richter-Menge, J.A.; Perovich, D.K.; Elder, B.C.; Claffey, K.; Rigor, I.; Ortmeyer, M. Ice mass-balance buoys: A tool for measuring and attributing changes in the thickness of the Arctic sea-ice cover. *Ann. Glaciol.* **2006**, *44*, 205–210.

23. Cox, G.F.N.; Weeks, W.F. Equations for determining the gas and brine volumes in sea ice samples. *J. Glaciol.* **1983**, *29*, 306–316.

24. Porter, K.G.; Feig, Y.S. The use of DAPI for identifying and counting aquatic microflora. *Limnol. Oceanogr.* **1980**, *25*, 943–948.

25. Bergmann, M.; Peters, R.H. A simple reflectance method for the measurement of particulate pigment in lake water and its application to phosphorus-chlorophyll-seston relationships. *Can. J. Fish. Aquat. Sci.* **1980**, *37*, 111–114.

26. Kulp, T.R.; Han, S.; Saltikov, C.W.; Lanoil, B.D.; Zargar, K.; Oremland, R.S. Effects of imposed salinity gradients on dissimilatory arsenate reduction, sulfate reduction, and other microbial processes in sediments from California soda lakes. *Appl. Environ. Microbiol.* **2008**, *74*, 3618–3618.

27. Myers, R.; Fischer, S.G.; Lerman, L.S.; Maniatis, T. Nearly all single base substitutions in DNA fragments joined to a GC-clamp can be detected by denaturing gradient gel electrophoresis. *Nucleic Acids Res.* **1985**, *139*, 3131–3145.

28. Díez, B.; Pedrós-Alió, C.; Marsh, T.L.; Massana, R. Application of denaturing gradient gel electrophoresis (DGGE) to study the diversity of marine picoeukaryotic assemblages and comparison of DGGE with other molecular techniques. *Appl. Environ. Microbiol.* **2001**, *67*, 2942–2951.

29. Bano, N.; Ruffin, S.; Ransom, B.; Hollibaugh, J.T. Phylogenetic Composition of Arctic Ocean Archaeal Assemblages and Comparison with Antarctic Assemblages. *Appl. Environ. Microbiol.* **2004**, *70*, 781–789.

30. Labrenz, M.; Sintes, E.; Toetzke, F.; Zumsteg, A.; Herndl, G.J.; Seidler, M.; Jürgens, K. Relevance of a crenarchaeotal subcluster related to Candidatus Nitrosopumilus maritimus to ammonia oxidation in the suboxic zone of the central Baltic Sea. *ISME J.* **2010**, *4*, 1496–1508.

31. Polz, M.F.; Cavanaugh, C.M. Bias in template-to-product ratios in multitemplate PCR. *Appl. Environ. Microbiol.* **1998**, *64*, 3724–3730.

32. Muyzer, G.; de Waal, E.C.; Uitterlinden, A.G. Profiling of complex microbial populations by denaturing gradient gel electrophoresis analysis of polymerase chain reaction-amplified genes coding for 16S rRNA. *Appl. Environ. Microbiol.* **1993**, *59*, 695–700.

33. Delong, E.F. Archaea in coastal marine environments. *PNAS* **1992**, *89*, 5685–5689.

34. Skidmore, M.; Anderson, S.P.; Sharp, M.; Foght, J.; Lanoil, B.D. Comparison of microbial community compositions of two subglacial environments reveals a possible role for microbes in chemical weathering processes. *Appl. Environ. Microbiol.* **2005**, *71*, 6986–6997.

35. Good, I.J. The population frequencies of species and the estimation of population parameters. *Biometrika* **1953**, *40*, 237.

36. Lanoil, B.D.; Sassen, R.; La Duc, M.T.; Sweet, S.T.; Nealson, K.H. Bacteria and Archaea Physically Associated with Gulf of Mexico Gas Hydrates. *Appl. Environ. Microbiol.* **2001**, *67*, 5143–5153.

37. Decipher. Available online: http://decipher.cee.wisc.edu/ (accessed on 12 September 2012).

38. Drummond, A.; Ashton, B.; Buxton, S.; Cheung, M.; Cooper, A.; Duran, C.; Field, M.; Heled, J.; Kearse, M.; Markowitz, S.; *et al.* Geneious Pro: Geneious v5.4.5; Biomatters. Available online: http://www.geneious.com/ (accessed on 12 December 2011).

39. Junge, K.; Eicken, H.; Deming, J.W. Bacterial Activity at −2 to −20 °C in Arctic Wintertime Sea Ice. *Appl. Environ. Microbiol.* **2004**, *70*, 550–557.

40. Murray, A.E.; Wu, K.Y.; Moyer, C.L.; Karl, D.M.; Delong, E.F. Evidence for circumpolar distribution of planktonic Archaea in the Southern Ocean. *Aquat. Microb. Ecol.* **1999**, *18*, 263–273.

Ecology of Subglacial Lake Vostok (Antarctica), Based on Metagenomic/Metatranscriptomic Analyses of Accretion Ice

Scott O. Rogers, Yury M. Shtarkman, Zeynep A. Koçer, Robyn Edgar, Ram Veerapaneni and Tom D'Elia

Abstract: Lake Vostok is the largest of the nearly 400 subglacial Antarctic lakes and has been continuously buried by glacial ice for 15 million years. Extreme cold, heat (from possible hydrothermal activity), pressure (from the overriding glacier) and dissolved oxygen (delivered by melting meteoric ice), in addition to limited nutrients and complete darkness, combine to produce one of the most extreme environments on Earth. Metagenomic/metatranscriptomic analyses of ice that accreted over a shallow embayment and over the southern main lake basin indicate the presence of thousands of species of organisms (94% Bacteria, 6% Eukarya, and two Archaea). The predominant bacterial sequences were closest to those from species of Firmicutes, Proteobacteria and Actinobacteria, while the predominant eukaryotic sequences were most similar to those from species of ascomycetous and basidiomycetous Fungi. Based on the sequence data, the lake appears to contain a mixture of autotrophs and heterotrophs capable of performing nitrogen fixation, nitrogen cycling, carbon fixation and nutrient recycling. Sequences closest to those of psychrophiles and thermophiles indicate a cold lake with possible hydrothermal activity. Sequences most similar to those from marine and aquatic species suggest the presence of marine and freshwater regions.

Reprinted from *Biology*. Cite as: Rogers, S.O.; Shtarkman, Y.M.; Koçer, Z.A.; Edgar, R.; Veerapaneni, R.; D'Elia, T. Ecology of Subglacial Lake Vostok (Antarctica), Based on Metagenomic/Metatranscriptomic Analyses of Accretion Ice. *Biology* **2013**, *2*, 629-650.

1. Introduction

Nearly 400 subglacial lakes have been discovered in Antarctica, the largest of which is Lake Vostok [1–5]. While Lake Vostok covers an area (15,690 km^2) that is about 80% of the size of the Laurentian Great Lake Ontario, it holds a larger volume of water (5,400 km^3), due to its depth (maximum depth = 510 m). It lies beneath 3,700 to 4,200 m of ice, and has been continuously ice-covered for the past 15 million years, with the only known influx of water originating from melting of the overriding glacier. Lake Vostok consists of a northern and the southern basin (Figure 1). Little is known about the northern basin, but more is known about the southern basin because of studies based on an ice core that was drilled over the southeastern corner of the lake. While glacial ice melts over portions of the lake, water from the lake freezes (*i.e.*, accretes) to the bottom of the glacier in other parts of the lake creating an accretion ice layer that is over 200 m thick in some locales [2,6–8]. Because the glacier moves across the lake at a rate of approximately 3 m per year, the accretion ice holds a temporal record that spans approximately 5,000 to 20,000 years [7], as well as a spatial record of the surface waters of the lake. The accretion ice from the core represents several parts of the southern portion of the lake, including a region near a shallow embayment on the southwestern corner of the lake and the southern portions of the southern main basin.

Accretion ice from the ice core has been analyzed to determine the concentrations of specific ions (e.g., Na^+, K^+, Ca^{2+}, Cl^-, SO_4^{2-}) [9–12] and biomass [13–19] originating from several locations in the lake. The ice that formed in the vicinity of the shallow embayment (3,538–3,608 m, termed type 1 accretion ice) contains fine particulate matter [7], as well as relatively high concentrations of ions [9], organisms and nucleic acids [9,16,17]. Conversely, accretion ice that formed over the southern basin (3,609–3,769 m, termed type 2 ice) contains low concentrations of particulates, ions, organisms and nucleic acids. Dozens of bacterial cells, fungal cells and sequences have been reported from several accretion ice core sections [12–19] with the highest numbers concentrated in the core sections that represent regions of the lake near the shallow embayment. They include many types of organisms that are common to other aquatic environments, as well as many that remain unidentified. Sequences from thermophilic bacteria have been reported, indicative of possible hydrothermal activity in the lake [20–22]. Autotrophic and heterotrophic species have also been reported from the accretion ice [16,17]. These reports indicate that Lake Vostok might be more biologically complex than previously concluded.

Another feature of Lake Vostok is that it lies entirely below current mean sea level. Its surface is more than 200 m below sea level, and the deepest point is almost 800 m below sea level. A recent study [23] concluded that Lake Vostok lies within a graben (similar to those in the Great Rift Valley in Africa) that formed more than 60 million years ago. A second study, based on radar data [24], reported that 35 million years ago when Antarctica was free of ice, the Southern Ocean was in the immediate vicinity of Lake Vostok. However, by 34 million years ago, ice had covered the lake and lowered sea level, which might have isolated it from a direct connection to the ocean. The fact that parts of Lake Vostok contain moderate levels of salt, and that sequences from marine organisms have been detected in the accretion ice indicate that this lake might have a complex history. In this study, we utilize metagenomic/metatranscriptomic sequence data ([22]; Supplementary Tables S1–S10) to reconstruct the possible ecology of Lake Vostok.

Figure 1. Source of ice core sections used in this study. (**a**) Location of Lake Vostok (small rectangle) in Antarctica; (**b**) Detail of the outline of Lake Vostok, as indicated by radar [1,8,23,24]; (**c**) Detail of the southern end of Lake Vostok, showing the locations of the shallow embayment, ridge, southern basin, track of glacier to the drill site (dashed line), and approximate locations where the accretion ice samples (V5 and V6) were formed.

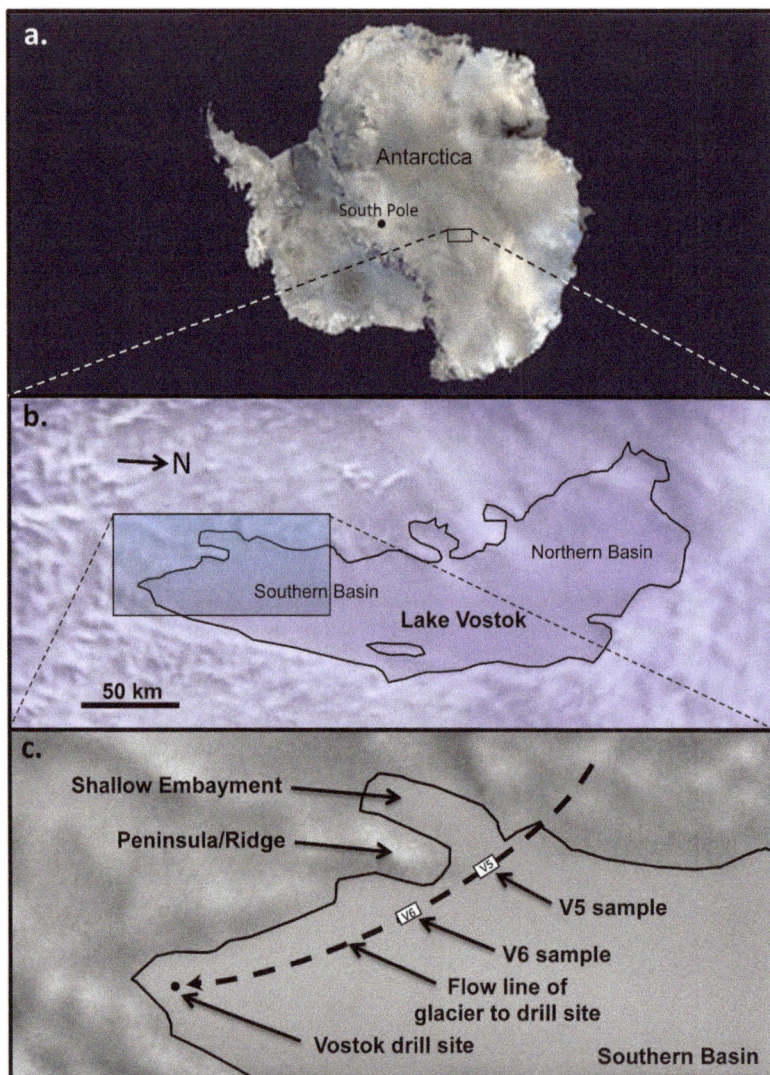

Table 1. Summary of sequence results for V5 (total of 1,863; or 3,718, including sequences that cannot be classified to species). The number of sequences in each taxon, ecology/physiology of each classified taxon and species characteristics are presented. Sequences are grouped by Domain and Phylum.

Taxon	Unique gene sequences	Unique rRNA gene sequences [a]		Ecology and physiology [b]	Species characteristics [b]
		≥200 nt	<200 nt		
BACTERIA	3495	2535	460		
Acidobacteria	2	1	0	acidophilic, soil, adaptable	chemoorganotrophic heterotrophs
Actinobacteria	228	151	24	thermophilic, halotolerant, psychrotolerant, alkalaitolerant, psychrophilic, Antarctic, deep sea sediments, lake sediments, some grow on limestone	nitrogen fixation, nitrite oxidation, ammonia oxidation, organic decomposition, heterotrophs
Bacterioidetes/ Chlorobi	88	61	8	aquatic, sediments, thermophilic, psychrophilic, alkalaiphilic, anaerobic	carbon fixation (use sulfide ions, hydrogen or ferrous ions), reductive TCA cycle
Chloroflexi	1	0	0	aerobic, thermophilic	carbon fixation using the 3-hydroxylpropionic bicycle
Cyanobacteria	228	144	60	common in Antarctic lakes, at least one is thermophilic (*Thermosynecoccus* sp.)	carbon fixation using the reductive pentose phosphate cycle, some are from anoxygenic ancestors
Deferribacteres	1	1	0	animal intestines, anaerobic	chemoorganotrophic heterotrophic
Deinococcus/ Thermus	5	1	1	thermophilic, radiophilic, aerobic some associate with cyanobacteria	chemoorganotrophic heterotrophic
Fibrobacteres	1	0	1	anaerobic, inhabit animal intestines	chemoorganotrophic heterotrophic
Firmicutes	602	401	40	Spore formers, common in extreme environments, thermophiles, mesophilic, psychrophilic, psychrotolerant, halophilic, hot springs, deep sea thermophilic, anaerobic, aerobic	heterotrophic
Fusobacteria	10	8	0	parasitic on animals, anaerobic	chemoorganotrophic heterotrophic
Planctomycetes	6	2	2	Fresh, brackish and saline lakes/ponds, anaerobic	chemoautolithotrophic anammox, nitrite reduction using ammonium as electron donor
Proteobacteria	474	265	46		
Alphaproteobacteria	91	45	7	Psychrophilic, mesophilic, thermophilic, Antarctic lakes, animal symbionts, aerobic, soil/sediments, aquatic, alkalaitolerant, require calcium, marine, halotolerant	nitrite reduction, nitrifying bacteria, denitrification (nitrate to nitrogen gas), methylotrophic, use inorganic sulfur, oxidize sulfate and thiosulfate, carbon fixation using the reductive pentose phosphate cycle, carbon fixation using the reductive TCA cycle

Table 1. *Cont.*

Taxon	Unique gene sequences	Unique rRNA gene sequences [a] ≥200 nt	<200 nt	Ecology and physiology [b]	Species characteristics [b]
Betaproteobacteria	105	31	7	thermophilic, mesophilic, psychrophilic, aquatic, aerobic, highly adaptable	nitrogen fixation, nitrate reduction, ammonia oxidation, carbon fixation using the reductive pentose phosphate cycle, manganese oxidation, iron oxidation, inorganic sulfur oxidation, arsenic oxidation
Deltaproteobacteria	10	5	0	aquatic, soil, mesophilic, anaerobic, aerobic, freshwater debris, predator of Gram-negative bacteria, halotolerant, marine	carbon fixation using the reductive TCA cycle, iron reduction, sulfur reduction, ethanol fermentation
Epsilonproteobacteria	6	3	2	Some animal associated, mesophilic, thermophilic, aerobic, anaerobic	carbon fixation using the reductive TCA cycle
Gammaproteobacteria	254	176	28	thermophilic, mesophilic, psychrophilic, psychrotolerant, aerobic, anaerobic, peizophilic, deep sea, halophilic, polar ice, soil, sediments, permafrost, 33 distinct sequences from species of *Psychrobacter*, 10 distinct sequences from species of *Halomonas* (halophilic), some produce intracellular gas vesicles, some are animal associated	nitrogen fixation, nitrate reduction, nitrite respiration, denitrification, sulfur oxidation, chemolithoautotrophs, iron oxidation, mineralization of aromatics, carbon fixation using the reductive pentose phosphate cycle
Uncultured Proteobacteria	8	5	2	unknown	unknown
Spirochaetes	3	3	0	animal pathogens	heterotrophic
Tenericutes	4	4	0	saprobes and arthropod pathogens/symbionts, anaerobic	heterotrophic
Verrucomicrobia	3	1	0	freshwater, soil, symbionts of protists and nematodes, aerobic	heterotrophic
Uncultured Bacteria	1839	1492	278	Sequences similar to those from uncultured and unidentified species, many from other environmental metagenomic studies	Unknown
ARCHAEA	**2**	**0**	**0**	deep hydrate-bearing sediment, peizotolerant, psychrotolerant	Methanotrophic, carbon fixation using the reductive acetyl-CoA pathway
EUKARYA	**221**	**124**	**27**		
Amoebozoa	1	1	0	*Nolandella* sp.; aquatic; feed on bacteria, diatoms, nematodes, fungi, protozoans and organic matter	Heterotrophic
Archaeplastida	74	28	9		
Chlorophyta	10	5	4	Antarctic and polar green algal species	carbon fixation using the reductive pentose phosphate cycle

Table 1. *Cont.*

Taxon	Unique gene sequences	Unique rRNA gene sequences [a] ≥200 nt	<200 nt	Ecology and physiology [b]	Species characteristics [b]
Rhodophyta	1	0	0	Antarctic red alga	carbon fixation using the reductive pentose phosphate cycle
Streptophyta	63	23	5	Pollen from lake sediments or from glacial deposition?	(carbon fixation using the reductive pentose phosphate cycle)—non-viable?
Chromalveolata	12	6	2	diatoms, heterokonts, predatory protists, dinoflagellates, ciliates, Antarctic, aquatic	carbon fixation using the reductive pentose phosphate cycle, heterotrophic
Excavata	2	0	0	freshwater species	heterotrophic
Opisthokonta	115	79	10		
Animalia	24	10	3		
Arthropoda	16	8	0	Arctic, Antarctic, aquatic. (e.g., *Daphnia* sp., Ellipura, Branchiopoda, Entomobryiadae).	heterotrophic
Bilateria	1	0	1	Deep sediment environmental sample	unknown
Chordata	3	1	0	Aves, from meteoric ice or contaminant?	heterotrophic
Cnideria	1	0	0	Small sea anemone, lives in soft sediment with water salinities of 9 to 52 ppt at temperatures from −1 to 28 °C.	heterotrophic
Mollusca	1	0	1	*Nutricola* sp., cold water marine bivalve that burrows into sediments.	heterotrophic
Rotifera	1	1	0	Survives under extreme conditions; feed on detritus, bacteria, algae and protists.	heterotrophic
Tardigrada	1	0	1	Hardy animal, eats rotifers and algae, can survive from approximately −270 to 150 °C	heterotrophic
Fungi	91	69	7		
Ascomycota	48	34	4	Antarctic, polar, aquatic, soil	heterotrophic
Basidiomycota	29	24	0	Antarctic, polar, psychrophilic, psychrotolerant	heterotrophic
Mucorales	1	0	1	Aquatic, parasitic on arthropods	heterotrophic
Uncultured fungi	13	11	2	unknown	unknown
Rhizaria	1	0	0	Freshwater, *Paulinella* sp.	heterotrophic
Uncultured eukaryotes	16	10	6	unknown	unknown

[a] Sequences ≥200 nt were submitted to NCBI GenBank and were assigned accession numbers, while those shorter than 200 nt could not be submitted to NCBI GenBank, and therefore do not have accession numbers. Totals based on BLAST searches using pyrosequencing reads; [b] Ecological, physiological and other characters were based on information from the sequenced organisms identified in the BLAST searches. Sources of information were NCBI descriptions, publications cited in the NCBI descriptions and web sources (see Supplementary Tables S1–S6).

2. Results and Discussion

2.1. Summary of Results

Two samples, each consisting of meltwater from two accretion ice core sections, were analyzed for this research. One (termed "V5") contained meltwater from two ice core sections (3,563 m and 3,585 m; Figure 1) that accreted in the vicinity of the shallow embayment on the southwestern corner of Lake Vostok. The other (termed "V6") contained meltwater from two ice core sections (3,606 m and 3,621 m) that accreted on the western side of the southern basin. Sequences have been deposited in the NCBI (National Center for Biotechnology) GenBank database (accession numbers are provided in the Experimental Section; BLAST results are presented in Supplementary Tables S1–S10 [22].). For the V5 sample, 36,754,464 bp of sequence data was obtained that included 94,728 high quality 454 sequence reads, with mean lengths of 388 bp. For the V6 sample, 1,170,900 bp of sequence data was obtained that included 5,204 high quality reads, with mean lengths of 225 bp. The lower quantity of sequence data for V6 is consistent with our previous results from the same core sections indicating much lower concentrations of cells and viable cells in the V6 ice core sections compared to the V5 ice core sections [16,17]. However, the lower number of sequences and the shorter average read lengths also might indicate that the nucleic acids in this sample were degraded to a greater extent than those in the V5 sample. Overall, approximately 15% of the sequences were unique, while the remaining 85% were additional copies from the unique set of sequences. A total of 3,718 unique sequences were retrieved from V5 (3,146 were rRNA gene sequences), of which 1,863 could be classified to species (Table 1; Supplementary Tables S1–S6 [22]), and 184 unique gene sequences were retrieved from V6 (111 were rRNA gene sequences), of which 133 could be classified to species (Table 2; Supplementary Tables S7–S10). Approximately 94% of the unique sequences in V5 and 85% in V6 were from Bacteria. Sequences closest to those from species of autotrophs and heterotrophs were present. Only two unique Archaea sequences were found (both in V5), and they were most similar to species of methanotrophs from deep-ocean sediments. The remaining sequences were closest to those from species of Eukarya (6% in V5 and 15% in V6), including more than 200 unique sequences from multicellular organisms, most of which were Fungi (primarily ascomycetes and basidiomycetes). In general, the species indicated by the sequence comparisons were organisms specific to lakes, brackish water, oceans/seas, soil, lake sediments, deep-sea sediments, deep-sea thermal vents, animals and plants.

Table 2. Summary of sequence results for V6 (total of 133 classified taxonomically; or 184, including sequences that cannot be classified to species). The number of sequences in each taxon, ecology/physiology of each classified taxon and species characteristics are presented. Sequences are grouped by Domain and Phylum.

Taxon	Unique gene sequences	Unique rRNA gene sequences [a] ≥200 nt	<200 nt	Ecology and physiology [b]	Species characteristics [b]
BACTERIA	**155**	**69**	**21**		
Actinobacteria	14	1	4	fish pathogen, psychrophilic, ocean/lake sediments	chemoorganotrophic heterotrophic
Bacterioidetes/Chlorobi	1	0	0	psychrophilic, alkalaiphilic, aerobic	heterotrophic
Chloroflexi	1	0	1		
Deinococcus/Thermus	1	0	0	thermophilic, radiophilic, some associate with cyanobacteria	chemoorganotrophic heterotrophic
Firmicutes	16	5	0	alkalaiphilic, thermophilic, mesophilic, psychrophilic, soil/sediments, anaerobic, some parasitic/symbiotic on animals	heterotrophic, nitrate reduction
Fusobacteria	1	0	0	mesophilic, parasitic on animals, anaerobic	heterotrophic
Proteobacteria	71	27	6		
Alphaproteobacteria	8	5	0	mesophilic, psychrophilic, aerobic, acid tolerance, aquatic, sediments, animal symbionts	nitrogen fixation, heterotrophic, carbon fixation using the reductive pentose phosphate cycle
Betaproteobacteria	22	6	3	annelid symbiont, annelid associated, Arctic soils, aquatic, Antarctic marine, intracellular gas vacuoles, high amounts of 16:1 ω7c fatty acids, psychrophilic, mesophilic, thermophilic, aerobic, highly adaptable, hot springs, (e.g., *Thiobacillus* sp., related to *Hydrogenophilus thermoluteus*, previously reported by Bulat *et al.* 2004 [20,21] Lake Vostok accretion ice at 3,607 m depth)	nitrogen fixation, chemoorganotrophic heterotrophic, aromatic hydrocarbon degradation, nitrous oxide reduction, arsenic oxidation, arsenic reduction, inorganic sulfur oxidation, chemolithoautotroph,, hydrogen oxidation, carbon fixation using the reductive pentose phosphate cycle
Gammaproteobacteria	39	14	3	fish intestinal symbionts (2 species), nematode associated, animal associated, plant associated, aquatic, soil/sediment, thermophilic, mesophilic, psychrophilic, anaerobic, aerobic, halotolerant	nitrogen fixation, nitrate reduction, nitrite respiration, heterotrophic, carbon fixation using the reductive pentose phosphate cycle
Uncultured Proteobacteria	2	2	0		

Table 2. *Cont.*

Taxon	Unique gene sequences	Unique rRNA gene sequences [a]		Ecology and physiology [b]	Species characteristics [b]
		≥200 nt	<200 nt		
Uncultured Bacteria	50	36	10	sequences similar to those from uncultured and unidentified species, many from other environmental metagenomic studies	unknown
EUKARYA	29	12	9		
Archaeplastida	2	1	1		
Streptophyta	2	1	1	pollen from lake sediments or from glacial deposition?	(carbon fixation using the reductive pentose phosphate cycle)—non-viable?
Opisthokonta	26	10	8		
Animalia	5	0	0		
Arthropoda	5	0	0	aquatic, Acari, parasitic	heterotrophic
Fungi	22	10	8		
Ascomycota	13	7	3	aquatic, one grows on marble and limestone, one isolated from mid-ocean hydrothermal vents, some from sediments, one can use methanol as a carbon source, Antarctic species	heterotrophic
Basidiomycota	4	0	4	Antarctic, marine, aquatic	heterotrophic
Uncultured fungi	4	3	1	unknown	unknown
Uncultured eukaryote	1	1	0	unknown	unknown

[a] Sequences ≥200 nt were submitted to NCBI GenBank and were assigned accession numbers, while those shorter than 200 nt could not be submitted to NCBI GenBank, and therefore do not have accession numbers; [b] Ecological, physiological and other characters were based on information from the sequenced organisms identified in the BLAST searches. Sources of information were NCBI descriptions, publications cited in the NCBI descriptions and web sources (see Supplementary Tables S7–S10).

2.2. Extremophiles

A large number of the sequences were most similar to those from psychrophilic and psychrotolerant species (Tables 1 and 2; Supplementary Tables S1–S10). Within the Gammaproteobacteria, there were 33 unique sequences closest to various *Psychrobacter* species (most with rRNA SSU gene identities of greater than 97%), all described as psychrophiles. Also present were sequences closest to psychrophilic or psychrotolerant species of Actinobacteria, Alphaproteobacteria, Archaea, Archaeplastida, Bacteroidetes, Betaproteobacteria, Firmicutes, Chromalveolata, and Opisthokonta (both Animalia and Fungi). Conversely, there were many sequences that were closest to those from several thermophilic species (many with rRNA SSU gene identities greater than 97%). While most were found in V5 ice (46 with rRNA sequence identities greater than 97% to known taxa), a few were found in V6 ice (2 with rRNA sequence identities greater than 97% to known taxa). A number of sequences most similar to those of sulfur oxidizing bacteria were found in V5 (Tables 1 and 2).

Previously, several gene sequences from a thermophilic bacterium, *Hydrogenophilus thermoluteolus* were reported from the Lake Vostok accretion ice [20,21].

Several sequences from marine species, as well as from halophilic and halotolerant species, were present in the metagenome/metatranscriptome data set (Tables 1 and 2). Sequences closest to *Jeotgalicoccus halotolerans*, *Nesterenkonia halotolerans*, and other halophilic bacteria were found in V5 accretion ice, several of which have rRNA SSU gene identities greater than 98%. Some of these species were alkalaitolerant. A number of sequences in V5 and V6 were most similar to sequences from marine species, including marine bacteria (many with rRNA identities above 97%), a sea squirt-associated bacterium (99% rRNA SSU gene identity to *Pseudomonas xanthomarina*), an oyster pathogen (100% identity to a hypothetical protein from *Perkinseus maratimus*), a sea anemone (78% identity to a hypothetical protein from *Nematostella vectensis*, a small sea anemone, related to *Hydra* spp.) and a marine mollusk (100% rRNA SSU identity to *Nutricola tantilla*). The presence of marine, halophilic and halotolerant species is suggestive of marine layers, or other regions with high ion concentrations, within the lake or lake sediments. Saltwater layers and submarine brine lakes have been reported at the bottom of the Mediterranean Sea and Gulf of Mexico [25,26]. The number of V5 sample sequences with ≥97% identities to psychrophiles and thermophiles were roughly equal (46 and 49, respectively).The combination of possible halophiles, psychrophiles and thermophiles (at the ≥97% identity levels; Tables 1 and 2; Figure 2), in addition to higher concentrations of ions and particulate matter in the V5 sample, all are suggestive of a diversity of conditions in the southwestern region of the lake. In the V6 samples, there were more sequences (sequence identities ≥97%) closest to psychrophiles (7) than thermophiles (2). This is consistent with the presence of hydrothermal activity in the vicinity of the shallow embayment, and colder conditions in the main basin.

Figure 2. Comparisons of percentages of sequences according to characteristics of the closest species. Bars represent the percentages of sequences in V5 (green bars, N = 338) and V6 (blue bars, N = 38) that could be categorized with sequence identities ≥97% to NCBI sequences. Halo- = halophilic or halotolerant; Sed. = from lake or ocean sediments; Psychro- = psychrophilic or psychrotolerant; Thermo- = thermophilic or thermotolerant.

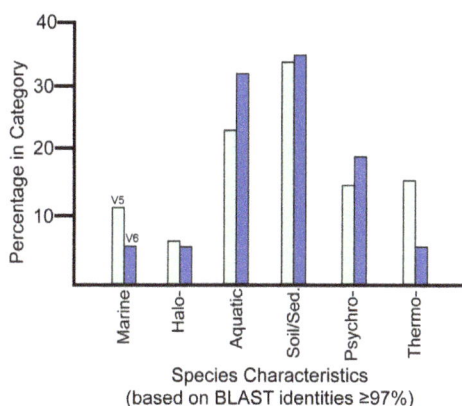

2.3. Metabolic Classification

We considered only sequences whose species and functional characteristics had been clearly identified. Genes for glycolysis, the TCA (tricarboxylic acid) cycle and genes for the synthesis of most amino acids were present (Supplementary Figures S1–S5). Incomplete pathways were found for the synthesis of arginine, histidine and proline. Alternate enzymes might be used for these processes, as determination of some sequences could not be made unambiguously. Some of the genes for each of these pathways were present, and therefore it is assumed that some of the unidentified sequences may encode for the missing enzymes in the pathways.

Sequences closest to those from bacteria capable of nitrogen fixation, nitrosification, nitrification, nitrate reduction, denitrification, anammox, assimilation and decomposition were present (Figure 3). Several types of nitrogen fixing bacteria were indicated by the sequences, including species of *Azospirillum, Azotobacter, Bacillus, Burkholderia,* Cyanobacteria, *Frankia, Klebsiella, Rhizobium, Rhodobacter, Rhodopseudomonas* and *Sinorhizobium.* Sequences of nitrifying bacteria included species of *Methylococcus, Nitrobacter, Nitrococcus, Notrosococcus* and *Nitrosomonas.* Species important in other parts of the nitrogen cycle were within the genera *Alkaligenes, Bacillus, Clostridium, Micrococcus, Paracoccus, Proteus, Pseudomonas, Streptomyces* and *Thiobacter.* Several sequences from planctomycetes were closest to sequences from species capable of anammox metabolic processes in marine environments.

Sequences of genes and organisms involved in many phases of carbon fixation cycles and pathways were found (Figure 3). Three forms of carbon fixation were indicated. The sequences indicated that most of the organisms utilize either the reductive TCA (rTCA) cycle (α-, δ- and ε-proteobacteria, and Chlorobi) (Tables 1 and 2) or the reductive pentose phosphate (rPP; Calvin-Benson) cycle (Archaeplastida, Chromalveolates, Cyanobacteria, and α-, β-, and γ-Proteobacteria) [27]. Based on the frequency of gene sequences, the most common mode of CO_2-fixation was the rTCA cycle (Tables 1 and 2), while the rPP cycle was the second most common. The two Archaea in V5 may fix carbon via the reductive acetyl-CoA (rACA) pathway [27]. However, mRNA gene sequences for enzymes in this pathway were not found in searches of the metagenome/metatranscriptome data set.

A large number and diversity of sequences from phototrophs were present in the accretion ice, including 228 cyanobacterial, 11 algal, 12 chromalveolate, and other unique sequences. Sequences for many of the genes involved in the light reactions of photosynthesis in cyanobacteria were found in the accretion ice. These included light-independent photochlorophyllide reductase and oxidase, phycocyanobilin oxidoreductase, a phycoerythrin subunit and several genes involved in carotenoid biosynthesis. However, gene function was not measured in this study, and it is possible that some of the gene sequences were from pseudogenes, that they are inactive genes, or they originated from organisms once entrapped in the meteoric ice.

Figure 3. Ecological characterization of organisms based on BLAST search results from the V5 and V6 metagenomic/metatranscriptomic sequences. The taxonomic classifications listed are based on the highest identities to sequences from species within the taxa that have been documented to have the functions specified in the boxes. These have been grouped primarily by Phylum for simplicity. Greater than 95% of the taxa are either primary producers (autotrophs), bacteria involved in the nitrogen cycle, or decomposers, including bacteria and fungi, and a few chromalveolates (shaded green box). Primary and secondary consumers comprise less than 5% of the species richness (*i.e.*, number of unique sequences). Organisms listed in black font were determined by sequences that exhibited ≥97% sequence identity with sequences in GenBank (NCBI). Organisms listed in red font exhibited <97% sequence identity or were suggested because the sequences were closest to symbionts, parasites or pathogens of those organisms (at ≥97% sequence identity).

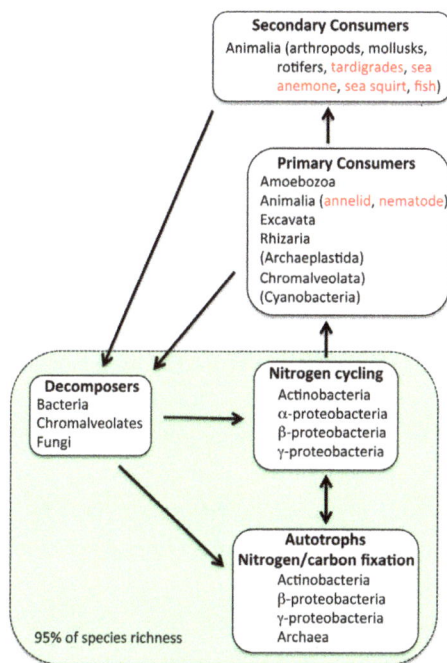

2.4. Eukaryotes

While only about 6% of the unique sequences were closest to eukaryotes (221 from V5, 87 of which have sequence identities of ≥97% with sequences on NCBI; and 29 from V6, 24 of which have identities ≥97%), diverse taxonomic groups were represented. Most of the sequences were most similar to those from Fungi (91 sequences in V5 and 22 in V6), including one rRNA SSU sequence that was 99% similar to a marine fungus sequence that previously had been recovered from a deep-sea thermal vent [28]. Several sequences from species of Animalia were found, including 21 sequences closest to those from arthropods (16 in V5 and 5 in V6), many of which are predatory or parasitic, including sequences closest to species of *Daphnia* (planctonic crustaceans; 98% identity), Ellipura (≥95% identity with species of springtails, some of which are aquatic or marine), Branchiopoda (fairy shrimp, primarily freshwater; 93% identity) and Entomobryidae (slender spingtails, some of which are aquatic; 89–98% identity). Additionally, V5 contained sequences closest to an unidentified bilaterian, a rotifer (closest to *Adineta* sp., which is a hardy, cosmopolitan, freshwater species; 98% identity), a tardigrade (closest to *Milnesium* sp., which is a hardy, predatory, cosmopolitan, freshwater species; 93% identity), a mollusk (*Nutricola tantilla*, a small [maximum diameter of 9 mm] marine bivalve that lives in sediments to about 120 m water depth; 100% identity) and a cnidarian (related to *Nematostella* sp., a small sea anemone; 78% identity). Several Archaeplastida sequences were found in V5 and V6 (most with sequence identities of ≥99%).

Sequences closest to an uncultured lobster gut bacterium (98% identity), *Verminephorbacter* sp. (an annelid nephridia symbiont; 92% identity), *Renibacter salmonarium* (a salmonid fish pathogen; 98% identity), a rainbow trout intestinal bacterium T1 (93% identity) and *Photorhabdus asymbiotica* (a nematode symbiont; 98% identity) all were found in the V6 accretion ice sample. Additionally, sequences closest to *Carnobacterium mobile* (associated with fish; 100% identity), *Macrococcus* sp. (a bivalve-associated bacterium; 97% identity), *Pseudomonas xanthomarina* (associated with sea squirts; 99% identity) and *Mycobacterium marinum* (associated with fish; 99% identity) were found in the V5 accretion ice sample. All of the species are dependent on intimate associations (symbiotic or parasitic) with their eukaryotic hosts, which are crustaceans, annelids, tunicates, nematodes or fish. Species of all of these animal groups have been found in the vicinity of deep-sea thermal vents and elsewhere in marine and aquatic environments [29–37]. Additional indications of animals in the lake came from sequences of several species in the Enterobacteriaceae, which were present in both the V5 and V6 samples. These included sequences closest to several strains/species of *E. coli*, *Erwinia*, *Klebsiella*, *Salmonella*, and *Shigella* (most with sequence identities ≥97%), all of which are found in the digestive systems of fish and other aquatic and marine animals. In addition, sequences closest to those from species of Fusobacteria (some with ≥97% identities) that are parasitic on animals, Alphaproteobacteria (some with ≥97% identities) that are animal symbionts, and Tenericutes that are arthropod symbionts and pathogens (up to 91% sequence identity) were found in the V5 sample.

In V5, there were 16 sequences closest to those from single-celled eukaryotic species. These included sequences closest to species of Excavata (≥98% identity), Rhizaria (closest to *Paulinella* sp., a freshwater phototroph; 94% identity), Amoebozoa (*Naeglaria gruberei*, 98% identity; *Nolandella* sp., marine, 88% identity) and Chromalveolata (10 unique sequences, including two Ciliophora, 99% identity each; three bacillariophytes, *Stephanodiscus* sp.—99%, *Stephanodiscus* sp.—100%, *Hatzschia* sp.—95% identity; three heterokonts, *Aphanomyces euteiches*—98% identity, *Botrydiopsis constricta*—99% identity, *Halosiphon tomentosus*—97% identity); a cryptophyte, *Cryptomonas paramecium*—100% identity; and a member of Perkinsea, obligate parasite of mollusks—100% identity. While some of these eukaryotes are free-living, many are symbionts, parasites or commensals on multicellular organisms, including animals and plants. Several multicellular organisms were indicated by the sequence data, but the sequences closest to symbionts and parasites suggest that a more diverse set of multicellular eukaryotes might exist in the lake.

2.5. Possible Marine Environment in Lake Vostok

Organisms in Lake Vostok have had millions of years to adapt and evolve. The lake has been continuously ice-covered for approximately 15 million years [1,4,10,23,24,38]. During parts of the Miocene (15 to 25 million years ago, mya), it was intermittently ice-free, and previous to that, a cooling period from 25–34 mya led to the lake to be ice-covered much of the time [38]. However, prior to 35 mya (during the Eocene), most of Antarctica (including the Lake Vostok region) was free of ice, sea levels were higher and extensive complex ecosystems were present on the continent, complete with lakes and streams, as well as diverse sets of microbes, plants (including extensive forests), fungi and animals. During these times, Lake Vostok probably contained a species-rich biota. Currently, Lake Vostok is below sea level. Currently, it is separated from regions that once were occupied by ocean water by low ridges that are approximately 20–50 m above current mean sea level [23,24,38]. However, 35 million years ago, sea levels were between 50–100 m higher. Therefore, this low area may have been a straight that connected Lake Vostok with the ocean, thus making it a large marine bay (Figure 4). Our metagenomic/metatranscriptomic data set included sequences that are most similar to sequences from marine organisms, including many Bacteria, halophiles, a marine mollusk, a sea anemone, a marine thermal vent fungus and two Archaea that previously were described from deep-ocean sediments. As temperatures in Antarctica began to cool and sea levels dropped about 34 million years ago, Lake Vostok become ice covered and isolated from the ocean [24]. Aerobic organisms that survive in cold and partially shaded aquatic environments could have survived in the epilimnion. Anaerobes would have been limited to the benthic and sediment regions. Hydrothermal regions may have been present much of the time, because rifting of the region began more than 60 million years ago [23]. As freshwater from the melting glacier flowed into the lake, ion gradients were likely to develop. These gradients are common in other ice-covered Antarctic lakes [39–42], and the differences in ion concentrations in the embayment compared to the main basin are consistent with this hypothesis. Hydrothermal activity in the vicinity of the embayment might cause mixing of any stratified layers in the lake in that region, leading to the higher ion and mineral inclusion concentrations in the accretion ice from that region.

Once current continuous ice-cover began approximately 15 million years ago, additional selection would have occurred as light intensities decreased, and oxygen concentrations and water pressures increased. Although extinctions would probably have been extensive, it is unlikely that all life would disappear from the lake and its sediments. Based on the number of unique species matches in BLAST searches of our accretion ice metagenomic/metatranscriptomic data set, we estimate the total number of species to be at least 1,996 in Lake Vostok, of which over 90% (approximately 1,800) are Bacteria. This converts to approximately 4 species mL^{-1}, assuming that the concentration of nucleic acids in the accretion ice is representative of the concentration of nucleic acids and organisms in the lake water, and that the samples are representative of the lake in general. However, as ice forms, it can force out many components, including some molecules, minerals and cells. Therefore, the concentrations of organisms and nucleic acids are probably higher in the lake water than in the accretion ice. The calculated concentration is below the number of species found in ocean water, sediments and most freshwater lakes [43–46]. Because the accretion ice represents primarily water at the surface of the lake, species density and richness in the lake could be much higher in some locales. In any case, a relatively high number of sequences matching sequences from a diverse set of species were found in Lake Vostok accretion ice. The 15–35 million years of coverage by ice appears to have been ample time for many organisms to adapt to the extreme conditions that developed in the lake, which has likely led to numerous speciation events.

Figure 4. Sea level changes relative to the current level of water in Lake Vostok (based on reference [24]). (**a**) Current mean sea level. Blue indicates areas below sea level if all of the ice was removed. Light green indicates areas above sea level. The upper surface of Lake Vostok (black region) is approximately 200 m below current sea level; (**b**) View of the same region with a 50 m sea level rise. Current outline of Lake Vostok is indicated by the dashed line; (**c**) View of the same region with a 100 m sea level rise.

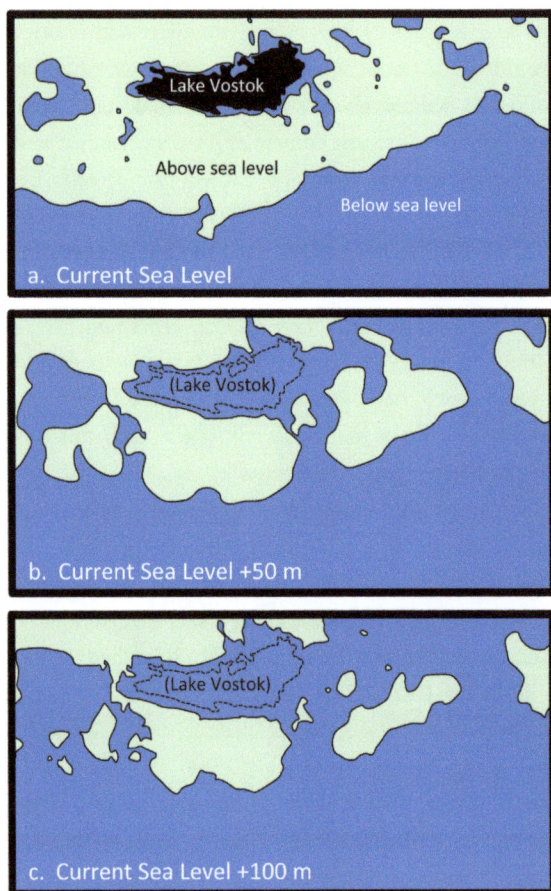

3. Experimental Section

3.1. Acquisition and Processing of Ice Core Sections

All ice core sections were selected from the USGS NICL (United States Geological Survey, National Ice Core Laboratory, Denver, CO, USA). They were selected based on desired depths and for the absence of cracks (to avoid possible external contamination). They were shipped frozen to our laboratory. Sections were surface sterilized according to a tested method that assures destruction and removal of all surface contaminating cells and nucleic acids, while preserving cells and nucleic acids frozen in the ice [16,17,47,48]. Briefly, quartered ice core sections, 6–16 cm in length (total volume approximately 125 mL), were warmed at 4 °C for 30 min (to avoid thermal shock and cracking) before surface decontamination. The work surfaces in a room (under positive pressure) separate from the main laboratory were treated with 0.5% sodium hypochlorite, 70% ethanol and UV irradiation for one hour prior to surface sterilizing and melting the ice core sections. Inside a sterile laminar flow hood, the ice core sections were surface decontaminated by total immersion in a 5.25% sodium hypochlorite solution (pre-chilled to 4 °C for at least 2 h) for 10 s followed by three rinses with 800 mL of sterile water (4 °C, 18.2 MΩ, <1 ppb [parts per billion] total organic carbon, TOC, and autoclaved). Then, the core section was transferred into a sterile funnel and melted at room temperature by collection of 25–50 mL aliquots. This protocol significantly reduces the risk of contamination of the ice core meltwater samples [47,48]. The meltwater was then frozen at −20 °C. Sample V5 included meltwater from Vostok 5G core sections at depths of 3,563 and 3,585 m, corresponding to type 1 ice that accreted in the vicinity of the embayment. Sample V6 included core sections 3,606 and 3,621 m, corresponding to type 2 ice that accreted over a portion of the southern main basin of Lake Vostok (Figure 1). A total of 250 mL of meltwater was used for each sample (approximately 125 mL from each ice core section). The meltwater samples were filtered sequentially through 1.2, 0.45 and 0.22 µm Durapore filters (Millipore, Billerica, MA, USA). The filters were stored at −80 °C for future reference. Then, the filtered meltwater was subjected to ultracentrifugation at 100,000 xg for 16 h to pellet cells and nucleic acids. Two control samples (purified water, 18.2 MΩ, <1 ppb TOC; and the same water, autoclaved and subjected to concentration by ultracentrifugation) also were processed using the same protocols. The V5, V6, and control samples were ultracentrifuged on different days to lessen potential cross-contamination. Pellets were rehydrated in 50 µL 0.1× TE (1 mM Tris [pH 7.5], 0.1 mM EDTA).

3.2. DNA and RNA Extraction

Nucleic acid extraction was performed using MinElute Virus Spin Kits (QIAGEN, Valencia, CA) and eluted in 150 µL AVE buffer (water with 0.04% sodium azide). This kit isolates both RNA and DNA. The eluted nucleic acids were further concentrated by precipitating overnight at −20 °C with 0.5 M NaCl in 80% ethanol. They were then pelleted by centrifugation at 16,000 × g for 15 min, washed with cold 80% ethanol and centrifuged at 16,000 × g for 5 min. They were dried under vacuum, and then they were resuspended in 15 µL 0.1× TE.

3.3. cDNA Synthesis and Amplification of cDNA and DNA

Complementary DNAs (cDNAs) were synthesized from the extracted RNAs. The procedure was performed using a SuperScript Choice cDNA kit (Invitrogen, Grand Island, NY, USA), according to the manufacturer's instructions, using 10 µL of the extracted RNA and 80 pmol of random hexamer primers. The cDNA was then mixed with 10 µL of extracted DNA (less than 1 ng/µL) from the same meltwater sample, and *Eco*RI (Not I) adapters (AATTCGCGGCCGCGTCGAC, dsDNA) were added using T4 DNA ligase. The final concentration of components in each reaction for addition of *Eco*RI adapters was: 66 mM Tris-HCl (pH 7.6), 10 mM MgCl$_2$, 1 mM ATP, 14 mM DTT, 100 pmols *Eco*RI (*Not* I) adapters and 0.5 units of T4 DNA ligase, in 50 µL total volume. The reaction was incubated at 15 °C for 20 h. Then, the reaction was heated to 70 °C for 10 min to inactive the ligase. [Note: The cDNAs and DNAs were mixed in order to maximize the biomass of nucleic acids, necessary for successful pyrosequencing. Thus, the cDNA comprised the metatranscriptomic fraction (of which most was from rRNA) and the DNA comprised the metagenomic fraction of each sample].

The products were size fractionated by column chromatography. Each 2 mL column contained 1 mL of Sephacryl® S-500 HR resin. TEN buffer (10 mM Tris-HCl [pH 7.5], 0.1 mM EDTA, 25 mM NaCl; autoclaved) was utilized to wash the columns and elute the samples through the columns. Fractions of approximately 40 µL were collected by chromatography. After measurement of the volume of each fraction, fractions 6–18 were precipitated to concentrate the DNA. Concentration was accomplished by adding 0.5 volumes (of the fraction size) of 1 M NaCl, and two volumes of −20 °C absolute ethanol. After gentle mixing, each was left to precipitate at −20 °C overnight. The fractions were centrifuged in a microfuge for 20 min at room temperature and decanted. Then, the pellets were washed with 0.5 mL of −20 °C 80% ethanol and centrifuged for 5 min. Finally each of the DNA pellets was dried under vacuum and rehydrated in 20 µL of 0.1× TE buffer.

After resuspension, fractions 6–18 were subjected to PCR amplification using *Eco*RI (*Not*I) adapter primers (AATTCGCGGCCGCGCTCGAC). The samples were amplified using a GeneAmp PCR Reagent Kit (Applied Biosystems, Carlsbad, CA, USA). Each reaction mixture contained: 10 mM Tris-HCl [pH 8.3], 50 mM KCl, 1.5 mM MgCl$_2$, 0.001% (w/v) gelatin, 5 pmol each dNTP (dATP, dCTP, dGTP, and dTTP), 1 U Ampli*Taq* DNA polymerase and 50 pmols *Eco*RI (*Not*I) adapter primers (AATTCGCGGCCGCGCTCGAC), each in 25 µL total volume. The thermal cycling program was: 94 °C for 4 min; then 40 cycles of 94 °C for 1 min, 55 °C for 2 min, 72 °C for 2 min; followed by an incubation for 10 min at 72 °C. A 1 µL aliquot of each was subjected to 1% agarose gel electrophoresis at 5 V/cm in TBE (89 mM tris, 89 mM borate, 2 mM EDTA [pH 8.0]), containing 0.5 µg/mL ethidium bromide, and visualized by UV irradiation. Fractions that excluded small (<200 bp) and large (>2.0 kb) fragments were used for further processing. Then, amplified products for each fraction of the desired size range (as above) were pooled (approximately 350 µL per sample), and were precipitated with NaCl and ethanol, washed, and dried (as above). Each was rehydrated in 35 µL of 0.1× TE.

3.4. Addition of 454 A and B Sequences by PCR Amplification

Each of the amplified samples was then reamplified using primers that contained *Eco*RI/*Not*I sequences on their 3' ends and 454-specific primers on their 5' ends (one primer with sequence A; underlined: CGTATCGCCTCCCTCGCGCCATCAGAATTCGCGGCCGCGTCGAC; and the other with sequence B; CTATGCGCCTTGCCAGCCCGCTCAGAATTCGCGGCCGCGTCGAC). The thermal cycling program was: 94 °C for 4 min; then 40 cycles of 94 °C for 1 min, 55 °C for 3 min, 72 °C for 3 min; followed by an incubation for 10 min at 72 °C. All PCR products were cleaned with a PCR purification kit (QIAGEN, Valencia, CA, USA). The amplicons were quantified on agarose gels (as above) to calculate concentrations (based on comparisons to plasmid pGEM4Z (Promega, Madison, WI, USA) standards on the same gel). After adjusting concentrations to approximately 1 µg/µL, 20 µg of each was sent to Roche 454 Life Sciences 454 Technologies (Roche, Branford, CT, USA) for 454 pyrosequencing using a 454 GS Junior System.

3.5. Sequence Analysis

The sequences were extracted from the data file and organized using Python (Python Software Foundation) on the Ohio Super Computer (OSC, Columbus, OH, USA). The sequences were deposited in the GenBank nucleotide database at the National Center for Biotechnology Information (NCBI; accession numbers: JQ997163–JQ997235; JQ997237–JQ997322; JQ997324–JQ997402; JQ997404–JQ997547; JQ997549–JQ998298; JQ998300–JQ998745; JQ998747–JQ999505; JQ999568–JQ999624; JQ999909–JQ999910; JQ997196–JQ997198; JQ997274; JQ997284; JQ997285; JQ997287; JQ997308; JQ997309; JQ997361; JQ997374; JQ997375; JQ997378; JQ997384; JQ997393; JQ997394; JQ997443; JQ997448; JQ997457; JQ997460; JQ997469; JQ997487–JQ997497; JQ997541; JQ997613; JQ997623; JQ997624; JQ997638; JQ997639; JQ997651; JQ997695; JQ997698; JQ997801; JQ997804; JQ997847; JQ998421; JQ998746; JQ999303; JQ999327; JQ999330; JQ999348; JQ999360; JQ999361; JJQ999365–JQ999369; JQ999371; JQ999492; JQ999493; JQ999635–JQ999829; JQ999837–JQ999897; JQ999899; JQ999901–JQ999905; JQ999507–JQ999509; JQ999512; JQ999515; JQ999518–JQ999521; JQ999523–JQ999526; JQ999529–JQ999530; JQ999533; JQ999538; JQ999540; JQ999545; JQ999549; JQ999552; JQ999554; JQ999556–JQ999564; JQ999567; JQ999629; JQ999631; JQ999830; JQ999831; JQ999833–JQ999835). They were assembled using MIRA 3.0.5 (Whole Genome Shotgun and EST Sequence Assembler [49]), using the following command line: job=denovo,genome,accurate,454. From the assembly, the average lengths for V5 and V6 were 539 and 318, respectively. The average quality scores were 39.7 and 40.2, respectively; maximum coverages were 3.3 and 6.2, respectively; and average coverages were 2.2 and 3.5, respectively. Initial taxonomic analyses were performed on MG-RAST [50] and Galaxy [51], Batch Mega-BLAST searches were performed to determine taxonomic and gene identities. The BLAST execution file was set up to retrieve the top 10 similar sequences, with e-value cutoffs of 10^{-10}. They were subjected to batch BLASTN similarity searches on the OSC, and then sorted according to gene, taxon, and similarity e-values using FileMaker (FileMaker, Inc., Santa Clara, CA, USA). The top BLASTN hit was used to determine taxonomic classification (when genus and species names were provided), also considering the lengths and percent similarities of the matches. The sequences were divided into four category files: V5 rRNA genes, V6 rRNA genes, V5 mRNA genes and V6 mRNA genes (Full list of sequences and descriptions in reference [22], and presented in Supplementary Tables S1–S10. The rRNA gene results were used primarily to determine taxonomic classifications. Each species was then categorized according to temperature requirements, growth requirements, metabolic functions and ecological niche, based on NCBI descriptions, publications cited in the NCBI descriptions, and internet sources. Some mRNA sequences were used to determine or confirm species identifications, where possible.

3.6. Metabolic Analysis

Sequences were uploaded onto the KAAS-KEGG (KEGG Automatic Annotation Server; KEGG—Kyoto Encyclopedia of Genes and Genomes, Kyoto, Japan) pathway website and blasted against the default set of bacterial and eukaryotic genes [52]. The sequences were compared to known sequences from 40 taxa (23 provided on the KAAS-KEGG site, and 17 additional taxa added manually) The additional taxa (with abbreviations used for searches) were: *Cryptococcus neoformans* JEC21; *Thalassiosira pseudonana*; *Dictyostelium discoideum*; *Burkholderia mallei* ATCC 23344; *Campylobacter jejuni* NCTC11168; *Desulfovibrio vulgaris* DP4; *Caulobacter crescentus* CB15; *Micrococcus luteus*; *Acidobacterium capsulatum*; *Flavobacterium johnsoniae*; *Fibrobacter succinogenes*; *Fusobacterium nucleatum*; *Opitutus terrae*; *Gemmatimonas aurantiaca*; *Rhodopirellula baltica*; *Chlorobium limicola*; *Chloroflexus aurantiacus*. Based on the results from the KAAS-KEGG analysis, metabolic pathways that were present in our dataset were identified. Tables of enzymes that matched our sequences from each pathway were retrieved. Subsets of these are presented in Supplementary Figures S1–S5.

4. Conclusions

Lake Vostok accretion ice contains nucleic acids from a diversity of species. While many were closest to those from psychrophilic species, a large number of sequences were closest to those from thermophilic species. Sequences from both anaerobes and aerobes were represented, as well as halophiles, aquatic and marine species. The list of taxa included approximately 94% Bacteria and 6% Eukarya, including over 100 species of multicellular Eukarya. While most were fungi (primarily ascomycetes and basidiomycetes), a number of animals also were indicated, including a rotifer, tardigrade, nematode, bivalves, sea anemone, crustaceans, and possibly fish (as suggested by the presence of sequences that were most similar to those from bacterial symbionts and pathogens of fish species). The species indicated by the sequences include those that participate in many parts of the nitrogen cycle, as well as those that fix, utilize and recycle carbon. Because of the higher concentrations of nucleic acids and viable organisms in accretion ice compared to the overriding meteoric ice [16,17], it is likely that the organisms and nucleic acids in the accretion ice originated in the lake water. The indications of large numbers of marine, halophilic and halotolerant organisms suggest that marine layers, or other saline regions, might exist in the lake. They may have originated millions of years ago at a time when the lake might have been physically connected with the surrounding ocean. Therefore, Lake Vostok might contain a complex interdependent set of organisms, zones and habitats that have developed over the tens of millions of years of its existence.

Acknowledgments

We thank NICL, NICL-SMO and the Ice Core Working group for allowing us access to the ice core sections used in this research. This research was partially funded by the National Science Foundation (ANT-0536870) and Bowling Green State University. We greatly appreciate the donation of sequencing services by Roche 454 Life Sciences for testing of their GS junior system. We thank Paul F. Morris for his valuable help with the data analyses.

References

1. Kapista, A.; Ridley, J.F.; Robin, G.Q.; Siegert, M.J.; Zotikov, I. Large deep freshwater lake beneath the ice of central Antarctica. *Nature* **1996**, *381*, 684–686.
2. MacGregor, J.A.; Matsuoka, K.; Studinger, M. Radar detection of accreted ice over Lake Vostok, Antarctica. *Earth Planet Sci. Lett.* **2009**, *282*, 222–233.
3. Masalov, V.N.; Lukin, V.V.; Shermetiev, A.N.; Popov, S.V. Geophysical investigation of the subglacial Lake Vostok in Eastern Antarctica. *Dokl. Earth Sci.* **2001**, *379A*, 734–738.
4. Studinger, M.; Karner, G.D.; Bell, R.E.; Levin, V.; Raymond, C.A.; Tikku, A. Geophysical models for the tectonic framework of the Lake Vostok region East Antarctica. *Earth Planet Sci. Lett.* **2003**, *216*, 663–677.
5. Wright, A.; Siegert, M.J. The identification and physiographical setting of Antarctic subglacial lakes: An update based on recent discoveries. *Geophys. Monogr. Ser.* **2011**, *192*, 9–26.
6. Jouzel, J.; Petit, J.R.; Souchez, R.; Barkov, N.I.; Lipenkov, V.Y.; Raynaud, D.; Stievenard, M.; Vassiliev, N.; Verbeke, V.; Vimeux, F. More than 200 meters of lake ice above subglacial Lake Vostok, Antarctica. *Science* **1999**, *286*, 2138–2141.
7. Bell, R.; Studinger, M.; Tikku, A.; Castello, J.D. Comparative biological analyses of accretion ice from subglacial Lake Vostok. In *Life in Ancient Ice*; Castello, J.D., Rogers, S.O., Eds.; Princeton University Press: Princeton, NJ, USA, 2005; pp. 251–267.
8. Gramling, C. A tiny window opens into Lake Vostok, while a vast continent awaits. *Science* **2012**, *335*, 788–789.
9. Siegert, M.J.; Ellis-Evans, J.C.; Tranter, M.; Mayer, C.; Petit, J.; Salamatin, A.; Priscu, J.C. Physical, chemical and biological processes in Lake Vostok and other Antarctic subglacial lakes. *Nature* **2001**, *414*, 603–609.
10. Siegert, M.J.; Tranter, M.; Ellis-Evans, J.C.; Priscu, J.C.; Lyons, W.B. The hydrochemistry of Lake Vostok and the potential for life in Antarctic subglacial lakes. *Hydrol. Process.* **2003**, *17*, 795–814.
11. Bulat, S.A.; Alekhina, I.A.; Lipenkov, V.Y.; Lukin, V.V.; Marie, D.; Petit, J.R. Cell concentrations of microorganisms in glacial and lake ice of the Vostok ice core, East Antarctica. *Microbiology* **2009**, *78*, 808–810.

240

12. Christner, B.C.; Royston-Bishop, G.; Foreman, C.M.; Arnold, B.R.; Tranter, M.; Welch, K.A.; Lyons, W.B.; Tsapin, A.I.; Studinger, M.; Priscu, J.C. Limnological conditions in subglacial Lake Vostok, Antarctica. *Limnol. Oceanogr.* **2006**, *51*, 2485–2501.

13. Abyzov, S.S.; Poglazova, M.N.; Mitskevich, J.N.; Ivanov, M.V. Common features of microorganisms in ancient layers of the Antarctic ice sheet. In *Life in Ancient Ice*; Castello, J.D., Rogers, S.O., Eds.; Princeton University Press: Princeton, NJ, USA, 2005; pp. 240–250.

14. Christner, B.C.; Mosley-Thompson, E.; Thompson, L.G.; Reeve, J.N. Isolation of bacteria and 16S rDNAs from Lake Vostok accretion ice. *Environ. Micrbiol.* **2001**, *3*, 570–577.

15. Christner, B.C.; Mosley-Thompson, E.; Thompson, L.G.; Reeve, J.N. Classification of bacteria in polar and nonpolar global ice. In *Life in Ancient Ice*; Castello, J.D., Rogers, S.O., Eds.; Princeton University Press: Princeton, NJ, USA, 2005; pp. 227–239.

16. D'Elia, T.; Veerapaneni, R.; Rogers, S.O. Isolation of microbes from Lake Vostok accretion ice. *Appl. Environ. Microbiol.* **2008**, *74*, 4962–4965.

17. D'Elia, T.; Veerapaneni, R.; Theraisnathan, V.; Rogers, S.O. Isolation of fungi from Lake Vostok accretion ice. *Mycologia* **2009**, *101*, 751–763.

18. Karl, D.M.; Bird, D.F.; Björkman, K.; Houlihan, T.; Shakelford, R.; Tupas, L. Microorganisms in the accreted ice of Lake Vostok, Antarctica. *Science* **1999**, *286*, 2144–2147.

19. Priscu, J.C.; Adams, E.E.; Lyons, W.B.; Voytek, M.A.; Mogk, D.W.; Brown, R.L.; McKay, C.P.; Takacs, C.D.; Welch, K.A.; Wolf, C.F. Geomicrobiology of subglacial ice above Lake Vostok, Antarctica. *Science* **1999**, *286*, 2141–2144.

20. Bulat, S.A.; Alekhina, I.A.; Blot, M.; Petit, J.R.; Waggenbach, D.; Lipenkov, V.Y.; Vasilyeva, L.P.; Wloch, D.M.; Raynaud, D. DNA signature of thermophilic bacteria from the aged accretion ice of Lake Vostok, Antarctica: Implications for searching for life in extreme icy environments. *Int. J. Astobiol.* **2004**, *1*, 1–12.

21. Lavire, C.; Normand, P.; Alekhina, I.; Bulat, S.; Prieur, D.; Birrien, J.L.; Fournier, P.; Hänni, C.; Petit, J.R. Presence of *Hydrogenophilus thermoluteolus* DNA in accretion ice in the subglacial Lake Vostok, Antarctica, assessed using *rrs, cbb* and *hox. Environ. Microbiol.* **2006**, *8*, 2106–2114.

22. Shtarkman, Y.M.; Koçer, Z.A.; Edgar, R.; Veerapaneni, R.; D'Elia, T.; Morris, P.F.; Rogers, S.O. Subglacial Lake Vostok (Antarctica) accretion ice contains a diverse set of sequences from aquatic, marine and sediment-inhabiting Bacteria and Eukarya. *PLoS One*, **2012**, in revision.

23. Ferracciolli, F.; Finn, C.A.; Jordan, T.A.; Bell, R.E.; Anderson, L.M.; Damaske, D. East Antarctic rifting triggers uplift of the Gamburtsev Mountains. *Nature* **2011**, *479*, 388–392.

24. Young, D.A.; Wright, A.P.; Roberts, J.L.; Warner, R.C.; Young, N.W.; Greenbaum, J.S.; Schroeder, D.M.; Holt, J.W.; Sugden, D.E.; Blankenship, D.D. A dynamic early East Antarctic Ice Sheet suggested by ice-covered fjord landscapes. *Nature* **2011**, *474*, 72–75.

25. Ferrer, M.; Werner, J.; Chernikova, T.N.; Barjiela, R.; Fernández, L.; La Cono, V.; Waldmann, J.; Teeling, H.; Golyshina, O.V.; Glöckner, F.O. Unveiling microbial life in the new deep-sea hypersaline Lake *Thetis*. Part II: A metagenomic study. *Environ. Microbiol.* **2011**, *14*, 268–281.

26. Toth, D.J.; Lerman, A. Stratified lake and ocean brines: Salt movement and time limits of existence. *Limnol. Oceanog.* **1975**, *20*, 715–728.

27. Bar-Even, A.; Noor, E.; Milo, R. A survey of carbon fixation pathways through a quantitative lens. *J. Exp. Bot.* **2012**, *63*, 2325–2342.

28. Burgaud, G.; Le Calvez, T.; Arzur, T.D.; Vandenkoornhuyse, P.; Barbier, G. Diversity of culturable marine filamentous fungi from deep-sea hydrothermal vents. *Environ. Microbiol.* **2009**, *11*, 1588–1600.

29. Bell, E. *Life at Extremes: Environments, Organisms and Strategies for Survival*; CABI: Cambridge, MA, USA, 2012; p. 576.

30. Daniel, M.; Cohen, D.M.; Richard, H.; Rosenblatt, R.H.; Moser, H.G. Biology and description of a bythitid fish from deep-sea thermal vents in the tropical eastern Pacific. *Deep Sea Res.* **1990**, *37*, 267–283.

31. Gaill, F.; Mann, K.; Wiedemann, H.; Engel, J.; Timpl, R. Structural comparison of cuticle and interstitial collagens from annelids living in shallow sea-water and at deep-sea hydrothermal vents. *J. Mol. Biol.* **1995**, *246*, 284–294.

32. Shank, T.M.; Black, M.B.; Halanych, K.M.; Lutz, R.A.; Vrijenhoek, R.C. Miocene radiation of deep-sea hydeothermal vent shrimp (Caridea: Bresiliidae): Evidence from mitochondrial cytochrome oxidase subunit I. *Mol. Phylogenet. Evol.* **1999**, *13*, 244–254.

33. Tunnicliffe, V. The biology of hydrothermal vents: Ecology and evolution. *Oceanogr. Mar. Biol.* **1991**, *29*, 319–407.

34. Tunnicliffe, V.; McArthur, A.G.; McHugh, D. A biogeographical perspective of the deep-sea hydrothermal vent fauna. *Adv. Mar. Biol.* **1998**, *34*, 353–442.

35. Vishnivetskaya, T.A.; Erokhina, L.G.; Spirina, E.V.; Shatilovich, A.V.; Vorobyova, E.A.; Tsapin, A.; Gilichinsky, D. Viable phototrophs: Cyanobacteria and green algae from the permafrost darkness. In *Life in Ancient Ice*; Castello, J.D., Rogers, S.O., Eds.; Princeton University Press: Princeton, NJ, USA, 2005; pp. 140–158.

36. Vrijenhoek, R.C. Gene flow and genetic diversity in naturally fragmented metapopulations of deep-sea thermal vent animals. *J. Hered.* **1997**, *88*, 285–293.

37. Tarasov, V.G.; Gebruk, A.V.; Mironov, A.N.; Moskalev, L.I. Deep-sea and shallow-water hydrothermal vent communities: Two different phenomena? *Chem. Geol.* **2005**, *224*, 5–39.

38. Zachos, J.; Pagani, M.; Sloan, L.; Thomas, E.; Billups, K. Trends, Rhythms, and Aberrations in Global Climate 65 Ma to Present. *Science* **2001**, *292*, 686–693.

39. Brambilla, E.; Hippe, H.; Hagelstein, A.; Tindall, B.J.; Stackebrandt, E. 16S rDNA diversity of cultured and uncultured prokaryotes of a mat sample from Lake Fryxell, McMurdo Dry Valleys, Antarctica. *Extremophiles* **2001**, *5*, 23–33.

40. Clocksin, K.M.; Jung, D.O.; Madigan, M.T. Cold-Active Chemoorganotrophic Bacteria from Permanently Ice-Covered Lake Hoare, McMurdo Dry Valleys, Antarctica. *Appl. Environ. Microb.* **2007**, *73*, 3077–3083.

41. Laybourn-Parry, J.; Pearce, D.A. The biodiversity and ecology of Antarctic lakes: Models for evolution. *Philos. Trans. Roy. Soc. B* **2007**, *362*, 2273–2289.

42. Mosier, A.C.; Murray, A.E.; Fritsen, C.H. Microbiota within the perennial ice cover of Lake Vida, Antarctica. *FEMS Microbiol. Ecol.* **2007**, *59*, 274–288.

43. Kalyuzhnaya, M.G.; Isapidus, A.; Ivanova, N.; Copeland, A.C.; McHardy, A.C.; Szeto, E.; Salamov, A.; Grigoriev, I.V.; Suciu, D.; Levine, S.R. High-resolution metagenomics targets specific functional types in complex microbial communities. *Nat. Biotechnol.* **2008**, *26*, 1029–1034.

44. Kemp, P.F.; Aller, J.Y. Estimating prokaryotic diversity: When are 16S rDNA libraries large enough? *Limnol. Oceanogr.* **2004**, *2*, 114–125.

45. Newton, R.J.; Jones, S.E.; Eiler, A.; McMahon, K.D.; Bertlisson, S. A guide to the natural history of freshwater lake bacteria. *Microbiol. Mol. Biol.* **2011**, *75*, 14–49.

46. Ravenschlag, K.; Sahm, K.; Pernthaler, J.; Amann, R. High Bacterial Diversity in Permanently Cold Marine Sediments. *Appl. Environ. Microb.* **1999**, *65*, 3982–3989.

47. Rogers, S.O.; Ma, L.J.; Zhao, Y.; Catranis, C.M.; Starmer, W.T.; Castello, J.D. Recommendations for elimination of contaminants and authentication of isolates in ancient ice cores. In *Life in Ancient Ice*; Castello, J.D., Rogers, S.O., Eds.; Princeton University Press: Princeton, NJ, USA, 2005; pp. 5–21.

48. Rogers, S.O.; Theraisnathan, V.; Ma, L.J.; Zhao, Y.; Zhang, G.; Shin, S.-G.; Castello, J.; Starmer, W. Comparisons of protocols to decontaminate environmental ice samples for biological and molecular examinations. *Appl. Environ. Microbiol.* **2005**, *70*, 2540–2544.

49. Chevreux, B.; Wetter, T.; Suhai, S. Genome sequence assembly using trace signals and additional sequence information, Computer Science and Biology. In Proceedings of the German Conference on Bioinformatics (GCB), Hannover, Germany, 4–6 October 1999; pp. 45–56.

50. Meyer, F.; Paarmann, D.; D'Souza, M.; Olson, R.; Glass, E.M.; Kubal, M.; Paczian, T.; Rodriguez, A.; Stevens, R.; Wilke, A. The Metagenomics RAST server—A public resource for the automatic phylogenetic and functional analysis of metagenomes. *BMC Bioinforma.* **2008**, *9*, 386.

51. Goecks, J.; Nekrutenko, A.; Taylor, J. The Galaxy Team. Galaxy: A comprehensive approach for supporting accessible, reproducible, and transparent computational research in the life sciences. *Genome Biol.* **2010**, *11*, R86.

52. Moriya, Y.; Itoh, M.; Okuda, S.; Yoshizawa, A.C.; Kanehisa, K. KAAS: An automatic genome annotation and pathway reconstruction server. *Nucleic Acids Res.* **2007**, *35*, W182–W185.

Sea Ice Microorganisms: Environmental Constraints and Extracellular Responses

Marcela Ewert and Jody W. Deming

Abstract: Inherent to sea ice, like other high latitude environments, is the strong seasonality driven by changes in insolation throughout the year. Sea-ice organisms are exposed to shifting, sometimes limiting, conditions of temperature and salinity. An array of adaptations to survive these and other challenges has been acquired by those organisms that inhabit the ice. One key adaptive response is the production of extracellular polymeric substances (EPS), which play multiple roles in the entrapment, retention and survival of microorganisms in sea ice. In this concept paper we consider two main areas of sea-ice microbiology: the physico-chemical properties that define sea ice as a microbial habitat, imparting particular advantages and limits; and extracellular responses elicited in microbial inhabitants as they exploit or survive these conditions. Emphasis is placed on protective strategies used in the face of fluctuating and extreme environmental conditions in sea ice. Gaps in knowledge and testable hypotheses are identified for future research.

Reprinted from *Biology*. Cite as: Ewert, M.; Deming, J.W. Sea Ice Microorganisms: Environmental Constraints and Extracellular Responses. *Biology* **2013**, *2*, 603–628.

1. Introduction

Sea ice is a dynamic, porous matrix that harbors within its interior network of brine pores and channels an active (e.g., [1,2]) and diverse [3–6] community. The sympagic (ice-associated) community has multiple trophic levels including photosynthetic bacteria and algae, chemoautotrophic bacteria and archaea, and heterotrophic bacteria, archaea, flagellates, fungi and small metazoans [5,7–11]. Members of this community, particularly the bacteria and algae, play important roles in cycling carbon [12,13] and nitrogen [14,15] in polar regions; selected bacteria also respond to pollutants such as crude oil [16,17] and mercury [18,19].

The seasonal (autumnal) decrease in temperature that leads to the formation of sea ice in polar waters progressively reduces the liquid phase of the ice—the brine volume fraction—and consequently increases the concentration of solutes and particles in the brine. Phase equations of sea ice [20–22] are frequently used to estimate brine salinity and brine volume fraction based on the temperature of the ice and its bulk salinity (salinity after melting). Temperature determines solute concentration such that when the ice reaches a temperature of -5 °C, just $\sim$3 °C below the freezing point of seawater, the estimated brine volume fraction has decreased below 0.3 (even as low as 0.05 for ice with low bulk salinity; Figure 1a) and the estimated brine salinity has increased to nearly 100 (Figure 1b). At extreme winter temperatures, salt precipitation within the brine phase adds complexity to these constraints (see deflection points at -22.9 °C, the eutectic for hydrohalite, in Figure 1b). Organisms, previously at seawater temperature and salinity, are thus exposed to much lower temperatures, higher salinities, and reduced habitable space soon after entrapment in sea ice.

Figure 1. Dependence of brine volume fraction (**a**) and brine salinity (**b**) on sea-ice temperature, according to phase equations from Cox and Weeks [21]. Contour lines in (**a**) indicate the effect of different bulk salinities on brine volume fraction. Brine salinity (**b**) is independent of bulk ice salinity, conventionally determined only by temperature; we suggest, by the shadowing of the line, that the presence of extracellular polymeric substances (EPS) produced by sea-ice organisms may influence brine salinity in as yet unpredictable ways (see Section 3.3).

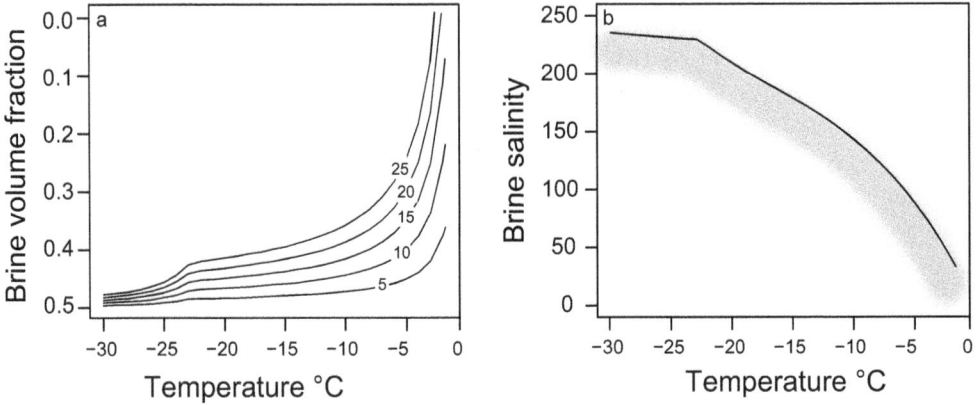

Other properties of sea-ice brines also show temperature-dependent changes. For instance, the solubility of biologically relevant gases, including CO_2 and O_2, decreases as the salinity of brines increases, translating to a general degassing effect throughout the winter season [23]. Subsequently, biological O_2 consumption can further reduce O_2 concentration in some sections of the sea-ice column leading to the development of microbial communities and processes that are favored by low oxygen conditions [14,15,24].

Summer brings a new set of changes. Melting of snow and then surface ice, and the consequent formation of melt ponds, exposes sea-ice microorganisms to salinities close to that of freshwater, or else flushes them back to the ocean through brine drainage [13,25]. Although disadvantageous to halophiles, this surface melting generates an environment suitable for the growth of a distinctive community of freshwater microorganisms [26,27]. Increased solar radiation in the spring and summer also promotes growth of photosynthetic organisms, further influencing sea-ice brine composition and prompting the secretion of protective screening and quenching compounds [28]. The onset of significant algal photosynthesis imposes additional challenges to the resident microbial community in the form of oxidative stress (by the increase of O_2) or decrease in pH (by consumption of CO_2 and consequent shifts in the carbonate chemistry of brine).

Some microorganisms have constitutive adaptations (expressed constantly) that allow them to thrive or survive under specific conditions of high salinity or low temperature. In a variable environment such as sea ice, however, acclimation mechanisms that allow a microorganism to function across a range of conditions may have an advantage over constitutive mechanisms finely tuned to an extreme, but relatively constant, environment. Microbial adaptations to sea ice may involve intracellular processes, membrane proteins or the secretion of extracellular polymeric substances (EPS). The latter can modify the immediate surroundings of the organism and its neighbors. The protective role of EPS, which tends to be gelatinous in nature, has been recognized since the earliest stages of sea-ice microbiology, as this account from 1918 by McLean [29] relates:

> "It is a curious fact, and yet a well-known experience, to find that bacteria may live dormant in ice for prolonged periods, and that infection may be carried through ice, but it is not so generally recognized that some bacteria prefer to grow on ice. Microorganisms, as a rule, are capable of resisting a low temperature when their ordinary activities cease, and they tend, either as single units or in clusters, to throw out a mucilaginous protein substance for their protection."

This review addresses some of the physico-chemical parameters that define the sea-ice environment on a seasonal basis and the protective responses they elicit in microorganisms, particularly bacteria, the category of microorganisms known to inhabit all depths, forms and seasons of sea ice (Figure 2). It diverges from prior reviews on sea-ice microorganisms and their environment (e.g., [7,30–35]) by giving particular attention to the protective role of EPS in the face of extreme fluctuations in key environmental parameters characteristic of the Arctic. In doing so, gaps in knowledge and new directions for research are identified.

2. Sea Ice as a Microbial Environment

2.1. Incorporation into Sea Ice

Sea ice formation starts in the fall. When water reaches its freezing point, ice crystals, known as frazil ice, form throughout the upper water column and rise to the surface accumulating in a slush. This surface slush continues to freeze, consolidating into an upper layer of ice. Ensuing sea-ice growth depends on the conditions of the ocean: calm conditions promote the formation of congelation ice; turbulent conditions lead to frazil ice growth [36]. The growth of sea ice produces a porous structure that, unlike glacial ice, retains abundant impurities that were present in the source water. Salts and other solutes, organic and inorganic particles, and microorganisms are rejected by the growing ice lattice into interconnected liquid inclusions within the ice. The liquid inclusions, brine channels and pores, form an interconnected network that accumulates high concentrations of solutes and constitutes the inhabited fraction of sea ice [37].

Figure 2. Schematic diagram of seasonal (fall through winter) processes influencing microorganisms in sea ice, including transport mechanisms (orange arrows) and some of the microbial adaptive responses (italics). During sea-ice formation, larger organisms ascend with rising frazil ice crystals; smaller bacteria and archaea likely attach to algae, particles or ice crystals. Once entrained in the ice, microorganisms inhabit a network of brine channels where they experience low temperature (T), high brine salinity (S) and reduced living space, but are protected from fluctuations in air temperature by the insulating properties of snow and ice. As sea ice consolidates, brines are expelled into the ocean (desalination) and onto the surface; a fraction of the microorganisms, EPS and other components of the brine are expelled, too. Surface-expelled brines and their contents form a skim layer that can be incorporated into frost flowers and snow, prone to wind dispersal. The skim layer and frost flowers, directly exposed to the atmosphere, experience more extreme fluctuations in temperature and brine salinity and, as the sun rises in late winter, greater UV exposure. From remaining areas of open water, including leads, wind can transport marine microorganisms in aerosols. Airborne microorganisms (including terrestrial bacteria) can nucleate snow and return to the ice/snow surface.

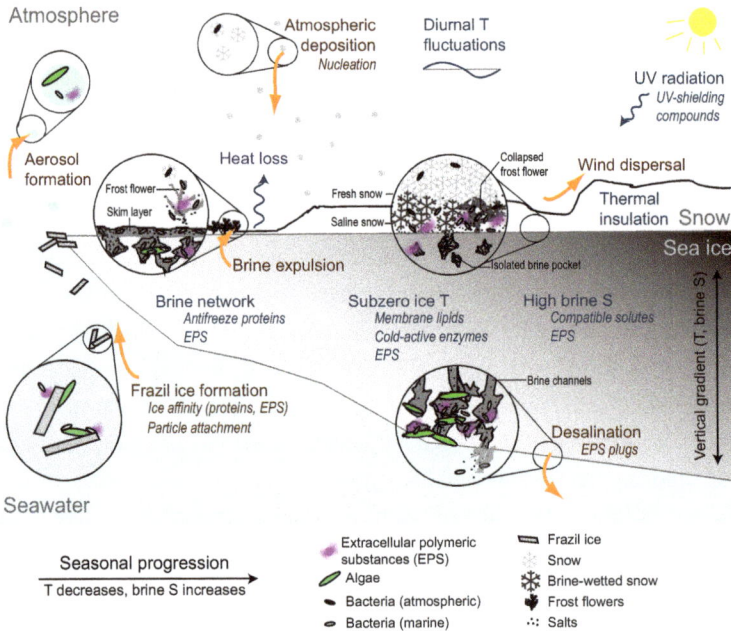

Different processes can contribute to the retention of some of the microorganisms, particles and organic substances entrained into the ice from the source water (Figure 2). Physical concentration can enrich algal cells, either through scavenging by frazil ice or by water circulation through the newly established ice layer [38]. Scavenging occurs when frazil ice crystals drift towards the surface, dragging in their path particles and algal cells from the water column that will later concentrate in the ice [39,40]. The physical concentration of algae has been observed in the field, with algal cells

>10 μm preferentially enriched in young Arctic sea ice [8]. The ability of marine algae to nucleate ice crystals [41] could also contribute to the enrichment process of sea-ice algae by favoring the formation of ice crystals in their immediate environment that can lift them towards the consolidating ice layer.

In contrast, enrichment of smaller bacterial (and archaeal) cells in sea ice does not appear to occur directly by physical processes. Bacterial incorporation can be facilitated by the presence of algae, through bacterial association with algal cells or aggregated algal EPS, that are then concentrated by physical processes [38,42,43]. Attachment of bacteria to algae in young Arctic sea-ice samples has been observed but, because insufficient data are available to consider it a widespread phenomenon, the potential role of EPS is highlighted instead [43]. Indeed, particulate EPS (pEPS; > 0.4 μm) of likely algal origin are rapidly enriched in newly forming sea ice [43,44]. Bacterial EPS also have the potential to play a role in the entrainment and retention of bacteria in ice either directly or by promoting attachment to algal cells or detrital particles amenable to physical entrainment. Dissolved EPS (dEPS; < 0.4 μm) produced by *Colwellia psychrerythraea* strain 34H, a model marine psychrophilic bacterium [45], was found to be selectively retained in saline ice under experimental conditions [46]. Marine bacteria with the ability to produce an EPS coating with similar properties could be retained in the ice by this means alone, independently of association with algal cells or their byproducts, an hypothesis that remains to be tested.

As sea ice consolidates, brines in the upper layers of the ice are expelled upwards to the ice surface forming a surface skim layer (Figure 2). Sea-ice microorganisms (primarily bacteria), EPS and dissolved organic compounds are carried with the brine to this even colder habitat at the ice-atmosphere interface. A fraction of the bacteria and EPS may be selectively retained in the ice, however, following the arguments and potential mechanisms outlined for initial entrainment into the ice (EPS coatings and attachment to larger particles or ice crystals). Brines in this skim layer, and the bacteria and organic substances within it, can subsequently be incorporated into frost flowers that form on the new ice surface [47] or into the saline snow layer, which represents a vast bacterial habitat in its own right [48]. Frost flowers and saline snow will have brine inclusions with properties similar to those of the ice, but exposed to more extreme environmental parameters.

2.2. The Low Temperature Constraint

A defining characteristic of the sea-ice environment is temperatures below, and sometimes well below, 0 °C. Meltponds, the accumulated meltwater from snow and surface ice during the summer season, are the only ice-associated environments with temperatures above 0 °C (between 0.4 and 1.5 °C, according to Lee and collaborators [13]). Sea-ice temperatures range typically between -2 °C and -30 °C with the coldest ones recorded in upper sea ice during Arctic winter [49] (winter lows for Antarctic sea ice are less extreme). Environments associated with the surface of new ice in winter, such as the brine skim layer and frost flowers, can be exposed to air temperatures below -30 °C. As a result of the insulating properties of ice and snow, environments experiencing the most severe fluctuations in temperature (and thus brine salinity) are those directly exposed to the winter atmosphere: the brine skim layer and frost flowers on the surface of new ice and, to a lesser degree, the saline snow layer when snow is blown to minimal thickness (as discussed by Ewert and collaborators [48]). To illustrate a typical range of temperatures and fluctuations seasonally experienced in Arctic sea ice and associated environments, Figure 3 presents detailed temperature measurements from the Mass Balance Observatory Site from the University of Alaska Fairbanks [50]. This observatory, located in landfast coastal sea ice in the coast of Barrow, Alaska (156.5° W, 71.4° N), measures air-snow-ice-water temperature profiles throughout the winter and spring seasons.

Figure 3. Temperature recorded at the Mass Balance Observatory Site (Barrow, AK, USA) during 2011 (days of year 25–158) at different depths above and below the ice surface. Dashed lines mark seasonal transitions. Spring equinox was on day 79, summer solstice on day 171.

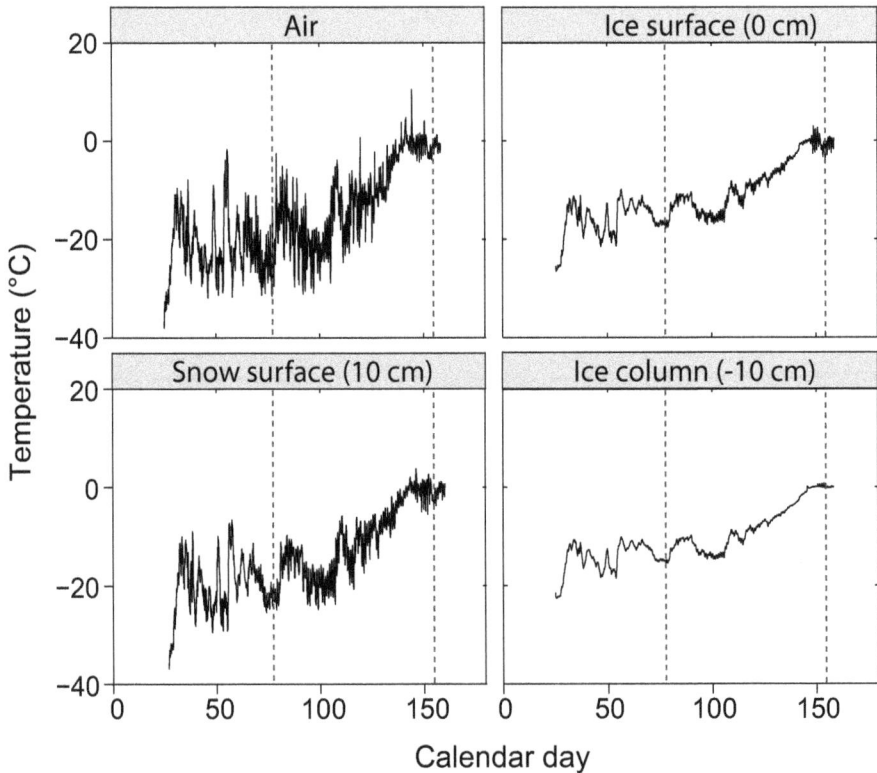

Low temperatures impose constraints at different levels in the single-cell microorganism and, consequently, elicit responses spanning different aspects of its physiology. Reaction and transport rates decrease with temperature, slowing most physiological processes. Protein folding is affected by a decrease in hydrophobic forces and changes in hydration [51]. Membranes become rigid. Nucleic acids become more stable, which hinders replication, transcription and translation processes. Microbial solutions to these constraints also span the gamut of possibilities. Cold-active enzymes remain functional at low temperatures by favoring amino acids that allow higher flexibility and structural modifications that provide ligands better access to the catalytic site [51]. Bakermans and collaborators [52] found that *Psychrobacter cryohalolentis* K5, a psychrotolerant bacterium isolated from permafrost, showed important changes in its proteome with up to 30% of the proteins having significantly different levels of expression when exposed to low temperatures. Among these proteins were cold-shock chaperones to facilitate translation, cold-adapted alleles that would allow a same function be performed at two different temperatures, and an increase in the expression of certain transporters. Bacteria also respond to low temperatures by changing the type of fatty acids and carotenoids present in their membranes and altering the membrane protein content [35,52,53]. In fact, some bacteria can perceive changes in temperature through a membrane-bound sensor that triggers the expression of cold-activated genes [53].

Given the prevailing low temperatures, organisms inhabiting sea ice and associated environments can be expected to use many of these strategies to cope with low-temperature constraints. When compared with the underlying water, the sea-ice bacterial community is enriched in culturable taxa considered psychrophilic, *i.e.*, uniquely adapted to low temperatures [54]. Also, psychrophilic organisms tend to be more abundant in the upper layers of the ice, which have been exposed to the coldest temperatures during the winter [54]. Extracellular enzymes from sea ice have also been recognized for their unusually low optimal temperatures [55], especially those assayed in winter ice [56].

2.3. The Brine Channel Network

The brine channel network containing the liquid fraction of the ice has a complex three-dimensional structure [57]. For relatively warm ice near the ice-water interface, brine channels range in diameter from a few to hundreds of micrometers; the network is dominated by the smallest channels (<40 μm) that account for about 50 % of the surface area [9]. Brine inclusions are characterized by their volume fraction and connectivity, temperature-dependent properties that determine the permeability of the ice. For ice with a given bulk salinity, the size and connectivity of the brine inclusions will decrease with temperature until the ice reaches a critical porosity where it is no longer permeable [58]. According to phase equations of sea ice, the threshold for fluid permeability in sea ice occurs when the porosity approaches 5%, which occurs at a temperature of about −5 °C for ice with a salt concentration of 5 ppt [59]. Once the ice reaches the permeability threshold, pockets of brine become isolated from the underlying seawater and from each other (though micrometer-scale connections remain possible [56]). Shrinking of brine inclusions leads to an increase in the concentration of salts and other solutes, and of organisms and other particles present in the brines. This temperature-dependent concentration of solutes exposes sea-ice organisms to seasonal changes in salinity. The uppermost section of the ice experiences the most drastic changes, where, as temperature decreases in the winter, the concentration of salts in the brine can reach a salinity of 220, or 6–7 times higher than that experienced by microorganisms in seawater before their entrapment in the ice. As salts in the brine concentrate above their saturation points, ice experiences the successive precipitation of ikaite ($CaCO_3 \cdot 6H_2O$) at −2.2 °C, mirabilite ($Na_2SO_4 \cdot 10H_2O$) at

−8.2 °C, and hydrohalite (NaCl·2H$_2$O) at −22.9 °C [20,60]. Salt precipitation changes the ionic composition of sea-ice brines with respect to the source seawater and generates additional solid surfaces (salt crystals) with which microorganisms can interact. Such interactions have not been explored, except conceptually (see Section 3.3).

Temperature-dependent reduction of brine volume also increases the percentage of brine channel area covered by organisms [9], as well as the concentration of bacteria, viruses and free DNA within the brine [61,62] simultaneously increasing their contact rates in exponential fashion [63]. High concentrations and contact rates with viruses and nucleic acids have been hypothesized to promote lateral gene transfer in sea ice [56,64]. Properties of the brine channel network, mainly its connectivity and volume fraction, also affect the type of predators and the predator-prey dynamics of the sea-ice community. In general, larger predators only access the lowermost sections of the ice [65], with pores of less than 200 μm considered a refuge for smaller organisms (microalgae, ciliates and bacteria [9]). Some metazoan predators such as rotifers and turbellaria, though, are flexible enough to squeeze into brine channels with diameters much smaller than their bodies, and also adjust their body size according to changes in ambient salinity [9]. In the smallest sea-ice brine inclusions, viruses take the role as main predators [11,61].

Observations by Krembs and collaborators [66] confirmed the role of EPS-producing microorganisms in directly influencing the properties of the brine channel network. The presence of EPS increased the abundance of pores in the ice by 15% (over EPS-free ice) and led to the formation of pores with convoluted irregular shapes [66]. The effects of EPS were also evident in the larger pores (> 250 μm), where perimeter-to-length ratios corresponded to a fractal geometry, as opposed to the Euclidean geometry characteristic of pores in artificially grown sea ice lacking EPS [66]. The presence of EPS thus affects the habitability of sea ice by increasing the volume of the habitable liquid phase and the interior ice-liquid surface area available for "colonization." These effects result from interactions between EPS and the ice, whether by clogging the brine channel network, changing the viscosity of the brine or directly associating with the ice crystals [66].

Sea-ice microorganisms can also modify the brine channel network through their antifreeze proteins, another type of extracellular substance produced by both sea-ice diatoms [67] and bacteria [68]. Extracellular antifreeze proteins secreted by the sea-ice diatom *Fragilariopsis cylindrus* can alter the microscopic and macroscopic structure of saline ice, opening the possibility for this protein and similar ones to play an important role in shaping the sea-ice microbial environment if produced in sufficient quantities [69,70]. The presence of extracellular organic substances with the ability to change macroscopic and microscopic structure of sea ice suggests a possible need to re-evaluate the applicability of Cox and Weeks [21] equations to describe brine salinity and brine volume fraction in natural, EPS-rich sea ice (Figure 1a). The issue is of particular relevance since the phase equations of sea ice are a common tool for estimating the brine salinities experienced by sea-ice organisms *in situ*.

2.4. The Brine Salinity Constraint

The high salinity characteristic of sea-ice brine imposes at least two types of constraints on resident microorganisms. First, high concentrations of salts tend to affect the functioning of proteins, including precipitating them. Bacteria and archaea inhabiting high salinity environments tend to have, as a response, acidic proteins that, given their abundant negative charges, remain soluble and functional at higher salinities than basic proteins. Second, high environmental salinity exposes organisms to high osmotic pressure that drives water out of the cell, resulting in potential dehydration, loss of turgor pressure and reduction of cell volume. To counteract this water efflux, microorganisms of all types compensate for excessive concentrations of external solutes by accumulating compatible solutes in the interior of the cell. The general microbial ability to tolerate and even thrive in sea-ice brines comes with the added benefit of refuge against metazoan predators more susceptible to increases in salinity, such as those reported by Krembs and collaborators [9].

Figure 4. Brine salinity estimated from temperature data in Figure 3. Depths and dashed lines as in Figure 3. Brine salinity calculated using air temperature represents the extreme situation in which expelled sea-ice brines are directly exposed to the atmosphere and in thermal equilibrium with it.

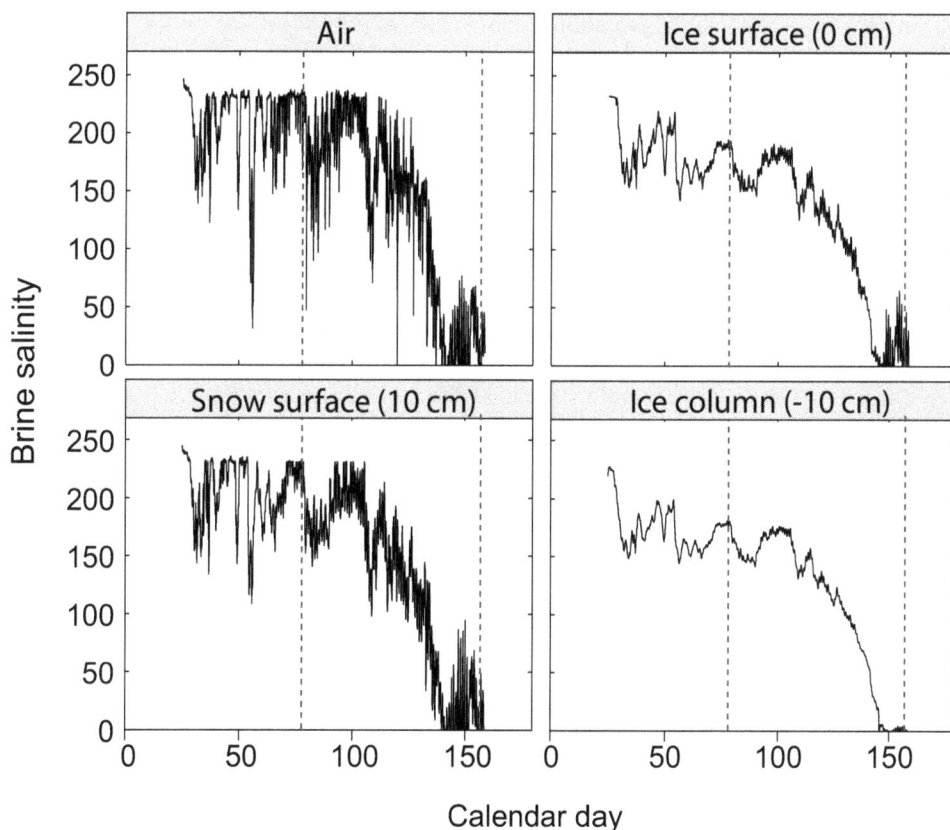

The osmotic up-shift that occurs with ice formation happens quickly as the temperature of the ice drops (Figure 1b). Sea-ice brines, though, are distinguished from other high-salinity environments not only by subzero temperature but also by extreme fluctuations in salinity (Figure 4). A common bacterial response to osmotic up-shift starts with the transient accumulation of K^+ and glutamate, accompanied by a release of putrescine to balance intracellular charges [71,72]. Avoiding growth limitations inherent to an intracellular accumulation of salts, microorganisms replace the accumulated K^+ with compatible solutes, which are either imported or synthesized directly in the cell [71]. Compatible solutes are small, water-soluble organic molecules that increase the osmolarity of the cytoplasm without the disruptive effect of salt ions [73]. Dozens of compatible solutes have been described for Bacteria and Archaea, including free amino acids and their derivatives, sugars and their derivatives, and polyols and their derivatives. Among the most common compatible solutes are betaine, ectoine, trehalose, α-glucosylglycerol and glutamate [73]. A suite of genes allows for the transport of compatible solutes from the environment and/or their synthesis in the cell [74].

Not all organisms accumulate compatible solutes. Some extremely halophilic microorganisms, such as *Halobacterium salinarum*, which grows optimally at 26% NaCl, compensate high external concentration of solutes by incorporating salts in their cytoplasm. To keep cytoplasmic proteins functioning after the accumulation of salts, *H. salinarum* expresses an unusually high ratio of acidic to basic proteins (4.9 for the complete proteome [75]). Membrane proteins, adapted to function when directly exposed to the high-salinity environment, present a similar tendency to be acidic independently of the intracellular accumulation of organic solutes or K^+ salts in the cytoplasm [76].

Partial proteomes available to examine for bacteria known from sea ice, when compared with bacteria and archaea from other saline and fresh-water environments (following [76]), also present the signature of a high acid-to-basic ratio in their membrane proteins (Table 1), consistent with the high salinities seasonally experienced in sea-ice brines. Their cytoplasmic ratios, however, do not compare with the extremely halophilic reference strain. Although the difference may be domain-specific (*Halobacterium*, contrary to its name, is an archaeal genus), we hypothesize that the salting-in strategy used by *H. salinarum*, with long term accumulation of K^+ ions in the cytoplasm and the majority of cytoplasmic proteins being acidic, has not been adopted by microorganisms from the sea-ice environment despite exposure to high brine salinities. Greater physiological flexibility will be provided by use of the compatible solutes strategy in the face of strong seasonal fluctuations in salinity inherent to sea-ice brines. Note that *Psychromonas ingrahamii*, which has the lowest ratio of acidic to basic proteins in either its membrane or cytoplasmic proteome compared with other marine isolates considered (Table 1), was isolated from the sea ice/water interface, which tends to have lower salinities and species not necessarily adapted to life in the brine channel network [77].

Extracellular polymeric substances are also used by sea-ice microorganisms as a response to elevated salinities. The sea-ice diatom *Fragilariopsis cylindrus* has been shown to increase the production of all types of EPS (soluble, insoluble and frustule-associated) when frozen under high-salinity conditions [77]. Likewise, high concentrations of EPS from a sea-ice isolate of the bacterial genus *Pseudoalteromonas* were shown to extend the range of salinities at which this strain could grow, while also providing protection against freeze-thaw cycles [78].

Osmotic down-shift will be experienced in the late spring and the summer as the ice warms and melts. The extent of the osmotic down-shift depends on summer brine drainage. If melting prompts the brine to be flushed back into the ocean, microorganisms may not likely experience salinities much lower than seawater. If melting results in the formation of surface meltponds, then one of two conditions will follow: meltponds connected to seawater will have salinities close to 29, similar to those in nearby surface water; unconnected meltponds will have salinities below 5, reaching values as low as 0.1 [13]. In the latter case, microorganisms will be exposed to a drastic down-shift in salinity, which could result in the lysis of a significant fraction of the population. For bacteria, even if not directly lysed, a down-shift may prompt lysogenic viruses (already carried by the cell) to enter the lytic stage in those cells with an active metabolism [79,80] and lead to bacterial loss by that mechanism. The possibility that an EPS coating may protect against a drastic down-shift in salinity or viral lysis has not been tested (see Section 3.2).

Table 1. Ratio of acidic to basic proteins in partial proteomes of selected microorganisms.

Organism	Membrane	Cytoplasmic	Environment
Extremely halophilic			
Halobacterium salinarum	3.88	16.8	Highly saline lakes
Halophilic			
Psychrobacter cryohalolentis	2.22	3.27	Cryopeg
Roseobacter denitrificans	2.22	3.00	Marine
Psychrobacter arcticus	2.14	3.34	Permafrost
Sphingopyxis alaskensis	1.90	2.23	Marine
Shewanella frigidimarina	1.52	2.90	Marine, sea ice
Colwellia psychrerythraea	1.47	3.17	Marine sediments, sea ice
Shewanella oneidensis	1.47	3.26	Anaerobic sediments
Marinobacter aquaeolei	1.42	3.13	Marine
Oceanobacillus iheyensis	1.15	3.78	Marine sediments
Psychromonas ingrahamii	1.10	2.00	Sea ice / water interface
Non-halophilic			
Flavobacterium psychrophilum	0.75	1.44	Freshwater fish
Lactococcus lactis subsp. *lactis*	0.74	4.03	Gut flora
Sphingomonas wittichii	0.69	2.28	River

Ratio of proteins with isoelectric point (pI) < 7 to proteins with pI > 7; pI calculated with the *Compute pI/Mw* tool from the ExPasy Bioinformatics Resource Portal [81]. All reviewed protein entries for each organism retrieved from the UniProtKB data base on November 2012 [82], annotated for location as either "membrane" or "cytoplasmic."

2.5. Insolation

Solar radiation drives numerous reactions, biotic and abiotic, including the alteration and destruction of biologically relevant molecules. Processes driven by solar radiation, and the responses they trigger in microorganisms, are of particular relevance in polar regions where strong seasonal changes in insolation occur. At the organism level, UV radiation (UVR) can damage DNA and other nucleic acids by the formation of thymine dimers. UVR is known to decrease viability in bacteria from aquatic ecosystems [83] and damage the photosynthetic potential of benthic algae [84].

Given the potential detrimental effects of high irradiation, microorganisms have developed multiple protective responses, including the production of shading pigments, antioxidant compounds, and the performance of rapid DNA repair. For instance, in sediment-associated diatoms, exposure to UV-B prompts motility (away from radiation) and the production of carotenoid pigments able to function as quenching agents [84]. The synthesis of mycosporine-like amino acids (MAA), a type of UVR-screening compound, is widespread in marine microscopic algae, especially those associated with surface blooms [85]. Similar responses to UVR are found in the sea-ice microbial community. Uusikivi and collaborators measured relatively high concentrations of MAA in Baltic sea ice, particularly in the surface layers [86]. Likewise, Mundy and collaborators reported the production of carotenoid pigments and mycosporine-like amino acids by algal communities associated with sea ice during the melting season of Arctic coastal first year ice, under high levels of UVR [28]. Motility by sea-ice algae in response to changing irradiance has been suggested [87] but, to the best of our knowledge, has not been confirmed as a mechanism of photoadaptation in sea ice. The general effects of UVR on EPS are less clear, with some (non-sea-ice) studies finding an increase [83] and others a decrease [84] in EPS content of the UV-exposed community.

Solar radiation further influences the sea-ice ecosystem by driving reactions that modify the dissolved organic carbon (DOC) pool. The potential of organic compounds to participate in photochemical reactions can be inferred from their absorption of visible radiation and UVR. DOC from spring sea ice, known to absorb UVR, has been shown to undergo varied photochemical reactions including changes in bioavailability and photooxidation to CO_2 [23,88]. EPS are also affected by solar radiation. Ortega-Retuerta and collaborators demonstrated that transparent exopolymer particles (an alternative descriptor for pEPS) from natural North Sea water and from cultures of the marine diatom, *Chaetoceros affinis*, can be photolysed by UV-B (290–315 nm), and to some extent by UV-A (315–400 nm) and photosynthetically active radiation (400–700 nm) [89].

Possible effects of solar radiation on EPS specific to sea-ice environments have been considered [47] but not tested to our knowledge. As part of a larger study [48], we examined the susceptibility of Arctic sea-ice EPS to photochemical reactions by measuring absorption spectra for pEPS samples from upper sea ice and saline snow. Absorption, converted to Napierian absorption coefficients (m^{-1}), was higher in the UV-B range (Figure 5), suggesting that EPS associated with the winter sea-ice bacterial community may be susceptible to photochemical changes during the spring and summer when the radiation level increases. Similar profiles to the one in Figure 5 have been observed for particulate organic matter from late-winter surface Baltic sea ice [86], except for a peak in the 320–345 range associated with MAA that was absent from our samples. Samples from Baltic sea ice [86] were collected after the snow melt and contained a community of microscopic algae likely responsible for the production of MAA; in contrast, our samples were collected before snow melt and dominated by a bacterial community, explaining the absence of a MAA signature. In fact, Cockell and collaborators found that a snow cover of 5–15 cm thickness could reduce the transmittance of UV by an order of magnitude and reduce the impact of radiation in bacterial spores [90]. The snow cover over sea ice may thus act as a seasonal shading agent, protecting surface sea-ice microorganisms against UV radiation. This protective cover, though, is highly heterogeneous in thickness and melts early in the season [91].

Figure 5. Absorption spectra for pEPS solution concentrated from surface samples of winter first year ice (open circles, 13 mg glu-eq mL^{-1}) and saline snow (filled circles, 9.3 mg glu-eq mL^{-1}). Samples were collected offshore Barrow, Alaska, in February 2010, filtered onto 0.4 μm polycarbonate filters as described by Ewert and collaborators [48], kept frozen in the dark at $-20\,^{\circ}$C for 20 months, and resuspended in 1.5 mL of distilled water for analysis.

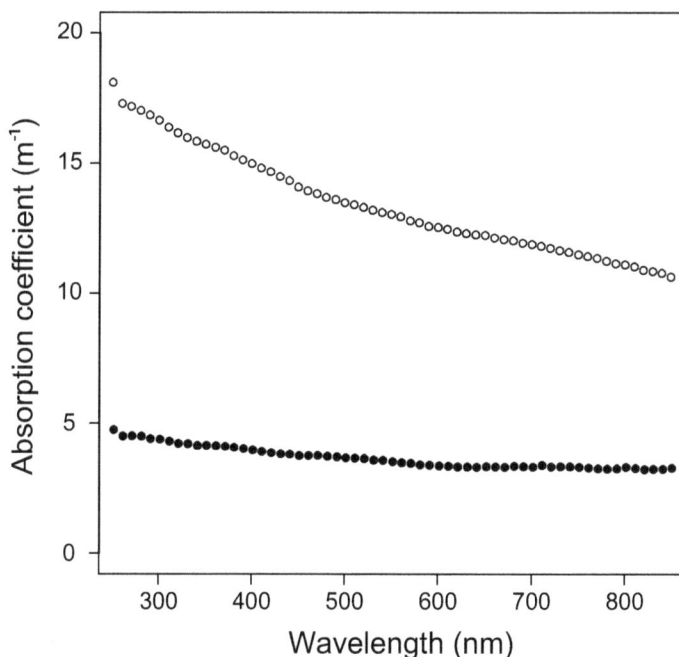

3. Extracellular Responses to Sea-Ice Environmental Constraints

3.1. Extracellular Polymeric Substances

Extracellular polymeric substances (EPS), composed primarily of polysaccharides, are commonly produced by a wide range of microorganisms from both terrestrial and marine environments. EPS differ amongst organisms and producing conditions in sugar chain length and branching, sugar composition, type of sugar linkages, and the presence of additional chemical groups such as sulfates, proteins, lipids and even nucleic acids [92,93]. Different bacterial strains can produce EPS of different chemical composition and structure [93,94], but a single strain can also produce more than one kind of EPS [95]. Likewise, the type and amount of EPS produced by a bacterial strain can be modified by exposure to certain environmental conditions such as salinity [96], temperature [97] or presence of heavy metals [98]. Hence, the term EPS does not refer to a single chemically defined molecule but a complex mixture of diverse polysaccharides and ancillary compounds. Because EPS measurements typically quantify only the polysaccharide fraction of these components, the term EPS has also been used to refer specifically to extracellular polysaccharide substances. We use EPS throughout this review in its broadest meaning, unless otherwise specified.

EPS can be either tightly bound to the cell surface, loosely attached, or cell free [99]. Cell-free hydrophobic EPS from mesophilic bacteria have been shown to self-assemble into polymer microgels and to accelerate the self-assembly of microgels in seawater, with implications for concentrating organic-rich substrates for bacterial degradation [100]. These properties, however, were inhibited at low temperature for the particular polysaccharides studied [100]. The extent to which such self-assembly may occur in the sea-ice environment, where solute concentrations are high and EPS may be derived from psychrophilic microorganisms, has not been fully explored. An initial analysis in winter ice indicated minimal self-assembly [101], yet EPS aggregates with spherical diameters between 2 and 50 μm have been observed in sea ice [44].

3.2. EPS in Sea Ice

Most of the EPS in sea ice can be attributed to production by ice algae, either before or after entrainment into the ice. Even in regions of the ice dominated by bacteria, the trail of algal EPS is expected to overwhelm the amount produced by sea-ice bacteria [11,66]. Diatoms may produce distinctive EPS depending on their particular sea-ice habitat. Two diatoms isolated from the sea-ice brine network, *Fragilariopsis curta* and *F. cylindrus*, produced complex polysaccharides of higher molecular mass, with low relative abundance of glucose but high relative content of galactose, xylose and fucose. In contrast, a species of *Synedropsis* from the ice-water interface produced EPS dominated by low-molecular weight polysaccharides with low complexity and high relative content of glucose [77].

EPS abundance in sea ice and associated environments has been quantified in numerous studies beginning with those by Krembs and collaborators in 2002 [101] and by Meiners and collaborators in 2003 [44]. During fall, the number of EPS particles in sea ice can be an order of magnitude higher than in underlying water and often correlates with the presence of sea-ice algae [44]. The dissolved EPS fraction is consistently more abundant in sea ice [66] and sea-ice associated environments such as frost flowers [102] and saline snow [48].

The EPS pool in sea ice is established during ice formation [43,46,102] but can be modified subsequently by the entrained microorganisms. For instance, sea-ice microorganisms can add EPS to the existing pool by producing it *in situ* as a stress response, a process inferred from the increase in EPS concentration in winter sea ice [49,101]. On the other hand, bacteria may selectively degrade and consume certain fractions of the EPS pool, changing its overall chemical composition and size fractionation, as suggested by the detailed analyses of Underwood and collaborators [103].

The widespread, yet heterogeneous (e.g., [103]), presence of EPS in sea ice and associated environments may reflect the varied functions these polymers perform at different ecosystem levels [99,104]. At the microorganism level EPS have been associated with cell adhesion and aggregation [105], motility [106], affinity for metals [107], and with providing a sticky framework to keep extracellular enzymes in the immediate vicinity of the cell [99]. EPS can also provide protection against toxic heavy metals [108] and desiccation [109]. All of these functions have relevance in sea-ice environments. In particular, recent experimental data have shown that EPS can play a role in protecting sea-ice bacteria [78] and diatoms [77] against the challenges of high-salinity brines. These results are in agreement with data from other environments where high-salinity stress triggered changes in the type and amount of EPS produced by microorganisms from anaerobic sludge [96] and by freshwater cyanobacteria [110]. Likewise, EPS could have a role in protection against low salinity shocks. The marine psychrophilic bacterium *Colwellia psychrerythraea* strain 34H, whose immediate relatives are found in sea ice, increased the amount of EPS produced per cell when exposed to low salinities not permissive of growth [111]. The survival benefit was implied but not directly tested.

3.3. Influence of EPS on Physical-Chemical Properties of Sea Ice

Further insight into the protective role of EPS comes from experiments by Krembs and collaborators [66], who observed that artificial ice formations containing algal EPS had higher bulk salinities than EPS-free counterparts. This result has been related to the potential of EPS to form "plugs" in the brine channels, increasing the amount of salts that are retained (Figure 2). Following the phase equations of sea ice, higher bulk salinities result in higher brine volume fractions under similar temperature regimes (Figure 1b), effectively increasing the available habitable space for microorganisms.

The salinity of the brine pockets, however, is conventionally described as a function of temperature only and does not depend on the bulk salinity of ice. Following earlier work [66,112], we suggested in Figure 1b that the presence of EPS may have an effect on the validity of traditional phase equations when applied to natural sea ice. Some possible mechanisms may involve extracellular polymers (whether EPS or proteins [46]) with ice activity interacting with the ice surface of the brine pores and channels. If an important fraction of the surface area is covered, the growth of ice crystals might be restricted, resulting in local areas with lower salinity than predicted. EPS may also partition the brine within an ice pore creating microscale salinity gradients that affect ice crystal growth in currently unpredictable ways [112]. Another option could be antifreeze proteins, whereby more water in the liquid state would mean lower salinities. An EPS plug physically decreasing the minimal size of the brine pocket would have a similar effect.

Divalent cations present in sea-ice brines can also interact with charged groups in the backbone of EPS. In the marine environment, this interaction has been suggested to play a role in the binding of key nutrients for the cell such as iron [113]. Likewise, the interaction of EPS with Ca^{2+} determines self-assembly of marine gels, which can in turn increase the availability of nutrients for the microbial population [114]. In the case of sea ice, the relationship between EPS and Ca^{2+} may figure in the fate of carbonates in sea ice [115]. Relationships between bulk measures of dissolved organic matter and $CaCO_3$ precipitation were not evident in Antarctic sea ice [116], but experiments specifically using EPS under ice-brine conditions have not been reported. If the dissolved organic matter measured in seawater by Chave and Suess [117] included EPS, then evidence exists for a role in delaying the onset of $CaCO_3$ precipitation. Data on the interactions between EPS and calcium in other environments [118] may inform first tests of this hypothesis for sea ice.

Bergmann and collaborators used conductometric titrations to estimate the amount of binding sites for divalent cations present in ionic and nonionic bacterial extracellular polysaccharides [119]. Conductometric titrations [120] measure changes in conductivity resulting from the addition of a saline solution to a solution of interest, and provide information on the charge density of polyelectrolytes such as ionic polysaccharides. A non-ionic polysaccharide such as dextran has a titration curve where no interaction with Ca^{2+} ions is evident. Xanthan, being an ionic polysaccharide, has a titration curve with a clear offset due to its conductive properties and the presence of associated counter-ions. Its curve also shows two segments with distinctive slopes, indicating that Ca^{2+} ions interacted with the polysaccharide until all binding sites were occupied [119].

Following this approach, conductometric titrations were performed on solutions (in de-ionized water) of dextran, xanthan, and EPS obtained from a culture of *Colwellia psychrerythraea* strain 34H. A blank with no polysaccharide added was also included (see Figure 6 for details). The resulting titration curves and the slopes of their respective linear regressions (Figure 6) agree with results from Bergmann and collaborators [119]. The titration curve of dextran and the blank closely resemble each other, whereas the titration curve of xanthan has an offset and two segments with distinctive slopes. The slope of the 34H EPS curve is the same as the slope of the first segment for xanthan, the segment where interaction with Ca^{2+} is expected; there is no change in the slope, however, and the offset is 6 times higher. EPS from strain 34H thus likely contains charged polysaccharides with abundant backbone charges and associated counter-ions (high curve offset) and multiple binding sites for Ca^{2+} (no change in slope over the tested range of $CaCl_2$ concentrations). The presence of charged EPS from this Arctic marine psychrophile has implications for the dynamics of carbonates in the sea-ice environment given that charged polysaccharides, unlike non-ionic polysaccharides, have known effects on the precipitation of carbonates [121].

Interactions between cations and polysaccharides also confer both algal and bacterial EPS with the potential to adsorb heavy metal contaminants such as Cd^{2+} [98], Pb^{2+} [122] and Hg^{2+} [123], which can then be incorporated into the food chain [124,125]. Disconcerting concentrations of these heavy metals have been found in the Arctic marine food web [126]; EPS from sea-ice organisms may be playing a role in the fate of these contaminants. Of special interest is the dependence of heavy metal adsorption to EPS on properties such as salinity, pH and Ca^{2+} concentrations [122–124], properties that undergo seasonal changes in sea-ice brines.

Figure 6. Conductometric titration of polysaccharide solutions with $CaCl_2$ (0.05 M). Each data point shows the effect of increasing concentration of $CaCl_2$ on preexisting solutions of polysaccharide (0.5 g L^{-1}). Value in parentheses is the slope of the titration curve. Slopes were calculated using linear regressions, all of which have R > 0.99 and p value < 0.001. Experiments were performed at room temperature, with less than 1 degree difference among treatments (blank, 22.0 °C ± 0.1; dextran, 22.1 °C ± 0.1; xanthan 22.0 °C ± 0.1; 34H EPS, 22.9 °C ± 0.1). Cell-free EPS from strain 34H was extracted by centrifugation and precipitation with ethanol as in [111], followed by freeze-drying.

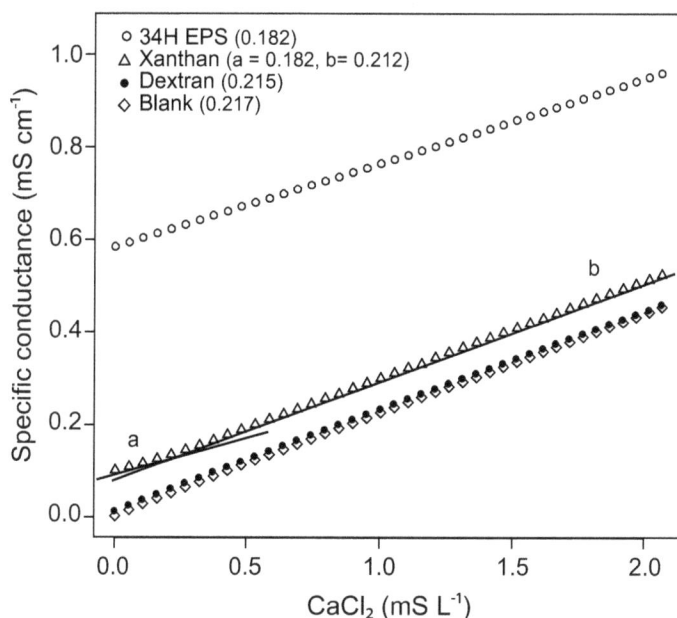

EPS can also interact with other extracellular macromolecules, most relevantly with proteins. Non-covalent interactions between EPS and proteins allow the formation of complexes, coacervates and aggregates, increasing the range of pH in which a protein is soluble [127]. In the case of cold-adapted marine organisms, EPS have been shown to increase the stability and half-life of a cold-active extracellular aminopeptidase from *C. psychrerythraea* 34H [128]. The interaction of EPS with antifreeze proteins is of particular interest for it could allow the accumulation of such proteins in the immediate vicinity of the cell, concentrating their antifreeze effects [70] to the benefit of sea-ice inhabitants as temperatures drop seasonally. Interactions between EPS and proteins under *in situ* conditions relevant to microbial life in sea ice remain largely unexplored.

4. Prospectus for Future Research

In focusing on the extracellular responses of sea-ice microorganisms to the sometimes severe and fluctuating environmental conditions of their habitat, our goal has been to highlight a number of features, particularly regarding EPS, that have not been fully explored or that raise new research directions. The extracellular products of microorganisms entrapped in sea ice are known to influence the microstructure of the ice, and thus its habitability, but how they may influence the effectiveness of traditional equations for calculating key parameters of sea ice or function at the micrometer scale within an ice pore is not clear. Unanswered questions involve the potential role of EPS in mitigating the salt concentration directly experienced by the cell, contributing to the adaptive strategy of compatible solutes, and blocking viral attack. EPS interactions with other exudates, including ice-active proteins and hydrolytic enzymes that serve a substrate-acquisition function for the cell, are poorly known. If the unexplored interactions of EPS with inorganic ions, particularly Ca^{2+}, were to be as significant in sea ice as they are in other environments, then the implications for carbon transport through the sea-ice cover could be quantitatively important. As much as has been learned over the past decades about sea-ice microorganisms and their self-protective responses to the constraints of their habitat, more awaits discovery.

Acknowledgments

This work was supported by NSF-OPP award 0908724 and the Walters Endowed Professorship to J.W.D. We thank our lab-group colleagues for insightful discussion and anonymous reviewers for their helpful comments.

References

1. Junge, K.; Eicken, H.; Deming, J.W. Bacterial activity at –2 to –20 °C in Arctic wintertime sea ice. *Appl. Environ. Microbiol.* **2004**, *70*, 550–557.
2. Søgaard, D.H.; Kristensen, M.; Rysgaard, S.; Glud, R.N.; Hansen, P.J.; Hilligsøe, K.M. Autotrophic and heterotrophic activity in Arctic first-year sea ice: Seasonal study from Malene Bight, SW Greenland. *Mar. Ecol. Prog. Ser.* **2010**, *419*, 31–45.
3. Brown, M.V.; Bowman, J.P. A molecular phylogenetic survey of sea-ice microbial communities (SIMCO). *FEMS Microbiol. Ecol.* **2001**, *35*, 267–275.
4. Brinkmeyer, R.; Knittel, K.; Jürgens, J.; Weyland, H.; Amann, R.; Helmke, E. Diversity and structure of bacterial communities in Arctic versus Antarctic pack ice. *Appl. Environ. Microbiol.* **2003**, *69*, 6610–6619.
5. Bowman, J.S.; Rasmussen, S.; Blom, N.; Deming, J.W.; Rysgaard, S.; Sicheritz-Ponten, T. Microbial community structure of Arctic multiyear sea ice and surface seawater by 454 sequencing of the 16S RNA gene. *ISME J.* **2012**, *6*, 11–20.
6. Maas, E.W.; Simpson, A.M.; Martin, A.; Thompson, S.; Koh, E.Y.; Davy, S.K.; Ryan, K.G.; O'Toole, R.F. Phylogenetic analyses of bacteria in sea ice at Cape Hallett, Antarctica. *N. Z. J. Mar. Freshw. Res.* **2012**, *46*, 3–12.

7. Horner, R.; Ackley, S.F.; Dieckmann, G.S.; Gulliksen, B.; Hoshiai, T.; Legendre, L.; Melnikov, I.A.; Reeburgh, W.S.; Spindler, M.; Sullivan, C.W. Ecology of sea ice biota. *Polar Biol.* **1992**, *12*, 417–427.

8. Gradinger, R.; Ikävalko, J. Organism incorporation into newly forming Arctic sea ice in the Greenland Sea. *J. Plankton Res.* **1998**, *20*, 871–886.

9. Krembs, C.; Gradinger, R.; Spindler, M. Implications of brine channel geometry and surface area for the interaction of sympagic organisms in Arctic sea ice. *J. Exp. Mar. Biol. Ecol.* **2000**, *243*, 55–80.

10. Lizotte, M.P. The microbiology of sea ice. In *Sea Ice: An Introduction to Its Physics, Chemistry, Biology and Geology*; Thomas, D.N., Dieckmann, G.S., Eds.; Wiley-Blackwell: Oxford, UK, 2003; pp. 184–210.

11. Collins, R.E.; Rocap, G.; Deming, J.W. Persistence of bacterial and archaeal communities in sea ice through an Arctic winter. *Environ. Microbiol.* **2010**, *12*, 1828–1841.

12. Miller, L.A.; Papakyriakou, T.N.; Collins, R.E.; Deming, J.W.; Ehn, J.K.; Macdonald, R.W.; Mucci, A.; Owens, O.; Raudsepp, M.; Sutherland, N. Carbon dynamics in sea ice: A winter flux time series. *J. Geophys. Res.* **2011**, *116*, C02028.

13. Lee, S.H.; Stockwell, D.A.; Joo, H.M.; Son, Y.B.; Kang, C.K.; Whitledge, T.E. Phytoplankton production from melting ponds on Arctic sea ice. *J. Geophys. Res.* **2012**, *117*, C04030.

14. Kaartokallio, H. Evidence for active microbial nitrogen transformations in sea ice (Gulf of Bothnia, Baltic Sea) in midwinter. *Polar Biol.* **2001**, *24*, 21–28.

15. Rysgaard, S.R.; Glud, R.N. Anaerobic N_2 production in Arctic sea ice. *Limnol. Oceanogr.* **2004**, *49*, 86–94.

16. Delille, D.; Basseres, A.; Dessommes, A. Seasonal variation of bacteria in sea ice contaminated by diesel fuel and dispersed crude oil. *Microb. Ecol.* **1997**, *33*, 97–105.

17. Brakstad, O.G.; Nonstad, I.; Faksness, L.G.; Brandvik, P.J. Responses of microbial communities in Arctic sea ice after contamination by crude petroleum oil. *Microb. Ecol.* **2008**, *55*, 540–552.

18. Barkay, T.; Poulain, A.J. Mercury (micro) biogeochemistry in polar environments. *FEMS Microbiol. Ecol.* **2006**, *59*, 232–241.

19. Møller, A.K.; Barkay, T.; Abu Al-Soud, W.; Sørensen, S.J.; Skov, H.; Kroer, N. Diversity and characterization of mercury-resistant bacteria in snow, freshwater and sea-ice brine from the High Arctic. *FEMS Microbiol. Ecol.* **2011**, *75*, 390–401.

20. Assur, A. *Composition of Sea Ice and Its Tensile Strength*; University of California Libraries: San Diego, CA, USA, 1960; pp. 106–138.

21. Cox, G.F.N.; Weeks, W.F. Equations for determining the gas and brine volumes in sea-ice samples. *J. Glaciol.* **1983**, *29*, 306–316.

22. Leppäranta, M.; Manninen, T. The brine and gas content of sea ice with attention to low salinities and high temperatures. *Finn. Inst. Mar. Res. Int. Rep.* **1988**, *2*, 1–14.

23. Thomas, D.N.; Papadimitriou, S.; Michel, C. Biogeochemistry of Sea Ice. In *Sea Ice*, 2nd ed.; Thomas, D.N., Dieckmann, G.S., Eds.; Wiley-Blackwell: Oxford, UK, 2010; pp. 425–467.

24. Petri, R.; Imhoff, J.F. Genetic analysis of sea-ice bacterial communities of the Western Baltic Sea using an improved double gradient method. *Polar Biol.* **2001**, *24*, 252–257.

25. Meiners, K.; Krembs, C.; Gradinger, R. Exopolymer particles: Microbial hotspots of enhanced bacterial activity in Arctic fast ice (Chuckchi Sea). *Aquat. Microb. Ecol.* **2008**, *52*, 195–207.

26. Brinkmeyer, R.; Glockner, F.O.; Helmke, E.; Amann, R. Predominance of β-proteobacteria in summer melt pools on Arctic pack ice. *Limnol. Oceanogr.* **2004**, *49*, 1013–1021.

27. Gradinger, R.; Meiners, K.; Plumley, G.; Zhang, Q.; Bluhm, B.A. Abundance and composition of the sea-ice meiofauna in off-shore pack ice of the Beaufort Gyre in summer 2002 and 2003. *Polar Biol.* **2005**, *28*, 171–181.

28. Mundy, C.J.; Gosselin, M.; Ehn, J.K.; Belzile, C.; Poulin, M.; Alou, E.; Roy, S.; Hop, H.; Lessard, S.; Papakyriakou, T.N.; *et al.* Characteristics of two distinct high-light acclimated algal communities during advanced stages of sea ice melt. *Polar Biol.* **2011**, *34*, 1869–1886.

29. McLean, A.L. Bacteria of ice and snow in Antarctica. *Nature* **1918**, *102*, 35–39.

30. Legendre, L.; Ackley, S.F.; Dieckmann, G.S.; Gulliksen, B.; Horner, R.; Hoshiai, T.; Melnikov, I.A.; Reeburgh, W.S.; Spindler, M.; Sullivan, C.W. Ecology of sea ice biota. *Polar Biol.* **1992**, *12*, 429–444.

31. Ackley, S.F.; Sullivan, C.W. Physical controls on the development and characteristics of Antarctic sea ice biological communities—A review and synthesis. *Deep Sea Res. Part I* **1994**, *41*, 1583–1604.

32. Staley, J.T.; Gosink, J.J. Poles apart: Biodiversity and biogeography of sea ice bacteria. *Annu. Rev. Microbiol.* **1999**, *53*, 189–215.

33. Thomas, D.N.; Dieckmann, G.S. Antarctic sea ice—A habitat for extremophiles. *Science* **2002**, *295*, 641–644.

34. Mock, T.; Thomas, D.N. Recent advances in sea-ice microbiology. *Environ. Microbiol.* **2005**, *7*, 605–619.

35. Deming, J.W. Extremophiles: Cold Environments. In *Encyclopedia of Microbiology*; Shaechter, M., Ed.; Elsevier: Oxford, UK, 2009; pp. 147–158.

36. Petrich, C.; Eicken, H. Growth, Structure and Properties of Sea Ice. In *Sea Ice*, 2nd ed.; Thomas, D.N., Dieckmann, G.S., Eds.; Wiley-Blackwell: Oxford, UK, 2010; pp. 247–282.

37. Junge, K.; Krembs, C.; Deming, J.W.; Stierle, A.; Eicken, H. A microscopic approach to investigate bacteria under in-situ conditions in sea-ice samples. *Ann. Glaciol.* **2001**, *33*, 304–310.

38. Weissenberger, J.; Grossmann, S. Experimental formation of sea ice: Importance of water circulation and wave action for incorporation of phytoplankton and bacteria. *Polar Biol.* **1998**, *20*, 178–188.

39. Garrison, D.L.; Ackley, S.F.; Buck, K.R. A physical mechanism for establishing algal populations in frazil ice. *Nature* **1983**, *306*, 363–365.

40. Garrison, D.L.; Close, A.R.; Reimnitz, E. Algae concentrated by frazil ice: Evidence from laboratory experiments and field measurements. *Antarct. Sci.* **1989**, *1*, 313–316.

41. Knopf, D.A.; Alpert, P.A.; Wang, B.; Aller, J.Y. Stimulation of ice nucleation by marine diatoms. *Nature Geosci.* **2011**, *4*, 88–90.

42. Grossmann, S.; Dieckmann, G.S. Bacterial standing stock, activity, and carbon production during formation and growth of sea ice in the Weddell Sea, Antarctica. *Appl. Environ. Microbiol.* **1994**, *60*, 2746–2753.

43. Riedel, A.; Michel, C.; Gosselin, M.; LeBlanc, B. Enrichment of nutrients, exopolymeric substances and microorganisms in newly formed sea ice on the Mackenzie shelf. *Mar. Ecol. Prog. Ser.* **2007**, *342*, 55–67.

44. Meiners, K.; Gradinger, R.; Fehling, J.; Civitarese, G.; Spindler, M. Vertical distribution of exopolymer particles in sea ice of the Fram Strait (Arctic) during autumn. *Mar. Ecol. Prog. Ser.* **2003**, *248*, 1–13.

45. Deming, J.W. Sea Ice Bacteria and Viruses. In *Sea Ice*, 2nd ed.; Thomas, D.N., Dieckmann, G.S., Eds.; Wiley-Blackwell: Oxford, UK, 2010; pp. 247–282.

46. Ewert, M.; Deming, J.W. Selective retention in saline ice of extracellular polysaccharides produced by the cold-adapted marine bacterium *Colwellia psychrerythraea* strain 34H. *Ann. Glaciol.* **2011**, *52*, 111–117.

47. Bowman, J.S.; Deming, J.W. Elevated bacterial abundance and exopolymers in saline frost flowers and implications for atmospheric chemistry and microbial dispersal. *Geophys. Res. Lett.* **2010**, *37*, L13501, doi:10.1029/2010GL043020.

48. Ewert, M.; Carpenter, S.D.; Colangelo-Lillis, J.; Deming, J.W. Bacterial and extracellular polysaccharide content of brine-wetted snow over Arctic winter first-year sea ice. *J. Geophys. Res.* **2013**, doi:10.1002/jgrc.20055.

49. Collins, R.E.; Carpenter, S.D.; Deming, J.W. Spatial heterogeneity and temporal dynamics of particles, bacteria, and pEPS in Arctic winter sea ice. *J. Mar. Syst.* **2008**, *74*, 902–917.

50. Druckenmiller, M.L.; Eicken, H.; Johnson, M.; Pringle, D.; Williams, C. Towards an integrated coastal sea-ice observatory: System components and a case study at Barrow, Alaska. *Cold Reg. Sci. Technol.* **2009**, *56*, 61–72.

51. Georlette, D.; Blaise, V.; Collins, T.; D'Amico, S.; Gratia, E.; Hoyoux, A.; Marx, J.C.; Sonan, G.; Feller, G.; Gerday, C. Some like it cold: Biocatalysis at low temperatures. *FEMS Microbiol. Rev.* **2004**, *28*, 25–42.

52. Bakermans, C.; Tollaksen, S.L.; Giometti, C.S.; Wilkerson, C.; Tiedje, J.M.; Thomashow, M.F. Proteomic analysis of *Psychrobacter cryohalolentis* K5 during growth at subzero temperatures. *Extremophiles* **2007**, *11*, 343–354.

53. Shivaji, S.; Prakash, J.S.S. How do bacteria sense and respond to low temperature? *Arch. Microbiol.* **2010**, *192*, 85–95.

54. Bowman, J.P.; McCammon, S.A.; Brown, M.V.; Nichols, D.S.; McMeekin, T.A. Diversity and association of psychrophilic bacteria in Antarctic sea ice. *Appl. Environ. Microbiol.* **1997**, *63*, 3068–3078.

55. Huston, A.L.; Krieger-Brockett, B.B.; Deming, J.W. Remarkably low temperature optima for extracellular enzyme activity from Arctic bacteria and sea ice. *Environ. Microbiol.* **2001**, *2*, 383–388.

56. Deming, J.W. Life in Ice Formations at Very Cold Temperatures. In *Physiology and Biochemistry of Extremophiles*; ASM Press: Washington, DC, USA, 2007; pp. 133–144.

57. Weissenberger, J.; Dieckmann, G.; Gradinger, R.; Spindler, M. Sea ice: A cast technique to examine and analyze brine pockets and channel structure. *Limnol. Oceanogr.* **1992**, *37*, 179–183.

58. Golden, K.M.; Eicken, H.; Heaton, A.L.; Miner, J.; Pringle, D.J.; Zhu, J. Thermal evolution of permeability and microstructure in sea ice. *Geophys. Res. Lett.* **2007**, *34*, L16501.

59. Golden, K.M.; Ackley, S.F.; Lytle, V.I. The percolation phase transition in sea ice. *Science* **1998**, *282*, 2238–2241.

60. Rysgaard, S.R.; Bendtsen, J.; Delille, B.; Dieckmann, G.S.; Glud, R.N.; Kennedy, H.; Mortensen, J.; Papadimitriou, S.; Thomas, D.N.; Tison, J.L. Sea ice contribution to the air-sea CO_2 exchange in the Arctic and Southern Oceans. *Tellus B* **2011**, *63*, 823–830.

61. Collins, R.E.; Deming, J.W. Abundant dissolved genetic material in Arctic sea ice Part II: Viral dynamics during autumn freeze-up. *Polar Biol.* **2011**, *34*, 1831–1841.

62. Collins, R.E.; Deming, J.W. Abundant dissolved genetic material in Arctic sea ice Part I: Extracellular DNA. *Polar Biol.* **2011**, *34*, 1819–1830.

63. Wells, L.E.; Deming, J.W. Modelled and measured dynamics of viruses in Arctic winter sea-ice brines. *Environ. Microbiol.* **2006**, *8*, 1115–1121.

64. Collins, R.E. Microbial evolution in sea ice: Communities to genes. PhD Thesis, University of Washington, Seattle, WA, USA, 2009.

65. Gradinger, R.; Friedrich, C.; Spindler, M. Abundance, biomass and composition of the sea ice biota of the Greenland Sea pack ice. *Deep Sea Res. Part II* **1999**, *46*, 1457–1472.

66. Krembs, C.; Eicken, H.; Deming, J.W. Exopolymer alteration of physical properties of sea ice and implications for ice habitability and biogeochemistry in a warmer Arctic. *Proc. Natl. Acad. Sci. USA* **2011**, *108*, 3653–3658.

67. Janech, M.G.; Krell, A.; Mock, T.; Kang, J.; Raymond, J.A. Ice-binding proteins from sea ice diatoms (Bacillarophyceae). *J. Phycol.* **2006**, *42*, 410–416.

68. Raymond, J.A.; Fritsen, C.; Shen, K. An ice-binding protein from an Antarctic sea ice bacterium. *FEMS Microbiol. Ecol.* **2007**, *61*, 214–221.

69. Raymond, J.A. Algal ice-binding proteins change the structure of sea ice. *Proc. Natl. Acad. Sci. USA* **2011**, *108*, E198.

70. Bayer-Giraldi, M.; Weikusat, I.; Besir, H.; Dieckmann, G. Characterization of an antifreeze protein from the polar diatom *Fragilariopsis cylindrus* and its relevance in sea ice. *Cryobiology* **2011**, *63*, 210–219.

71. Dinnbier, U.; Limpinsel, E.; Schmid, R.; Bakker, E.P. Transient accumulation of potassium glutamate and its replacement by trehalose during adaptation of growing cells of *Escherichia coli* K-12 to elevated sodium chloride concentrations. *Arch. Microbiol.* **1988**, *150*, 348–357.

72. Wood, J.M. Osmosensing by bacteria: Signals and membrane-based sensors. *Microbiol. Mol. Biol. Rev.* **1999**, *63*, 230–262.

73. Roberts, M. Organic compatible solutes of halotolerant and halophilic microorganisms. *Saline Syst.* **2005**, *1*, doi:10.1186/1746-1448-1-5.

74. Mao, X.; Olman, V.; Stuart, R.; Paulsen, I.T.; Palenik, B.; Xu, Y. Computational prediction of the osmoregulation network in *Synechococcus* sp. WH8102. *BMC Genomics* **2010**, *11*, 291.

75. Kennedy, S.P.; Ng, W.V.; Salzberg, S.L.; Hood, L.; DasSarma, S. Understanding the adaptation of *Halobacterium* species NRC-1 to its extreme environment through computational analysis of its genome sequence. *Genome Res.* **2001**, *11*, 1641–1650.

76. Saum, S.H.; Pfeiffer, F.; Palm, P.; Rampp, M.; Schuster, S.C.; Müller, V.; Oesterhelt, D. Chloride and organic osmolytes: A hybrid strategy to cope with elevated salinities by the moderately halophilic, chloride-dependent bacterium *Halobacillus halophilus*. *Environ. Microbiol.* **2012**, doi:10.1111/j.1462-2920.2012.02770.x.

77. Aslam, S.; Cresswell-Maynard, T.; Thomas, D.N.; Underwood, G.J.C. Production and characterization of the intra- and extracellular carbohydrates and polymeric substances (EPS) of three sea-ice diatom species, and evidence for a cryoprotective role for EPS. *J. Phycol.* **2012**, *48*, 1494–1509.

78. Liu, S.B.; Chen, X.L.; He, H.L.; Zhang, X.Y.; Xie, B.B.; Yu, Y.; Chen, B.; Zhou, B.C.; Zhang, Y.Z. Structure and ecological roles of a novel exopolysaccharide from the Arctic sea ice bacterium *Pseudoalteromonas* sp. Strain SM20310. *Appl. Environ. Microbiol.* **2013**, *79*, 224–230.

79. Ghosh, D.; Roy, K.; Williamson, K.E.; Srinivasiah, S.; Wommack, K.E.; Radosevich, M. Acyl-homoserine lactones can induce virus production in lysogenic bacteria: An alternative paradigm for prophage induction. *Appl. Environ. Microbiol.* **2009**, *75*, 7142–7152.

80. Shkilnyj, P.; Koudelka, G.B. Effect of salt shock on stability of γ^{imm433} lysogens. *J. Bacteriol.* **2007**, *189*, 3115–3123.

81. ExPASy. SIB Bioinformatics Resource Portal, Compute pI/Mw. Available online: http://web.expasy.org/compute_pi/ (accessed on 1 November 2012).

82. Universal Protein Resource (UniProt). Available online: http://www.uniprot.org/uniprot/ (accessed on 1 November 2012).

83. Thomas, V.K.; Kuehn, K.A.; Francoeur, S.N. Effects of UV radiation on wetland periphyton: Algae, bacteria, and extracellular polysaccharides. *J. Freshw. Ecol.* **2009**, *24*, 315–326.

84. Underwood, G.J.C.; Nilsson, C.; Sundbäck, K.; Wulff, A. Short-term effects of UVB radiation on chlorophyll fluorescence, biomass, pigments, and carbohydrate fractions in a benthic diatom mat. *J. Phycol.* **1999**, *35*, 656–666.

85. Jeffrey, S.W.; MacTavish, H.S.; Dunlap, W.C.; Vesk, M.; Groenewoud, K. Occurrence of UVA-and UVB-absorbing compounds in 152 species (206 strains) of marine microalgae. *Mar. Ecol. Prog. Ser.* **1999**, *189*, 35–51.

86. Uusikivi, J.; Vähätalo, A.V.; Granskog, M.A.; Sommaruga, R. Contribution of mycosporine-like amino acids and colored dissolved and particulate matter to sea ice optical properties and ultraviolet attenuation. *Limnol. Oceanogr.* **2010**, *55*, 703–713.

87. Horner, R.; Schrader, G.C. Relative contributions of ice algae, phytoplankton, and benthic microalgae to primary production in nearshore regions of the Beaufort Sea. *Arctic* **1982**, *4*, 485–503.

88. Norman, L.; Thomas, D.N.; Stedmon, C.A.; Granskog, M.A.; Papadimitriou, S.; Krapp, R.H.; Meiners, K.M.; Lannuzel, D.; van der Merwe, P.; Dieckmann, G.S. The characteristics of dissolved organic matter (DOM) and chromophoric dissolved organic matter (CDOM) in Antarctic sea ice. *Deep Sea Res. Part II* **2011**, *58*, 1075–1091.

89. Ortega-Retuerta, E.; Passow, U.; Duarte, C.M.; Reche, I. Effects of ultraviolet B radiation on (not so) transparent exopolymer particles. *Biogeosci. Discuss.* **2009**, *6*, 7599–7625.

90. Cockell, C.; Rettberg, P.; Horneck, G.; Scherer, K.; Stokes, D.M. Measurements of microbial protection from ultraviolet radiation in polar terrestrial microhabitats. *Polar Biol.* **2003**, *26*, 62–69.

91. Sturm, M.; Massom, R.A. Snow and Sea Ice. In *Sea Ice*, 2nd ed.; Thomas, D.N., Dieckmann, G.S., Eds.; Wiley-Blackwell: Oxford, UK, 2010; pp. 153–204.

92. Ruas-Madiedo, P.; de los Reyes-Gavilán, C.G. Methods for the screening, isolation and characterization of exopolysaccharides produced by lactic acid bacteria. *J. Dairy Sci.* **2005**, *88*, 843–856.

93. Mancuso Nichols, C.; Garon Lardière, S.; Bowman, J.P.; Nichols, P.D.; Gibson, J.A.E.; Guézennec, J. Chemical characterization of exopolysaccharides from Antarctic marine bacteria. *Microb. Ecol.* **2005**, *49*, 578–589.

94. Lemoine, J.; Chirat, F.; Wieruszeski, J.M.; Strecker, G.; Favre, N.; Neeser, J.R. Structural characterization of the exocellular polysaccharides produced by *Streptococcus thermophilus* SFi39 and SFi12. *Appl. Environ. Microbiol.* **1997**, *63*, 3512–3518.

95. Schiano Moriello, V.; Lama, L.; Poli, A.; Gugliandolo, C.; Maugeri, T.L.; Gambacorta, A.; Nicolaus, B. Production of exopolysaccharides from a thermophilic microorganism isolated from a marine hot spring in flegrean areas. *J. Ind. Microbiol. Biotechnol.* **2003**, *30*, 95–101.

96. Vyrides, I.; Stuckey, D. Adaptation of anaerobic biomass to saline conditions: Role of compatible solutes and extracellular polysaccharides. *Enzyme Microb. Technol.* **2009**, *44*, 46–51.

97. Mancuso Nichols, C.; Bowman, J.P.; Guezennec, J. Effects of incubation temperature on growth and production of exopolysaccharides by an Antarctic sea ice bacterium grown in batch culture. *Appl. Environ. Microbiol.* **2005**, *71*, 3519–3523.

98. Guibaud, G.; Comte, S.; Bordas, F.; Dupuy, S.; Baudu, M. Comparison of the complexation potential of extracellular polymeric substances (EPS), extracted from activated sludge and produced by pure bacterial strains, for cadmium, lead and nickel. *Chemosphere* **2005**, *59*, 629–638.

99. Decho, A.W. Microbial Exopolymer Secretions in Ocean Environments: Their Role(s) in Food Webs and Marine Processes. In *Oceanography and Marine Biology, an Annual Review*; Barnes, H., Barnes, M., Eds.; Aberdeen University Press: Aberdeen, UK, 1990; Volume 28, pp. 73–153.

100. Ding, Y.X.; Chin, W.C.; Rodriguez, A.; Hung, C.C.; Santschi, P.H.; Verdugo, P. Amphiphilic exopolymers from *Sagittula stellata* induce DOM self-assembly and formation of marine microgels. *Mar. Chem.* **2008**, *112*, 11–19.

101. Krembs, C.; Eicken, H.; Junge, K.; Deming, J.W. High concentrations of exopolymeric substances in Arctic winter sea ice: Implications for the polar ocean carbon cycle and cryoprotection of diatoms. *Deep Sea Res. Part I* **2002**, *49*, 2163–2181.

102. Aslam, S.; Underwood, G.J.C.; Kaartokallio, H.; Norman, L.; Autio, R.; Fischer, M.; Kuosa, H.; Dieckmann, G.S.; Thomas, D.N. Dissolved extracellular polymeric substances (dEPS) dynamics and bacterial growth during sea ice formation in an ice tank study. *Polar Biol.* **2012**, *35*, 661–676.

103. Underwood, G.J.C.; Fietz, S.; Papadimitriou, S.; Thomas, D.N.; Dieckmann, G. Distribution and composition of dissolved extracellular polymeric substances (EPS) in Antarctic sea ice. *Mar. Ecol. Prog. Ser.* **2010**, *404*, 1–19.

104. Wolfaardt, G.M.; Lawrence, J.R.; Korber, D.R. Function of EPS. In *Microbial Extracellular Polymeric Substances*; Wingender, J., Neu, T.R., Flemming, H.-C., Eds.; Springer Verlag: Berlin, Germany, 1999; pp. 171–200.

105. Mora, P.; Rosconi, F.; Franco Fraguas, L.; Castro-Sowinski, S. *Azospirillum brasilense* Sp7 produces an outer-membrane lectin that specifically binds to surface-exposed extracellular polysaccharide produced by the bacterium. *Arch. Microbiol.* **2008**, *189*, 519–524.

106. Lind, J.L.; Heimann, K.; Miller, E.A.; van Vliet, C.; Hoogenraad, N.J.; Wetherbee, R. Substratum adhesion and gliding in a diatom are mediated by extracellular proteoglycans. *Planta* **1997**, *203*, 213–221.

107. Baker, M.G.; Lalonde, S.V.; Konhauser, K.O.; Foght, J.M. Role of extracellular polymeric substances in the surface chemical reactivity of *Hymenobacter aerophilus*, a psychrotolerant bacterium. *Appl. Environ. Microbiol.* **2010**, *76*, 102–109.

108. Bitton, G.; Freihofer, V. Influence of extracellular polysaccharides on the toxicity of copper and cadmium toward *Klebsiella aerogenes*. *Microb. Ecol.* **1977**, *4*, 119–125.

109. Knowes, E.J.; Castenholz, R.W. Effect of exogenous extracellular polysaccharides on the dessication and freezing tolerance of rock-inhabiting phototrophic microorganisms. *FEMS Microbiol. Ecol.* **2008**, *66*, 261–270.

110. Ozturk, S.; Aslim, B. Modification of exopolysaccharide composition and production by three cyanobacterial isolates under salt stress. *Environ. Sci. Pollut. R.* **2010**, *17*, 595–602.

111. Marx, J.G.; Carpenter, S.D.; Deming, J.W. Production of cryoprotectant extracellular polysaccharide substances (EPS) by the marine psychrophilic bacterium *Colwellia psychrerythraea* strain 34H under extreme conditions. *Can. J. Microbiol.* **2009**, *55*, 63–72.

112. Krembs, C.; Deming, J.W. The role of exopolymers in microbial adaptation to sea ice. In *Psychrophiles: From Biodiversity to Biotechnology*; Margesin, R., Schinner, F., Marx, J.C., Gerda, C., Eds.; Springer-Verlag: Berlin, Germany, 2008; pp. 247–264.

113. Mancuso Nichols, C.; Guezennec, J.; Bowman, J.P. Bacterial exopolysaccharides from extreme marine environments with special consideration of the southern ocean, sea ice, and deep-sea hydrothermal vents: A review. *Mar. Biotechnol.* **2005**, *7*, 253–271.

114. Verdugo, P. Marine microgels. *Annu. Rev. Mar. Sci.* **2012**, *4*, 375–400.

115. Rysgaard, S.R.; Søgaard, D.S.; Cooper, M.; Pucko, M.; Lennert, K.; Papakyriakou, T.; Wang, F.; Geilfus, N.; Glud, R.N.; Ehn, J.; *et al.* Ikaite crystal distribution in Arctic winter sea ice and implications for CO_2 system dynamics. *Cryosphere* **2013**, in press.

116. Fischer, M.; Thomas, D.N.; Krell, A.; Nehrke, G.; Göttlicher, J.; Norma, L.; Riaux-Gobin, C.; Dieckmann, G. Quantification of ikaite in Antarctic sea ice. *Cryosphere Discuss.* **2012**, *6*, 505–530.

117. Chave, K.E.; Suess, E. Calcium carbonate saturation in seawater: Effects of dissolved organic matter. *Limnol. Oceanogr.* **1970**, *15*, 633–637.

118. Braissant, O.; Cailleau, G.; Dupraz, C.; Verrecchia, E.P. Bacterially induced mineralization of calcium carbonate in terrestrial environments: The role of exopolysaccharides and amino acids. *J. Sediment. Res.* **2003**, *73*, 485–490.

119. Bergmann, D.; Furth, G.; Mayer, C. Binding of bivalent cations by xanthan in aqueous solution. *Int. J. Biol. Macromol.* **2008**, *43*, 245–251.

120. Farris, S.; Mora, L.; Capretti, G.; Piergiovanni, L. Charge density quantification of polyelectrolyte polysaccharides by conductometric titration: An analytical chemistry experiment. *J. Chem. Educ.* **2011**, *89*, 121–124.

121. Hardikar, V.V.; Matijević, E. Influence of ionic and nonionic dextrans on the formation of calcium hydroxide and calcium carbonate particles. *Colloid. Surf. A* **2001**, *186*, 23–31.

122. Comte, S.; Guibaud, G.; Baudu, M. Biosorption properties of extracellular polymeric substances (EPS) towards Cd, Cu and Pb for different pH values. *J. Hazard Mater.* **2008**, *151*, 185–193.

123. Zhang, D.; Lee, D.J.; Pan, X. Desorption of Hg (II) and Sb (V) on extracellular polymeric substances: Effects of pH, EDTA, Ca (II) and temperature shocks. *Bioresour. Technol.* **2013**, *128*, 711–715.

124. Bhaskar, P.V.; Bhosle, N.B. Bacterial extracellular polymeric substance (EPS): A carrier of heavy metals in the marine food-chain. *Environ. Int.* **2006**, *32*, 191–198.

125. Schlekat, C.E.; Decho, A.W.; Chandler, G.T. Bioavailability of particle-associated silver, cadmium, and zinc to the estuarine amphipod *Leptocheirus plumulosus* through dietary ingestion. *Limnol. Oceanogr.* **2000**, *45*, 11–21.

126. Campbell, L.M.; Norstrom, R.J.; Hobson, K.A.; Muir, D.C.G.; Backus, S.; Fisk, A.T. Mercury and other trace elements in a pelagic Arctic marine food web (Northwater Polynya, Baffin Bay). *Sci. Total Environ.* **2005**, *351*, 247–263.

127. Turgeon, S.L.; Schmitt, C.; Sanchez, C. Protein–polysaccharide complexes and coacervates. *Curr. Opin. Colloid Interface Sci.* **2007**, *12*, 166–178.
128. Huston, A.L.; Methe, B.; Deming, J.W. Purification, characterization, and sequencing of an extracellular cold-active aminopeptidase produced by marine psychrophile *Colwellia psychrerythraea* strain 34H. *Appl. Environ. Microbiol.* **2004**, *70*, 3321–3328.

Endolithic Microbial Life in Extreme Cold Climate: Snow Is Required, but Perhaps Less Is More

Henry J. Sun

Abstract: Cyanobacteria and lichens living under sandstone surfaces in the McMurdo Dry Valleys require snow for moisture. Snow accumulated beyond a thin layer, however, is counterproductive, interfering with rock insolation, snow melting, and photosynthetic access to light. With this in mind, the facts that rock slope and direction control colonization, and that climate change results in regional extinctions, can be explained. Vertical cliffs, which lack snow cover and are perpetually dry, are devoid of organisms. Boulder tops and edges can trap snow, but gravity and wind prevent excessive buildup. There, the organisms flourish. In places where snow-thinning cannot occur and snow drifts collect, rocks may contain living or dead communities. In light of these observations, the possibility of finding extraterrestrial endolithic communities on Mars cannot be eliminated.

Reprinted from *Biology*. Cite as: Sun, H.J. Endolithic Microbial Life in Extreme Cold Climate: Snow Is Required, but Perhaps Less Is More. *Biology* **2013**, *2*, 693-701.

1. Introduction

The Ross Desert, an unofficial geographic name referring to high-altitude (>1000 m) areas of the McMurdo Dry Valleys, is one of coldest environments on Earth. Here, the air temperature does not rise much above 0 °C in the peak of summer [1]. The year-round low temperatures create a secondary challenge for life: low water activity, or high aridity. While snow—the only form of precipitation in the region—falls regularly during the summer months, most of the snow is either blown away or sublimates without melting. Together, these two extremes—low temperatures and high aridity—create a desert environment where life is restricted to a few protected niches. Pioneering work by Imre Friedmann and his colleagues showed that the interior of sandstone is one such niche, occupied by cryptoendolithic cyanobacteria and lichens [2,3]. During the summer, the rocks are warmed by solar radiation or insolation, intermittently reaching temperatures high enough to melt snow and support biological activity [1,4,5]. In addition, the sandstones are translucent, especially when wet, with the outer centimeter of the rock, where the organisms reside, receiving 0.1%–1% of incident sunlight [6]. The organisms also actively improve the optical properties of the surrounding sandstone by leaching iron from it [3,7].

The Ross Desert cryptoendoliths do not reside under all available sandstone surfaces, and they don't survive under all rock surfaces or at all locations. On Mount Fleming, for example, the community is mostly dead and fossilized [8]. Early researchers attributed the absence of life in these locations to the absence of warm temperatures. North-facing slopes, which receive direct solar radiation and are, therefore, warm, are nearly always colonized [3]. In contrast, south-facing slopes, which receive less insolation, are generally devoid of colonization. Taking this logic further, it was suggested that minor changes in temperature during periods of glaciation and global cooling can cause the endolithic community in an entire region to go extinct [7].

In this article, I present an alternative hypothesis, which emphasizes the volume of snow that a rock surface actually receives, or the effective snow condition. In an extreme cold climate where snow is the sole moisture source, photosynthetic microorganisms living within rocks are faced with unique ecological challenges. For instance, snow, unlike rain, cannot wet vertical surfaces. Hence, cliffs are perpetually dry. At the other extreme, a rock can be covered by too much snow. Under a thick snow cover, a rock may no longer receive sufficient insolation to reach temperatures high enough to melt snow. In addition, the light level within the rock may no longer be adequate to support photosynthesis. An ideal effective snow condition occurs on rocks that can trap some snow, but where gravity or frequent gusty winds can prevent excessive buildup. As shown below, all biological variations on Battleship Promontory, which were previously attributed to temperature, can be explained by variations in effective snow condition.

In light of these new observations, the generally-held notion that the surface of Mars is too cold to support extant life [7–10] should be revisited. Given the recent evidence that suitable rock types, frost formation, and conditions for stable liquid water all occur on Mars, in equatorial lowlands, the possibility of finding living endolithic microorganisms there cannot be eliminated.

2. Results and Discussion

2.1. Battleship Promontory: Correlation between Biology and Snow

On Battleship Promontory (76°55′S, 161°58′E, elevation 1294 m), in the Convoy Range, sandstone rocks vary widely in size and shape, from outcrops tens of meters across, to boulders a few meters high, and to small stones forming a part of the rubble field (Figure 1). An opportunity to observe the effective snow condition presented itself during a field trip in late January, 2005. Following a significant snowfall, the snow covering the rocks was drastically re-arranged by wind.

Direct contact with snow is not always necessary for a rock to be colonized, and the presence of moisture is not the sole criterion for colonization. For instance, the feet of boulders and stones in loose rubble fields—kept moist by contact with damp soil—are uniformly colonized. Endolithic organisms can also exist within the lower surface of a thin overhang, apparently sustained by downward movement of moisture penetrating the upper surface. In the Dry Valleys, where winds frequently gust up to 15 meters per second, mostly from the southeast [1], some sandstone surfaces are heavily abraded and undergo grain-by-grain disintegration [7]. Under these conditions, slow-growing endolithic organisms are unable to establish a foothold.

These special situations aside, contact with snow is essential for colonization. Hence, vertical cliffs, which cannot trap snow, are devoid of organisms. This is true for both north- and south-facing cliffs (Figure 1). These "abiotic" surfaces are covered by a relatively uniform dark red coating [7]. This coating is the consequence, not the cause, of the rock's abiotic condition. Where such surfaces have access to moisture, for example, if they lie next to a colonized corner, the coating is destroyed by biological activity and recedes (Figure 1).

Figure 1. Relationship between rock slope, orientation, and biological activity on Battleship Promontory, Convoy Range, Antarctica. North is to the lower left corner of the photograph (note shadow on the ground). Vertical surfaces are devoid of organisms, as is evident from their dark red coloration, because they cannot trap snow. Sloped surfaces are colonized and show exfoliation, regardless of orientation (boulders in foreground), because they trap some snow, but excess snow is removed by gravity and wind. Heavily-scoured surfaces are devoid of colonization. The rate of scouring is such that the organisms are unable to establish a foothold.

Moderately-sloped surfaces at the tops of boulders have the ideal effective snow condition. They can trap some snow, but excess snow either falls off or is blown away by strong winds. As a result, snow covers on these surfaces are thin, especially around the edges (Figure 2). North- and south-facing slopes are equally well colonized, suggesting that snow, not temperature, controls where the organisms can or cannot exist. The presence of microorganisms under these surfaces was confirmed both in the field (Figure 3) and by examining returned samples using scanning electron microscopy (Figure 4). These relatively dry surfaces are colonized primarily by the lichen-dominated community, while the permanently moist rocks on the ground generally harbor cyanobacteria.

Figure 2. Snow melting on a moderately-sloped boulder top on Battleship Promontory. Melting occurs in summer around noon, when insolation is maximal. (Scale 10 cm).

Figure 3. Fractured sandstone showing the presence of microorganisms just below the surface. The leaching of iron from the surrounding sandstone improves photosynthetic access to sunlight. (Scale 5 cm).

Figure 4. Scanning electron micrograph showing endolithic lichens in sandstone, with soredia, characteristic reproductive structures consisting of small groups of algal cells surrounded by fungal cells (arrows). (Scale 10 μm).

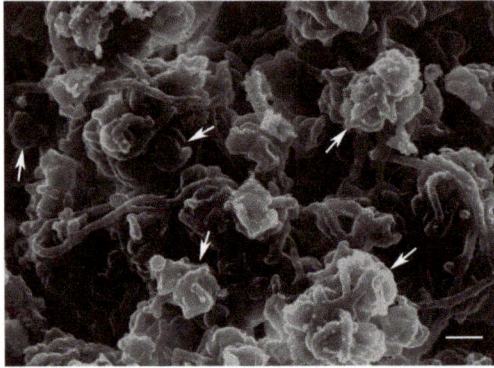

Figure 5. Evidence that persistent thick snow covers are detrimental. (**a**) Partially-covered sandstone surface in the lee of a boulder (red arrow) on Battleship Promontory after winds cleared the snow off the area around the edge. (**b**) View of the exposed area from above, showing that it is heavily colonized. (**c**) View after the snow was deliberately removed, showing that the covered area (below the added black line) is largely devoid of biological activity. (Scale 20 cm).

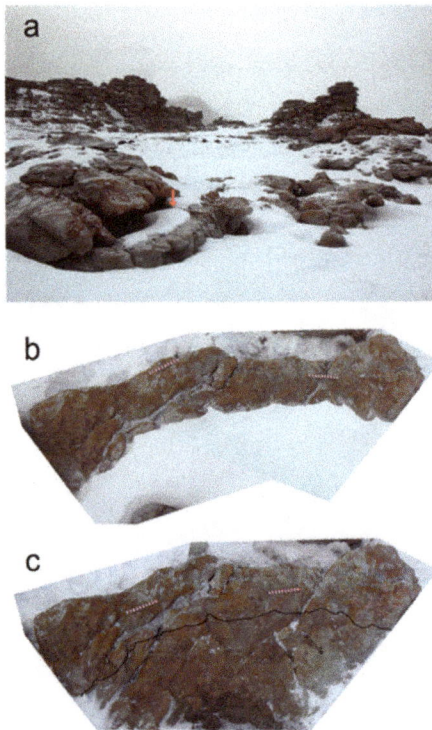

Perhaps the strongest evidence that snow, not temperature, controls colonization on Battleship Promontory comes from flat, horizontal surfaces. Despite uniform insolation, these surfaces are not always uniformly colonized. Where the colonization is not uniform, it is correlated with the distribution of snow. It seems that the organisms prefer less, not more snow. An example of this observation is shown in Figure 5. This sandstone slab, situated in the lee of a boulder, was partially covered by 8–10 cm of snow (Figure 5a). The photographs in Figure 5b and 5c show a bird's-eye view of the slab before and after the snow was deliberately removed for observation. The snow-free outer edge is actively colonized, but the snow-covered area shows little evidence of biological activity.

The colonized sandstone rocks on Battleship Promontory can be divided into two categories. First, there are surfaces elevated above the ground. Due to gravity- and wind-assisted snow removal, the effective snow condition of these rocks stays relatively constant and optimal regardless of snowfall volume. These communities appear to be all viable. Second, there are surfaces at ground level and surfaces in a topographic low (e.g., gullies), where snow removal cannot occur and drift snow accumulates. On these surfaces, the effective snow condition can vary considerably and change through time. A surface that is favorable in a climate with low annual precipitation may become unfavorable in a climate with high annual precipitation, and vice versa. In an extreme environment, where the organisms grow slowly [11], extinction occurs relatively quickly, but re-colonization would be slow. As a result, repeated episodes of colonization, death, and re-colonization may occur in these rocks.

2.2. Cause of Death on Mount Fleming and Horseshoe Mountain: Climate Cooling or Fluctuations in Precipitation?

Mount Fleming and Horseshoe Mountain, in the Asgard Range, are two sites where sandstone outcrops contain mostly dead and entirely dead microbial communities, respectively. Relative to Battleship Promontory, these sites are located farther south, slightly more inland, and at a higher elevation (2200 m). Accordingly, they have a colder climate. The mean January air temperature on Battleship Promontory is −16.6 °C. In contrast, the values for Mount Fleming and Horseshoe Mountain are −18.7 °C and −22.4 °C, respectively [8]. Based on these data, Friedmann and his colleagues concluded that the cold limit separating hostile from life-supporting environments runs roughly through the area of Mount Fleming [8]. This conclusion was based on the assumption that, at those locations, the communities went extinct because the temperatures no longer rose sufficiently to melt whatever snow there was. In light of the effective snow condition, it is possible that the Mount Fleming and Horseshoe Mountain communities went extinct from changes in annual precipitation, not from low temperature. The landscapes on Mount Fleming and Horseshoe Mountain are relatively flat, with little opportunity to trap snow. During periods of relatively high precipitation, it is possible that exposed surfaces were covered by a thin film of snow that permitted the rocks to warm up to a degree sufficient to produce meltwater. During drier periods with less snow, however, whatever snow there was could have been more effectively removed by the wind, thereby reducing the overall period of metabolic activity and increasing the probability of the death of the community. In this case, the isolated occurrence of colonies on Mount Fleming could be a sign of recovery of the ecosystem during what is now a wetter period.

2.3. Extraterrestrial Endolithic Microorganisms on Mars?

The surface of Mars is generally considered uninhabitable because of its low atmospheric pressure, somewhat less than 10 mbars. This pressure is below the triple point of water and so does not allow for the presence of stable liquid water. Numerical calculations by Lobitz and colleagues indicate, however, that this may not be the case across the entire planet [12]. Specifically, in low-lying regions between equator and 40°N, temperature and pressure conditions for stable liquid water may occur during summer months. In Utopia Planitia, favorable conditions may last for up to one third of the year. Frost formation in this region is well-documented by images returned by the Viking 2 lander (Figure 6). Furthermore, recent missions indicate that soil sulfate and gypsum, which are suitable for colonization by endolithic organisms [13], are widespread on Mars [14]. Until we definitively establish the cold limit of life on Earth, the possibility that rocks in Utopia Planitia contain live microorganisms cannot be eliminated.

Figure 6. Possible extraterrestrial endolithic habitat on Utopia Planitia, Mars, where water-ice frost formation (arrow), conditions for stable liquid water, and suitable rock types exist.

Image credit: NASA/JPL.

3. Experimental Section

Field surveys of biological activity relied on macroscopic biosignatures visible on the rock surface. Where possible, observations were documented by photography. The following protocol was used to prepare samples for electron microscopy. Specimens were rehydrated in saline phosphate buffer and then fixed in 0.5% formaldehyde and 1% glutaraldehyde for 15 minutes. Fixed specimens were dehydrated in an ethanol series: 15%, 30%, 50%, 75%, 95%, 100%, each for 15 minutes. After two additional changes in anhydrous ethanol, the specimens were placed in hexamethyldisilazane (HMDS) twice, each time for 30 minutes. After the second wash, the HMDS was decanted, and the specimens were air-dried. Specimens were carbon-coated and viewed using a scanning electron microscope (JSM-5610).

4. Conclusions

On Battleship Promontory, colonization barriers of microbial communities under sandstone surfaces are imposed by the uneven distribution of snow, not temperature. The presence of dead communities under some rock surfaces may be attributed to fluctuations in annual precipitation. On Mount Fleming and Horseshoe Mountain, the communities may have died during a period of climatic cooling or during a period when changes in annual precipitation caused the effective snow condition to become unfavorable. On Mars, extant endolithic communities may exist in equatorial lowlands.

Acknowledgement

Thanks to P. Conrad and R. Carlson for leading the field expedition, to C. McKay for discussions, To J. Nienow and R. Kreidberg for editing, to L. Wable for graphic assistance, and to three anonymous reviewers for comments that improved the manuscript. This work was in part supported by a grant from the NASA Astrobiology Program (NNX08AO45G).

References

1. Friedmann, E.I.; McKay, C.P.; Nienow, J.A. The cryptoendolithic microbial environment in the Ross Desert of Antarctica: Satellite-transmitted continous nanoclimate data, 1984–1986. *Polar Biol.* **1987**, *7*, 273–287.
2. Friedmann, E.I.; Ocampo, R. Endolithic blue-green algae in the dry valleys: Primary producers in the Antarctic desert ecosystem. *Science* **1976**, *193*, 1247–1249.
3. Friedmann, E.I. Endolithic microorganisms in the Antarctic cold desert. *Science* **1982**, *215*, 1045–1053.
4. Friedmann, E.I. Melting snow in the dry valleys is a source of water for endolithic microorganisms. *Antarctic J.* **1978**, *13*, 162–163.
5. McKay, C.P.; Friedmann, E.I. The cryptoendolithic microbial environment in the Antarctica cold desert: Temperature variations in nature. *Polar Biol.* **1985**, *4*, 19–25.

6. Nienow, J.A.; McKay, C.P.; Friedmann, E.I. The cryptoendolithic microbial environment in the Ross Desert of Antarctica: light in the photosynthetically active region. *Microb. Ecol.* **1988**, *16*, 271–289.

7. Friedmann, E.I.; Weed, R. Microbial trace-fossil formation, biogenous, and abiotic weathering in the Antarctic cold desert. *Science* **1987**, *236*, 703–705.

8. Friedmann, E.I.; Druk, A.Y.; McKay, C.P. Limits of life and microbial extinction in the antarctic desert. *Antarctic J.* **1994**, *29*, 176–179.

9. Friedmann, E.I. The Antarctic cold desert and the search for traces of life on Mars. *Adv. Space Res.* **1986**, *6*, 265–268.

10. McKay, C.P.; Friedmann, E.I.; Wharton, R.A.; Davies, W.L. History of water on Mars: A biological perspective. *Adv. Space Res.* **1992**, *12*, 231–238.

11. Sun, H.J.; Friedmann, E.I. Growth on geological time scales in the Antarctic cryptoendolithic microbial community. *Geomicrobiol. J.* **1999**, *16*, 193–202.

12. Lobitz, B.; Wood, B.L.; Averner, M.M.; McKay, C.P. Use of spacecraft data to derive regions on Mars where liquid water would be stable. *Proc. Natl. Acad. Sci. USA.* **2001**, *98*, 2132–2137.

13. Dong, H.; Rech, J.A.; Jiang, H.; Sun, H.; Buck, B.J. Endolithic cyanobacteria in soil gypsum: Occurrences in Atacama (Chile), Mojave (USA), and Al-Jafr Basin (Jordan) Deserts. *J. Geophys. Res.* **2007**, *112*, G02030.

14. Gendrin, A.; Mangold, N.; Bibring, J.P.; Langevin, Y.; Gondet, B.; Poulet, F.; Bonello, G.; Quantin, C.; Mustard, J.; Arvidson, R.; LeMouélic, S. Sulfates in Martian layered terrains: The OMEGA/Mars Express view. *Science* **2005**, *307*, 1587–1591.

Psychrophily and Catalysis

Charles Gerday

Abstract: Polar and other low temperature environments are characterized by a low content in energy and this factor has a strong incidence on living organisms which populate these rather common habitats. Indeed, low temperatures have a negative effect on ectothermic populations since they can affect their growth, reaction rates of biochemical reactions, membrane permeability, diffusion rates, action potentials, protein folding, nucleic acids dynamics and other temperature-dependent biochemical processes. Since the discovery that these ecosystems, contrary to what was initially expected, sustain a rather high density and broad diversity of living organisms, increasing efforts have been dedicated to the understanding of the molecular mechanisms involved in their successful adaptation to apparently unfavorable physical conditions. The first question that comes to mind is: How do these organisms compensate for the exponential decrease of reaction rate when temperature is lowered? As most of the chemical reactions that occur in living organisms are catalyzed by enzymes, the kinetic and thermodynamic properties of cold-adapted enzymes have been investigated. Presently, many crystallographic structures of these enzymes have been elucidated and allowed for a rather clear view of their adaptation to cold. They are characterized by a high specific activity at low and moderate temperatures and a rather low thermal stability, which induces a high flexibility that prevents the freezing effect of low temperatures on structure dynamics. These enzymes also display a low activation enthalpy that renders them less dependent on temperature fluctuations. This is accompanied by a larger negative value of the activation entropy, thus giving evidence of a more disordered ground state. Appropriate folding kinetics is apparently secured through a large expression of trigger factors and peptidyl–prolyl *cis*/*trans*-isomerases.

Reprinted from *Biology*. Cite as: Gerday, C. Psychrophily and Catalysis. *Biology* **2013**, *2*, 719-741.

1. Introduction

Polar ecosystems are characterized by a high diversity and abundance of microorganisms. Indeed, cell densities from 0.9 to 14.9×10^5 mL^{-1} have been recorded in Arctic pack ice with similar figures in adjacent seawater. Thirty-three phylotypes were identified in these environments; they belong to the γ- and α-proteobacteria with less than 1% being cultivable [1]. In Arctic tundra soils, and in winter, the total bacterial cell counts can be as high as 5×10^9 cells per g of soil [2]. Even Arctic permafrost, characterized by temperatures below the freezing point of water, is highly populated in metabolically active or possibly dormant microorganisms with cell counts of 3.56×10^7 per g of soil of Canadian permafrost [3]. High cell densities are also present in the Antarctic with figures of 5.4 to 7.9×10^7 cells per g of lake sediment [4], whereas in free waters, at Terra Nova Bay (Ross Sea) for example, cell counts vary from 0.1 to 15.7×10^5 cells mL^{-1} [5]. These cell densities, both in the Arctic and Antarctic oceans, are similar to those recorded in temperate habitats and correspond to microbial diversities much greater than those initially expected [6]. Such data testify to successful adaptations of microbial communities to extremely cold environments. Knowing that low temperatures have usually a negative effect on population growth, one has to conclude that a complete set of molecular adaptations has taken place to notably compensate for the freezing effect of low temperatures on reaction rate, diffusion rate, membrane permeability and nucleic acids dynamics, for instance. Reaction rates are clearly crucial for the survival of microorganisms at low temperatures, since they vary in an exponential way as a function of temperature according to the Arrhenius law, in which the rate constant, $k = A \cdot e^{-Ea/RT}$, depends on the pre-exponential factor A, also called frequency factor, which, in the reaction rate expression, derived from the transition state theory, takes the form of $A = k_B(T/h)\exp(+\Delta S^*/R)\exp(1)$; k_B is the Boltzmann constant, h, the Planck constant, T, the temperature in Kelvin and ΔS^* the activation entropy of the reaction [7]. One can see that the frequency factor A strongly depends on the activation entropy and is directly dependent on temperature. E_a is called the activation energy; it is equal to the term, $\Delta H^* + RT$, and the activation enthalpy of the reaction can be easily determined from Arrhenius plots in which lnk is expressed as a function of 1/T. In enzyme-catalyzed reactions, these Arrhenius plots usually give a straight line of slope—E_a/R over a more or less wide range of temperatures provided that the temperature conditions do not alter the enzyme structure or the enzyme–substrate complex. In the case of psychrophiles, one has also to take into consideration the viscosity of the environment which can have a strong effect on reaction rates. In 2004, Garcia-Viloca and coworkers [8] proposed, in the case of enzyme-catalyzed reactions, a generalized expression of reaction rate, in which $k_{cat} = \gamma$ $_{(T)}k_B(T/h)\exp(-\Delta G^*/RT)$. The factor γ is, in the context of viscosity, an extended expression of the old transmission factor κ that takes into account the probability that some of the activated molecules will return to the ground state rather than be transformed into product; in other words, they can re-cross the energy barrier. This factor is usually neglected, but at low temperature, it can significantly differ from unity. Some works have been devoted to this problem and we can mention that of Demchenko *et al.* [9] who studied the influence of the viscosity of the medium on the catalytic properties of lactate dehydrogenase. They, for example, demonstrated that the V_{max}, for lactate oxidation in the presence of NAD^+, decreases from 8.5 units in low-viscosity buffer to 1.5 in a 44% sucrose solution equivalent to a viscosity of about 6cP. It is worth noting that at 20 °C the average viscosity of the intracellular space is 2.5cP, whereas at 0 °C, this viscosity raises to 5cP [10]. Also, if at 20 °C the viscosity of pure water is close to 1, it subsequently raises to 1.787 at 0 °C. Thus, clearly the high viscosity of aqueous media at low temperatures should also have a depressive effect on reaction rates. This problem has been also addressed by Siddiqui and coworkers [11].

2. General Properties of Cold-Adapted Enzymes

Many cold-adapted enzymes have now been fully characterized in terms of catalytic properties and the main characteristic of these enzymes is that the thermo-dependent activity curve is always displaced towards low temperatures as illustrated in Figure 1. The left curve corresponds to the evolution of the activity as a function of temperature of a cold-adapted α-amylase from the Antarctic strain *Pseudoalteromonas haloplanktis*. The right curve illustrates a similar curve recorded for the homologous α-amylase from a thermophilic microorganism, *Bacillus amyloliquefaciens* [12]. Three main differences can be observed:

(1). The apparent optimum temperature of the cold-adapted enzyme is displaced towards low temperatures by as much as 30 °C.
(2). The cold-adapted enzyme displays a much higher catalytic efficiency than the thermophilic enzyme up to approximately its apparent optimum.
(3). The cold-adapted enzyme is, in contrast with the thermophilic one, rapidly inactivated at temperatures above 25 °C.

Figure 1. Specific activity as a function of temperature of the α-amylase from the Antarctic strain *Pseudoalteromonas haloplanktis* (black dots) and of the thermophilic counterpart from *Bacillus amyloliquefaciens* (open circles). Worth noting is the important shift of the apparent optimum towards low temperature. Adapted from [12].

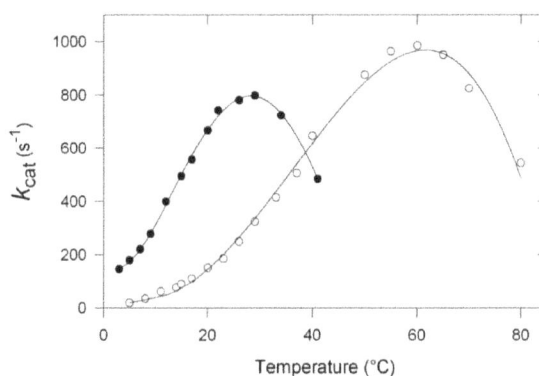

Some additional commentaries are necessary to fully appreciate the significance of these curves. First, one has to consider that the so-called "optimum temperature" still reported as such in many papers, is only an apparent optimum since, at this temperature, the enzyme is already under severe thermal stress and cannot be exposed at this temperature for a long time. A more appropriate term would be "critical temperature," to indicate that a partial inactivation has already taken place. Second, as far as the shift towards low temperature is concerned, it is worth mentioning that its amplitude strongly depends on the enzyme investigated. Third, in the case of many cold-adapted enzymes, the activity that corresponds to the apparent optimum is somewhat lower than that recorded for homologous mesophilic or thermophilic enzymes, and this can possibly reflect an incomplete adaptation to low temperatures. Also, in the same context, the higher activity, observed at low and moderate temperatures, considerably varies with the enzyme under investigation. In the present cases, at 10 °C, the specific activity of the cold α-amylase is about 10 times as high as that of the thermophilic enzyme.

The first significant report on the properties of cold-adapted enzymes was made in 1984 and concerned a heat labile alkaline phosphatase isolated from Antarctic seawater bacteria [13]. Not only the three main properties of cold-adapted enzymes were correctly described, but the authors also suggested that these enzymes could present significant advantages over mesophilic counterparts for biotechnological purposes. Recent investigations devoted to the study of new cold-adapted enzymes have strictly confirmed the properties of these enzymes. We can mention the hormone-sensitive lipase isolated from the Antarctic strain *Psychrobacter* sp. TA144 [14], the periplasmic nitrate reductase from the Antarctic bacterium *Shewanella gelidimarina* [15], and the serine hydroxymethyltransferase from a cold-adapted *Psychromonas ingramii* isolated from Arctic polar sea ice [16]. Another parameter which can be influenced by temperature is the K_m which, in general, is related to the affinity of the enzyme for the substrate, provided that the rate constants that could interfere with the constants directly involved in the true dissociation constant of the enzyme–susbtrate complex could be neglected. The comparison of the K_m values of enzymes from orthologous species, differently adapted to temperature (Table 1), reveals three main features:

Table 1. K_m values of some psychrophilic, mesophilic and thermophilic enzymes.

Enzyme-organism	T (°C)	K_m	Ref
Alpha-amylase			
P: *Pseudoalteromonas haloplanktis*	25	234.00 uM	
M: Pig pancreatic	25	65.00	[17]
Aspartate aminotransferase			
P: *P. haloplanktis*	07	5.82 mM	
	25	8.34	
	25	7.31	[18]
M: *E.coli*	35	21.04	
Aspartate transcarbamylase			
P: Gram-TAD1	11	20.00 mM	
M: *E.coli*	30	0.014	[19]
Citrate synthase			
P: Antarctic bacterium DS2-3R		230 uM	
M: mesophiles	23	<50 uM	[20]
DNA ligase			
P: *Pseudoalteromonas haloplanktis*	04	0.165 uM	
M: *E.coli*	18	0.179	
T: *Thermus scotoductus*	18	0.179	[21]
	30	0.702	
T: *Thermus scotoductus*	45	0.236	
Elongation factor TU			
P: *Moraxella* TAC II 25	15	0.36 uM	
M: *E.coli*	15	0.13	[22]
Endonuclease I			
P: *Vibrio salmonicido*	05	246.00 mM	
M: *Vibrio cholerae*	05	118.00	[23]
Isocitrate dehydrogenase			
P: *Colwellia maris*	15	62.00 mM	
M: *E. coli*	15	3.30	[24]

Table 1. *Cont.*

Enzyme-organism	T (°C)	K_m	Ref
Lactate dehydrogenase			
P: *Champsocephalus gunnarii*	00	0.16 mM	
M: *Deinococcus radiodurans*	48	0.21	[25]
T: *Thermus thermophilus*	90	0.16	
Ornithine transcarbamylase			
P: *Moritella abyssi*	05	1.78 mM	
M: *E. coli*	37	2.40	[26]
T: *Thermus thermophilus*	55	0.10	
RNA-dependent ATPase			
P: *Pseudoalteromonas haloplanktis*	10	0.60 mM	
	25	0.9	[27]
M: *E.coli*	10	0.02	
	25	0.06	
Subtilisin			
P: *Bacillus* (Antarctic)	05	26.00 uM	
	25	37.00	[28]
M: *Bacillus licheniformis*	05	6.00	
	25	17.00	
Triose phosphate isomerase			
P: *Vibrio marinus*	10	1.90 mM	[29]
M: *E. coli*	25	1.09	

First, an increase in the temperature over a threshold value causes an increase in K_m; in other words, a lower affinity of the enzyme for the substrate; second, the lowest K_m values are observed in the temperature range usually experimented by the organism [30] third, with two exceptions, aspartate aminotransferase and DNA ligase, the K_m values of cold-adapted enzymes are usually higher than those of mesophilic counterparts but are closer to each other at the respective temperature of their environment. These data suggest that enzymes from species adapted to different temperatures have evolved molecular adaptations in order to maintain a conformation enabling an appropriate interaction between enzymes and substrates at the usual temperature of the environment. In certain cases, these adaptations are not likely to be complete, as previously mentioned.

3. Activity and Stability

Various techniques have been used to evaluate the thermal stability of proteins. The thermal unfolding of a protein is accompanied by a positive modification of the enthalpy and the heat absorption can be followed in a microcalorimeter that directly provides the Tm values of the respective domains of a protein or, if the unfolding is highly cooperative, the Tm value of the whole molecular edifice. Fluorescence signals can also be used because the exposure to the solvent of tryptophane residues, usually buried into the protein, is associated with a red shift of the emission wavelength, which is easily followed as a function of temperature. Circular dichroism signals are also altered as a function of unfolding; in some cases, the far U-V region has been used, but in this range of wavelength, around 220 nm, only the secondary structure is concerned and the ellipticity values are related to the percentage of helical and beta-structures. This does not allow the detection of the modifications of the tertiary structure, which can be evaluated only if the circular dichroïsm measurements are carried out in the near U-V region, in the absorption bands of aromatic residues. The signals are, however, rather weak, and the ellipticity changes can be positive or negative upon unfolding. In many cases, also the thermal unfolding has been determined using the measurement of the residual activity of the enzyme after exposure for a certain time at a given temperature followed by cooling. This technique is not suitable due to the possible refolding on cooling. To validate this technique, one has first to demonstrate that the thermal unfolding is truly irreversible at all temperatures tested. In this context, it is worth knowing that psychrophilic enzymes are particularly prone to rapid spontaneous refolding. One of the characteristic features of cold-adapted enzymes is that, often, the thermal inactivation of the enzyme precedes any detectable changes in tertiary structure by the techniques mentioned above. This is clearly illustrated in Figure 2. This figure shows the profiles of the thermal inactivation and structural transition curves of orthologous α-amylases (Figure 2A,C) and glycoside hydrolases, xylanases and cellulase, (Figure 2B,D) adapted to different temperatures. In the case of α-amylases, the thermal unfolding has been followed by fluorescence spectroscopy at an emission wavelength of 350 nm, and one can see that for the cold-adapted enzyme (AHA) a significant inactivation is reached before any detectable changes in the three-dimensional structure of the protein. On the contrary, in the case of mesophilic and thermophilic enzymes, the inactivation strictly corresponds to detectable changes in the three-dimensional structure. Similar data are observed in the case of psychrophilic, mesophilic and thermophilic glycoside hydrolases (Figure 2B,D). It is worth mentioning that thermograms in panel 2D have been obtained by differential scanning calorimetry. They correspond to the measure of the heat absorbed during the thermal unfolding of the proteins. These data suggest either that the active site of these psychrophilic enzymes are more heat-labile than the protein structure or that the substrate-enzyme complex becomes highly unstable and that its thermal dissociation precedes any change in the structure of the protein. This hypothesis is supported by the fact that the K_m values of cold-adapted enzymes are, in general, higher than that of their mesophilic or thermophilic counterparts. For these latter, the loss of activity is concomitant with unfolding. The higher specific activity of cold-adapted enzymes, at low and moderate temperatures, can be attributed to activation energies lower than those of their mesophilic and thermophilic counterparts, as shown in Table 2. This lower activation energy is the result of a drastic reduction of the activation enthalpy. This corresponds to a lower temperature dependence of the activity and suggests that less enthalpy-driven bonds have to be broken to secure an appropriate interaction between the enzyme and the substrate which, in general, is an induced process. The Eyring equation, $k_{cat} = \kappa \cdot k_B T/h \cdot e^{-\Delta G^*/RT}$ which is another form of the Arrhenius equation, and in which κ is the transmission coefficient; k_B, the Boltzmann constant; h, the Planck constant and ΔG^*, the free energy of activation, indicates that the catalytic constant is not only exponentially dependent on temperature, but also on the free activation energy. Table 2 shows that this free energy of activation is, as expected from the values of the activation enthalpies, lower than that of the mesophilic counterparts, with two exceptions: arginine kinase and chitinase. The small amplitude of the differences between the respective activation energies is the result of larger negative values of the activation entropies in the case of cold-adapted enzymes that act as a compensating factor. Indeed, the differences observed in the catalytic constants would have been much larger if similar values had been recorded for the activation entropies of psychrophilic and mesophilic enzymes. The more negative values of the activation entropies of cold-adapted enzymes also suggest that the ground state of these enzymes displays a larger disorder than their mesophilic homologues. As mentioned earlier, there are two exceptions to these observations: arginine kinase and chitinase, in these cases, the activation energy is apparently higher than those of the mesophilic counterpart, despite the fact that both the activation enthalpy and activation entropy strictly follow the usual trend of psychrophilic enzymes. As ΔG^* derives from the difference between the enthalpic and entropic terms, it is probable that the higher activation energy of the cold-adapted arginine kinase is due to experimental errors on these terms. In the case of chitinase, the situation is possibly different since it has been argued that the lower specific activity of this psychrophilic enzyme at 15 °C originates from the fact that a soluble preparation of chitin from crabs was used as substrate and this may not be a good substrate for the cold-adapted enzyme, since chitins from different origins can be structurally very different [31]. The analysis of Table 2 also shows that, in some cases, the activation entropy is positive, meaning that the activated state displays a higher disorder than the ground state. This is only recorded in the case of mesophilic enzymes and can be either attributed to a particularly high rigidity of these mesophilic enzymes or, alternatively, to a difference in the redistribution of water molecules associated with the enzyme. Although the differences observed in the free energy of activation could seem rather weak, they are however high enough to explain the higher specific activity of cold-adapted enzymes. Indeed, and as an example, the difference in the activation energy of mesophilic and psychrophilic α-amylases is only of 0.8kJ/mole at 10 °C but this is enough to secure a threefold increase of the k_{cat} for the cold-adapted enzyme.

Table 2. Catalytic constants and activation parameters of a few cold-adapted enzymes as compared with mesophilic counterparts.

Enzyme	Type	T (°C)	$k_{cat}(s^{-1})$	ΔG^*	ΔH^*	$T\Delta S^*$	Reference
					kJ/mole		
Amylase	Psy	10	294.0	57.7	34.7	−23.0	[32]
	Mes		97.0	58.5	46.4	−12.1	
Arginine kinase	Psy	25	3.3	69.4	18.8	−50.6	[33]
	Mes		13.4	66.6	41.9	−4.7	
Cellulase	Psy	4	0.18	71.6	46.2	−25.4	[34]
	Mes		0.01	78.2	65.8	−12.4	
Chitinase	Psy	15	1.7	69.2	60.2	−9.0	[31]
	Mes		3.9	67.2	74.3	+7.1	
Chitobiase	Psy	15	3.8	59.5	44.7	−14.8	[35]
	Mes		0.9	63.5	71.5	+8.0	

Table 2. *Cont.*

Enzyme	Type	T (°C)	$k_{cat}(s^{-1})$	ΔG*	ΔH* kJ/mole	TΔS*	Reference
Citrate synthase	Psy				7.4	ΔS* = 22.7	[20]
	Mes				11.5	ΔS* = 9.7	
Endonuclease	Psy	5	9.41	62.8	33.4	−29.4	[23]
	Mes		1.03	67.9	74.0	+6.1	
LDH	Psy	0	250.0	75.0	22.0	−53.0	[25]
	Mes		72.0	75.0	45.0	−30.0	
Lysozyme	Psy			45.1	31.9	−13.2	[36]
	Mes			46.2	49.4	+3.2	
Subtilisin	Psy	15	25.4	62.0	36.0	−26.5	[37]
	Mes		5.4	66.0	46.0	−20.2	
Xylanase (bact)	Psy	10	515.5	54.0	21.0	−33.0	[38]
	Mes		59.5	60.0	58.0	−2.0	
Xylanase (yeast)	Psy	5	14.8	52.3	45.3	−7.0	[39]
	Mes		4.9	54.6	49.9	−4.7	

Figure 2. (A) Percentages of specific activities of psychrophilic (AHA), mesophilic (PPA) and thermophilic (BAA) α-amylases as a function of temperature and **(C)** concomitant thermal transitions as observed by fluorescence spectroscopy. **(B)**. Percentages of specific activities as a function of temperature of psychrophilic xylanase from the Antarctic strain *Pseudoalteromonas haloplanktis* (pXyl), mesophilic xylanase from *Streptomyces* sp. S38 (Xyl 1), and thermophilic endoglucanase from *Clostridium thermocellum* (Cel A) and their **(D)** concomitant thermal unfolding as recorded by differential scanning calorimetry. Reproduced with permission from [40].

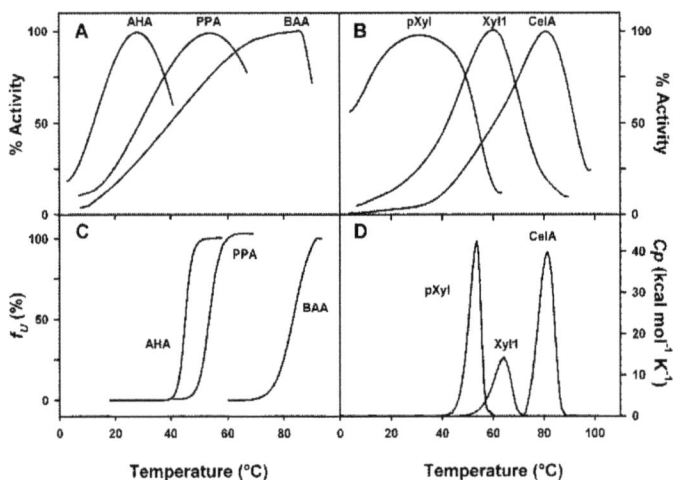

4. Thermodynamic Stability

The thermodynamic stability of a protein can be easily evaluated at a given temperature by differential scanning calorimetry provided that the unfolded form is in a two-state reversible thermodynamic equilibrium with the folded structure according to the equation: $N \rightleftharpoons U$, defined by a thermodynamic equilibrium constant K. Then, the Gibbs free energy of unfolding, also known as thermodynamic stability, can be calculated from the Gibbs–Helmholtz equation: $\Delta G_{N-U} = \Delta H_{N-U} - T\Delta S_{N-U} = -RT\ln K$. This equation can be rewritten as a function of the information derived from a differential scanning calorimetry profile, as shown in Figure 2, panel D.

$$\Delta G_{N-U(T)} = \Delta H_{cal} (1 - T/T_m) + \Delta C_p(T - T_m) - T\Delta C_p\ln(T/T_m)$$

ΔH_{cal}, is the heat absorbed on unfolding; it is given by the area limited by the curve; T is the temperature investigated; T_m is the temperature of half-unfolding for $\Delta G_{N-U} = 0$ when $U/N = 1$; ΔC_p, is the heat capacity change from the native to the unfolded state and is mainly due to the exposure of hydrophobic groups followed by their hydration [41,42]. It corresponds to the amount of surface exposed to the solvent upon unfolding, in other words, to the accessible surface area (ASA). This factor can also be determined experimentally using a microcalorimeter or can be calculated if the three-dimensional structure is known; it is a positive value. A typical thermodynamic stability curve is shown in Figure 3, in which are also described the thermodynamic components of the above mentioned equation. One can see that the maximum thermodynamic stability is obtained around 20 °C, the enthalpic and entropic terms vary with the temperature and one can predict an unfolding induced by cold around −10 °C. Experimentally, such stability curves have been obtained in the case of psychrophilic, mesophilic and thermophilic α-amylases, as shown in Figure 4. Several interesting features can be deduced from the analysis of these curves. First of all, the maximum stability, which corresponds to the highest value of ΔG_{N-U}, is for the three orthologous enzymes recorded around 25 °C, even so, the melting temperatures are very different, around 42 °C for AHA, 62 °C for the mesophilic enzyme, and 85 °C in the case of the thermophilic α-amylase. This has been attributed to the hydrophobic effect that plays a crucial role in the folding and stability of proteins [43]. Second, in the case of mesophilic (PPA) and thermophilic enzymes (BAA), the usual environmental temperature of these organisms lies on the right limb of the stability curve and therefore does not correspond to the maximum stability of these enzymes. This low stability at the environmental temperature is in fact required to secure an appropriate flexibility of the molecular edifice that allows a good interaction of these enzymes with the substrates. This flexibility is secured through the increase of temperature of the unfavorable stabilization entropy, ΔS_{N-U}. At the maximum stability, $\Delta S_{N-U} = 0$, this term becomes negative at low temperatures due to the propensity of hydrophobic groups to favor hydration rather than association with similar groups.

Therefore, on the right side of the curve, mesophilic and thermophilic enzymes are stabilized by the enthalpic term. Conversely, in the case of the cold-adapted enzyme, the environmental temperature lies on the left side of the curve; this part of the curve is characterized by a negative value of the entropic term, which is therefore the stabilizing factor. The negative value of the stabilization enthalpy, chiefly the result of a weakening of hydrophobic and electrostatic interactions induced by the hydration of these groups, appears, on the other hand, to be essential in conferring the appropriate flexibility of the psychrophilic enzyme at low temperatures.

Figure 3. A conformational stability curve displaying the change in stabilization energy, ΔG, as a function of temperature. It also shows the concomitant change of the enthalpic, ΔH, and entropic, ΔS, contributions. It is worth noting that the relatively small values of ΔG result from the difference between rather large figures of ΔH and $T\Delta S$. Reproduced with permission from [44].

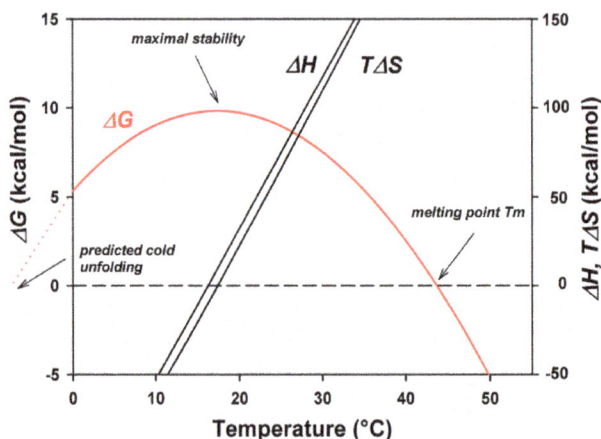

Figure 4. Stabilization energy, ΔG, of psychrophilic (AHA), mesophilic (PPA) and thermophilic (BAA) α-amylases as a function of temperature. Note that the maximum stability is for the three enzymes close to each other, around 25 °C, despite the large difference in the melting temperatures of these enzymes. AHA (*Pseudoalteromonas haloplanktis* α-amylase); PPA (Pig pancreatic α-amylase); BAA (*Bacillus amyloliquefaciens* α-amylase). Adapted from [32].

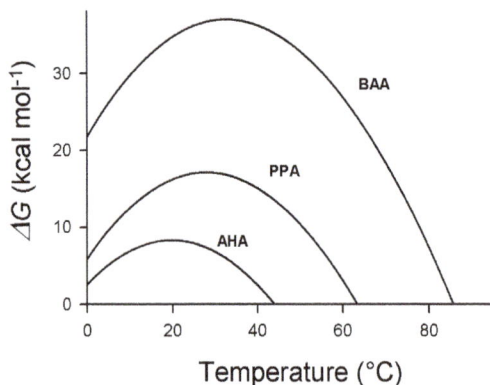

Ultimately, the hydration of these groups leads to cold denaturation and, paradoxically, the cold-adapted enzyme is more sensitive to low temperatures than the homologous mesophilic and thermophilic enzymes. It will be shown that this is due to the fact that the structure of the cold-adapted enzyme is stabilized by a lower number of interactions. In the case of the orthologous α-amylases, one can see that the higher stability of the mesophilic and thermophilic enzymes, leading to higher melting points, is obtained by lifting the stability curve, that is, by increasing the amplitude of the enthalpic contribution to the stability [32]. This is mainly achieved through reinforcing hydrophobic interactions and increasing the number of hydrogen and ionic interactions next to a non-systematic decrease of the entropic contribution. This latter factor can result from an increase in the number of proline residues and a decrease in the number of glycine residues that both decrease the entropy of the denatured form. As a consequence, notably of the decrease of hydrophobic interactions in cold-adapted enzymes, many of them show a reversible thermal unfolding due to their lower propensity to form aggregates when the hydrophobic core is exposed to the solvent.

5. Engineering Cold-Adapted Enzymes

About thirty-three dimensional structures of cold-adapted enzymes have been solved by X-ray crystallography at high resolution (Table 3) and the analysis of these structures has largely contributed to defining the structural determinants of the lower stability of these enzymes. As already stated, these factors include: a lowering of the strength and/or number of the usual weak interactions; the hydrophobic, ionic and hydrogen bonds; a higher proportion of glycine residues, providing an increase of the mobility of certain regions; a decrease in the number of proline residues, also acting in this direction (there is, indeed, a tendency toward an increase in proline residues from psychrophiles to thermophiles); the weakening of the hydrophobic clusters that contributes to reduce the compactness of the protein; an increase in the number of hydrophobic residues at the surface of the molecule (this process induces a clustering of water molecules around these groups and induces a decrease in the entropy and a concomitant increase in free energy); the increase of the number of charged groups at the surface that favors the interaction with the solvent and the solubility at low temperatures; the substitution of the N- and C-caps of α-helices by uncharged residues that reduce the interactions between these residues and the dipoles formed by the helices that indeed carry a net positive charge at the N-terminal part and a net negative charge near the C-terminal end [45]; the increase of the length of the loops connecting the secondary structures; a decrease in the metal-binding affinity in metalloproteins; and, finally, psychrophilic proteins can exhibit loose protein extremities, which are preferentially tightly bound or buried in thermophilic proteins. Each protein adapted to cold uses a few of these factors in various combinations for the adaptation, and this seriously complicates the identification of the factors that are truly involved, even when the three-dimensional structure is known. A first approach to shed some light on the molecular characteristics of cold adaptation is site-directed mutagenesis. Amino acids substitutions can be introduced on the basis of hypotheses resulting from the analysis of crystallographic or of model structures of these enzymes. This rational approach is rendered hazardous by the fact that, in the absence of thermal selective pressure, some genetic drift probably occurs that tends to give rise to structural modifications not really implicated in the adaptation to cold. In one of the pioneering studies [17], numerous single mutations were introduced in the cold-adapted α-amylase isolated from the Antarctic strain *Pseudoalteromonas haloplanktis*. This enzyme displays a severe reduction of stability when compared to mesophilic and thermophilic counterparts (Figure 4). The effect of the mutations, supposed to restore structural specificities found in mesophilic and thermophilic counterparts, were evaluated both at the level of thermal stability and specific activity. Fourteen mutations, including double mutants, were introduced to reproduce weak bonds found in the highly similar structure of the homologous mesophilic enzyme from pig which displays eight additional ion pairs, fifteen additional arginine residues, and ten additional aromatic interactions. The introduced mutations aimed to restore hydrogen bonds, salt bridges, helix dipole stabilization, and various hydrophobic interactions. All these mutations were located far from the catalytic site. Some interesting conclusions were drawn: the additional selected electrostatic interactions provided the highest contribution to stability, increasing both the Tm and ΔH_{cal}, the latter by values as high as 4.4 kcal/mol at 20 °C; a double mutant, consisting in an additional salt bridge and double aromatic interactions, increased the stabilization energy of the cold-adapted enzyme by a factor of two at 10 °C; and, the reinforcement of the hydrophobic core structure induced the appearance of multiple calorimetric domains absent in the cold-adapted enzyme characterized by a two-state unfolding process. Stabilized mutants were also found to refold with less efficiency, and most of the mutations that provided a mesophilic character by increasing the stability of the cold-adapted enzyme also led to a decrease in both k_{cat} and K_m. This can be explained by the reduction of the conformational entropy of the binding species. It was also concluded that the cold-adapted enzyme has reached a limit in thermal stability precluding any further improvement of the specific activity through a decrease in stability. In another similar work carried out on this enzyme, an additional disulfide bond was also introduced to mimic the situation occurring not only in pig pancreatic α-amylase, but also in all chloride-dependent α-amylases from mammals and birds [46]. In these organisms, this disulfide bridge connects domains A and B and is located near the active site. As expected from the previous study, the introduction of this disulfide bridge decreases the specific activity by a factor higher than two, as well as the K_m both at 5 °C and 20 °C. The mutant also shows a higher compactness as demonstrated by fluorescence-quenching experiments, whereas its microcalorimetric characterization displays, contrarily to the wild cold-adapted enzyme, a profile typical of the pig pancreatic enzyme with multiple transitions. However, the T_m of the first transition was lower than that of the cold-adapted enzyme. This is indicative of an unfavorable constraint created by the introduction of the disulfide bridge. Stabilizing effects are, however, recorded in other parts of the enzyme since the overall calorimetric enthalpy increases from 214 kcal mol^{-1} for the cold-adapted enzyme to 241 kcal mol^{-1} for the mutant CC. These values have to be compared to 295 kcal mol^{-1} found for the pig pancreatic α-amylase. In a recent work, two multiple mutants of this cold-adapted α- amylase, derived from the data recorded from single mutations, were also investigated [47]. The first, Mut5, bears five mutations, previously described, that correspond, as stated above, to structural peculiarities existing in the mesophilic pig pancreatic α-amylase: N150D introduces a salt bridge, V196F restores an aromatic interaction, K300R provides a bidentate interaction with the chloride ion, and T232V reinforces a hydrophobic cluster, as well as Q164I. The second mutant, Mut5CC, is identical to Mut5 with the addition of the disulfide bridge Q58C/A99C found in chloride dependent α-amylases from mammals and birds. First of all, the specificity of the two multiple mutants, towards various natural and chromogenic substrates, was strictly similar to that of the parent cold-adapted enzyme. By contrast, the mutants showed a specific activity similar to that of the mesophilic enzyme and a very significant increase of the thermal stability as recorded by fluorescence spectroscopy and differential scanning calorimetry. Contrarily again to the mesophilic enzyme, the thermal inactivation above 35 °C occurs without any detectable structural changes supporting the idea that the active site remains the most thermal sensitive structural element. Fluorescence quenching, using acrylamide as quencher of the tryptophane residues, shows that Mut5 and Mut5CC display a reduced structural permeability. The reversibility of the unfolding was also strongly affected and the rate constants for irreversible unfolding differed by several order of magnitude at 43 °C. Reversible unfolding was only possible in the presence of non-detergent sulfobetaine, 3-(1-pyridinio)-1-propane-sulfonate. It was concluded that these two mutants, as well as the single mutants from which they derive, can be considered as structural intermediates between the psychrophilic and the mesophilic enzymes. These data also strongly supports the prevailing hypothesis that the reduction of the force of the weak interactions that stabilize the cold-adapted enzyme induces an increase in the flexibility of the enzyme. This provides an appropriate accessibility of the substrate at low temperature that leads to a higher specific activity of cold-adapted enzymes at low and moderate temperatures to the expense of thermal stability. This view was further supported by a recent and independent work related to the molecular dynamics of these mutants. The aim of this study was to identify in atomic details the effect of these mutations, especially the long-distance effects, and to identify the structural determinants that led to the incomplete conversion of the cold-adapted enzyme into a mesophilic-like edifice [48]. Multiple MD simulations were applied to the seven mutants (single and multiple) and it was shown that these mutants display a reduced flexibility in various regions of the protein, especially near the active site and substrate-binding groove. They also elicit, in some cases, unexpected long-range effects that even affect the flexibility of domain C that did not carry any mutation. This confirms the ability of these mutations to modulate the dynamic properties of AHA in conferring to it a mesophilic-like behavior, both in terms of activity and substrate binding. The current view of the relationship between enzyme dynamics, flexibility, and catalytic properties was contested in a study that submitted a cold-adapted protease to directed evolution [49]. After several cycles, a multiple mutant of the cold-adapted subtilisin S41 [28] was generated associating a high thermostability to a high specific activity at low temperatures. The selection of the mutants was made on the basis of a higher stability coupled with a specific activity higher or equal to that of the wild-type enzyme towards a synthetic substrate, s-AAPF-pNa. It was concluded that, at least *in vitro*, it was quite possible to simultaneously increase the stability of an enzyme concomitantly with its specific activity. The fact that, in nature, these two parameters seem to evolve in opposite directions could be due to the absence of selective pressure on thermal stability in low temperature environments. Later, however, it was reported that the mutant was poorly active towards macromolecular substrates. The importance of the choice of the substrate in this type of experiment making use of multi-substrates enzymes was underlined in similar experiments tending to confer to a mesophilic subtilisin BPN' a cold-adapted character. Indeed, mutants associating a reduced stability to a higher specific activity were obtained only when synthetic substrates were used but not when casein, a natural substrate, was used [50]. Other attempts to confer a higher activity at low temperature were carried out on a thermophilic protease, WF146, showing a high proportion (>60%) of identical amino acids with the aforementioned cold-adapted protease [51]. Here again, a combination of random and site-directed mutagenesis was used using casein as substrate. The selected mutants were found to be more active than the wild-type enzyme and were all less stable, especially the mutant R29 that includes four single mutations. It had a specific activity at 25 °C, three times as high as that of the wild-type enzyme and showed a half-life of 4.5 min at 80 °C to be compared to 60 min for the wild-type enzyme. In this experiment, it was also shown that this multiple mutant and the other mutants were found to be less active than the wild-type enzyme over the synthetic substrate s-AAPF-pNA and displayed higher Km. This confirms that, generally, the high specific activity of cold-adapted enzymes is gained to the detriment of the thermal stability. This is necessary to confer an appropriate flexibility of crucial parts of the molecular structure at low temperature that facilitates the accommodation of the substrate and probably also the release of products. The difference in the catalytic properties of the afore-mentioned mutants towards low and high molecular weight substrates reflects this trend. A mutant well adapted to small-sized substrates will be too rigid to properly interact with larger substrates and, conversely, mutants well adapted to large-sized substrates will be relatively less efficient in the case of small-sized substrates due to an excess of flexibility reflected in a Km increase, as observed previously. That does not necessarily mean that, *in vitro*, it is not possible to both increase, to a certain extent, the thermal stability and specific activity of a cold-adapted enzyme. Nature has not probably tested all the possibilities and has evolved enzymes up to the point where, in a specific environment, an appropriate compromise between stability–flexibility and activity was reached [44].

Table 3. Crystallographic structures of cold-adapted enzymes obtained at high resolution.

Cold-Adapted Enzyme	Host Organism	Reference
Adenylate kinase	*Bacillus globisporus*	[52]
Adenylate kinase	*Marinibacillus marinus*	[53]
Alkaline metalloprotease	*Pseudomonas* sp.	[54]
Alkaline phosphatase	*Pandalus borealis*	[55]
Alkaline phosphatase	*Bacterial strain* TAB5	[56]
Anionic trypsin	*Salmo salar*	[57]
Alpha-amylase	*Pseudoalteromonas haloplanktis*	[58]
Aspartate carbamoyltransferase	*Moritella profunda*	[59]
Beta-galactosidase	*Arthrobacter* sp. C2-2	[60]
Catalase	*Vibrio salmonicida*	[61]
Cellulase	*Pseudoalteromonas haloplanktis*	[62]
Citrate synthase	*Arthrobacter* sp. strain DS2-3R	[63]
Elastase	*Salmo salar*	[64]
Endonuclease I	*Vibrio salmonicida*	[23]
Esterase	*Pseudoalteromonas* sp. 643A	[65]
Lactate dehydrogenase	*Champsocephalus gunnari*	[25]
Lipase B	*Candida antarctica*	[66]

Table 3. *Cont.*

Cold-Adapted Enzyme	Host Organism	Reference
Malate dehydrogenase	*Aquaspirillium articum*	[67]
Pepsin	*Gadus morhua*	[68]
Phenylalanine hydroxylase	*Colwellia psychrerythtaea*	[69]
Proteinase K-like	*Serratia* sp.	[70]
Protein-tyrosinephosphatase	*Shewanella* sp.	[71]
S-formylglutathione hydrolase	*Pseudoalteromonas haloplanktis*	[72]
Subtilisin-like protease	*Vibrio* sp. PA-44	[73]
Subtilisin S41	*Bacillus* sp.	[74]
Superoxide dismutase	*Aliivibrio salmonicida*	[75]
Triose phosphate isomerase	*Vibrio marinus*	[29]
Trypsin	*Oncorhynchus ketav*	[76]
Uracil-DNA N-glycosylase	*Gadus morhua*	[77]
Xylanase	*Pseudoalteromonas* sp.	[78]

6. Folding at Low Temperature

Protein synthesis consists in the production, at the ribosome level, of unfolded polypeptide chainsthat should either properly fold in the complex environment of the intracellular space or be conditioned in order to safely reach their final destination, in organites, cellular envelopes, or extracellular space. Although some proteins fold spontaneously without problem, others need assistance by proteins known as chaperones. These, indeed, help the protein to adopt a favorable equilibrium between the unfolded and folded state; they assist the protein up to their specific localization, prevent misfolding, aggregation, and actively participate in the appropriate turnover by controlling the hydrolysis of non-functional proteins. They also play an active role in the assembly of multimeric proteins [79,80]. Psychrophilic, mesophilic and thermophilic proteins presumably fold according to similar but very complex processes that involve the nucleation-condensation mechanism and/or parallel routes. The first mechanism occurs through the production of native-like secondary structures, involving residues separated by short sequences, stabilized by hydrogen bonds that induce a condensation of the structure around this nucleus and the formation of intermediate tertiary structures stabilized by other bonds and characterized by various transition state barriers [81]. The second mechanism, known as parallel routes, involves different folding channels in which misfolding is a natural consequence of hierarchical folding and that productive folding intermediates are also stabilized by non-native interactions [82]. The folding process can be viewed as a folding funnel populated by various and transient energy-distinct intermediates converging in a more or less defined native form characterized by a lower free energy level. Psychrophilic and mesophilic–thermophilic proteins, however, display distinct folding funnels as proposed by D'Amico *et al.* [32] in the case of cold-adapted α-amylases. This is illustrated in Figure 5. As cold-adapted enzymes are less stable than their mesophilic counterparts, the average energy level of the native state is higher than that of the mesophilic counterpart, whereas the difference in dynamical properties is reflected by the number of local minima and in their possible inter-conversion between each other on the free energy landscape. In this context, one can see that the native state of the cold-adapted enzyme is characterized by a large population of conformations differing by a low energy level, phenomenon that allows the rapid conversion of one conformation into another. This can explain the high structural flexibility of psychrophilic enzymes. By contrast, in the mesophilic counterpart, the number of local free energy minima is limited. They display quite distinct energy levels that limit the conversion between the conformations represented on the conformational coordinates. That explains the lower flexibility of mesophilic enzymes at a given temperature. This model was found to be consistent with a cold-adapted zinc metalloprotease studied by molecular dynamics [83] and by similar comparative studies on the molecular dynamics of cold-adapted and mesophilic elastases and uracil DNA glycosylases [84]. It was concluded that psychrophilic enzymes present a greater number of metastable states at relevant temperatures and can explore several conformational basins that favor the activity at low temperatures, whereas mesophilic enzymes tend to be trapped into the main conformational basin. The lower energy level of the unfolded form of psychrophilic enzymes reflects an increase in the entropy of the unfolded form of some psychrophilic enzymes when compared to their mesophilic counterparts due to their higher proportion of glycine residues and to the reduction of their number of proline residues. It has already been mentioned that low temperatures, in favoring the hydration of individual groups, weaken some intra-molecular interactions, such as hydrophobic and ionic interactions, and also counteract the proper conversion of the *trans* configuration of proline residues into a *cis* configuration, a process which is a crucial rate-limiting step in the folding of proteins .These facts can be detrimental to the correct folding of proteins at low temperature, and it is essential to understand how folding is regulated in psychrophiles. From the limited number of studies devoted to this problem that make use of the differential expression of proteins at various growth temperatures, one can conclude that psychrophiles, although they are able to express all the chaperones discovered in mesophilic counterparts, fail to adopt a generalized strategy. Furthermore, depending on the species, even contrasting data were recorded. For example, in *Oleispira antarctica*, it was shown that the main chaperones GroEL and GroES had an optimum refolding activity at low temperature and were also found, by contrast to those of *E. coli*, to confer to the mesophilic bacterium the ability to grow at low temperatures [85,86]. However, in *Methanococcus burtonii* [87], as well as in the Antarctic bacterium *Pseudoalteromonas haloplanktis* [88], the expression of GroEL was strongly repressed and, by contrast, overexpressed in *Sphingopyxis alaskensis* [89] and *Acidithiobacillus ferrooxidans* [90]. For those microorganisms that repress, at low temperatures, the expression of the main chaperone GroEL, it has been hypothesized that, in psychrophiles, the hydrophobic core is often weaker than in mesophilic counterparts and, therefore, the risk of aggregation and misfolding is rather limited and the expression of GroEL is not generally required. This is supported by the fact that many cold-adapted proteins display a reversible thermal unfolding [44]. Also, in *E. coli*, the overexpression of the trigger factor at low temperature represses the expression of other chaperones [91]. The discrepancy observed in these data can possibly originate from the choice of the cardinal temperatures selected for comparison. Some of them correspond to the so-called optimum growth temperature based on the shortest doubling time, and it is well known that these temperatures induce a severe thermal stress in bacteria. This can significantly modify the data. There is, however, a common phenomenon observed at low temperatures; it is the systematic overexpression of the peptidyl–prolyl cis/trans isomerase and of the trigger factor which was found to embark a PPIase domain [92]. This clearly indicates that the *cis-trans* isomerization of prolyl residues is a particularly significant limiting step of the folding process of proteins at low temperatures.

Figure 5. Schematic representation of funnel-shaped folding energy landscapes of mesophilic (left diagram) and psychrophilic (right diagram) enzymes. Adapted from [32].

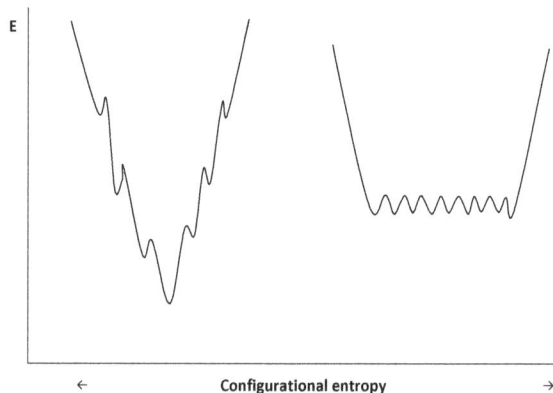

7. Conclusions

Life at low temperature requires the expression of enzymes that can compensate for the freezing effect of these temperatures on molecular structures. Indeed, cold can prevent the appropriate interaction with the substrate and is exponentially detrimental to the rate of the chemical reactions occurring in living organisms. Psychrophiles, during the course of evolution, have slightly altered the amino acid sequence of their enzymes in order to produce molecular structures characterized by a higher flexibility than their mesophilic counterparts at low temperatures. This was achieved through a decrease of the strength and number of classical weak interactions such as hydrogen and ionic bonds, as well as hydrophobic interactions, but also by an increase of the entropy of the unfolded state through a decrease in proline residues and an increase in glycine residues. Depending on the enzyme and on its capacity to tolerate some structural modifications, only a few of these destabilizing factors are mobilized so that the strategy in the adaptation appears specific to each enzyme. This has led to a spectacular increase of the specific activity of these enzymes at low and moderate temperatures that reaches more or less that of their mesophilic counterparts at their usual environmental temperatures. This strategy has been, however, generally detrimental to their thermal stability, which is drastically depressed with no consequence for these organisms as no selective pressure on thermal stability is exerted in their low temperature environments. Depending on their specific evolution history and on the physico–chemical characteristics of their permanent habitats each psychrophile has developed on this basis a specific strategy that led to a sort of continuum in the adaptation to cold of these enzymes. Indeed, some of these organisms are able to tolerate temperatures around 40 °C, while others display more accentuated characteristics with a limit of survival below 15 °C. The folding process of proteins is strongly dependent on temperature, and the production of cold-adapted enzymes necessitates some adjustments at this level. In this case also, the strategy adopted by cold-adapted organisms seems to be specific with a modulated expression of the various chaperones that assist folding. However, one consistency seems to be the overexpression of factors that facilitate the conversion of *cis* form of proline residues into their *trans* conformers, such as the enzyme peptidyl–prolyl *cis/trans* isomerase and the related trigger factor.

References

1. Brinkmeyer, R.; Knittel, K.; Jürgens, J.; Weyland, H.; Amann, R.; Helmke, E. Diversity and structure of bacterial communities in Arctic *versus* Antarctic pack ice. *Appl. Envirom. Microbiol.* **2003**, *69*, 6610–6619.
2. Buckeridge, K.M.; Grogan, P. Deepened snow alters soil microbial nutrient limitations in arctic birch hummock tundra. *Appl. Soil. Ecol.* **2008**, *39*, 210–222.
3. Steven, B.; Briggs, S.; McKay, C.P.; Pollard, W.H.; Greer, C.W.; Whyte, L.G. Characterization of the microbial diversity in a permafrost sample from the Canadian high Arctic using culture-dependent and culture-independent methods. *FEMS Microbiol. Ecol.* **2007**, *59*, 513–523.
4. Shivagi, S.; Kumari, K.; Kishore, K.H.; Pindi, P.K.; Rao, P.S.; Srivinas, T.N.R.; Asthana, R.; Ravindra, R. Vertical distribution of bacteria in a lake sediment from Antarctica by culture-independent and culture-dependent approaches. *Res. Microbiol.* **2011**, *162*, 191–203.
5. Lo Giudice, A.; Caruso, C.; Mangano, S.; Bruni, V.; De Domenico, M.; Michaud, L. Marine bacterioplankton diversity and community composition in an Antarctic coastal environment. *Microbiol. Ecol.* **2012**, *63*, 210–223.
6. Kirby, B.M.; Easton, S.; Tuffin, I.M.; Cowan, D.A. Bacterial diversity in polar habitats. In *Polar Microbiology. Life in a Deep Freeze*; Miller, R.V., Whyte, L.G., Eds.; ASM Press: Washington DC, USA, 2012; Chapter 1, p.3.
7. Collins, T.; D'Amico, S.; Marx, J.C.; Feller, G.; Gerday, C. Cold-adapted enzymes. In *Physiology and Biochemistry of Extremophiles*; Gerday, C.H., Glansdorff, N., Eds.; ASM Press: Washington DC, USA, 2007; Chapter 13, p.165.
8. Garcia-Viloca, M.; Gao, J.; Karplus, M.; Truhlar, D.G. How enzymes work: Analysis by modern rate theory and computer simulation. *Science* **2004**, *303*, 186–195.
9. Demchenko, A.P.; Rusyn, O.I.; Saburova, E.A. Kinetics of the lactate dehydrogenase reaction in high viscosity media. *Biochim. Biophys. Acta* **1989**, *998*, 196–203.
10. Mastro, A.M.; Keith, A.D. Diffusion in the aqueous compartment. *J. Cell Biol.* **1984**, *99*, 180–187.
11. Siddiqui, K.S.; Bokhari, S.A.; Afzal, A.J.; Singh, S. A novel thermodynamic relationship based on Kramers theory for studying enzyme kinetics under high viscosity. *IUBMB Life* **2004**, *56*, 403–407.
12. Feller, G.; Lonhienne, T.; Deroanne, C.; Libioulle, C.; Van Beeumen, J.; Gerday, C. Purification, characterization, and nucleotide sequence of the thermolabile α-amylase from the Antarctic psychrotroph *Alteromonas. haloplanktis* A23. *J. Biol. Chem.* **1992**, *267*, 5217–5221.

13. Kobori, H.; Sullivan, C.W.; Shizuya, H. Heat-labile alkaline phosphatase from Antarctic bacteria: Rapid 5' end-labeling of nucleic acid. *Proc. Natl. Acad. Sci.* USA **1984**, *81*, 6691–6695.

14. De Santi, C.; Tutino, M.L.; Mandrich, L.; Giuliani, M.; Parrilli, E.; Del Vecchio, P.; de Pascale, D. The hormone-sensitive lipase from *Psychrobacter.* sp. TA144: New insight in the structural/functional characterization. *Biochimie* **2010**, *92*, 949–957.

15. Simpson, P.J.L.; Codd, R. Cold adaptation of the mononuclear molybdoenzyme periplasmic nitrate reductase from the Antarctic bacterium *Shewanella. gelidimarina. Biochem. Biophys. Res. Comm.* **2011**, *414*, 783–788.

16. Angelaccio, S.; Florio, R.; Consalvi, V.; Festa, G.; Pascarella, P. Serine hydroxymethyltransferase from the cold adapted microorganism *Psychromonas. ingrahamii*: A low temperature active enzyme with broad specificity. *Int. Mol. Sci.* **2012**, *13*, 1314–1326.

17. D'Amico, S.; Gerday, C.; Feller, G. Structural determinants of cold adaptation and stability in a large protein. *J. Biol. Chem.* **2001**, *276*, 25791–25796.

18. Birolo, M.; Tutino, L. Aspartate aminotransferase from the Antarctic bacterium *Pseudoalteromonas. haloplanktis* TAC 125. Cloning, expression, properties, and molecular modelling. *Eur. J. Biochem.* **2000**, *267*, 2790–2802.

19. Sun, K.; Camardella, L.; Di Prisco, G.; Hervé, G. Properties of aspartate transcarbamylase from TAD1, a psychrophilic bacterial strain isolated from Antarctica. *FEMS Microbiol. Lett.* **1998**, *164*, 375–382.

20. Gerike, U.; Danson, M.J.; Russell, N.J.; Hough, D.W. Sequencing and expression of the gene encoding a cold-active citrate synthase from an Antarctic bacterium strain DS-3R. *Eur. J. Biochem.* **1997**, *248*, 49–57.

21. Georlette, D.; Jonsson, Z.O.; Van Petegem, F.; Chessa, J.; Van Beeumen, J.; Hubscher, U.; Gerday, C. A DNA ligase from the psychrophile *Pseudoalteromonas. haloplanktis* gives insights into the adaptation of proteins to low temperatures. *Eur. J. Biochem.* **2000**, *267*, 3502–3512.

22. Masullo, M.; Arcari, P. Psychrophilic elongation factor TU from the Antarctic *Moraxella.* sp. TAC II 125: Biochemical characterization and cloning of the encoding gene. *Biochemistry* **2000**, *39*, 15531–15539.

23. Altermak, B.; Niiranen, L.; Willassen, N.P.; Smalas, A.O.; Moe, E. Comparative studies of endonuclease I from cold-adapted *Vibrio. salmonicida* and mesophilic *Vibrio. cholerae. FEBS J.* **2007**, *274*, 252–263.

24. Watanabe, S.; Yasutake, Y.; Tanaka, I.; Takada, Y. Elucidation of stability determinants of cold-adapted monomeric isocitrate dehydrogenase from a psychrophilic bazcterium, *Colwellia. maris*, by construction of chimeric enzymes. *Microbiology* **2005**, *151*, 1083–1094.

25. Coquelle, N.; Fioravanti, E.; Weik, M.; Vellieux, F.; Madern, D. Activity, stability and structural studies of lactate dehydrogenases adapted to extreme thermal environments. *J. Mol. Biol.* **2007**, *374*, 547–562.

26. Xu, Y.; Feller, G.; Gerday, C.; Glansdorff, N. Metabolic enzymes from psychrophilic bacteria: Challenge of adaptation to low temperatures in ornithine carbamoyltransferase from *Moritella. abyssi. J. Bacteriol.* **2003**, *185*, 2161–2168.

27. Cartier, G.; Lorieux, F.; Allemand, F.; Dreyfus, M.; Bizebard, T. Cold adaptation in DEAD-box proteins. *Biochemistry* **2010**, *49*, 2636–2646.

28. Narinx, E.; Baise, E.; Gerday, C. Subtilisin from antarctic bacteria: Characterization and site-directed mutagenesis of residues possibly involved in the adaptation to cold. *Prot. Engineer.* **1997**, *10*, 1271–1279.

29. Alvarez, M.; Zeelen, J.P.; Mainfroid, V.; Rentier-Delrue, F.; Martial, J.A.; Wyns, L.; Wierenga R.K., Maes, D. Triose-phosphate isomerase (TIM) of the psychrophilic bacterium *Vibrio. marinus. J. Biol. Chem.* **1998**, *273*, 2199–2206.

30. Hochachka, P.W.; Somero, G.N. Temperature. In *Biochemical Adaptation*; Hochachka, P.W., Domero, G.N., Eds.; Oxford University Press: New York, NY, USA, 2002; Chapter 7, p.290.

31. Lonhienne, T.; Baise, E.; Feller, G.; Bouriotis, V.; Gerday, C. Enzyme activity determination on macromolecular substrates by isothermal titration calorimetry: Application to mesophilic and psychrophilic chitinases. *Biochim. Biophys. Acta* **2001**, *1545*, 349–356.

32. D'Amico, S.; Marx, J.-C.; Gerday, C.; Feller, G. Activity-stability relationship in extremophilic enzymes. *J. Biol. Chem.* **2003**, *278*, 7891–7896.

33. Suzuki, T.; Yamamoto, K.; Tada, H.; Uda, K. Cold-adapted features of arginine kinase from the deep-sea. *Calyptogena. kaikoi. Mar. Biotechnol.* **2012**, *14*, 294–303.

34. Garsoux, G.; Lamotte-Brasseur, J. ; Gerday, C.; Feller, G. Kinetic and structural optimisation to catalysis at low temperatures in a psychrophilic cellulase from the Antarctic bacterium *Pseudoalteromonas. haloplanktis. Biochem. J.* **2004**, *384*, 247–253.

35. Lonhienne, T.; Zoidakis, J.; Vorgias, E.; Feller, G.; Gerday, C.; Bouriotis, V. Modular structure, local flexibility and cold-activity of a novel chitobiase from a psychrophilic Antarctic bacterium. *J. Mol. Biol.* **2001**, *310*, 291–297.

36. Sotelo-Mundo, R.R.; Lopez-Zavala, A.A.; Garcia-Orozco, K.D.; Arvizu-Flores, A.A.; Velazquez-Conteras, E.F.; Vaalenzuela-Soto, E.M.; Rojo-Dominguez, A.; Kanost, M.R. The lysozyme from insect (*Manduca. sexta*) is a cold-adapted enzyme. *Protein Pept. Lett.* **2007**, *14*, 774–778.

37. Davail, S.; Feller, G.; Narinx, E.; Gerday, C. Cold adaptation of proteins. Purification, characterization and sequence of the heat labile subtilisin from the Antarctic psychrophile *Bacillus* TA41. *J. Biol. Chem.* **1994**, *269*, 17448–17453.

38. Collins, T.; Meuwis M-A. Gerday, C.; Feller, G. Activity, stability and flexibility in glycosidases adapted to extreme thermal environments. *J. Mol. Biol.* **2003**, *328*, 419–428.

39. Petrescu, J.; Lamotte-Brasseur, J.; Chessa, J.-P.; Ntarima, P.; Claeyssens, M.; Devreese, B.; Marino, G.; Gerday, C. Xylanase from psychrophilic yeast *Cryptococcus. adeliae. Extremophiles* **2000**, *4*, 137–144.

40. Marx, J.-C.; Collins, S.; D'Amico, S.; Feller, G.; Gerday, C. Cold-adapted enzymes from marine Antarctic microorganisms. *Mar. Biotechnol.* **2006**, *9*, 293–304.

41. Robindon, G.W.; Cho, C.H. Role of hydration water in protein unfolding. *Biophys. J.* **1999**, *77*, 3331–3318.

42. Loladze, V.V.; Ermolenko, D.N.; Makhatadze, G.I. Heat capacity changes upon burial of polar and non polar groups in proteins. *Protein Sci.* **2001**, *10*, 1343–1352.

43. Kumar, S.; Tsai, C.J.; Nussinov, R. Maximal stabilities of reversible two-state proteins. *Biochemistry.* **2002**, *41*, 5359–5374.

44. Feller, G. Protein stability and enzyme activity at extreme biological temperatures. *J. Phys. Condens. Matter* **2010**, 22, 32–49.

45. Chakravarty, S.; Varadarajan, R. Elucidation of factors responsible for enhanced thermal stabilitiy of protein: A structural genomics based study. *Biochemistry* **2002**, *41*, 8152–8161.

46. D'Amico, S.; Gerday, C.; Feller, G. Dual effects of an extra disulfide bond on the activity and stability of a cold-adapted α-amylase. *J. Biol. Chem.* **2002**, *277*, 46110–46115.

47. Cipolla, A.; D'Amico, S.; Barumandzadeh, R.; Matagne, A.; Feller, G. Stepwise adaptations to low temperature as revealed by multiple muatnts of psychrophilic α-amylase from Antarctic bacterium. *J. Biol. Chem.* **2011**, *286*, 38348–38555.

48. Papaleo, E.; Pasi, M.; Tiberti, M.; De Gioia, L. Molecular dynamics of mesophilic-like mutants of a cold-adapted enzyme: Insights into distal effects induced by the mutations. *PLoS One* **2011**, *6*, e24214.

49. Miyazaki, K.; Wintrode, P.L.; Grayling, R.A.; Rubingh, D.N.; Arnold, F.H. Directed evolution study of temperature adaptation in a psychrophilic enzyme. *J. Mol. Biol.* **2000**, *297*, 1015–1026.

50. Taguchi, S.; Komada, S.; Momose, H. The complete amino acid substitutions at position 131 that are positively involved in cold adaptation of subtilisin BPN'. *Appl. Environ. Microbiol.* **2000**, *66*, 1410–1415.

51. Zong, C.Q.; Song, S.; Fang, N.; Liang, X.; Zhu, H.; Tang, X.; Tang, B. Improvement of low-temperature caseinolytic activity of a thermophilic subtilase by directed evolution and site-directed mutagenesis. *Biotechnol. Bioeng.* **2009**, *104*, 862–870.

52. Bae, E.; Phillips, G.N. Structure and analysis of highly homologous psychrophilic, mesophilic and thermophilic adenylate kinases. *J. Biol. Chem.* **2004**, *279*, 28202–28208.

53. Davlieva, M.; Shammo, Y. Structure and biochemical characterization of an adenylate kinase originating from the psychrophilic organism *Marinibacillus. marinus. Acta Crystallogr. Sect. F Struct. Biol. Cryst. Commun.* **2009**, *65*, 751–756.

54. Aghajari, N.; Van Petegem, F.; Villeret, V.; Chessa, J.P.; Gerday, C.; Haser, R.; Van Beeumen, J. Crystal structures of a psychrophilic metalloprotease reveal new insights into catalysis by cold-adapted proteases. *Proteins* **2003**, *50*, 636–647.

55. De Baker, M.; McSweeney, S.; Rasmussen, H.B.; Riise, B.W.; Lindley, P.; Hough, E. The 1.9 Å crystal structure of heat-labile shrimp alkaline phosphatase. *J. Mol. Biol.* **2002**, *318*, 1265–1274.

56. Wang, E.; Koutsioulis, D.; Leiros, H.K.; Andersen, O.A.; Bouriotis, V.; Hough, E.; Heikinheimo, P. Crystal structure of alkaline phosphatase from the Antarctic bacterium TAB5. *J. Mol. Biol.* **2007**, *366*, 1318–1331.

57. Helland, R.; Leiros, I.; Berglund, G.I.; Willassen, N.P.; Smalas, A.O. The crystral structure of anionic salmon trypsin in complex with bovine pancreatic trypsin inhibitor. *Eur. J. Biochem.* **1998**, *256*, 317–324.

58. Aghajari, N.; Feller, G.; Gerday, C.; Haser, R. Structures of the psychrophilic *Alteromonas. haloplanctis* alpha-amylase give insights into cold adaptation at a molecular level. *Structure* **1998**, *6*, 1503–1516.

59. De Vos, D.; Xu, Y.; Hulpiau, P.; Vergauwen, B.; Van Beeumen, J.J. Structural investigation of cold activity and regulation of aspartate carbamoyltransferase from the extreme psychrophilic bacterium *Moritella. profunda. J. Mol. Biol.* **2007**, *365*, 379–395.

60. Skalova, T.; Dohnalek, J.; Spiwok, V.; Lipovova, P.; Vondrackova, E.; Petrokova, H.; Duskova, J.; Strnad, H.; Kralova, B.; Hasek, J. Cold-active beta-galactosidase from *Arthrobacter.* sp. C2–2 forms compact 660 kDa hexamers: Crystal structure at 1.9 Å resolution. *J. Mol. Biol.* **2005**, *353*, 282–294.

61. Riise, E.K.; Lorentzen, M.S.; Helland, R.; Smalas, A.O.; Leiros, H.-K.S.; Willassen, N.P. The first structure of a cold-active catalase from *Vibrio. salmo*nicida at 1.96 Å reveals structural aspects of cold adaptation. *Acta Crystallogr. D Biol. Crystallogr.* **2007**, *63*, 135–148.

62. Violot, S.; Aghajari, N.; Czjzek, M.; Feller, G.; Sonan, G.; Gouet, P.; Gerday, C.; Haser, R.; Receveur-Bréchot, V. Structure of a full length psychrophilic cellulase from *Pseudoalteromonas. haloplanktis* revealed by X-ray diffraction and small angle X-ray scattering. *J. Mol. Biol.* **2005**, *348*, 1211–1224.

63. Russell, R.J.; Gerike, U.; Danson, M.J.; Hough, D.W.; Taylor, G.L. Structural adaptations of the cold-active citrate synthase from an Antarctic bacterium. *Structure* **1998**, *6*, 351–361.

64. Berglund, G.I.; Willassen, N.P.; Hordvik, A.; Smalas, A.O. Structure of native pancreatic elastase from North Atlantic salmon at 1.61 Å resolution. *Acta Crystallogr. D Biol. Crystallogr.* **1995**, *51*, 925–937.

65. Brzuszkiewicz, A.; Nowak, E.; Dauter, Z.; Dauter, M.; Cieslinski, H.; Dlugolecka, A.; Kur, J. Structure of EstA esterase from psychrotrophic *Pseudoalteromonas.* sp. 643A covanently inhibited by monoethylphosphonate. *Acta Crystallogr. Sect. F Struct. Biol. Cryst. Commun.* **2009**, *65*, 862–865.

66. Uppenberg, J.; Hansen, M.T.; Paktar, S.; Jones, T.A. The sequence, crystal structure determination and refinement of two crystal forms of lipase B from *Candida antarctica*. *Structure* **1994**, *2*, 293–308.

67. Kim, S.Y.; Hwang, K.Y.; Kim, S.H.; Sung, H.C.; Han, Y.S.; Cho, Y. Structural basis for cold adaptation. Sequence, biochemical properties, and crystal structure of malate dehydrogenase from a psychrophile *Aquaspirillium. arcticum. J. Biol. Chem.* **1999**, *274*, 11761–11767.

68. Karlsen, S.; Hough, E.; Olsen, R.L. Structure and proposed amino-acid sequence of a trypsin from Atlantic cod (*Gadus. morhua*). *Acta Crystallogr. D Biol. Crystallogr.* **1998**, *54*, 32–46.

69. Leiros, H.K.; Pey, A.L.; Innselset, M.; Moe, E.; Leiros, I.; Steen, I.H.; Martinez, A. Structure of phenylalanine hydroxylase from *Colwellia. psychrerythraea* 34H, a monomeric cold active enzyme with local flexibility around the active site and high overall stability. *J. Biol. Chem.* **1997**, *282*, 21973–21986.

70. Helland, R.; Larsen, A.N.; Smalas, A.O.; Willassen, N.P. The 1.8 Å crystal structure of a proteinase K-like enzyme from a psychrotroph *Serratia.* species. *FEBS. J.* **2006**, *273*, 61–71.

71. Tsuruta, H.; Mikami, B.; Aizono, Y. Crystal structure of cold-active protein-tyrosine phosphatase from a psychrophile, *Shewanella.* sp. *J. Biochem.* **2005**, *137*, 69–77.

72. Alterio, V.; Aurilia, V.; Romanelli, A.; Parracino, A.; Saviano, M.; D'Auria, S.; De Simone, G. Crystal structure of an S-formyl glutathione hydrolase from *Pseudoalteromonas. haloplanktis* TAC 125. *Biopolymers* **2010**, *93*, 669–677.

73. Arnorsdottir, J.; Kristjansson, M.M.; Ficner, R. Crystal structure of a subtilisin-like serine proteinase from a psychrotrophic *Vibrio.* species reveals structural aspects of cold adaptation. *FEBS. J.* **2005**, *272*, 832–845.

74. Almog, O.; Gonzalez, A.; Godin, N.; de Leeuw, M.; Mekel, M.J.; Klein, D.; Braun, S.; Shoham, G.; Walter, R.L. The crystal structure of the psychrophilic subtilisin S41 and the mesophilic subtilisin Sph reveal the same calcium-loaded state. *Proteins* **2009**, *74*, 489–496.

75. Pedersen, H.L.; Willassen, N.P.; Leiros, I. The first structure of a cold-adapted superoxide dismutase (SOD): Biochemical and structural characterization of iron SOD from *Aliivibrio. samonicida. Acta Crystallogr. Sect. F Biol. Cryst. Commun.* **2009**, *65*, 84–92.

76. Toyota, E.; Ng, K.K.; Kuninaga, S.; Sekizaki, H.; Itoh, K.; Tanizawa, K.; James, M.N. Crystal structure and nucleotide sequence of an anionic trypsin from chum salmon (*Oncorrhynchus. keta*) in comparison with Atlantic salmon (*Salmo. salar*) and bovine trypsin. *J. Mol. Biol.* **2002**, *324*, 391–397.

77. Leiros, I.; Moe, E.; Lanes, O.; Smalas, A.O.; Willassen, N.P. The structure of uracil-DNA glycosylase from Atlantic cod (*Gadus. morhua*) reveals cold-adaptation features. *Acta Crystallogr. D Biol. Crystallogr.* **2003**, *59*, 1357–1365.

78. Van Petegem, F.; Collins, T.; Meuwis, M.A.; Gerday, C.; Feller, G.; Van Beeumen, J. The structure of a cold-adapted family 8 xylanase at 1.3 Å resolution. Structural adaptations to cold and investigation of the active site. *J. Biol. Chem.* **2003**, *278*, 7531–7539.

79. Hoffman, A.; Bukau, B.; Kramer, G. Structure and function of the molecular chaperone, trigger factor. *Biophys. Biochim. Acta* **2010**, *1803*, 650–661.

80. Tartaglia, G.G.; Dobson, C.M.; Hartl, F.U. Physicochemical determinants of chaperone requirements. *J. Mol. Biol.* **2010**, *400*, 579–588.

81. Nölting, B.; Salimi, N.; Guth, U. Protein folding forces. *J. Theor. Biol.* **2008**, *251*, 331–347.

82. Gianni, S.; Ivarsson, Y.; Jemth, P.; Brunori, M.; Travaglini-Allocatelli, C. Identification and characterization of protein folding intermediates. *Biophys. Chem.* **2007**, *128*, 105–113.

83. Xie, B.; Bian, F.; Chen, X.; He, H.; Guo, J.; Gao, X.; Zeng, Y.; Chen, B.; Zhou, B.; Zhang, Y. Cold adaptation of zinc metalloprotease in the thermolysin family from deep sea and Arctic sea ice bacteria revealed by catalytic and structural properties and molecular dynamics. *J. Biol. Chem.* **2009**, *284*, 9257–9269.

84. Mereghetti, P.; Riccardi, L.; Brandsdal, B.O.; Fantucci, P.; De Gioia, L.; Papaleo, E. Near native-state conformational landscape of psychrophilic and mesophilic enzymes: Probing the folding funnel model. *J. Phys. Chem. B.* **2010**, *114*, 7609–7619.

85. Ferrer, M.; Chernikova, T.N.; Yakimov, M.M.; Golyshin, P.N.; Timmis, K.N. Chaperonins govern growth of *Escherichia coli* at low temperatures. *Nat. Biotechnol.* **2003**, *21*, 1266–1267.

86. Ferrer, M.; Lunsdorf, H.; Chernikova, T.N.; Yakimov, M.; Timmis, K.N.; Golyshin, P.N. Functional consequences of single: Double ring transitions in chaperonins: Life in the cold. *Mol. Microbiol.* **2004**, *53*, 167–182.

87. Goodchild, A.; Saunders, N.F.; Erlan, H.; Raftery, M.; Guilhaus, M.; Curmi, P.M.; Cavicchioli, R. A proteomic determination of cold-adaptation in the Antarctic archaeon, *Methanococcoides. burtonii. Mol. Microbiol.* **2004**, *53*, 309–321.

88. Piette, F.; D'Amico, S.; Struvay, C.; Mazzuchelli, G.; Renaut, J.; Tutino, M.L.; Danchin, A.; Leprince, P.; Feller, G. Proteomics of life of low temperatures: Trigger factor is the primary chaperone in the Antarctic bacterium *Pseudoalteromonas. haloplanktis* TAC 125. *Mol. Microbiol.* **2010**, *76*, 120–132.

89. Ting, L.; Williams, T.J.; Cowley, M.J.; Lauro, F.M.; Guilhaus, M.; Raftery, M.J.; Cavicchioli, R. Cold adaptation in the marine bacterium *Sphingopyxis. alaskensis* assessed using quantitative proteomics. *Environ. Microbiol.* **2010**, *12*, 2658–2676.

90. Mykytczuk, N.C.; Trevors, J.T.; Foote, S.J.; Leduc, L.G.; Ferroni, G.D.; Twine, S.M. Proteomics insights into cold adaptation of psychrotrophic and mesophilic *Acidothiobacillus. ferrooxidans* strains *Antonie. Van Leeuwenhoek* **2011**, *100*, 259–277.

91. Kandror, O.; Goldberg, A.L. Trigger factor is induced upon cold shock and enhances viability of *Escherichia coli* at low temperatures. *Proc. Natl. Acad. Sci. USA* **1997**, *94*, 4978–4981.

92. Kramer, G.; Patzelt, H.; Rauch, T.; Kurtz, T.A.; Vorderwulbecke, S.; Bukau, B.; Deurling, E. Trigger factor peptidyl-prolyl *cis/trans* isomerase activity is not essential for the folding of cytosolic proteins in *Escherichia coli. J. Biol. Chem.* **2004**, *14*, 14165–14170.

Biotechnology of Cold-Active Proteases

Swati Joshi and Tulasi Satyanarayana

Abstract: The bulk of Earth's biosphere is cold (<5 °C) and inhabited by psychrophiles. Biocatalysts from psychrophilic organisms (psychrozymes) have attracted attention because of their application in the ongoing efforts to decrease energy consumption. Proteinases as a class represent the largest category of industrial enzymes. There has been an emphasis on employing cold-active proteases in detergents because this allows laundry operations at ambient temperatures. Proteases have been used in environmental bioremediation, food industry and molecular biology. In view of the present limited understanding and availability of cold-active proteases with diverse characteristics, it is essential to explore Earth's surface more in search of an ideal cold-active protease. The understanding of molecular and mechanistic details of these proteases will open up new avenues to tailor proteases with the desired properties. A detailed account of the developments in the production and applications of cold-active proteases is presented in this review.

Reprinted from *Biology*. Cite as: Joshi, S.; Satyanarayana, T. Biotechnology of Cold-Active Proteases. *Biology* **2013**, *2*, 755-783.

1. Introduction

Life exists at temperatures as low as −20 °C in the permafrost soil and as high as 122 °C in thermal environments [1,2]. Almost 70% of our planet's surface is covered by oceans, and thus, is a major environment where the temperature is around 4 °C. Polar regions constitute 15% of the Earth's surface and 20% of the terrestrial region of Earth is permafrost. Thus 80% of earth's surface is permanently cold with the temperatures below 5 °C [3]. All the cold geographical regions of the Earth harbor cold-adapted microorganisms, which are known as psychrophiles [4]. Modern biotech industry requires macromolecules that can function under extreme conditions. Psychrophilic and psychrotolerant microorganisms and their cold-adapted proteins and enzymes have a host of biotechnological applications [5]. Microorganisms get adapted to different niches, and thus, lead to evolution in their molecular machinery. The cold-adapted microbes are known to produce cold-active enzymes [6]. Among cold-active enzymes (α-amylase [7,8], lipase [9,10], aspartate transcarbamylase [11], Ca^+Zn^{+2} protease, [12], citrate synthetase, [13], α-lactamase [14], malate dehydrogenase [15], triose-phosphate isomerase [16], DNA ligase [17], xylanase [18], citrate synthase [19], metalloprotease [12], polygalacturonase [20], cellulases and xylanase [21], chitinase [22], endo-arabinanase [23], and pectinase [24]), proteases constitute an important group which have high catalytic efficiencies at lower temperatures. Proteases constitute an important class of hydrolytic enzymes that are found in all life forms as they are essential in physiological, metabolic and regulatory functions [25]. Nowadays, approximately 60% of the total enzyme market is shared by proteases in various industries, and according to a recent report from Business Communications Company (BCC 2008), the global market for industrial enzymes had been estimated to reach US $ 4.9 billion by 2013 [26,27]. Proteases have found applications in diverse fields such as detergent industry, leather processing, silk degumming, food and dairy, baking, pharmaceutical industries, silver recovery from x-ray films, waste management and others. Cold-active proteases have been reported from various microorganisms, but detailed investigations on their adaptation to cold environments and structure and bioenergetics are scarce. Their application potential has not yet been exploited fully for the benefit of mankind. Microbes with high potential are still waiting in the cold and harsh niches. This review attempts to summarize the developments in cold-active proteases, and strategies that can be adapted to search for more potent and versatile cold-active proteases to suit industrial requirements.

2. Microbes Producing Cold-Active Proteases

Cold-active proteases are mainly sourced from microorganisms from cold habitats such as arctic regions, polar regions, deep sea and glacier soils, glacier ice, permafrost, cold desert soil, sub-Antarctic sediments, sub-glacial water, alpine regions and other cold regions on earth. The potential of psychrophiles and enzymes produced by them have been reviewed from time to time [17,28–31]. Morita [32] defined psychrophiles based on their optimal growth temperature. Organisms growing optimally at about 15 °C or below with a maximum temperature of growth at about 20 °C and the ability to survive at 0 °C are known as psychrophiles. In contrast, psychrotolerant microbes generally have optimum and maximum temperatures of growth at 20 °C or above. Psychrotolerant microbes have an optimum growth temperature between 20 and 40 °C, but are also capable of growth at 0 °C [33].

Oceans cover more than 70% of Earth's surface and it is a major ecosystem with an average temperature of around 5 °C, and hence, this is one of the habitats for psychrophiles. A diverse range of psychrophilic microorganism have been isolated from sea belonging to different microbial groups such as gram-negative (e.g., *Pseudoalteromonas, Moraxella, Psychrobacter, Polaromonas, Psychroflexus, etc.*) and gram-positive (e.g., *Arthrobacter, Bacillus, Micrococcus*] bacteria, archaea [e.g., *Methanogenium, Halorubrum*), yeasts (e.g., *Candida, Cryptococcus*) and fungi (e.g., *Penicillium, Cladosporium*).

Cold-active protease producing microorganisms have been isolated from different geographical regions such as *Azospirillum* sp. from mountain soil [34], *Bacillus licheniformis* from glacier soil [35] *Clostridium* sp. from Antarctic region [36], *Colwellia* sp. from sea ice [37] and sub-antarctic sediments [38], *Curtobacterium luteum* from glacier soil [39], *Exiguobacterium* sp. from cold desert soil [40], *Pedobacter cryoconitis* from glacier ice [41], *Penicillium chrysogenum* from cold marine environment [42], *Pseudomonas* sp. from deap sea [43], *Psychrobacter proteolyticus* from Antarctic krill *Euphasia superba* Dana [44], *Serratia* sp. from coastal water [45], *Vibrio* sp. from marine water [46] and *Xanthomonas maltophilia* from alpine environment [47]. Yu *et al.* [48] screened organisms from the sandy sediment of Nella Fjord, Eastern Antarctica [69°22'6" S, 76°21'45" E] for the cold-active hydrolytic enzymes. Out of 33 isolates screened, *Sulfitobacter* sp NF1-26, *Photobacterium* NF1-15, *Pseudomonas* NF1-39-1, *Shewanella* NF1-3, *Bizionia* NF1-21, *Flavobacterim* NF1-9, *Salinibacterium* NF2-5 were found to secrete proteolytic enzymes. While Kuddus *et al.* [49] isolated cold-active alkaline protease producing *Stenotrophomonas* sp. from the soil of Gangotri glacier (western Himalaya, India). Some microorganisms that are known to produce cold-active alkaline proteases are listed in Table 1.

Table 1. Microorganisms producing cold-active alkaline protease.

S. No.	Organisms	Properties of the proteases			Reference
		Mol. weight (kDa)	T_{Opt} (°C)	$pH_{Opt.}$	
1	*Alcaligenes faecalis*	-	30	8.8	[50]
2	*Alkaliphilus transvaalensis*	30	40	12.6	[51]
3	*Alteromonas haloplanktis*	74–76	20	8–9	[52]
4	*Aspergillus ustus*	45	32	9	[53]
5	*Azospirillum* sp.	48.6	40	8.5	[34]
6	*Bacillus* sp.	-	30	9.6	[54]
7	*Bacillus* spp.	-	40	10.5–11	[55]
8	*Bacillus amyloliquefaciens* S94	45	-	10	[56]
9	*Bacillus cereus*	-	20	9	[57]
10	*Bacillus licheniformis* RKK-04	31	50	10	[58]
11	*Bacillus pumilus*	-	30	11.5	[59]
12	*Beauveria bassiana*	-	37	10	[60]
13	*Candida humicola*	-	37	10	[61]
14	*Clostridium* sp.	46	37	7	[36]
15	*Colwellia* sp.	60	35	8–9	[62]
16	*Colwellia psychrerythraea* strain 34H	71	19	6–8.5	[63]
17	*Curtobacterium luteum*	115	20	7	[39]
18	*Engyodontium album*	-	25	11	[64]
19	*Escherichia freundii*	55	25	10	[65]
20	*Exiguobacterium* sp.SKPB5	36	40	8	[40]
21	*Flavobacterium* YS-80	49	30	8–11	[66]
22	*Flavobacterium balustinum* P104	70	40	7–9	[67]
23	*Leucosporidium antarcticum* 171	34.4	30	8	[68]
24	*Pedobacter cryoconitis,*	27	40	8	[41]
25	*Penicillium chrysogenum* FS010	41	35	9	[42]
26	*Planomicrobium* sp. 547	-	35	9	[69]
27	*Pseudoalteromonas* sp. D12-004	34	35	7–8	[70]
28	*Pseudoalteromonas* sp. NJ276	28	30	8	[37]
29	*Pseudoalteromonas* sp. P96-47	-	20	8	[71]
30	*Pseudoalteromonas* sp. SM9913	65.84	25	9	[72]
31	*Pseudomonas* sp Ele-2	45	40	-	[73]
32	*Pseudomonas* sp.	-	20		[74]
33	*Pseudomonas* strain DY-A	-	40	10	[43]
34	*Pseudomonas aerugenosa* MTCC 7926	-	40	9	[75]
35	*Pseudomonas lundensis*	48	30	10.5	[76]
36	*Pseudomonas fluorescens*	-	35	5	[77]
37	*Pseudomonas fluorescens* 114.	47	35-40	8	[78]

Table 1. *Cont.*

| S. No. | Organisms | Properties of the Proteases | | | Reference |
		Mol. Weight (kDa)	T_{Opt} (°C)	$pH_{Opt.}$	
38	*Pycnoporus cinnabarinus* ss3	-	30	4	[79]
39	*Roseobacter* sp. [MMD040]	-	37-40	8–9	[80]
40	*Serratia marcescens* AP3801	58	40	6.5–8.0	[81]
41	*Serratia marcescens* TS1	56	40	8	[82]
42	*Serratia proteamaculans* 94	50	4-30	8	[83]
43	*Shewanella* strain Ac10	44	5-15	9	[84]
44	*Stenotrophomonas* sp.	55	15	10	[85]
45	*Stenotrophomonas maltophilia* MTCC 7528	75	20	10	[49]
46	*Streptomyces* sp.	-	30	10	[86]
47	*Streptomyces alboniger*	-	37	9–11	[87]
48	*Teredinobacter turnirae*	-	25	7	[88]
49	*Trichoderma atroviride*	24	25	6.2	[89]
50	*Vibrio* sp.	35	40	8.5–9.0	[90]
51	*Vibrio* sp. PA-44	47	25	8.6	[46]

3. Classification of Proteases

According to the Enzyme Commission [EC] classification, proteases are members of the group 3 [Hydrolases], and sub-group 4 [hydrolyzing peptide bonds]. Proteases have been divided into two broad groups on the basis of their ability to hydrolyze N- or C- terminal peptide bonds [exopeptidases] or internal peptide bonds [endopeptidases]. Although exopeptidases are used in some commercial applications, endopeptidases are industrially more important than the former. Exopeptidases are subdivided as aminopeptidases that cleave the N-terminal peptide linkage and carboxypeptidases that cleave the C-terminal peptide bond.Several other features have also been used in classifying proteases into different groups such as occurrence of charged moieties at sites relative to susceptible bond [91], their pH optima [as acidic, neutral or alkaline], substrate specificity [collagenase, keratinase, elastase], or their homology to previously characterized proteases such as trypsin, pepsin and others [trypsin-like, pepsin-like]. Morihara [92] classified serine proteases as trypsin-like proteinases, alkaline proteinases, *Myxobacter* α-lytic proteinases and staphylococcal proteinases. Hartley *et al.* [93] classified endoproteases into four groups on the basis of their active site and sensitivity to various inhibitors. The properties of the enzymes are summarized in Table 2.

4. Optimization of Fermentation Conditions for Production of Cold-Active Proteases

Proteases produced by microorganisms are predominantly extracellular in nature and are greatly affected by nutritional and physicochemical factors. Optimization of different media components can greatly affect the production cost and can lead either to profit or loss in an industry based on production of bioactive compounds by microorganisms. Proper balance of various media components determines the utilization of each component. In order to have a cost effective method of enzyme production, optimization of various media components is needed. Importance of this step is revealed by the fact that 30%–40% of production cost of industrial enzymes is estimated to be the cost of the growth medium [94]. No single medium can be used for production of protease from different psychrotrophic microbes. Each microorganism has its own specific idiosyncratic, physicochemical and nutritional requirements for the production of maximum enzyme titer. Therefore, it is necessary to optimize the production conditions for the strain of interest. Protease production by psychrotrophic microorganisms is affected by media components such as changes in C/N ratio, presence or absence of some easily metabolizable sugars such as glucose and sucrose in the production medium. Casein was the best nitrogen source, but the presence of carbohydrates like glucose, sucrose and lactose led to catabolic repression of protease production in *Colwellia* sp. [37]. Metal ions in the surrounding environment affect the growth of the organisms. Some having positive effect and some inhibits the growth of the organism. It is critical to find out which metal ion supports both the growth of the organism under study and the protease production. In *Stenotrophomona* sp., the enzyme production was enhanced by Cu^{2+} [126.8%] and Cr^{2+} [134.6%], but Co^{2+} reduced it [43.5%]. The other heavy metals such as Hg^{2+}, Cd^{2+} and Zn^{2+} had no significant effect [49]. Vazquez *et al.* [95] reported that increasing concentrations of calcium chloride [0 to 0.3 g l^{-1}] in culture media enhanced protease production in *Stenotrophomonas maltophilia*; the highest titre was attained after 36 h of growth.

Most of the proteolytic enzymes are produced and secreted in late exponential growth phase [96]. *Stenotrophomona* sp. has been reported to secrete maximum enzyme at 120 h [49]. While the enzyme production by *Pedobacter cryoconitis* attained a peak in 72 h, and thereafter, there was plateau in enzyme production [41]. *Pseudomonas* sp. strain DY-A produced maximum protease after 30 h incubation [43]. In case of *Pseudomonas* sp. strain DY-A protease, 10 °C was found optimum both for growth and protease production. Temperature change to 25 °C reduced both the growth and protease production [43]. In *Stenotrophomona* sp., a high protease titre [56.2 U/ml] was attained at 20 °C. This observation suggested that high enzyme titers could be produced in the temperature range between 15 and 25 °C [49]. *Pedobacter cryoconitis* produced maximum enzyme at 15 °C, although 44% of the maximum enzyme titer was also attained at 1 °C [41].

Table 2. Classification and biochemical characteristics of endoproteases.

Endoprotease	EC No.	Mol. Mas Range (kDa)	pH$_{Opt.}$	T$_{Opt.}$ (°C)	Metal Ion Required	Active Site a Residues	Major Inhibitor(s)
Aspartic or Carboxyl proteases	3.4.23	30–45	3–5	40–55	Ca^{2+}	Aspartate or cysteine	Pepstatin
Cysteine or thiol proteases	3.4.22	34–35	2–3	40–55	-	Aspartate or cysteine	Indoacetamide, p-CMB
Metallo- proteases	3.4.24	19–37	5–7	65–85	Zn^{2+}, Ca^{2+}	Phenylalanine or leucine	Chelating agents such as EDTA, EGTA
Serine proteases	3.4.21	18–35	6–11	50–70	Ca^{2+}	Serine, histidine and aspartate	PMSF, DIFP, EDTA, soybean trypsin inhibitor, phosphate buffers, indole, phenol, triamino acetic acid

5. Purification of Cold-Active Proteases

Proteases have been screened and purified from different sources. Different strategies have been employed for purifying cold-active proteases from diverse sources. The purification strategies used for purifying cold-active proteases from different sources are presented in Table 3. Proteases secreted into the medium are first concentrated by using methods such as ultrafiltration [96–98], ammonium sulphate [36,39,42,43] or acetone precipitation [68,81]. A few methods involve use of PEG [72] and lyophilization [69]. After concentrating protein, further purification is achieved either by single technique or by combining two different methods. Ion exchange chromatography is a method of choice in maximum cases. DEAE [diethyl amino ethyl] and CM [carboxy methyl] group containing matrices are mainly used to which protein molecules get adsorbed and can be eluted either by pH change or change in ionic strength of the eluent buffer.

Affinity chromatography technique is also a successful method of purification but labile nature of affinity ligands and higher cost are limiting factors. Hydrophobic interaction chromatography [HIC] and gel filtration chromatography have also been used extensively for protease purification either at an early to middle stage or in the final stage. Sephacyl, Superdex, Superose and Topopearl gels are most commonly used for filtration purpose. Zambare *et al.* [99] used various chromatographic techniques to purify protease from *P. aeruginosa* MCM B-327 and determined its molecular weight (Figure 1).

Figure 1. Native-PAGE of crude and purified protease from *P. aeruginosa* MCM B-327. (A) silver stained gel; (B) zymogram of protease with casein; (C) plot of Rf values *versus* standard molecular weights [99].

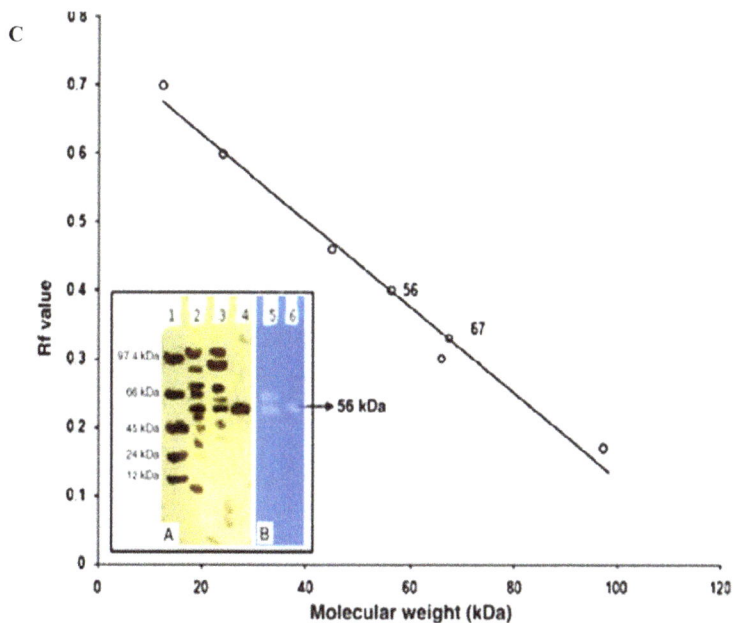

Table 3. Strategies used for purification of cold-active proteases.

Protease Source	Protease Type	Concentration Method	Column Matrices	Fold Purification	References
Alkaliphilus transvaalensis	Serine protease	Amicon Ultra-15	DEAE Toyopearl 650M resin,CM-Toyopearl 650M	96	[58]
Clostridium species	Serine-type metalloenzyme	Ammonium sulfate precipitation	Sephadex G-100	12.7	[36]
Colwellia psychrerythraea strain 34H.	Aminopeptidase	-	Sepharose Q,Hydroxyapatite,Resource Q	460	[63]
Curtobacterium luteum MTCC 7529	Metalloprotease	Ammonium sulphate precipitation	DEAE- Cellulose	34.1	[39]
Escherichia freundii,	Neutral serine protease	Ammonium sulfate precipitate	CM-cellulose, DEAE-Sephadex A-50, Sephadex G-100	-	[65]
Leucosporidium antarcticum 171	Serine proteinase	Acetone precipitation	Sephadex G-75,Diethylaminoethyl-Sephacel,Sephacryl S-100	1,568	[68]
Oerskovia xanthineolytica TK-1	Serine protease	Ultrafiltration	Phenyl-Sepharose CL-4B, DEAE-Sephacel	39.6	[97]
Pedobacter cryoconitis	Metalloprotease	-	SP Sepharose,Syn-Chropak CM300	-	[39]
Penicillium chrysogenum FS010	Serine protease	Ammoniumsulfate precipitation	DEAE Sepharose,Sephadex G-100	103.2	[42]
Planomicrobium species	Serine protease	Ammonium sulfate precipitation, Lyophilization	DEAE-52	-	[69]
Pseudoalteromonas sp. NJ276	Serine protease	Ammonium sulfate precipitation	DEAE-Sephadex A50,Sephadex G-75	22.5	[37]
Pseudoalteromonas sp. SM9913	Serine protease.	Ammonium sulfate precipitation, PEG 2000.	Sephadex G100	-	[53]
Pseudomonas aeruginosa IFO 3455	Metalloprotease	-	QAE-agarose	-	[74]
P. fluorescence 114	Neutral metalloprotease	Ammoniumsulfate precipitation	DEAE Toyopearl 650 M, Superdex 200 HR 10/30	-	[78]

Table 3. *Cont.*

Protease Source	Protease Type	Concentration Method	Column Matrices	Fold Purification	References
Pseudomonas strain DY-A	Serine protease	Ammonium sulfate precipitation,	DEAE Sepharose CL-6B, Sephadex G-100	84.2	[43]
Serratia marcescens AP3801	Metalloprotease	Ammonium sulfate precipitation	Sephacryl S-100, Q Sepharose	0.48	[62]
S. marcescens TS1.	Metalloprotease	Ammonium sulphateation, acetone precipitation	DEAE-cellulose	-	[82]
S. proteamaculans	Trypsin-like protease	Ultrafiltration	Q-Sepharose,BPTl-Sepharose	-	[98]
S. proteamaculans	Serine trypsin-like and Zn-dependent protease.	-	BPTl-Sepharose	-	[100]
S. proteamaculans 94	Cysteine protease	-	Arg-Silochrom Z-Gly-*DL*-Pro-Gly-Silochrom, Superise 12 HR 10/30 column	3433	[83]
Shewanella strain Ac10	Alkaline serine protease	-	Bacitracin-Sepharose column		[84]
Stenotrophomonas maltophilia	Serine proteases	Ultrafiltration	S-Sepharose	-	[101]
Stenotrophomonas sp.	Alkaline protease	Ammonium sulfate precipitation	DEAE-Sepharose	18.45	[85]
Marine psychrophilic strainPA-43	Serine peptidase	-	Q Sepharose, Sephacryl S-300, PBE 94	25.0	[102]
Vibrio sp. PA-44	subtilisin-like proteinase	Ammonium sulfate precipitation	N-carbobenzoxy-d-phenylalanyl-triethylenetetramine-Sepharose, phenyl-sepharose	-	[46]

6. Properties of Cold-Active Proteases

6.1. Temperature

Protease produced from *Stenotrophomonas maltophilia* MTCC 7528 is optimally active at 20 °C and the activity of the enzyme is retained even after repeated freez-thaw cycles [49]. M1 aminopeptidase (designated ColAP) produced by the marine psychrophilic bacterium *Colwellia psychrerythraea* strain 34H exhibited optimum activity at 19 °C [63]. Zhu *et al.* [42] reported optimal temperature for protease produced by *Penicillium chrysogenum* FS010 was 35 °C, about 10–15 °C lower than normally used industrial protease. The enzyme showed relatively high activity between15 and 35 °C. It retained 41% of proteolytic activity at 0 °C. Protease produced by *Pedobacter cryoconitis* showed optimum temperature for activity was 40 °C. Activity was significantly reduced at 50 °C, and total inactivation occurred at 60 °C. High activity (28%–79% of the maximum activity) was detected at 20–35 °C [41].

6.2. pH

The activity of any enzyme is greatly affected by the pH of the reaction mixture. Each enzyme has its own optimum pH at which is shows maximum activity. Protease fall into different classes based on their pH optima. An alkaline protease isolated from *Stenotrophomonas maltophilia* MTCC 7528 has been shown to be optimally active at pH10 [49]. Protease produced by *Pseudoalteromonas* sp. NJ276 has optimum enzyme activity at pH 8.0. About 31 and 38% of its optimum activity is detectable at pH 5.0 and 11.0, respectively [35]. Protease of *Pseudomonas* sp. strain DY-A showed a broad pH profile (pH 6.0–12.0) for casein hydrolysis and highest activity between pH 8.0 and 10.0. The highest stability of this protease was recorded at pH 10.0 [43].

6.3. Metal Ions

Cold-active protease from *Pseudoalteromonas* sp. NJ276 was partially inhibited by metals such as Mg^{2+}, Ca^{2+}, Cu^{2+}, Zn^{2+}, Ba^{2+}, Fe^{2+}, Pb^{2+} and Mn^{2+}. The enzyme was stable after incubation for 1 h in the presence of 2M NaCl, moreover, 56.5% of the maximum activity was detected in the presence of high-salt concentrations (up to 3M NaCl) [37]. Zeng *et al.* [43] reported that total activity of protease from *Pseudomonas* sp. strain DY-A production medium was increased by 30% in the presence of Ca^{2+} and Mg^{2+} (10 mM). These metal ions enhanced the enzymatic activity slightly (8%) and had an important role in enzyme stability. Among the cations tested, Co^{2+}, Cu^{2+} and Zn^{2+} inhibited the enzymatic activity, while Fe^{3+}, Mn^{2+}, K^+, Li^+, Hg^+, Ag^+ had no observable effect on enzymatic activity [43]. *Penicillium chrysogenum* FS010 protease activity was increased by the addition of Ca^{2+}, Na^+, Mg^{2+}, K^+, $NH4^+$, while Cu^{2+}, Co^{2+}, Fe^{3+} and EDTA inhibited the enzyme activity. The inhibitory effects of Fe^{3+} and EDTA were the strongest [43]. Cu^{2+} and Fe^{3+} showed strong inhibitory effect on Ps5 metalloprotease of *Pseudomonas lundensis,* while Co^{2+}, Fe^{2+}, Mn^{2+}, Al^{3+} reduced enzyme activity to 32%–14% [76]. Yang *et al.* [76] suggested that there could be a new catalytic pathway for reaction mechanism of Ps5 protease as contrary to other metalloproteases it showed inhibition by Zn^{2+}. The activity of ColAP was strongly inhibited by

Zn^{2+} and Mn^{2+}, while Ca^{2+} was slightly inhibitory, Mg^{2+} stimulated activity at 10 mM or higher, which is almost equivalent to the concentration found in seawater [63].

6.4. Effect of Inhibitors and Other Reagents

Cold-active protease of *Pseudoalteromonas* sp. NJ276 was inhibited by phenylmethylsulfonylfluoride [PMSF], sodium dodecyl sulfate [SDS], urea, thiourea, dithiothreitol [DTT], ethylenediaminetetraacetic acid [EDTA] and ethylene glycol tetraacetic acid [EGTA] but cystein protease inhibitor E-64 had no effect on the activity [37]. *Pseudomonas* sp. strain DY-A protease was completely inhibited by 1 mM of diisopropyl fluorophosphate (DFP), PMSF and 4-(2-Aminoethyl) benzenesulfonyl fluoride hydrochloride [AEBSF] which indicates that it was a serine protease. The enzyme was found resistant to thiol reducing agents such as DTT (10 mM) and β-mercaptoethanol [β-ME] (5%), suggesting that disulfide bonds were not involved in preserving proteolytic activity. The enzyme was found sensitive to urea (4 M), SDS (1%) and guanidine-HCl (1 M), indicating that hydrogen bonds played an important role in preserving the enzyme activity [43]. Seventy percent activity of cold-active metalloprotease Ps5 of *Pseudomonas lundensis* HW08 was inhibited by 10mM PMSF [76], indicating that it was a serine protease. Only 40% of the maximum activity was left after treatment with 1mM EDTA while EGTA had no significant effect on activity of Ps5 protease. Ps5 protease exhibited an interesting property of stimulation of activity 124% at 1.0% of H_2O_2, and retained 73% activity on increasing H_2O_2 concentration to 10%. Moreover 10% urea stimulated enzyme activity, but it lost the activity completely in the presence of anionic detergent SDS even at 1%. Hustan *et al.* [63] reported that the activity of ColAP was not affected significantly by PMSF, but inhibited by 10 mM DTT (a reducing agent) and EDTA (a metal-chelating agent).

6.5. Catalytic Efficiencies

Although the environmental and physiological effects on microorganisms dwelling in cold environments have been understood fairly well, the mechanistic details that allow enzymatic reactions in the cold niches have not been understood adequately. Cold temperatures lead to exponential decreases in rates of chemical reactions, as clear by the Arrhenius equation, and also lead to increase in the compact folding of proteins, and thus restricting the conformational ease needed for catalysis [102]. Despite these setbacks, cold-active enzymes have evolved in nature. In contrast to mesophilic enzymes, these enzymes exhibit three general distinguishing features: a higher specific activity [k_{cat}] or catalytic efficiency [*i.e.*, k_{cat}/K_m] at temperatures between 0 and 30 °C, a lower optimal temperature for activity, and reduced stability due to temperature rise and denaturing agents [103]. Kinetic parameters like K_m and k_{cat} have been studied in several psychrophilic enzymes and compared with their mesophilic and thermophilic counterparts [14,103]. Enzymes usually buried in huge amount of substrate tend to optimize their k_{cat} rather than K_m [14] for efficient functioning at low temperatures. Cold-active protease from *Clostridium* species isolated from Schirmacher oasis, Antarctica showed an increase in K_m with decrease in temperature which appears to be a characteristic that indicates a weak substrate binding which in turn lowers the activation energy [36]. This strategy to improve catalytic efficiency at low temperature has been adopted by many cold-active enzymes from Antarctic bacteria including alpha amylase from *Alteromonas haloplanctis* [104] and β-lactamase from *Psychrobacter immobilis* [14]. The *Clostridium* protease also exhibited a $Q10$ value (1.91) which is much lower than that observed for most mesophilic enzymes, which are in the order of 2 to 3 [14]. True psychrophilic enzymes are more flexible in structure and are invariably thermolabile. Despite displaying adaptational features of a cold-active enzyme like low $Q10$ value and increased K_m at low temperatures, protease from *Clostridium* has been moderately thermostable [36].

The Michaelis-Menten constant [K_m] and catalytic efficiency [k_{cat}/K_m] values for protease from *Pseudoalteromonas* sp. NJ276 were 0.41 mM and 45 $s^{-1}mM^{-1}$, respectively at 35 °C. The enzyme retained 54% of k_{cat} and k_{cat}/K_m at 0 °C. The activation energy of the protease was 34.8 kJ mol^{-1} [37].

Huston *el al.* [63] reported the highest specific activity [k_{cat}] for *Colwellia psychrerythraea* strain 34H cold-active protease (ColAP) at 19 °C. The highest k_{cat}/K_m value for ColAP was also recorded at 19 °C (5.0 $s^{-1}mM^{-1}$), and 44% of this was retained at 9 °C which is the optimum growth temperature for the strain 34H.

6.6. Substrate Spectrum

Proteases generally exhibit broad substrate specificity and are found active against both synthetic substrates and native proteins. Zeng *et al.* [43] reported *Pseudomonas* sp. strain DY-A protease displayed high activity towards N-succinyl-Ala-Ala-Pro-Phe-p-nitroanilide and N-succinyl-Ala-Ala-Pro-Leu-p-nitroanilide , which are well-known substrates for chymotrypsin but showed no activity towards N-Succinyl-Ala-Ala-Pro-Asp-p-nitroanilide, N-Succinyl-Ala-Ala-Ala-p-nitroanilide, N-Succinyl-Gly-Phe-p-nitroanilide. MCP-3 protease from *Pseudoalteromonas* sp. SM9913 displayed a broad substrate specificity and hydrolysed AAPF [N-Succinyl-Ala-Ala-Pro-Phe-p-nitroanilide] and partially hydrolyzed AAPL[N-Succinyl-Ala-Ala-Pro-Leu-p-nitroanilide], AAPK [N-Succinyl-Ala-Ala-Pro-Lys-p-nitroanilide], AAPR [N-Succinyl-Ala-Ala-Pro-Arg-p-nitroanilide], FAAF [N-Succinyl-Phe-Ala-Ala-Phe-p-nitroanilide]and FVR [N-Benzoyl-Phe-Val-Arg-p-nitroanilide] [105]. The protease from *Pseudomonas fluorescens* hydrolyzed various proteins with preference for milk proteins, caseins, as substrates [106]. Urea-Hb (urea-denatured haemoglobin) is the preferred substrate of the Antarctic yeast *Leucosporidium antarcticum* 171 proteinase lap2. The activity against native Hb and casein was found 60%–70% lower. The subtilase showed poor activity on elastin [68]. Psychrotrops have been reported to preferentially use synthetic substrates with proline at position P2 and an aromatic residue rather than an aliphatic residue at position P1 [63,102,107]. Protease from *Shewanella* strain PA- 43 displayed high activity towards N-succinyl AAPF p-nitroanilide. N-succinyl AAPL p-nitroanilide was also a relatively good substrate but showed reduced activity for N-succinyl AAVA p-nitroanilide. N-succinyl GGF p-nitroanilide, N-succinyl AAA p-nitroanilide and N-succinyl AAPD p-nitroanilide were poor substrates, and there was no activity with N-succinyl AAV p-nitroanilide, N-succinyl GFG p-nitroanilide, N-succinyl L-Phe p-nitroanilide, or N-benzoyl DL-Arg p-nitroanilide [102]. Acidic protease from psychrotrophic yeast *Candida humicola* hydrolyzed poly-L-Ala, poly-L-Ser, poly-L-Phe, and poly-L-Glu but was more active on native proteins such as BSA, casein, gelatin, and melittin [61]. *Flavobacterium balustinum* protease displayed endoprotease activity with N Suc- AAPL-p-nitroanilides, whereas, there was no exoprotease activity of protease when L-Ala-pNA and L-Phe-pNA was used as substrate [50]. As proteases from psychrotrophs have been found to exhibit wide range of substrate specificity, they can be utilized for bioremediation of industrial and domestic waste at ambient temperature.

7. Cloning and Expression of Cold-Active Proteases

Recombinant DNA technology has revolutionized the enzyme industry by providing a means for making the enzyme production process very economical. Till date very few attempts have been made in cloning of cold-active proteases and expression in heterologous or homologous hosts. Ni *et al.* [108] reported cloning of a 1248 bp long alkaline protease gene ORF (encoding a 42.9 kDa protein) from marine yeast *Aureobasidium pullulans* HN2-3 into surface display vector pINA1317-YlCWP110. This gene (ALP2) was heterologously expressed in cells of *Yarrowia lipolytica*. The recombinant *Y. lipolytica* cells displayed protease on its surface and were able to produce bioactive peptides from different sources of proteins. Khairullin *et al.* [98] sequenced and cloned a 78kDa novel trypsin like protease (PSP) from *Serratia proteamaculans.* The protease was cloned in *E. coli* expression vector pET23b (+) and trsansformed into *E. coli* BL21 [DE3] expression strain. Recombinant protein was purified using Ni^{2+}-NTA agarose column. The yield of expressed His6-PSP was 150 mg from 100 g of biomass. Yan *et al.* [105] cloned cold-adapted halophilic proteases from deep-sea psychrotolerant bacterium *Pseudoalteromonas* sp. SM9913. Yan *et al.* cloned the protease gene into pET22b (+) and expressed the gene as active protein in *E. coli* BL21 [DE3] cells. The recombinant protein was purified from fermentation broth as a multidomain protein containing one catalytic domain and two PPC domains which were further characterized using purified recombinant protein. Wintrode *et al.* [109] used cloning and expression technique to study the reversal of the properties of a mesophilic subtilisin like protease from *Bacillus sphaericus* to a protease resembling more to a psychrophilic protease. Directed laboratory evolution approach used by Wintrode resulted in generation of a protease which showed rate constant [k_{cat}] at 10 °C 6.6 times and a catalytic efficiency [k_{cat}/K_m] 9.6 times that of wild type. Its half-life at 70 °C is 3.3 times less than wild type. Wintrode *et al.* (109) used *E. coli-* *Bacillus* shuttle vector pSPH2R and a protease deficient *Bacillus* strain DB 428 for their study. DNA library was prepared in *E. coli* HB101 cells and then transformation of *Bacillus* competent cells was performed. Taguchi *et al.* [110] also took assistance of cloning and expression technology to improve psychrophilic features of a cold-active protease. The mutant subtilisin m-63 showed k_{cat}/K_m value 100% higher than that of the wild type at 10 °C when *N*-succinyl-L-Ala- L-Ala-L-Pro-L-Phe-p-nitroanilide was used as a substrate. This cold adaptation resulted due to three mutations, Val to Ile at position 72 [V72I], Ala to Thr at position 92 [A92T], and Gly to Asp at position 131 [G131D], and it was observed that an enhancement in substrate affinity was mostly responsible for the increased activity. Taguchi *et al.* [110] used pUC18 and pHY300PLK expression vectors for *E. coli* and *Bacillus* respectively and *E. coli* JM 109 and *Bacillus subtilis* UOT0999 as expression hosts. Kulakova *et al.* [84] prepared genomic library using genomic DNA of *Shewanella* sp. strain Ac10 in pUC118 and then selected positive clones expressing serine alkaline protease [SapSh]. Positive clones were used for subsequent retrieval of protease ORF from the pUC118 clone and then sub-cloning in pET21a for cloning the SapSh under T7 lac promoter. Recombinant protein was functionally active but shorter in size (44 kDa) than expected (85 kDa) which indicated removal of some protein sequence during protein processing. Sheng *et al.* [69] also reported cloning of alkaline protease gene from psychrophilic *Planomicrobium* sp. 547 into pTA2 vector. It has been felt since long that a different expression host should be developed for expressing cold-active proteins, since this will assist the proteins is several ways like proper folding and thus retaining activity. Parrilli *et al.* [111] developed *Pseudoalteromonas haloplanktis* TAC125 as a versatile psychrophilic host for recombinant protein production by disrupting its *gspE* gene. Other psychrophilic hosts should be generated so that a diverse array of cold-active proteins along with industrially important protease can be heterologously expressed.

8. Crystal Structure of Cold-Active Proteases

X-ray crystallography has revolutionized the study of proteins. So far structure-function relationship of many enzymes has been deduced using this technique. Understanding of tertiary structure of enzymes and spatial arrangement of catalytically important functional groups must be known in order to get information about possible changes in enzyme structure resulting from binding of substrate, products, stabilizers, inhibitors or effector molecules. 3D structure of proteases can provide an insight into the mechanism of the enzymatic action and also provide a template for the design of novel drugs if the studied protease is involved in pathogenesis, e.g., viral ptoreases. Crystallographic studies of proteins can also assist in understanding molecular basis of structure-environment adaptation relationships. Thus proteins in cold environments can be understood in further detail by this technique. Comparative investigations of numerous protein models and crystal structures revealed that cold-adapted enzymes tend to exhibit an attenuation of the strength and number of structural factors known to stabilize protein molecules [112].

Recently Almog *et al.* [113] reported that the calcium-loaded state is not responsible for the cold adaptation of psychrophilic cold-adapted subtilisin S41 from Antarctic *Bacilllus* sp. TA41. This conclusion was reached based on comparison of crystal structure of S41 with a mesophilic subtilisin Sph from *Bacillus sphaericus* 22973. These two subtilisins were highly similar in their calcium binding mode but differed in cold adaptability.

Many cold-active enzymes have been studied so far using crystallographic technique. Some of them include *Arthrobacter* β-galactosidase [114], Lipase [115] of *Moraxella* TA144 a strain from Antarctica, Lipase from *Psychrobactor immobilis* [116]. Dong *et al.* [117] solved crystal structure of subtilisin like and psychrophilic protease Apa1 from Antarctic *Pseudoalteromonas* sp. strain AS11. An aminopeptidase from *Aeromonas proteolytica* was solved at 1.8 Å by Chevrier *et al.* [118]. Zhang *et al.* [119] solved the crystal structure of psychrophilic protease from *Flavobacterium* YS-80 at 2 Å. The marine protease from *Flavobacterium* acquires a two domain structure. N-terminal domain includes amino acid residues 37–264 and C-terminal comprise of residues from 265 to 480. Zhang *et al.* (66) compared the structural feature of the marine protease MP of *Flavobacterium* YS-80 with another psychrophilic protease PAP from Antarctic *Pseudomonas* species and mesophilic counterpart AP from *Pseudomonas aerugenosa* and SMP from *S. marcescens* (Figure 2). It was found that the Zn^{2+}-Tyr-OH bond in PAP is more flexible in order to facilitate substrate accessibility and to maintain its activity in very low temperatures. Marine protease MP contains seven glycine residues which are not present in AP. Glycine is responsible for flexibility to the protein thus indicating that MP is more flexible in nature than AP [19]. Eight Ca^{2+} ions and one Zn^{2+} ions have been positioned in the electron density map. The MP has a comparable overall structure to PAP, AP and SMP with as it has a two domain structure as described earlier. After overlapping the overall structure of MP with PAP, AP and SMP using all the Ca atoms, the main-chain RMSDs were 0.87 Å [with PAP form1], 1.08 Å [with PAP form2], 1.03 Å [with PAP form3], 1.00 Å [with AP] and 1.28 Å [with SMP] [66].

Figure 2. (a) superimposed image of MP [in yellow] with PAP forms 2 [in magenta color] and AP [in cyan] and SMP [in pale green]. Zn and Ca ions shown in figure are from the structure of MP. **(b)** additional N-terminal Ca^{2+} binding site is shown in superimposed image of MP [in cornflower blue] with PAP form 1 [in purple] and AP [in cyan] demonstrating a stabilized loop formation shaped in MP. Amino acids of MP which are coordinating to the Ca ions are depicted as sticks [66].

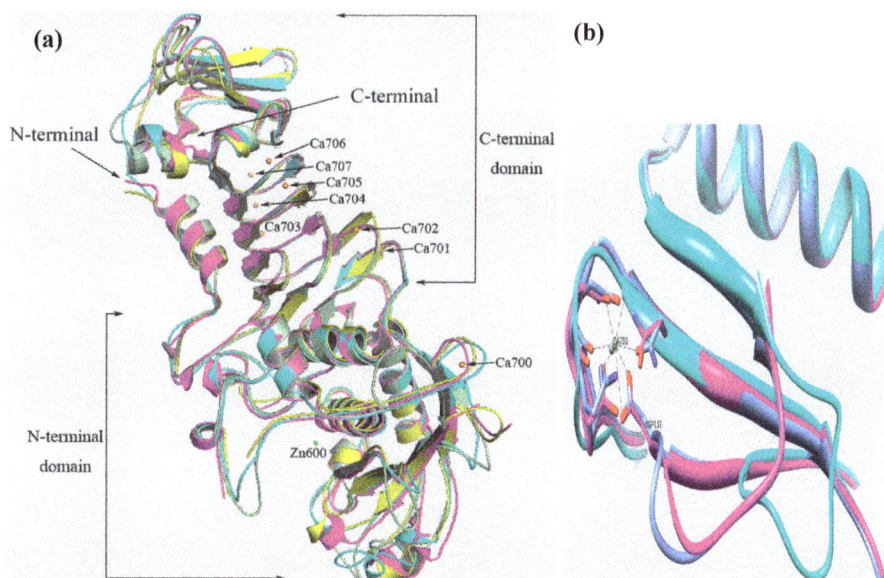

Papaleo *et al.* [119] studied seven mutant enzyme structures generated by digital effects to know whether substitution of few chosen amino acid residue confer a change in overall shift in optimum temperature or thermal stability of cold-adapted α-amylase. The mutation of a few residues alone would not change the thermal behavior of the particular enzyme, and other features also must be addressed to alter the thermal stability of the protein. Such a study can also be carried out for protease also to deduce if thermal stability of the enzyme can be altered. The limited knowledge of the protein structure is quite inadequate to reach the any thumb rule for deciding temperature range of enzymes.

However a general model proposed to explain higher activity at low temperature is that the enzyme possesses a more flexible conformation, metaphorically like an open hand, than their mesophilic and thermophilic counterparts. As a result of this increased flexibility, protein would be thermolabile often observed with cold-active enzymes [120]. In contrast, thermostable enzymes have more rigid and compact conformation more like a fist protecting them against destabilizing forces occurring at higher temperatures. A goal in comparing proteins from extremophiles is to test this and other proposed models for confirming whether such changes result in flexibility. Directed evolution studies have recently indicated that there is not a direct correlation between increased activity at low temperature and decreased thermostability [121]. In order to understand structural basis of cold adaptation, more crystal structures of similar mesophilic and thermophilic counterparts, along with the rationalized mutation studies are required. The most frequently reported structural differences between cold-adapted psychrozymes and their thermophilic and mesophilic counterparts involve interactions like fewer intra- or inter- subunit salt-bridges, loosely held hydrophobic packing in the protein core, longer surface loops and fewer prolines in such loops, increased number of glycine clusters, reduced number of arginines, improved solvent interactions through additional surface (mostly negative) charges, increased solvent exposure of apolar surface and a better accessibility of the active site [122].

9. Cold Environment Metagenomics: Tapping Biodiversity

Geographical regions with low temperature harbor psychrophiles and psychrotolerants. Diverse environments form extremely diverse niches, and the microorganisms are exposed to various extremes like pressure, temperature, nutrient availability and light. These organisms are a treasure of potentially unique biochemical and molecular profiles that might have the enzyme or molecule of enormous biotechnological interest and industrial application. The microbial enzymes from such environments are expected to have quite diverse biochemical and molecular properties. Isolation of both the microbe of interest and the molecule or enzyme of interest from these niches encounters obstacles mainly due to two reasons: first, despite the recent advances in the development of new culturing methodologies, most extremophiles could not be cultured using available technologies, and second, the problem of very low amount of biomass and thus the yield of DNA is very low for molecular analysis. Environmental genomics provides an answer for exploitation of the wealth offered by nature in extreme environments. Basic steps in this approach involve sample collection from the niche of interest such as cold environment and then this sample is processed for isolation of total environmental DNA. This environmental DNA is used either for cloning into suitable vector for genomic library construction or directly for sequencing or amplification using universal primers. These libraries are screened for the presence of enzymes of interest or for biomolecules of interest. The use of high throughput screening techniques and robotic systems make the screening process much faster and useful as large number of clones and libraries can be screened in relatively shorter time.

Environmental genomics approach has been used for isolating many cold-adapted enzymes like lipases [123,124], cellulases [125], amylases [126], xylanases [127] from the microflora existing in the cold environments. The isolation and characterization of various novel cold-adapted enzymes highlights and supports the potential of the cold environment metagenomics in future for the discovery of psychrophilic proteases too. Berlemont *et al.* [128] isolated three proteases along with other commercially important enzymes from Antarctic soil metagenome. This approach will accelerate the biotechnological exploitation of microbial diversity present in cold environments.

10. Enhancing Thermo-Stability of Cold-Active Proteases

The cold-active proteases from different microorganisms vary in their thermostability and alkalistability [41,43,73,65,63,101,105,129–131]. High proteolytic activity at lower temperatures shown by cold-active proteases is important in the commercial usage of proteases, but their low thermal stability is a common drawback that hinders their use in industries. To overcome this problem, various strategies have been used, among which reinforcement of the overall rigidity of the enzyme structure by increasing the number of disulfide bridges, intra-molecular salt bridges and shortening the length of loop regions are most commonly used [132,133]. Several ideas have been put forward for explaining the thermostability of proteins. In addition to providing insight into structural modifications, protein fluctuations can provide a mechanism of thermal stability too [134]. Experimental techniques such as nuclear magnetic resonance (NMR) [135,136], neutron diffraction methods [137] and theoretical approaches based on computer simulations on protein dynamics in solution [138,139] have supported the proposed idea. Molecular dynamics simulation has also been used to provide detailed atomic models of the protein stability and dynamics [140,141]. Attempts have been made to tailor the psychrophilic enzymes to have properties of industrial interest such as increased thermostability, tolerance to bleaches and detergents and to different organic solvents so that the proteins can become process friendly. The protein engineering has been used to alter the properties of the proteins by making changes in their primary structures. Mainly two engineering techniques have been used in attempts to create thermostable proteases: one is random mutagenesis and second is site-directed mutagenesis [SDM]. Pantoliano *et al.* [142] reported the improved thermostability and extreme alkalinity of subtilisin BPN by substitution of six individual amino acids [N218S, G169A, Y217K, M50F, Q206C, N76D]. The inactivation rate decreased several times as compared to the wild type BPN subtilisin. Strausberg *et al.* [143] reported 1000 times increase in $t_{1/2}$ of subtilisin BPN of *B. subtilis* by loop removal, cassette mutagenesis and screening procedure. Shao *et al.* [144] reported 8-fold increase in $t_{1/2}$ of subtilisin E by random priming and screening methods.

A mutant subtilisin E with enhanced thermostability at 60 °C was generated using SDM [145]. The thermostability and activity of subtilisin-like serine proteinase [VPR] had been improved by SDM approach [146]. Such studies clearly indicate that it is possible to improve one character [activity/stability] without affecting the other. Narinx *et al.* [147] performed SDM for introduction of an additional salt bridge, disulfide bonds, and increasing the affinity of the enzyme for calcium, and found that stability of the molecular structure was achieved by a modification of a calcium ligand T85D. The mutated enzyme was thermostable like mesophilic subtilisin.

Directed evolution of proteins involves recombinant DNA techniques such as DNA shuffling, random priming recombination and the staggered extension process [StEP]. Zhao and Arnold [148] reported increase in thermostability of subtilisin E by converting it to thermitase using directed evolution.

While studying the effect of trimethylamine N-oxide (TMAO) on the structure, activity, and stability of a psychrophilic protease (deseasin MCP-01), He *et al.* [149] suggested the possibility of using TMAO as an effective stabilizer to enhance the thermostability of a cold-adapted enzyme without compromising with its psychrophilic characters such as its overall structural flexibility and high catalytic efficiency at low temperature. The isolation temperature plays an important role in determining the cold adaptability of the enzyme of interest, as isolation temperatures is found to affect the enzyme properties. Vazquez and Mac Cormack [150] reported that the lower the strain isolation temperatures better the cold-adapted proteases in terms of optimal temperature and activation energy. Thus by using the recombineering and classical methods, thermostability of cold-active proteases can be improved.

11. Applications of Cold-Active Proteases

Economic benefits can be achieved by using cold-active proteases as they allow working at low temperatures even in an industrial scale. For example instead of heating and bringing the temperature during the industrial peeling process of leather by conventional protease from mesophilic or thermophilic microbes, the process can be performed at the temperature of tap water by using cold-active proteases. With the use of cold-active protease, energy saving is possible.

Proteases as a group found application in various fields such as baking, brewing, cheese making, in preparation of protein concentrates, leather industry, silk degumming, detergent industry, pharmaceutical industry, bioremediation, silver recovery from X-ray film and photographic industry are few to name the areas. The cold-active proteases find application in household processes, where they can be used for removal of macromolecular stains from fabrics along with other detergent components. As the whole process would be done at low temperature, the colors of the clothes will remain protected exposure to higher temperature. The treatment of wool and silk by protease can bring new and unique finishing to the surface of wool and silk fibers. Nowadays in textile industry, the synthetic fiber is being used. Some of the synthetic fibers cannot tolerate temperatures above 50–60 °C, and hence, require varied washing procedures [151]. During the past few years, a trend of lower washing temperature has gained popularity. Cold-active protease from *B. subtilis* showed stability in the presence of SDS and exhibited enhanced activity in Tween 80 and Wheel detergent, pH and detergent compatibility at low temperature, and thus, suits application in detergent formulation [35]. Protease from *Bacillus* sp.158 has found application in contact lens cleaning, thus increasing the transmittance of the lenses [152]. The protease of *P. aeruginosa* MCM B-327 was found to be useful in dehairing hides (Figure 3) [99].

Figure 3. Buffalo hide dehairing by PA02 protease of *P. aeruginosa* MCM B-327. (a) chemical treatment, (**b**) crude enzyme treatment, (**c**) control-water treatment [99].

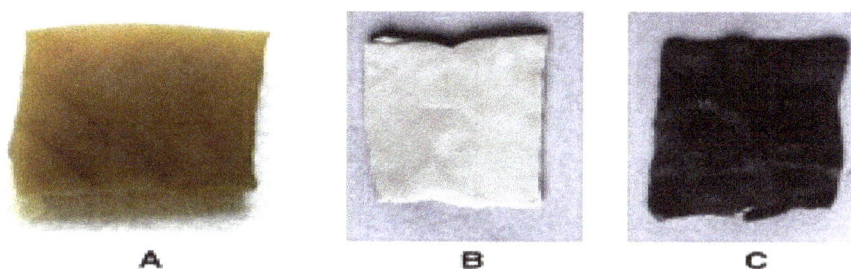

In the food industry, the property of having high catalytic activity at low-temperature allows transformation of heat labile products. They can be used in processes such as fermentation of fish or soy sauce with no spoilage and alterations in flavor and nutritional value. Cold-active proteases along with lipases can be used as rennet substitutes to accelerate the ripening of slow-ripening cheeses. Additionally cold-active proteases can find utility in softening and taste development of frozen or refrigerated meat products. Apart from this, thermal lability of such proteases can result in rapid inactivation by mild heat treatment [30]. This feature will prove beneficial in preserving the quality in the food industry.

Cold-adapted or low temperature tolerant enzymes suit well in waste management in cold environments, where the degradation capabilities of endogenous microflora are reduced due to low temperatures. Cold-adapted proteases thus can be used to optimize present day industrial processes and for developing future technologies with less energy inputs and process cost by removing the cost of heat inactivation step [28,30].

12. Conclusions and Future Perspectives

A wide range of microorganisms from diverse habitats, permanently cold as well as those exposed to cold during a part of the year, are known to produce cold-active proteases. Metagenomic culture-independent approaches have also been initiated for obtaining novel cold-active biocatalysts including proteases. A few attempts have been made to engineer and manipulate the cold-active proteases, but much success has not yet been achieved. Cloning of genes encoding cold-active protease from the wild strains and their over expression in suitable hosts is another area of research for cost effective production of these enzymes. The field of cold-active protease research is still wide open and expected to achieve spectacular success in the nearest future.

Acknowledgements

One of us (SJ) is grateful to the Council of Scientific & Industrial Research, Govt. of India, New Delhi for the award of fellowship while writing this review.

References

1. D'Amico, S.; Collins, T.; Marx, J.; Feller, G.; Gerday, C. Psychrophilic microorganisms: challenges for life. *EMBO Rep.* **2006**, *7*, 385–389.

2. Gomes, J.; Steiner, W. The biocatalytic potential of extremophiles and extremozymes. *Food Technol. Biotechnol.* **2004**, *42*, 223–235.

3. Rodrigues, D.F.; Tiedje, J.M. Coping with our cold planet. *Appl. Environ. Microbiol.* **2008**, *74*, 1677–1686.

4. Margesin, R.; Schinner, F.; Marx, J.C.; Gerday, C. *Psychrophiles: from Biodiversity to Biotechnology*; Springer: Berlin, Heidelberg, 2008; pp. 211–224.

5. Gounot, A. Bacterial life at low temperature: physiological aspects and biotechnological implications. *J. Appl. Bacteriol.* **1991**, *71*, 386–397.

6. Huston, A.L.; Krieger-Brockett, B.B.; Deming, J.W. Remarkably low temperature optima for extracellular enzyme activity from Arctic bacteria and sea ice. *Environ. Microbiol.* **2000**, *2*, 383–388.

7. Feller, G.; Bussy, O.L.; Gerday, C. Expression of psychrophilic genes in mesophilic hosts: assessment of the folding state of a recombinant α-amylase. *Appl. Environ. Microbiol.* **1998**, *64*, 1163–1165.

8. Aghajari, N.; Feller, G.; Gerday, C.; Haser, R. Crystallization and preliminary X-ray diffraction studies of α-amylase from the Antarctic psychrophile *Alteromonas haloplanctis* A23. *Prot. Sci.* **1996**, *5*, 2128–2129.

9. Jeon, J.H.; Kim, J.T.; Kim, Y.J.; Kim, H.K.; Lee, H.S.; Kang, S.G.; Kim, S.J.; Lee, J.H. Cloning and characterization of a new cold-active lipase from a deep-sea sediment metagenome. *Appl. Microbiol. Biotechnol.* **2009**, *81*, 865–874.

10. Suzuki, T.; Nakayama, T.; Kurihara, T.; Nishino, T.; Esaki, N. Cold-active lipolytic activity of psychrotrophic *Acinetobacter* sp. strain no. 6. *J. Biosci. Bioeng.* **2001**, *92*, 144–148.

11. Feller, G.; Amico, D.S.; Benotmane, A.M.; Joly, F.; Van Beeumen, J.; Gerday, C. Purification, characterization of nucleotide sequence of the thermolabile α–amylase from Antarctic psychrotroph *Alteromonas haloplanktis* A23. *J. Biol. Chem.* **1992**, *267*, 5217–5221.

12. Villeret, V.; Chessa, J.P.; Gerday, C.; Van Beeumen, J. Preliminary crystal structure determination of the alkaline protease from Antarctic psychrophile *Pseudomonas aeruginosa*. *Prot. Sci.* **1997**, *6*, 2462–2464.

13. Gerike, U.; Danson, M.J.; Hough, D.W. Sequencing and expression of the gene encoding a cold-active citrate synthase from an antarctic bacterium strain DS2–3R. *Eur. J. Biochem.* **1997**, *248*, 49–57.

14. Feller, G.; Gerday, C. Psychrophilic enzymes: molecular basis of cold adaptation. *Cell. Mol. Life Sci.* **1997**, *53*, 830–841.

15. Kim, S.Y.; Hwang, K.Y.; Kim, S.H.; Sung, H.C.; Han, Y.S.; Cho, Y. Structural basis of cold adaptation. Sequence, biochemical properties and crystal structure of malate dehydrogenase from a psychrophilic *Aquaspirillum articum*. *J. Biol. Chem.* **1999**, *274*, 11761–11767.

16. Alvarez, M.; Johan, P.H.; Zeelen, J.P.; Veronique Mainfroid, V.; Joseph, A.; Martial, J.A. Triose phosphate isomerase (TIM) of the psychrophilic bacterium *Vibrio marinus*. *J. Biol. Chem.* **1998**, *273*, 2199–2206.

17. Georlette, D.; Blaise, V.; Collins, T.; D'Amico, S. Some like it cold: biocatalysis at low temperatures. *FEMS Microbiol. Rev.* **2004**, *28*, 25–42.

18. Collins, T.; Meuwis, M.A.; Stals, I.; Claeyssens, M.; Feller, G.; Gerday, C. A novel family 8 xylanase, functional and physicochemical characterization. *J. Biol. Chem.* **2002**, *277*, 35133–35139.

19. Russell, R.J.; Gericke, U.; Danson, M.J.; Hough, D.W.; Taylor, G.L. Structural adaptations of the cold-active citrate synthase from an Antarctic bacterium. *Structure (Lond.)* **1998**, *6*, 351–361.

20. Birgisson, H.; Delgado, O.; Arroyo, L.G.; Hatti-Kaul, R.; Mattiasson, B. Cold-adapted yeasts as producers of cold-active polygalacturonases. *Extremophiles* **2003**, *7*, 185–193.

21. Akila, G.; Chandra, T.S. A novel cold-tolerant *Clostridium* strain PXYL1 isolated from a psychrophilic cattle manure digester that secretes thermolabile xylanase and cellulose. *FEMS Microbiol. Lett.* **2003**, *219*, 63–67.

22. Mavromatis, K.; Lorito, M.; Woo, S.L.; Bouriotis, V. Mode of action and antifungal properties of two cold-adapted chitinases. *Extremophiles* **2003**, *7*, 385–390.

23. Sakamoto, T.; Ihara, H.; Kozakic, S.; Kawasaki, H. A cold-adapted endo-arabinanase from *Penicillium chrysogenum. Biochim. Biophys. Acta* **2003**, *1624*, 70–75.

24. Nakagawa, T.; Nagaoka, T.; Taniguchi, S.; Miyaji, T.; Tomizuka, N. Isolation and characterization of psychrophilic yeasts producing cold-adapted pectinolytic enzymes. *Lett. Appl. Microbiol.* **2004**, *38*, 383–387.

25. Rao, M.B.; Tanksale, A.M.; Ghatge, M.S.; Deshpande, V.V. Molecular and biotechnological aspects of microbial proteases. *Microbiol. Mol. Biol. Rev.* **1998**, *62*, 597–635.

26. Godfrey, T.; West, S. Introduction to industrial enzymology. In *Industrial Enzymology*, 2nd ed.; Godfrey, W., Ed.; Macmillan Press: London, UK, 1996; pp. 1–8.

27. Gaur, S.; Agrahari, S.; Wadhwa, N. Purification of Protease from *Pseudomonas thermaerum* GW1 Isolated from poultry waste site. *Open Microbiol. J.* **2010**, *4*, 67–74.

28. Cavicchioli, R.; Siddiqui, K.S.; Andrews, D.; Sowers, K.R. Low-temperature extremophiles and their applications. *Curr. Opin. Biotechnol.* **2002**, *13*, 253–161.

29. Deming, J.W. Psychrophiles and polar regions. *Curr. Opin. Biotechnol.* **2002**, *5*, 301–309.

30. Margesin, R.; Feller, G.; Gerday, C.; Russell, N. Cold-adapted microorganisms: Adaptation strategies and biotechnological potential. In *The Encyclopedia of Environmental Microbiology*, Bitton Eds.; Wiley: New York, 2002; pp. 871–885.

31. Gerday, C.; Aittaleb, M.; Bentahir, M.; Chessa, J.P.; Claverie, P.; Collins. T.; D'Amico, S.; Dumont, J.; Garsoux, G.; Georiette, D.; Hoyoux, A.; Lonhience, T.; Meuwis, M.A;. Feller, G. Cold-adapted enzymes, from fundamentals to biotechnology. *Trends Biotechnol.* **2000**, *18*, 103–107.

32. Morita, R.J. Psychrophilic bacteria. *Bacteriol. Rev.* **1975**, *39*, 144–167.

33. Feller, G. Molecular adaptations to cold in psychrophilic enzymes. *Cell. Mol. Life Sci.* **2003**, *60*, 648–662.

34. Oh, K.H.; Seong, C.S.; Lee, S.W.; Kwon, O.S.; Park, Y.S. Isolation of a psychrotrophic *Azospirillum* sp. and characterization of its extracellular protease. *FEMS Microbiol. Lett.* **1999**, *174*, 173–178.

35. Baghel, V.S.; Tripathi, R.D.; Ramteke, R.W.; Gopal, K.; Dwivedi, S.; Jain, R.K.; Rai, U.N.; Singh, S.N. Psychrotrophic proteolytic bacteria from cold environments of Gangotri glacier, Westren Himalaya India. *Enzyme Microbial. Technol.* **2005**, *36*, 654–659.

36. Alam, S.I.; Dube, S.; Reddy, G.S.N.; Bhattacharya, B.K.; Shivaji, S.; Singh, L. Purification and characterization of extracellular protease produced by *Clostridium* sp. from Schirmacher oasis, Antarctica. *Enzyme Microbial. Technol.* **2005**, *36*, 824–831.

37. Wang, Q.; Hou, Y.; Xu, Z.; Miao, J.; Li, G. Optimization of cold-active protease production by the psychrophilic bacterium *Colwellia* sp NJ341 with response surface methodology. *Biores. Technol.* **2008**, *99*, 1926–1931.

38. Olivera, N.L.; Sequeiros, C.; Nievas, M.L. Diversity and enzyme properties of protease-producing bacteria isolated from sub- Antarctic sediments of Isla de Los Estados, Argentina. *Extremophiles* **2007**, *11*, 517–526.

39. Kuddus, M.; Ramteke, P.W. A cold–active extracellular metalloprotease from *Curtobacterium luteum.* (MTCC 7529), enzyme production and characterization. *J. Gen. Appl. Microbiol.* **2008**, *54*, 385–392.

40. Kasana, R.C.; Yadav, S.K. Isolation of a psychrotrophic *Exiguobacterium* sp SKPB5 (MTCC 7803) and characterization of its alkaline protease. *Curr. Microbiol.* **2007**, *54*, 224–229.

41. Margesin, R.; Dieplinger, H.; Hofmann, J.; Sarg, B.; Lindner, H. A cold-active extracellular metalloprotease from *Pedobacter cryoconitis*-production and properties. *Res. Microbiol.* **2005**, *156*, 499–505.

42. Zhu, H.Y.; Tian, Y.; Hou, Y.H.; Wang, T.H. Purification and characterization of the cold-active alkaline protease from marine cold-adaptive *Penicillium chrysogenum* FS010. *Mol. Biol. Rep.* **2009**, *36*, 2169–2174.

43. Zeng, R.; Zhang, R.; Zhao, J.; Lin, N. Cold-active serine alkaline protease from the psychrophilic bacterium *Pseudomonas* strain DY-A: enzyme purification and characterization. *Extremophiles* **2003**, *7*, 335–337.

44. Denner, E.B.; Mark, B.; Busse, H.J.; Turkiewicz, M.; Lubitz, W. *Psychrobacter proteolyticus* sp. Nov., a psychrotrophic, halotolerant bacterium isolated from the Antarctic krill *Euphausia superba* Dana, excreting a cold-adapted metalloprotease. *Syst. Appl. Microbiol.* **2001**, *24*, 44–53.

45. Larsen, A.L.; Moe, E.; Helland, R.; Gjellesvik, D.R.; Willassen, N.P. Characterization of a recombinantly expressed proteinase K-like enzyme from a psychrotrophic *Serratia* sp. *FEBS J.* **2006**, *273*, 47–60.

46. Kristjansson, M.M.; Magnusson, O.T.; Gudmundsson, H.M.; Alfredsson, G.A.; Matsuzawa, H. Properties of a subtilisin-like proteinase from a psychrotrophic *Vibrio* species comparison with proteinase K and aqualysin I. *Eur. J. Biochem.* **1999**, *260*, 752–760.

47. Margesin, R.; Schinner, F. Characterization of a metalloprotease from psychrophilic *Xanthomonas maltophilia*. *FEMS Microbiol. Lett.* **1991**, *79*, 257–262.

48. Yu, Y.; Li, H.R.; Zeng, Y.X.; Chen, B. Bacterial diversity and bioprospecting for cold-active hydrolytic enzymes from culturable bacteria associated with sediment from Nella Fjord, Eastern Antarctica. *Mar. Drugs* **2011**, *9*, 184–195.

49. Kuddus, M.; Ramteke, P.W. Production optimization of an extracellular cold-active alkaline protease from *Stenotrophomonas maltophilia* MTCC 7528 and its application in detergent industry. *Afr. J. Microbiol. Res.* **2011**, *7*, 809–816.

50. Thangam, E.B.; Rajkumar, G.S. Studies on the production of extracellular protease by *Alcaligenes faecalis*. *World J. Microb. Biot.* **2000**, *16*, 663–666.

51. Kobayashi, T.; Lu, J.; Li, Z.; Hung, V.S.; Kurata, A.; Hatada, Y.; Takai, K.; Ito, S.; Horikoshi, K. Extremely high alkaline protease from a deep-subsurface bacterium, *Alkaliphilus transvaalensis*. *Appl. Microbiol. Biotechnol.* **2007**, *75*, 71–80.

52. Suzuki, S.; Odagami, T. Low-temperature-active thiol protease from marine bacterium *Alteromonas haloplanktis*. *J.Biotechnol.***1997**, *l5*, 230–233.

53. Damare, C.; Raghukuma, C.; Muraleedharan, U.D.; Raghukumar, S. Deep-sea fungi as a source of alkaline and cold-tolerant proteases. *Enzyme Microb. Tech.* **2006**, *39*, 172–181.

54. Kaur, S.; Vohra, R.M.; Kapoor, M.; Beg, Q.K.; Hoondal, G.S. Enhanced production and characterization of a highly thermostable alkaline protease from *Bacillus* sp. P-2. *World J. Microb. Biot.* **2001**, *17*, 125–129.

55. Okuda, M.; Sumitomo, N.; Takimura, Y.; Ogawa, A.; Saeki, K.; Kawai, S.; Kobayashi, T.; Ito, S. A new subtilisin family: nucleotide and deduced amino acid sequences of new high-molecular-mass alkaline proteases from *Bacillus* spp. *Extremophiles* **2008**, *4*, 229–235.

56. Son, E.S.; Kim, J.I. Multicatalytic alkaline serine protease from the psychrotrophic *Bacillus amyloliquefaciens* S94. *J. Microbiol.* **2003**, *41*, 58–62.

57. Joshi, G.K.; Kumar, S.; Sharma, V. Production of moderately halotolerant, SDS stable alkaline protease from *Bacillus cereus* MTCC 6840 isolated from lake Nainital, Uttaranchal state, India. *Braz. J. Microbiol.* **2007**, *38*, 773–779.

58. Toyokawa, Y.; Takahara, H.; Reungsang, A.; Fukuta, M.; Hachimine, Y.; Tachibana, S.; Yasuda, M. Purification and characterization of a halotolerant serine proteinase from thermotolerant *Bacillus licheniformis* RKK-04 isolated from Thai fish sauce. *Appl. Microbiol. Biotechnol.* **2010**, *86*, 1867–1875.

59. Kumar, C.G. Purification and characterization of a thermostable alkaline protease from alkalophilic *Bacillus pumilus*. *Lett. Appl. Microbiol.* **2002**, *34*, 13–17.

60. Rao, Y.K.; Lu, S.C.; Liu, B.L.; Tzeng, Y.M. Enhanced production of an extracellular protease from *Beauveria bassiana* by optimization of cultivation processes. *Biochem. Eng. J.* **2006**, *28*, 57–66.

61. Ray, M.K.; Devi, K.U.; Kumar, G.S.; Shivaji, S. Extracellular protease from the Antarctic yeast *Candida humicola*. *Appl. Environ. Microbiol.* **1992**, *58*, 1918–1923.

62. Wang, Q.; Miao, J.L.; Hou, Y.H.; Ding, Y.; Wang, G.D.; Li, G.Y. Purification and characterization of an extracellular cold–active serine protease from the psychrophilic bacterium *Colwellia* sp. NJ341. *Biotech. Lett.* **2005**, *27*, 1195–1198.

63. Huston, A.L.; Methe, B.; Deming, J.W. Purification, characterization, sequencing of an extracellular cold–active aminopeptidase produced by marine psychrophile *Colwellia psychrerythraea* strain 34H. *Appl. Environ. Microbiol.* **2004**, *70*, 2321–2328.

64. Chellappan, S.; Jasmin, C.; Basheer, S.M.; Elyas, K.K.; Bhat, S.G.; Chandrasekaran, M. Production, purification and partial characterization of a novel protease from marine *Engyodontium album* BTMFS10 under solid state fermentation. *Process Biochem.* **2006**, *41*, 956–961.

65. Nakajima, M.; Mizusawa, K.; Yoshida, F. Purification and properties of an extracellular proteinase of psychrophilic *Escherichia freundii. Eur. J. Biochem.* **1974**, *44*, 87–96.

66. Zhang, S.C.; Sun, M.; Li, T.; Wang, Q.H.; Hao, J.H. Structure analysis of a new psychrophilic marine protease. *PLoS One* **2011**, doi:10.1371/ journal.pone.0026939.

67. Morita, Y.; Hasan, Q.; Sakaguchi, T.; Murakami, Y.; Yokoyama, K.; Tamiya, E. Properties of a cold–active protease from psychrotrophic *Flavobacterium balustinum* P104. *Appl. Microbiol. Biotechnol.* **1998**, *50*, 669–675.

68. Turkiewicz, M.; Pazgier, M.; Kalinowska, H.; Bielecki, S. A cold adapted extracellular serine protease of the yeast. *Leucosporidium antarcticum. Extremophiles* **2003**, *7*, 435–442.

69. Sheng, Y.X.; Lin, C.X.; Zhong, X.U.X.; Ying, Z.R. Cold-adaptive alkaline protease from the psychrophilic *Planomicrobium* sp. 547: Enzyme characterization and gene cloning. *Adv. Polar Sci.* **2011**, *22*, 49–54.

70. Xiong, H.; Song, S.; Xu, Y.; Tsoi, M.Y.; Dobretsov, S.; Qian, P.Y. Characterization of proteolytic bacteria from the Aleutian deep-sea and their proteases. *J. Ind. Microbiol. Biot.* **2007**, *34*, 63–71.

71. Vazquez, S.C.; Hernández, E.; Mac Cormack, W.P. Extracellular proteases from the Antarctic marine Pseudoalteromonas sp. P96–47 strain. *Rev. Argent. Microbiol.* **2008**, *40*, 63–71.

72. Chen, X.L.; Xie, B.B.; Lu, J.T.; He, H.L.; Zhang, Y. A novel type of subtilase from the psychrotolerant bacterium *Pseudoalteromonas* sp. SM9913: Catalytic and structural properties of deseasin MCP-01. *Microbiology* **2007**, *153*, 2116–2125.

73. Vazquez, S.C.; Coria, S.H.; Mac Cormack, W.P. Extracellular proteases from eight psychrotolerant Antarctic strains. *Microbiol. Res.* **2004**, *159*, 157–166.

74. Chessa, J.P.; Petrescu, I.; Bentahir, M.; Beeumen, J.V.; Gerday, C. Purification, physico-chemical characterization and sequence of a heat labile alkaline metalloprotease isolated from a psychrophilic *Pseudomonas* species. *Biochim. Biophys. Acta (BBA) - Protein Structure and Molecular Enzymology* **2000**, *1–2*, 265–274.

75. Patil, U.; Chaudhari, A. Optimal production of alkaline protease from solvent- tolerant alkaliphilic *Pseudomonas aeruginosa* MTCC 7926. *Indian J. Biotechnol.* **2011**, *10*, 329–339.

76. Yang, C.; Yang, F.; Hao, J.; Zhang, K.; Yuan, N.; Sun, M. Identification of a proteolytic bacterium HW08 and characterization of its extracllular cold-Active alkaline metalloprotease ps5. *Biosci. Biotechnol. Biochem.* **2010**, *74*, 1220–1225.

77. Koka, R.; Weimer, B.C. Isolation and characterization of a protease from *Pseudomonas fluorescens* RO98. *J. Appl. Microbiol.* **2000**, *89*, 280–288.

78. Hamamato, T.; Kaneda, M.; Horikoshi, K.; Kudo, T. Characterization of a Protease from a psychrotroph, *Pseudomonas fluorescens* 114. *Appl. Environ. Microbiol.* **1994**, *60*, 3878–3880.

79. Meza, J.C.; Auria, R.; Lomascolo, A.; Sigoillot, J.C.; Casalot, L. Role of ethanol on growth, laccase production and protease activity in *Pycnoporus cinnabarinus* ss3. *Enzyme Microb. Tech.* **2007**, *41*,162–168.

80. Shanmughapriya, S.; Krishnaveni, J.; Selvin, J.; Gandhimathi, R.; Arunkumar, M.; Thangavelu, T.; Kiran, G.S.; Natarajaseenivasan, K. Optimization of extracellular thermotolerant alkaline protease produced by marine *Roseobacter* sp. (MMD040). *Bioprocess Biosyst. Eng.* **2008**, *31*, 427–433.

81. Tariq, A.L.; Reyaz, A.L.; Prabakaran, J.J. Purification and characterization of 56 kDa cold-active protease from *Serratia marcescens*. *Afr. J. Microbiol. Res.* **2011**, *5*, 5841–5847.

82. Morita, Y.; Kondoha, K.; Hasanb, Q.; Sakaguchia, T.; Murakamia, Y.; Yokoyamaa, K.; Tamiyaa, E. Purification and characterization of a cold-Active protease from psychrotrophic *Serratia marcescens* AP3801. *J. Am. Oil Chem. Soc.* **1997**, *11*, 1377–1383.

83. Mozhina, N.V.; Burmistrova, O.A.; Pupov, D.V.; Rudenskaya, G.N.; Dunaevsky, Y.E.; Demiduk, I.V.; Kostrov, S.V. Isolation and properties of *Serratia proteamaculans* 94 cysteine protease. *Russ. J. Bioorg. Chem.* **2008**, *34*, 274–279.

84. Kulakova, L.; Galkin, A.; Kurihara, T.; Yoshimura, T.; Esaki, N. Coldactive serine alkaline protease from the psychrotrophic bacterium *Shewanella strain* ac10, gene cloning and enzyme purification and characterization. *Appl. Environ. Microbiol.* **1999**, *65*, 611–617.

85. Saba, I.; Qazi, P.H.; Rather, S.A.; Dar, R.A.; Qadri, Q.A.; Ahmad, N.; Johri, S.; Taneja, S.S.S. Purification and characterization of a cold-active alkaline protease from *Stenotrophomonas* sp., isolated from Kashmir, India. *World J. Microbiol. Biotechnol.* **2012**, *28*, 1071–1079.

86. Tokiwa, Y.; Kitagawa, M.; Fan, H.; Raku, T.; Hiraguri, Y.; Shibatani, S.; Kurane, R. Synthesis of vinyl arabinose ester catalyzed by protease from *Streptomyces* sp. *Biotechnol. Tech.* **1999**, *13*, 173–176.

87. Lopes, A.; Coelho, R.R.R.; Meirelles, M.N.L.; Branquinha, M.H.; Vermelho, A.B. Extracellular serine proteinase isolated from *Streptomyces alboniger*: Partial characterization and effect of aprotinin on cellular structure. *Mem. Inst. Oswaldo Cruz.* **1999**, *94*, 763–770.

88. Elibol, M.; Moreira, A.R. Optimizing some factors affecting alkaline protease production by a marine bacterium *Teredinobacter turnirae* under solid state fermentation. *Process Biochem.* **2005**, *40*, 1951–1956.

89. Kredics, L.; Terecskei, K.; Antal, Z.; Szekeres, A.; Hatvani, L.; Manczinger, L.; Vagvolgyi, C. Purification and preliminary characterization of a cold–adapted extracellular proteinase from *Trichoderma atroviride*. *Acta Biol. Hung.* **2008**, *59*, 259–268.

90. Hamamato, T.; Kaneda, M.; Kudo, T.; Horikoshi, K. Characterization of a protease from a psychrophilic *Vibrio* sp Strain 5709. *J. Mar. Biotechnol.* **1995**, *2*, 219–222.

91. Ward, O.P. Proteolytic Enzymes. In: *Comprehensive Biotechnology*; Moo-Young Ed.; Pergamon Press: Oxford, UK, 1985; Volume 3, pp. 789–818.

92. Morihara, K. Comparative specificity of microbial proteinases. *Adv. Enzymol.* **1974**, *41*, 179–243.

93. Hartley, B.S. Proteolytic enzymes. *Annu. Rev. Biochem.* **1960**, *29*, 45–72.

94. Joo, H.S.; Kumar, C.G.; Park, G.C.; Paik, S.R.; Chang, C.S. Oxidant and SDS-stable alkaline protease from *Bacillus clausii* I–52, production and some properties. *J. Appl. Microbiol.* **2003**, *95*, 267–272.

95. Vazquez, S.C.; MacCormack, W.P.; Rios Merino, L.N.; Fraile, E.R. Factors influencing protease production by two Antarctic strains of *Stenotrophomonas maltophilia*. *Rev. Argent. Microbiol.* **2000**, *32*, 53–62.

96. Dube, S.; Singh, L.; Alam, S.I. Proteolytic anaerobic bacteria from lake sediment of Antarctica. *Enzyme Microb. Tech.* **2001**, *20*, 114–118.

97. Saeki, K.; Iwata, J.; Watanabe, Y.; Tamai, Y. Purification and characterization of an alkaline protease from *Oerskovia xanthineolytica* TK-1. *J. Ferment. Bioeng.* **1994**, *77*, 554–556.

98. Khairullin, R.F.; Mikhailova, A.G.; Sebyakina, T.Y.; Lubenets, N.L.; Ziganshin, R.H.; Demidyuk, I.V.; Gromova, T.Y.; Kostrov, S.V.; Rumsh, L.D. Oligopeptidase B from *Serratia proteamaculans*. I. Determination of primary structure, isolation, and purification of wild-type and recombinant enzyme variants. *Biochem. (Moscow)* **2009**, *74*, 1164–1172.

99. Zambare, V.; Nilegaonkar, S.; Kanekar, P. A novel extracellular protease from *Pseudomonas aeruginosa* MCM B-327: Enzyme production and its partial characterization. *New Biotechnol.* **2011**, *28*, 173–181.

100. Mikhailova, A.G.; Likhareva, V.V.; Khairullin, R.F.; Lubenets, N.L.; Rumsh, L.D.; Demidyuk, I.V.; Kostrov, S.V. Psychrophilic trypsin-type protease from *Serratia proteamaculans*. *Biochem. (Moscow)* **2006**, *71*, 563–570.

101. Vazquez, S.; Ruberto, L.; Cormack, W.M. Properties of extracellular proteases from three psychrotolerant *Stenotrophomonas maltophilia* isolated from Antarctic soil. *Polar Biol.* **2005**, *28*, 319–325.

102. Irwin, J.A.; Alfredesson, G.A.; Lanzetti, A.J.; Haflidi, M.; Gudmundsson, H.M.; Engel, P.C. Purification and characterization of a serine peptidase from the marine psychrophile strain PA-43. *FEMS Microbiol. Lett.* **2001**, *201*, 285–290.

103. Feller, G.; Narinx, E.; Arpingy, J.L.; Aittaleb, M.; Baise, E.; Genicot, S.; Gerday, C. Enzymes from psychrophilic organisms. *FEMS Microbiol. Rev.* **1996**, *18*, 189–202.

104. Feller, G.; Payan, F.; Theys, F.; Qian, M.; Haser. R.; Gerday, C. Stability and structural analysis of α-amylase from the Antarctic psychrophile *Alteromonas haloplanctis* A23. *Eur. J. Biochem.* **1994**, *222*, 441–447.

105. Yan, B.Q.; Chen, X.L.; Hou, X.Y.; He, H.L.; Zhou, B.C.; Zhang, Y.Z. Molecular analysis of the gene encoding a cold-adapted halophilic subtilase from deep-sea psychrotolerant bacterium *Pseudoalteromonas* sp. SM9913: cloning, expression, characterization and function analysis of the C-terminal PPC domains. *Extremophiles* **2009**, *13*, 725–733.

106. Patel, T.R.; Jackman, D.M.; Bartlett, F.M. Heat-Stable protease from *Pseudomonas fluorescens* T16, purification by affinity column chromatography and characterization. *Appl. Environ. Microbiol.* **1983**. *46*, 333–337.

107. Kim, J.; Lee, S.M.; Jung, H.J. Characterization of calcium-activated bifunctional peptidase of the psychrotrophic *Bacillus cereus*. *J. Microbiol.* **2005**, *43*, 237–243.

108. Ni, X.; Yue, L.; Chi, Z.; Li, J.; Wang, X.; Madzak, C. Alkaline protease gene cloning from the marine yeast *Aureobasidium pullulans* HN2–3 and the protease surface display on *Yarrowia lipolytica* for bioactive peptide production. *J. Mar. Biotechnol.* **2009**, *11*, 81–89.

109. Wintrode, P.L.; Miyazaki, K.; Arnold, F.H. Cold adaptation of a mesophilic subtilisin-like protease by laboratory evolution. *J. Biochem.* **2000**, *275*, 31635–31640.

110. Taguchi, S.; Ozaki, A.; Momose, H. Engineering of a Cold-adapted protease by sequential random mutagenesis and a screening system. *Appl. Environ. Microbiol.* **1998**, *64*, 492–495.

111. Parrilli, E.; Vizio, D.D.; Cirulli, C.; Tutino, M.L. Development of an improved Pseudoalteromonas haloplanktis TAC125 strain for recombinant protein secretion at low temperature. *Microb. Cell Fact.* **2008**, doi:10.1186/1475-2859-7-2.

112. Huston, A.L. *Psychrophiles: From Biodiversity to Biotechnology*; Springer: Heidelberg, Germany, 2008; pp. 347–363.

113. Almog, O.; González, A.; Godin, N.; de Leeuw, M.; Mekel, M.J.; Klein, D.; Braun, S. The crystal structures of the *psychrophilic subtilisin* S41 and the mesophilic subtilisin Sph reveal the same calcium-loaded state. *Proteins* **2009**, *74*, 489–496.

114. Trimbur, D.E.; Gutshall, K.R.; Prema, P.; Brenchley, J.E. Characterization of a psychrotrophic *Arthrobacter* gene and its cold-active beta galactosidase. *Appl. Environ. Microbiol.* **1994**, *60*, 4544–4552.

115. Feller, G.; Thiry, M.; Gerday, C. Nucleotide sequence of the lipase gene lip2 from the Antarctic psychrotroph *Moraxella* TA 144 and site-specific mutagenesis of the conserved serine and histidine residues. *DNA Cell Biol.* **1991**, *10*, 381–388.

116. Arpigny, J.L.; Feller, G.; Gerday, C. Cloning, sequence and structural features of a lipase from the Antarctic facultative psychrophilic *Psychrobacter immobilis* B10. *Biochim. Biophys. Acta* **1993**, *1171*, 331–333.

117. Dong, D.; Ihara, T.; Motoshima, H.; Watanabe, K. Crystallization and preliminary X-ray crystallographic studies of a psychrophilic subtilisin-like protease Apa1 from Antarctic *Pseudoalteromonas* sp. strain AS-11. *Acta Crystallogr. Sect. F Struct. Biol. Cryst. Commun.* **2005**, *61*, 308–311.

118. Chevrier, B.; Schalk, C.; D'Orchymont, H.; Rondeau , J.M.; Moras, D.; Tarnus, C. Crystal structure of *Aeromonas proteolytica* aminopeptidase: A prototypical member of the co-catalytic zinc enzyme family. *Structure* **1994**, *2*, 283–291.

119. Papaleo, E.; Pasi, M.; Tiberti, M.; De Gioia, L. Molecular dynamics of mesophilic-like mutants of a cold-adapted enzyme: Insights into distal effects induced by the mutations. *PLoS ONE* **2011**, doi:10.1371/journal.pone.0024214.

120. Spiwok, V.; Lipovova, P.; Skalova, T.; Duskova, J.; Dohnalek, J.; Haaek, J.; Russell, N.J.; Kralova, B. Cold-active enzymes studied by comparative molecular dynamics simulation. *J. Mol. Model.* **2007**, *13*,485–497.

121. Miyazaki, K.; Wintrode, P.L.; Grayling, R.A.; Rubingh, D.N.; Arnold, F.H. Directed evolution study of temperature adaptation in a psychrophilic enzyme. *J. Mol. Biol.* **2000**, *297*, 1015–1026.

122. Kasana, R.C. Proteases from psychrotrophs: An overview. *Crit. Rev. Microbiol.* **2010**, *36*, 134–145.

123. Couto, G.H.; Glogauer, A.; Faoro, H.; Chubatsu, L.S.; Souza, E.M.; Margesin, F.O. Isolation of a novel lipase from a metagenomic library derived from mangrove sediment from the south Brazilian coast. *Genet. Mol. Res.* **2010**, *9*, 514–523.

124. Roh, C.; Villatte, F. Isolation of a low-temperature adapted lipolytic enzyme from uncultivated micro-organism. *J. Appl. Microbiol.* **2008**, *105*, 116–123.

125. Voget, S.; Steele, H.L.; Streit, W.R. Characterization of a metagenome-derived halotolerant cellulase. *J. Biotechnol.* **2006**, *126*, 26–36.

126. Sharma, S.; Khan, F.G.; Qazi, G.N. Molecular cloning and characterization of amylase from soil metagenomic library derived from Northwestern Himalayas. *Appl. Microbiol. Biotechnol.* **2010**, *86*, 1821–1828.

127. Lee, C.C.; Kibblewhite-Accinelli, R.E.; Wagschal, K.; Robertson, G.H.; Wong, D.W.S. Cloning and characterization of a cold-active xylanase enzyme from an environmental DNA library. *Extremophiles* **2006**, *10*, 295–300.

128. Berlemont, R.; Pipers, R.; Delsaute, M.; Angiono, F.; Feller, G.; Galleni, M.; Power, P. Exploring the Antarctic soil metagenome as a source of novel cold-adapted enzymes and genetic mobile elements. *Rev. Argent. Microbiol.* **2011**, *43*, 94–103.

129. Tondo, E.C.; Lakus, F.R.; Oliveira, F.A.; Brandelli, A. Identification of heat stable protease of *Klebsiella oxytoca* isolated from raw milk. *Lett. Appl. Microbiol.* **2004**, *38*, 146–150.

130. Secades, P.; Alvarez, B.; Guijarro, J.A. Purification and characterization of a psychrophilic calcium induced, growth-phase dependent metalloprotease from the fish pathogen *Flavobacterium psychrophilum*. *Appl. Environ. Microbiol.* **2001**, *67*, 2436–2444.

131. Matta, H.; Punj, V. Isolation and partial characterization of a thermostable extracellular protease of *Bacillus polymyxa* B–17. *Int. J. Food Microbiol.* **1998**, *42*, 139–145.

132. Zhang, Y.; Porcelli, M.; Cacciapuoti, G.; Ealick, S.E. The crystal structure of 5'-deoxy-5'-methylthioadenosine phosphorylase II from Sulfolobus solfataricus, a thermophilic enzyme stabilized by intramolecular disulfide bonds. *J. Mol. Biol.* **2006**, *357*, 252–262.

133. Storch, E.M.; Daggett, V.; Atkins, W.M. Engineering out motion: A surface disulfide bond alters the mobility of tryptophan 22 in cytochrome b5 as probed by time-resolved fluorescence and 1H NMR experiments. *Biochem.* **1999**, *38*, 5054–5064.

134. Matthews, B.W.; Nicholson, H.; Becktel, W.J. Enhanced protein thermostability from site-directed mutations that decrease the entropy of unfolding. *P. Natl. Acad. Sci. USA* **1987**, *84*, 6663–6667.

135. D'Amico, S.; Claverie, P.; Collins, T.; Georlette, D.; Gratia, E.; Hoyoux, A.; Meuwis, M.A.; Feller, G.; Gerday, C. Molecular basis of cold adaptation. *Philos. T. Roy. Soc. B.* **2002**, *357*, 917–925.

136. Boehr, D.D.; Dyson, H.J.; Wright, P.E. An NMR perspective on enzyme dynamics. *Chem. Rev.* **2006**, *106*, 3055–3079.

137. Henzler-Wildman, K.; Kern, D. Dynamic personalities of proteins. *Nature* **2007**, *450*, 964–972.

138. Tehei, M.; Zaccai, G. Adaptation to high temperatures through macromolecular dynamics by neutron scattering. *FEBS J.* **2007**, *274*, 4034–4043.

139. Adcock, S.A.; McCammon, J.A. Molecular Dynamics: Survey of methods for simulating the activity of proteins. *Chem. Rev.* **2006**, *106*, 1589–1615.

140. Van Gunsteren, W.F.; Bakowies, D.; Baron, R.; Chandrasekhar, I.; Christen, M.; Daura, X.; Gee, P.; Geerke, D.P.; Glattli, A.; Hunenberger, P.H. Biomolecular modeling: Goals, problems, perspectives. *Angewandte Chemie International Edition in English* **2006**, *45*, 4064–4092.

141. Van Gunsteren, W.F.; Dolenc, J.; Mark, A.E. Molecular simulation as an aid to experimentalists. *Curr. Opin. Struc. Biol.* **2008**, *18*, 149–153.

142. Pantoliano, M.W.; Whitlow, M.; Wood, J.F.; Dodd, S.W.; Hardman, K.D.; Rollence, M.L.; Bryan, P.N. Large increases in general stability for the subtilisin BPN' through incremental changes in the free energy of unfolding. *Biochem.* **1989**, *28*, 7205–7213.

143. Strausberg, S.L.; Alexander, P.A.; Gallagher, D.T.; Gilliland, G.L.; Barnett, B.L.; Bryan, P.N. Directed evolution of a subtilisin with calcium-independent stability. *Biotechnology* **1995**, *13*, 669–673.

144. Shao, Z.; Zhao, H.; Giver, L.; Arnold, F.H. Random-priming *in vitro* recombination: An effective tool for directed evolution. *Nucleic Acids Res.* **1998**, *26*, 681–683.

145. Yang, Y.; Jiang, L.; Yang, S.; Zhu, L.; Wu, Y.; Li, Z. A mutant subtilisin E with enhanced thermostability. *World J. Microb. Biot.* **2000**, *16*, 249–251.

146. Siguroardottir, A.G.; Arnorsdottir, J.; Thorbjarnardottir, S.H.; Eggertsson, G.; Suhre, K.; Kristjansson, M.M. Characteristics of mutants designed to incorporate a new ion pair into the structure of a cold adapted subtilisin-like serine proteinase. *Biochim. Biophys. Acta.* **2009**, *1794*, 512–518.

147. Narinx, E.; Baise, E.; Gerday, C. Subtilisin from psychrophilic antarctic bacteria: characterization and site-directed mutagenesis of residues possibly involved in the adaptation to cold. *Prot. Eng.* **1997**, *10*, 1271–1279.

148. Zhao, H.; Arnold, F.H. Directed evolution converts subtilisin E into a functional equivalent of thermitase. *Protein Engi.* **1999**, *12*, 47–53.

149. He, H.L.; Chen, X.L.; Zhang, X.Y.; Sun, C.Y.; Zou, B.C.; Zhang, Y.Z. Novel Use for the Osmolyte Trimethylamine N-oxide, Retaining the psychrophilic characters of cold–adapted protease *Deseasin* MCP-01 and simultaneously improving its thermostability. *Mar. Biotechnol.* **2009**, *11*, 710–716.

150. Vazquez, S.C.; MacCormack, W.P. Effect of isolation temperature on the characteristics of extracellular proteases produced by Antarctic bacteria. *Polar Res.* **2002**, *21*, 63–71.

151. Nielson, M.H.; Jepsen, S.J.; Outrup, H. Enzymes for lower temperature washing. *J. Am. Oil Chem. Soc.* **1981**, *58*, 644–649.

152. Pawar, R.; Zambare, V.; Barve, Z.; Paratkar, G. Application of protease isolated from *Bacillus* sp158 in enzymatic cleansing of contact lenses. *Biotechnology* **2009**, *8*, 276–280.

Antarctic Epilithic Lichens as Niches for Black Meristematic Fungi

Laura Selbmann, Martin Grube, Silvano Onofri, Daniela Isola and Laura Zucconi

Abstract: Sixteen epilithic lichen samples (13 species), collected from seven locations in Northern and Southern Victoria Land in Antarctica, were investigated for the presence of black fungi. Thirteen fungal strains isolated were studied by both morphological and molecular methods. Nuclear ribosomal 18S gene sequences were used together with the most similar published and unpublished sequences of fungi from other sources, to reconstruct an ML tree. Most of the studied fungi could be grouped together with described or still unnamed rock-inhabiting species in lichen dominated Antarctic cryptoendolithic communities. At the edge of life, epilithic lichens withdraw inside the airspaces of rocks to find conditions still compatible with life; this study provides evidence, for the first time, that the same microbes associated to epilithic thalli also have the same fate and chose endolithic life. These results support the concept of lichens being complex symbiotic systems, which offer attractive and sheltered habitats for other microbes.

Reprinted from *Biology*. Cite as: Selbmann, L.; Grube, M.; Onofri, S.; Isola, D.; Zucconi, L. Antarctic Epilithic Lichens as Niches for Black Meristematic Fungi. *Biology* **2013**, *2*, 784-797.

1. Introduction

Black meristematic fungi are known to be tolerant to extreme environmental conditions. The term black fungi embraces a polyphyletic group of fungi that share some phenotypic characters such as melanized cell walls and meristematic development, which seem to support survival and persistence in hostile environmental conditions. They are commonly isolated from environments that are almost devoid of other eukaryotic life-forms, including saltpans [1], acidic and contaminated sites [2–4], exposed rocks in dry and extremely hot or cold climates, ranging from hot deserts [5], the Mediterranean [6] to the Antarctic [7] and on monuments [8–12]. Owing to the stress pressure of the sites where they normally occur, black meristematic fungi are rarely found in complex microbial populations, rather they occur alone or in association with similar stress resistant organisms such as lichens [13,14] and cyanobacteria [15]. In the Antarctic, black meristematic fungi are recurrent members of endolithic microbial communities of ice free areas, including the lichen-dominated cryptoendolithic communities of the McMurdo Dry Valleys, one of the most inhospitable environments on Earth [16,17]. In these sites the limits for life are reached; since the conditions are too harsh to sustain epilithic settlement, mosses almost completely disappear and lichens grow protected in cracks and fissures or move inside the rocks, giving rise to well structured communities [16]. Together with lichens, other organisms can participate in these communities, in particular bacteria, cyanobacteria and non-lichenized fungi [18–20], but their biodiversity, their role and interactions are still scarcely investigated. Among these, the rock black fungi represent a peculiar group of colonizers [7,17].

Lichens are commonly described as a mutualistic symbiosis between fungi and "algae" (*Chlorophyta* or *Cyanobacteria*); however, recent studies revealed that they host a number of other microbes. Several culture-dependent and -independent studies have deepened our understanding of diverse populations of bacteria associated with lichens and their potential functional roles within the symbiosis [21–25]. Lichens also host numerous fungal species. The mycobiont is the dominant fungal species but other fungi may be present. These include lichenicolous fungi, which expresses symptoms [26,27] and endolichenic fungi, which grow without symptoms in the interior of lichens [28–31]. These studies have fueled the concept that, in addition to being symbiotic systems, where symbiotic partners may interact, lichens can also be considered miniature ecosystems [22,32]. However, despite a growing body of literature on organisms associated with lichens, we still have limited knowledge of the extent of eukaryotic diversity that may be associated with individual lichen thalli [31].

In this study we focused on extremotolerant black fungi associated with cold-loving lichens from the Antarctic, including the endemic species *Lecanora fuscobrunnea* Dodge & Baker. Antarctic lichens are still an unexplored niche for these organisms and we aimed to compare the black fungi diversity among different lichen species distributed in diverse ecosystems.

2. Experimental Section

Sampling sites and lichen identification: lichen thalli were collected using a sterile chisel and preserved in sterile plastic bags at −20 °C until processed for isolation of associated fungi. Lichens were identified using the key by Castello [33]. All data concerning the sampling sites and the identifications are reported in Table 1.

Isolation: in order to remove any potential contaminant before isolation, lichens were treated with H_2O_2 (8%) for 5 min; H_2O_2 was removed by washing with distilled sterile water for 5 min. The solution was filtered using 500 μm porosity filters. All fragments were collected and seeded on MEA (Malt Extract Agar, Oxoid, Ltd. Basingstoke, Hampshire, UK) in Petri Dishes and incubated at 5 °C and 15 °C. Plates were inspected weekly and as soon as new black colonies appeared they were transferred on fresh agar slant. Pure cultures were deposited in the CCFEE (Culture Collection of Fungi from Extreme Environments, DEB, Università degli Studi della Tuscia, Viterbo, Italy).

Morphology and temperature preferences: hyphal maturation was studied using light microscope. Slide cultures were seeded onto MEA, incubated for 10 w and mounted in lactic acid. Temperature preferences were performed in triplicate on MEA, in Petri dishes in the range 0–30 °C ± 1, with 5 °C intervals. Colony diameters were recorded monthly.

Molecular analysis: DNA was extracted from 6-months-old mycelium grown on MEA at 10 °C, using Nucleospin Plant kit (Macherey-Nagel, Düren, Germany) following the protocol optimized for fungi. Target gene for our analysis was the nuclear ribosomal 18S and ITS genes. PCR reactions were performed using BioMix (BioLine, Luckenwalde, Germany) and primers NS1-NS24 and ITS1-ITS4 to amplify 18S and ITS respectively [34]. Reaction mixtures were prepared by adding 5 pmol of each primer and 40 ng of template DNA in a final volume of 25 μL. For amplification, a MyCycler™ Thermal Cycler (Bio-Rad Laboratories, Munich, Germany) was used. The protocol used for amplification of the nuclear ribosomal 18S was as follows: 3 min at 95 °C for a first denaturation step, a denaturation step at 95 °C for 45 s, annealing at 52 °C for 30 s. Cycles were repeated 35 times, with a last extension at 72 °C for 5 min. ITS portion was amplified as previously described [7]. Products were purified using Nucleospin Extract kit (Macherey-Nagel, Düren, Germany). Sequencing reactions were performed according to the dideoxynucleotide method using the TF BigDye Terminator 1,1 RR kit (Applied Biosystems). Fragments were analyzed by Macrogen Inc. (Seoul, Korea). Sequence assembly was done using the software ChromasPro (version 1.32, Technelysium, Conor McCarthy School of Health Science, Griffith University, Southport, Queensland, Australia).

Table 1. Lichen species analyzed collection data and fungal strains isolated.

Lichen species	Location	Coordinates	Sampling date	Fungal strains (CCFEE)
Acarospora sp.	Ford Peak, NVL	75°41'26.3"S 160°26'25.3"E	28/01/2004	-
Acarospora flavocordia Castello & Nimis	Kay Island, NVL	75°04'13.7"S 165°19'02.0"E	30/01/2004	5324
Buellia frigida Darb.	Inexpressible Island, NVL	75°52'23.2"S 163°42'16.5"E	17/01/2004	-
Lecanora fuscobrunnea Dodge & Baker	Edmonson Point, NVL	74°19'43.7"S 165°08'00.7"E	29/01/2004	5320 *
Lecanora fuscobrunnea Dodge & Baker	Convoy Range, Terra SVL	76°54'33.0"S 160°50'00.0"E	25/01/2004	5303
Lecanora sp.	Inexpressible Island, NVL	75°52'23.2"S 163°42'16.5"E	17/01/2004	5319 *, 5323
Lecidea sp.	Starr Nunatak, NVL	75°53'55.7"S 162°35'31.3"E	15/02/2004	5318
Lecidea sp.	Starr Nunatak, Terra Vittoria del Nord	75°53'55.7"S 162°35'31.3"E	15/02/2004	5326
Lecidea cancriformis Dodge & Baker	Widowmaker Pass, NVL	74°55'23.5"S 162°24'17.0"E	12/02/2004	5321 **
Rhizocarpon sp.	Vegetation Island, NVL	74°47'05.2"S 163°38'40.3"E	16/01/2004	5312
Umbilicaria aprina Nyl.	Kay Island, NVL	75°04'13.7"S 165°19'02.0"E	30/01/2004	-
Umbilicaria decussata (Vill.) Zahlbr.	Kay Island, NVL	75°04'13.7"S 165°19'02.0"E	02/02/2004	-
Umbilicaria decussata (Vill.) Zahlbr.	Vegetation Island, NVL	74°47'05.2"S 163°38'40.3"E	16/01/2004	5317
Usnea antarctica Du Rietz	Kay Island, NVL	75°04'13.7"S 165°19'02.0"E	30/01/2004	-
Usnea antarctica Du Rietz	Vegetation Island, NVL	74°47'05.2"S 163°38'40.3"E	16/01/2004	5313 *
Xanthoria elegans (Link) th. Fr.	Kay Island, NVL	75°04'13.7"S 165°19'02.0"E	30/01/2004	5314, 5322

CCFEE—Culture Collection of fungi From Extreme Environments; NVL—Northern Victoria Land; SVL—Southern Victoria Land. Identified strains: * *Elasticomyces elasticus*; ** *Friedmanniomyces endolithicus*.

The alignment based on nuclear ribosomal 18S included 79 sequences of strains belonging to the class *Dothideomycetes* and *Eurotiomycetes* in the public domain chosen on the base of the Blastn results. Additional sequences of black fungi deposited in the database of the CCFEE (Culture Collection of Fungi from Extreme Environments, Università degli Studi della Tuscia, Viterbo, Italy) were analyzed (Table 2). Sequences were aligned iteratively with ClustalX [35], exported in Mega5 [36] for a manual improvement. The best-fit substitution model and Maximum Likelihood phylogenetic tree reconstruction was performed as previously described [17]. The robustness of the phylogenetic inference was estimated using the bootstrap method [37] with 1000 pseudoreplicates.

Table 2. List of strains and sequences analyzed.

Species	Strains no.	Source	Location	SSU
Acidomyces acidophilum	C2	acid mine drainage	CA, USA	AY374300
Acidomyces acidophilum	A3-7	acid mine drainage	CA, USA	AY374299
Acidomyces acidophilum	B1	acid mine drainage	CA, USA	AY374298
Aureobasidium pullulans	28v1	-	-	AY137505
Aureobasidium pullulans	30v4	-	-	AY137507
Botryosphaeria ribis	CBS 121.26	*Ribes rubrum*	-	U42477
Botryosphaeria ribis	CBS 115475	*Ribes*	-	DQ678000
Capnobotryella renispora	CBS 214.90	*Abies*	Japan	EF137360
Capnobotryella renispora	CBS 215.90	*Sphagnum*	Japan	AY220613
Capnobotryella renispora	CBS 572.89	Roof tile	Sweeden	AY220614
Capnobotryella renispora	UAMH 9870	*Sphagnum*	-	AY220611
Capronia coronata	CBS 617.96	Decorticated wood	New Zealand	AJ232939
Capronia semiimmersa	CBS 840.69	Decaying timber	Finland	AY554291
Catenulostroma abietis	CBS 459.93	*Abies*	Germany	DQ678040
Cladophialophora carrionii	CBS 260.83	Skin lesion	-	AY554285
Cladophialophora sp.	CBS 985.96	Brain	USA	AJ232953
Coccodinium bartschii	UME30232	-	-	U77668
Coniosporium sp.	MA 4597	Marble	Turkey	AJ972863
Cyphellophora laciniata	MUCL 9569	-	-	AY342010
Cryomyces antarcticus	CCFEE 514	Rock	Antarctica	GU250319
Cryomyces antarcticus	CCFEE 515	Rock	Antarctica	GU250320
Cryomyces antarcticus	CBS 116301T; CCFEE 534	Sandstone	Antarctica	DQ028269
Cryomyces minteri	CBS 116302; CCFEE 5187	Sandstone	Antarctica	DQ028270
Discosphaerina fagi	CBS 171.93	*Populus* leaf	UK	AY016342
Elasticomyces elasticus	**CBS 122538; CCFEE 5313**	**Lichen**	**Antarctica**	**FJ415474**
Elasticomyces elasticus	**CBS 122539; CCFEE 5319**	**Lichen**	**Antarctica**	**GU250332**
Elasticomyces elasticus	**CBS 122540; CCFEE 5320**	**Lichen**	**Antarctica**	**GU250333**
Elsinoe centrolobii	CBS 222.50	*Centrolobium robustum*	Brazil	DQ678041
Exophiala salmonis	CBS 157.67	*Salmo clarkii*	Canada	JN856020
Exophiala salmonis	AFTOL-ID 671	-	-	EF413608
Friedmanniomyces endolithicus	CCFEE 670	Rock	Antarctica	GU250322
Friedmanniomyces endolithicus	CCFEE 5208	Rock	Antarctica	Unpublished

Table 2. *Cont.*

Species	Strains no.	Source	Location	SSU
Friedmanniomyces endolithicus	**CCFEE 5321**	Lichen	Antarctic	Unpublished
Fonsecaea pedrosoi	CBS 272.37	-	-	AY554290
Guignardia mangiferae	IFO 33119	*Rhododendron pulchrum*	-	AB041247
Guignardia mangiferae	CBS 226.77	*Paphiopedilum callosum*	-	AB041248
Guignardia mangiferae	CBS 398.80	Orchid	-	AB041249
Hobsonia santessonii	-	-	-	AF289658
Hortaea werneckii	dH10921	Marble	-	Y18700
Hortaea werneckii	CBS 107.67	human *Tinea nigra*	-	Y18693
Knufia chersonesos	CBS 600.93; dH16058	Marble	Greece	Y18702
Knufia chersonesos	CBS 726.95	Marble	Italy	Unpublished
Knufia perforans	CBS 885.95	Marble	Delos, Greece	Y11714
Knufia perforans	CBS 665.80	Marble	Delos, Greece	Y11712
Mycocalicium victoriae	CBS 109863	Soil	Italy	Unpublished
Myriangium duriaei	CBS 260.36	*Chrysomphalus*	Argentina	NG_013129
Pseudotaeniolina globosa	CBS 109889	Rock	Italy	GU214576
Saxomyces alpinus	CCFEE 5466	Rock	Alps, Italy	GU250350
Saxomyces alpinus	CCFEE 5469	Rock	Alps, Italy	KC315860
Saxomyces alpinus	CCFEE 5470	Rock	Alps, Italy	KC315861
Saxomyces penninicus	CCFEE 5495	Rock	Alps, Italy	KC315864
Recurvomyces mirabilis	CBS 119434; CCFEE 5264	Rock	Antarctica	GU250329
Rhinocladiella atrovirens	CBS 688.76	*Pinus*	Australia	AJ232937
Rock black fungus	CCFEE 451	Rock	Antarctic	GU250314
Rock black fungus	CCFEE 457	Rock	Antarctic	GU250317
Rock black fungus	CCFEE 507	Rock	Antarctic	Unpublished
Rock black fungus	CCFEE 5176	Rock	Antarctic	GU250325
Rock black fungus	CCFEE 5177	Rock	Antarctic	Unpublished
Rock black fungus	CCFEE 5205	Rock	Antarctic	GU250327
Rock black fungus	CCFEE 5207	Rock	Antarctic	Unpublished
Rock black fungus	CCFEE 5267	Rock	Antarctic	Unpublished
Rock black fungus	CCFEE 5284	Rock	Antarctic	GU250330
Rock black fungus	CCFEE 5303	Rock	Antarctic	GU250331
Rock black fungus	CCFEE 5329	Rock	Antarctic	Unpublished
Teratosphaeria microspora	CBS 101951; STE-U 1960	Leaf	South Africa	EU167572
Teratosphaeria molleriana	CPC 1214	*Eucalyptus globulus*	Portugal	GU214606
Teratosphaeria molleriana	CPC 4577	*Eucalyptus*	Australia	GU214582
Teratosphaeria molleriana	CPC 10397	*Eucalyptus globulus*	Spain	GU214607
Teratosphaeria nubilosa	CPC 933	*Eucalyptus nitens*	South Africa	GU214608
Teratosphaeria nubilosa	CPC 937	*Eucalyptus globulus*	Australia	GU214609

Table 2. *Cont.*

Species	Strains no.	Source	Location	SSU
Unknown black fungus	**CCFEE 5304**	Lichen	Antarctic	Unpublished
Unknown black fungus	**CCFEE 5312**	Lichen	Antarctic	Unpublished
Unknown black fungus	**CCFEE 5314**	Lichen	Antarctic	Unpublished
Unknown black fungus	**CCFEE 5317**	Lichen	Antarctic	Unpublished
Unknown black fungus	**CCFEE 5318**	Lichen	Antarctic	Unpublished
Unknown black fungus	**CCFEE 5322**	Lichen	Antarctic	GU250334
Unknown black fungus	**CCFEE 5323**	Lichen	Antarctic	Unpublished
Unknown black fungus	**CCFEE 5324**	Lichen	Antarctic	Unpublished
Unknown black fungus	**CCFEE 5326**	Lichen	Antarctic	Unpublished

AFTOL—Assembling Fungal Tree Of Life project; CBS—Centraalbureau voor Schimmelcultures; CCFEE—Culture Collection of Fungi From Extreme Environments; CPC—Culture collection of P Crous, housed at the CBS; dH—de Hoog private collection housed at the CBS; IFO—Institute for Fermentation Culture Collection, Japan; MUCL—Belgian Co-ordinated Collections of micro-organisms; STE-U—University of Stellenbosch fungal culture collection, Stellenbosch, South Africa; UAMH—The University of Alberta Microfungus Collection and Herbarium, Edmonton, AB, Canada; UME—Herbarium Department of Ecology and Environmental Sciences (EMG) University of Umeå, Sweden. Strains isolated in this study are reported in bold.

3. Results and Discussion

Data concerning lichen sample (Figure 1), collection sites and black fungi isolated are reported in Table 1. The epilithic vegetation is rather rare in the Dry Valleys, it is therefore not surprising that only one lichen sample, *Lecanora fuscobrunnea*, out of 16 studied, was collected in Southern Victoria Land. Black fungi (Figure 2) were recovered from 11 out of 16 lichens examined.

Temperature relations are given in Table 3. All the strains tested were able to grow at 0 °C and none of the strains grew at 30 °C. Strains 5303, 5314, 5317, 5321, 5324, 5326 had their optimal growth temperature at 15 °C and did not show any growth above that temperature. All these strains can therefore be classified as psychrophilic, as defined for yeasts and other eukaryotic microorganisms [38]. Strain 5323, with optimal temperature and upper limit for growth at 20 °C, also may be defined as psychrophilic. Peculiar temperature relations, highlighting a more eurythermic behavior, were observed for strains 5313, 5319, 5320 with optimum at 15 °C but 25 °C as upper limit, too high for a true psychrophilic fungus. A similar profile was observed for strain CCFEE 5318 but with an optimal temperature at 20 °C.

Figure 1. Some of the lichen thalli examined for black fungi.

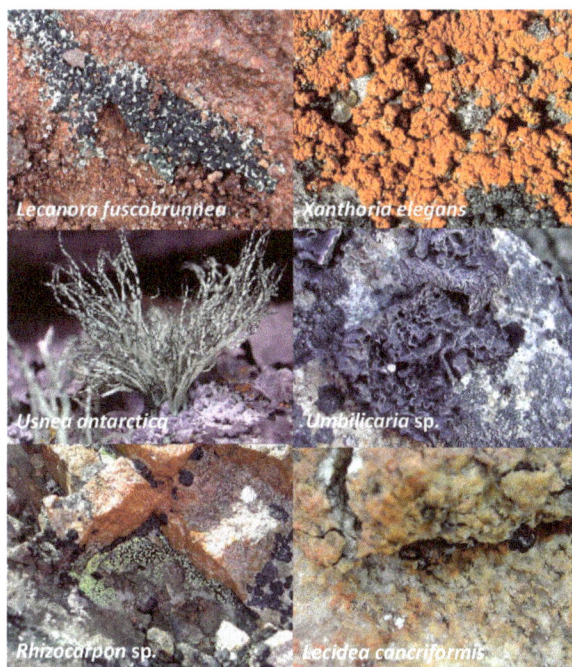

Figure 2. Some of the black fungi isolated from the lichens examined. This is a selection made based on morphological and phylogenetic characteristics.

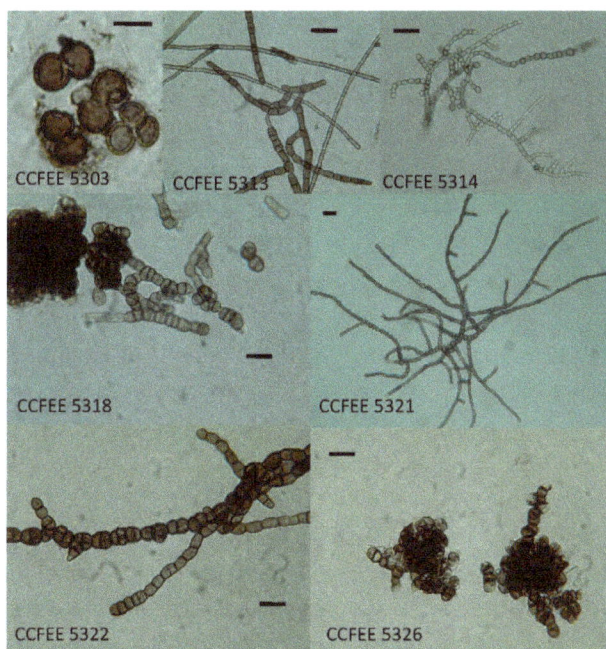

Table 3. Temperature relations.

Isolates	Temperature (°C)						
	0	5	10	15	20	25	30
CCFEE 5303	6.8 ± 1.8	4.3 ± 2.5	11.3 ± 0.4	**13.8 ± 1.8**	-	-	-
CCFEE 5312	6.9 ± 1.3	7.8 ± 1.1	11.8 ± 0.4	**18.2 ± 1.9**	5.2 ± 1.6	-	-
CCFEE 5313	16.4 ± 0.8	12.9 ± 0.6	25.8 ± 0.4	**31.9 ± 3.7**	30.8 ± 1.1	19.5 ± 3.5	-
CCFEE 5314	4.4 ± 0.6	5.2 ± 1.6	7 ± 0	**12.3 ± 2.5**	-	-	-
CCFEE 5317	4.8 ± 0.4	3.3 ± 0.4	8 ± 1.4	**12 ± 0**	-	-	-
CCFEE 5318	7.3 ± 0.4	4.1 ± 1.8	13.3 ± 0.4	13 ± 0.4	**13.5 ± 0.7**	11.3 ± 3.2	-
CCFEE 5319	15.5 ± 0.7	11.9 ± 0.1	22 ± 0	**30 ± 0.7**	24 ± 0.7	11.8 ± 1.1	-
CCFEE 5320	14 ± 0.7	10.8 ± 1.1	23 ± 1.4	**27.7 ± 0.9**	22.5 ± 0.7	16.8 ± 1.8	-
CCFEE 5321	3.5 ± 0.7	4.7 ± 0.9	8.8 ± 0	**10.3 ± 1.8**	-	-	-
CCFEE 5322	10.3 ± 1.8	9.8 ± 1.1	14 ± 1.4	**18.4 ± 2.3**	4 ± 0	-	-
CCFEE 5323	7 ± 0.7	4.7 ± 0.9	7 ± 2.1	18 ± 0.7	**20.5 ± 0.7**	-	-
CCFEE 5324	3.5 ± 0.7	4.1 ± 1.8	5.3 ± 0.4	**8.8 ± 0**	-	-	-
CCFEE 5326	2.5 ± 0	3.5 ± 0.7	4.1 ± 1.8	**7 ± 0.7**	-	-	-

Growth are reported as diameter of the colonies (mm) after 3 months of incubation. Highest growth values are reported in bold.

Most of the ITS sequences obtained showed too low identities in the GenBank and were not used for the phylogenetic inference. Figure 3 shows the ML phylogenetic tree, generated using a GTR+IG model, which was selected using the Akaike's information criterion with a Maximum likelihood approach. The alignment was based on 79 nuclear ribosomal 18S gene sequences and 1707 positions, including gaps, belonging to strains of both plant pathogenic and rock fungi, some of which were still unidentified. The tree includes two classes within the Ascomycota: Dothideomycetes (Orders Capnodiales, Dothideales, Myriangiales and Botryosphaeriales) and Eurotiomycetes (Order Chaetothyriales). The tree was rooted with *Debaryomyces hansenii* MUCL 29826.

The backbone remains uncertain, but orders in the class Dothideomycetes, are resolved although the 18S gene only was compared. The tree is in agreement with the most recent phylogenetic analyses with the Order Botryosphaeriales separated from Capnodiales [39,40]. Two sister clades are segregated in the Order Chaetothyriales: the group comprising most of the human opportunists of the family Herpotrichiellaceae as *Cladophialophora carrionii* (Trejos) de Hoog, Kwon-Chung & McGinnis, and the clade composed of mostly rock fungi, including the genus *Knufia* [41].

Seven of the strains here studied were grouped in the order Capnodiales placed in lineages purely constituted of fungi from rocks. Strains CCFEE 5312 and 5318 are included in a wide clade of rock fungi [40]; here only a selection of strains from the Antarctic is included, but the clade comprises rock fungi from the Mediterranean and Alps too, as well as the melanised micro-filamentous lichen *Cystocoleus ebeneus* (Dillwyn) Thwaites [42]. The strain CCFEE 5322 groups with rock black fungi exclusively from the Antarctic. The remaining strains in the Capnodiales belong to the rock fungal species *Elasticomyces elasticus* Zucconi & Selbmann and *Friedmanniomyces endolithicus* Onofri [43], the last one exclusively from the Antarctic continent [7,17]. The strain CCFEE 5304 as included in a well separated and supported clade of rock fungi from Antarctic rocks collected both in Northern and Southern Victoria land colonized with endolithic communities. This group remains without a clear assignment at any known fungal order. The remaining five strains were in the order Chaetothyriales (class Eurotiomycetes). Strains CCFEE 5326 and 5317 grouped together in a separated position with high bootstrap value and do not show clear relations with any described or undescribed species in the tree the ITS sequences were only 88% similar with the closest deposited in GenBank: this is not uncommon for black fungi from locations where genetic and geographic isolation, coupled with environmental pressure, promoted adaptive radiation [7]. Yet, their long branches indicate that these strains are distantly related to each other. Strains CCFEE 5314, 5323 and 5324 cluster with a rock Antarctic fungus, CCFEE 457, isolated from sandstone collected in the Dry Valleys; this group of Antarctic black fungi is sister of a clade represented by the recently formalized genus *Knufia* [41], including species mainly isolated from monuments.

All the strains examined here are not related to groups that contain known lichenicolous species. Rather, they show strict phylogenetic relations with fungi occurring on and in Antarctic rocks. Likewise, the rock black fungi included in this study were found to belong to two classes of Ascomycota: Dothideomycetes and Eurotiomycetes, in this last case specifically in the order Chaetothyriales (Figure 3). Dothideomycetous rock black fungi prefer natural, non-contaminated environments, while chaetotyriomycetous rock black fungi are recurrent particularly in areas influenced by human activities, rich in pollutants [44] probably as consequence of their ability to metabolize aromatic compounds [45].

Lichens can host a wide range of associated fungi with varied ecologies, specificities, and biological traits [26]. Some fast-growing lichenicolous species (e.g., *Athelia*, *Marchandiomyces*), with often low host specificity, can rapidly eradicate lichen vegetation, whereas many others grow slowly without expressing any or only local pathogenic symptoms on their specific hosts, apparently as a long-term result of evolutionary adaptation [46]. These lichenicolous fungi are not found to express their phenotypes without their hosts. Some groups of black fungi have also been observed to colonize a wide range of lichens, as lichenicolous fungi. Some species in the genus *Lichenothelia*, a cosmopolitan genus of rock-inhabiting melanised fungi in the superclass dothideomyceta [47] have been found in association with algae or with lichen thalli, where they produce fertile structures with asci and ascospores. However, species reported in this study were not related to *Lichenothelia* nor with any of the groups comprising known lichenicolous fungi. Moreover, they do not produce visible symptoms on thalli. Several melanized fungi were isolated from lichens from Armenia and the Alps with obscure discolourations [14] belonging to the genera *Mycosphaerella*, *Rhinocladiella*, *Capnobotryella* (class Dothideomycetes) and *Coniosporium*, in this last case related to *Knufia perforans* (Sterflinger) Tsuneda (class Eurotiomycetes). The strains CCFEE 5314, 5323, 5324 isolated during this study may be related to this last species, but the above mentioned isolates were not included in our tree since the SSU sequences are not available. Comparing the ITS sequences of our isolates we found that they were 10% distant from the sequences FJ265756 (*Coniosporium* sp. h6) and FJ265754 (*Coniosporium* sp. c-SH-2009a) isolated form *Caloplaca saxicola* (Hoffm.) Nordin and *Protoparmeliopsis muralis* (Scherb.) M. Choisy respectively, both from Armenia.

Figure 3. SSU ML tree indicating the phylogenetic position of the black meristematic fungi isolated from lichens (reported in bold in the tree). The strains reported as Rock black fungus are still unidentified rock fungi deposited in the Culture Collection of Fungi From Extreme Environments. Bootstrap values are the results of 1,000 pseudoreplicates. Values below 70 are not shown.

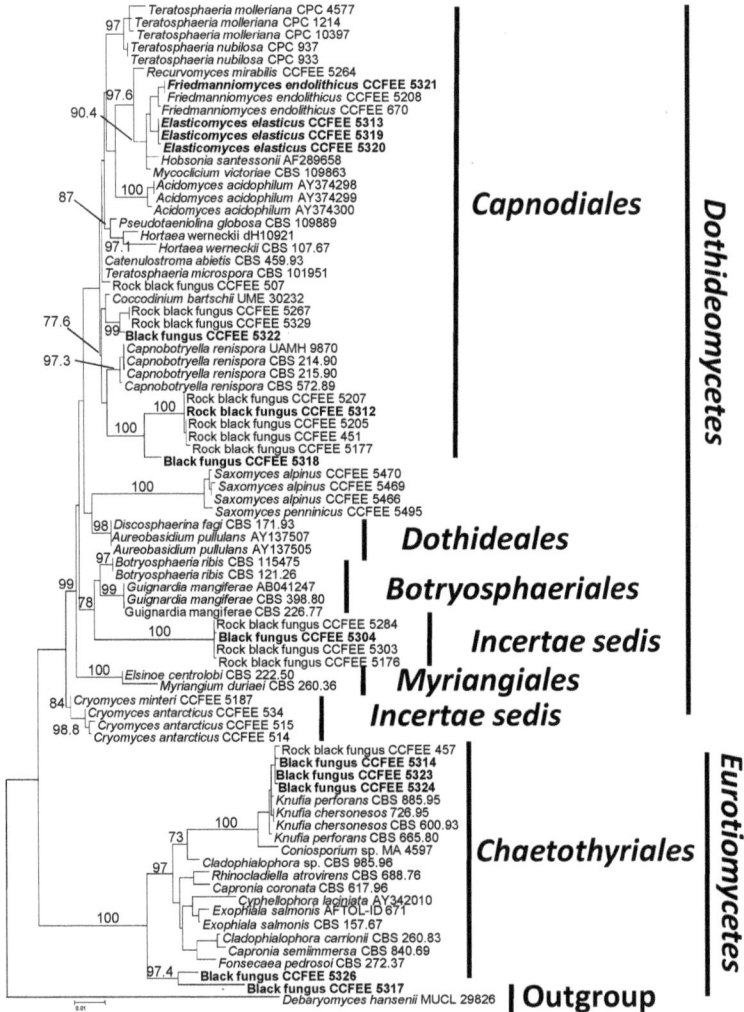

Association of black fungi with primary producers could be interpreted from a nutrition-ecological point of view. Oligotrophy is important adaptation for life on rocks and these fungi may often rely only on the sparse, airborne nutrients available, as pollutants in urban environments [45]. In natural environments, with low anthropogenic impact and scarce nutrient availability, they could more conveniently obtain nutrients, by living in association with lichens and other microbial primary producers, such as algae and cyanobacteria, which reside in the endolithic microbial communities of the highest mountain peaks and Antarctica [7,17,48].

Interestingly, some rock-inhabiting species were previously observed to develop lichen-like structures in axenic cultures with phototrophic algae [49,50]. This ability to develop symbiotic interactions with unicellular free-living algae might have allowed some rock-inhabiting fungal lineages to evolve lichenisation and a common link between rock-inhabiting meristematic and lichen-forming lifestyles of ascomycetous fungi has been recently hypothesized [47]. Some studies suggest that some rock-inhabiting fungi constitute early diverging lineages for lichenized fungal groups as Verrucariales and Arthoniomycetes [51].

4. Conclusions

This study represents the first contribution regarding black fungi associated with lichen thalli in the Antarctic. All strains isolated were either closely related or conspecific with black fungi previously found associated with Antarctic endolithic microbial communities. These are mostly cryptoendolithic lichen dominated communities, on the borderlines of what life can tolerate [16]. Even from cosmopolitan lichen species, we isolated endemic Antarctic fungi as *F. endolithicus*. Data obtained in this study give new insights into the biology of lichens: they are particularly well adapted to survive in extreme conditions and the ability to vary microbial communities associated according with the location may give further advantage in adaptation and survival of the whole community.

It is still unclear whether or not black fungi may supply benefits to epilithic lichens as well. It was suggested previously [7] that black fungi may play a role in in hydration or protection of photobionts by dissipating excessive sunlight. Cryptoendolitic lichens are melanized fungi that form a black barrier just above the photobionts stratification [14] making this a plausible scenario. The presence of black fungi may therefore play a crucial role to allow survival in these highly stressful conditions.

Apparently, at the cold edge of life, lichens, together with their associated microbes could find a solution to survive inside the rock by taking advantage of the suite of traits from each microbial partner in order to improve stress resistance and allow the whole community to survive in new conditions.

Acknowledgements

This work was carried out in the framework of the PNRA (Italian National Program for Antarctic Research). The Italian National Antarctic Museum "Felice Ippolito" for funding CCFEE (Culture Collection of Fungi from Extreme Environments).

References

1. Plemenitaš, A.; Gunde-Cimerman, N. Cellular responses in the halophilic black yeast *Hortaea werneckii* to high environmental salinity. In *Adaptation to Life at High Salt Concentrations in Archaea, Bacteria, and Eukarya*; Gunde-Cimerman, N., Oren, A., Plemenitaš, A., Eds.; Springer: Dordrecht, The Netherlands, 2005; pp. 455–470.

2. Baker, B.J.; Lutz, M.A.; Dawson, S.C.; Bond, P.L.; Banfield, J.F. Metabolically active eukaryotic communities in extremely acidic mine drainage. *Appl. Environ. Microbiol.* **2004**, *70*, 6264–6271.

3. Selbmann, L.; Egidi, E.; Isola, D.; Onofri, S.; Zucconi, Z.; de Hoog, G.S.; Chinaglia, S.; Testa, L.; Tosi, S.; Balestrazzi, A.; *et al.* Biodiversity, evolution and adaptation of fungi in extreme environments. *Plant Biosyst.* **2012**, *147*, doi:10.1080/11263504.2012.753134.

4. Isola, D.; Selbmann, L.; de Hoog, G.S.; Fenice, M.; Onofri, S.; Prenafeta-Boldú, F.X.; Zucconi, L. Isolation and screening of black fungi as degraders of volatile aromatic hydrocarbons. *Mycopathologia* **2013**, doi:10.1007/s11046-013-9635-2.

5. Staley, J.T.; Palmer, F.; Adams, J.B. Microcolonial fungi: Common inhabitants on desert rocks? *Science* **1982**, *215*, 1093–1095.

6. Ruibal, C.; Gonzalo, P.; Bills, G.F. Isolation and characterization of melanized fungi from limestone formations in Mallorca. *Mycol. Progress* **2005**, *4*, 23–38.

7. Selbmann, L.; de Hoog, G.S.; Mazzaglia, A.; Friedmann, E.I.; Onofri, S. Fungi at the edge of life: Cryptoendolithic black fungi from Antarctic desert. *Stud. Mycol.* **2005**, *51*, 1–32.

8. Sert, H.B.; Sümbül, H.; Sterflinger, K. Microcolonial fungi from antique marbles in Perge/Side/Termessos (Antalya/Turkey). *Antonie Leeuwenhoek* **2007**, *91*, 217–227.

9. Sert, H.B.; Sümbül, H.; Sterflinger, K. *Sarcinomyces sideticae*, a new black yeast from historical marble monuments in Side (Antalya, Turkey). *Bot. J. Linnean Soc.* **2007**, *154*, 373–380.

10. Sert, H.B.; Sümbül, H.; Sterflinger, K. A new species of *Capnobotryella* from monument surfaces. *Mycol. Res.* **2007**, *111*, 1235–1241.

11. Marvasi, M.; Donnarumma, F.; Frandi, A.; Mastromei, G.; Sterflinger, K.; Tiano, P.; Perito, B. Black microcolonial fungi as deteriogens of two famous marble statues in Florence, Italy. *Int. Biodeterior. Biodegrad.* **2012**, *68*, 36–44.

12. Zucconi, L.; Gagliardi, M.; Isola, D.; Onofri, S.; Andaloro, M.C.; Pelosi, C.; Pogliani, P.; Selbmann, L. Biodeteriorigenous agents dwelling the wall paintings of the Holy Saviour's Cave (Vallerano, Italy). *Int. Biodeterior. Biodegrad.* **2012**, *70*, 40–46.

13. Onofri, S.; Selbmann, L.; Zucconi, L.; de Hoog, G.S.; de los Rios, A.; Ruisi, S.; Grube, M. Fungal associations at the cold edge of life. In *Algae and Cyanobacteria in Extreme Environments*; Seckbach, J., Ed.; Springer: Dordrecht, The Netherlands, 2007; pp. 735–757.

14. Harutyunyan, S.; Muggia, L.; Grube, M. Black fungi in lichens from seasonally arid habitats. *Stud. Mycol.* **2008**, *61*, 83–90.

15. Sterflinger, K. Black Yeast and Meristematic Fungi: Ecology, Diversity and Identification. In *The Yeast Handbook. Biodiversity and Ecophysiology of Yeasts*; Péter, G., Rosa, C., Eds.; Springer: Berlin, Germany, 2006; pp. 501–514.

16. Nienow, J.A.; Friedmann, E.I. Terrestrial Litophytic (Rock) Communities. In *Antarctic Microbiology*; Friedmann, E.I., Ed.; Wiley-Liss: New York, NY, USA, 1993; pp. 343–412.

17. Selbmann, L.; de Hoog, G.S.; Gerrits van den Ende, A.H.G.; Ruibal, C.; de Leo, F.; Zucconi, L.; Isola, D.; Ruisi, S.; Onofri, S. Drought meets acid: Three new genera in a Dothidealean clade of extremotolerant fungi. *Stud. Mycol.* **2008**, *61*, 1–20.

18. Friedmann, E.I. Endolithic microorganisms in the Antarctic Cold Desert. *Science* **1982**, *215*, 1045–1053.

19. De la Torre, J.R.; Goebel, B.M.; Friedmann, E.I.; Pace, N.R. Microbial diversity of cryptoendolithic communities from the McMurdo Dry Valleys, Antarctica. *Appl. Environ. Microbiol.* **2003**, *215*, 3858–3867.

20. Selbmann, L.; Zucconi, L.; Ruisi, S.; Grube, M.; Cardinale, M.; Onofri, S. Culturable bacteria associated with Antarctic lichens: Affiliation and psychrotolerance. *Polar Biol.* **2010**, *33*, 71–83.

21. Cardinale, M.; Vieira de Castro, J., Jr.; Müller, H.; Berg, G.; Grube, M. *In situ* analysis of the bacterial community associated with the reindeer lichen *Cladonia arbuscula* reveals predominance of Alphaproteobacteria. *FEMS Microbiol. Ecol.* **2008**, *66*, 63–71.

22. Grube, M.; Cardinale, M.; Vieira de Castro, J.; Müller, H.; Berg, G. Species-specificstructural and functional diversity of bacterial communitiesin lichen symbioses. *ISME J.* **2009**, *3*, 1105–1115.

23. Hodkinson, B.; Lutzoni, F. A microbiotic survey of lichen-associated bacteria reveals a new lineage from the Rhizobiales. *Symbiosis* **2009**, *49*, 163–180.

24. Bates, S.T.; Cropsey, G.W.G.; Caporaso, J.G.; Knight, R.; Fierer, N. Bacterial communities associated with the lichen symbiosis. *Appl. Environ. Microbiol.* **2011**, *77*, 1309–1314.

25. Grube, M.; Berg, G. Microbial consortia of bacteria and fungi with focus on the lichen symbiosis. *Fungal Biol. Rev.* **2009**, *23*, 72–85.

26. Lawrey, J.D.; Diederich, P. Lichenicolous fungi: Interactions, evolution, and biodiversity. *Bryologist* **2003**, *106*, 80–120.

27. Lawrey, J.D.; Binder, M.; Diederich, P.; Molina, M.C.; Sikaroodi, M.; Ertz, D. Phylogenetic diversity of lichen-associated homobasidiomycetes. *Mol. Phylogenetics Evol.* **2007**, *44*, 778–789.

28. Girlanda, M.; Isocrono, D.; Bianco, C.; Luppi-Mosca, A.M. Two foliose lichens as microfungal ecological niches. *Mycologia* **1997**, *89*, 531–536.

29. Suryanarayanan, T.S.; Thirunavukkarasu, N.; Hariharan, G.N.; Balaji, P. Occurrence of non-obligate microfungi inside lichen thalli. *Sydowia* **2005**, *57*, 120–130.

30. U'Ren, J.; Lutzoni, F.; Miadlikowska, J.; Arnold, A.E. Community analysis reveals close affinities between endophytic and endolichenic fungi in mosses and lichens. *Microb. Ecol.* **2010**, *60*, 340–353.

31. Bates, S.T.; Berg-Lyons, D.; Lauber, C.L.; Walters, W.A.; Knight, R.; Fierer, N. A preliminary survey of lichen associated eukaryotes using Pyrosequencing. *Lichenologist* **2012**, *44*, 137–146.

32. Farrar, J.F. The Lichen as an Ecosystem: Observation and Experiment. In *Lichenology: Progress and Problems*; Brown, D.H., Hawksworth, D.L., Bailey, R.H., Eds.; Academic Press: London, UK, 1976; pp. 385–406.

33. Castello, M. Lichens of the Terra Nova Bay area, Northern Victoria Land (Continental Antactica). *Stud. Geobot.* **2003**, *22*, 3–54.

34. White, T.J.; Bruns, T.; Lee, S.B.; Taylor, J.W. Amplification and Direct Sequencing of Fungal Ribosomal RNA Genes for Phylogenetics. In *PCR Protocols, a Guide to Methods and Applications*; Innis, M.A., Gelfand, D.H., Sninsky, J.J., White, T.J., Eds.; Academic Press: San Diego, CA, USA, 1990; pp. 315–322.

35. Thompson, J.D.; Gibson, T.J.; Plewniak, F.; Jeanmougin, F.; Higgins, D.G. The ClustalX windows interface: Flexible strategies for multiple sequence alignment aided by quality analysis tools. *Nucleic Acids Res.* **1997**, *24*, 4876–4882.

36. Tamura, K.; Peterson, D.; Peterson, N.; Stecher, G.; Nei, M.; Kumar, S. MEGA5: Molecular evolutionary genetics analysis using maximum likelihood, evolutionary distance, and maximum parsimony methods. *Mol. Biol. Evol.* **2011**, *28*, 2731–2739.

37. Felsenstein, J. Confidence limits on phylogenies: An approach using the bootstrap. *Evolution* **1985**, *40*, 783–791.

38. Van Uden, N. Temperature profiles of yeasts. *Adv. Microbiol. Physiol.* **1984**, *25*, 195–251.

39. Schoch, C.L.; Shoemaker, R.A.; Seifert, K.A.; Hambleton, S.; Spatafora, J.W.; Crous, P.W. A multigene phylogeny of the *Dothideomycetes* using four nuclear loci. *Mycologia* **2006**, *98*, 1041–1052.

40. Ruibal, C.; Gueidan, C.; Selbmann, L.; Gorbushina, A.A.; Crous, P.W.; Groenewald, J.Z.; Muggia, L.; Grube, M.; Isola, D.; Schoch, C.L.; *et al.* Phylogeny of rock-inhabiting fungi related to *Dothideomycetes*. *Stud. Mycol.* **2009**, *64*, 123–133.

41. Tsuneda, A.; Hambleton, S.; Currah, R.S. The anamorph genus *Knufia* and its phylogenetically allied species in *Coniosporium*, *Sarcinomyces*, and *Phaeococcomyces*. *Botany* **2011**, *89*, 523–536.

42. Muggia, L.; Hafellner, J.; Wirtz, N.; Hawksworth, D.L.; Grube, M. The sterile microfilamentous lichenized fungi *Cystocoleus ebeneus* and *Racodium rupestre* are relatives of plant pathogens and clinically important dothidealean fungi. *Mycol. Res.* **2008**, *112*, 50–56.

43. Onofri, S.; Pagano, S.; Zucconi, L.; Tosi, S. *Friedmanniomyces endolithicus* (Fungi, Hyphomycetes), anam.-gen. and sp.nov., from continental Antarctica. *Nova Hedwig.* **1999**, *68*, 175–181.

44. Onofri, S.; Anastasi, A.; Del Frate, G.; di Piazza, S.; Garnero, N.; Guglielminetti, M.; Isola, D.; Panno, L.; Ripa, C.; Selbmann, L.; *et al.* Biodiversity of rock, beach and water fungi in Italy. *Plant Biosyst.* **2011**, *45*, 978–987.

45. Prenafeta-Boldú, F.X.; Summerbell, R.C.; de Hoog, G.S. Fungi growing on aromatic hydrocarbons: Biotechnology's unexpected encounter with biohazard. *FEMS Microbiol. Rev.* **2006**, *30*, 109–130.

46. Diederich, P.; Lawrey, J.D. New lichenicolous, muscicolous, corticolous and lignicolous taxa of *Burgoa* s. l. and *Marchandiomyces* s. l. (anamorphic Basidiomycota), a new genus for *Omphalina foliacea*, and a catalogue and a key to the non-lichenized, bulbilliferous basidiomycetes, *Mycol. Progress* **2007**, *6*, 61–80.

47. Muggia, L.; Gueidan, C.; Knudsen, K.; Perlmutter, G.; Grube, M. The lichen connections of black fungi. *Mycopathologia* **2012**, doi:10.1007/s11046-012-9598-8.

48. Selbmann, L.; Isola, D.; Egidi, E.; Zucconi, L.; Gueidan, C.; de Hoog, G.S.; Onofri, S. Rock inhabiting fungi: *Saxomyces* gen. nov. and four new species from the Alps. *Fungal Diver.* **2013**, in press.

49. Gorbushina, A.A.; Beck, A.; Shulte, A. Microcolonial rock inhabiting fungi and lichen photobionts: Evidence for mutualistic interactions. *Mycol. Res.* **2005**, *109*, 1288–1296.

50. Brunauer, G.; Blaha, J.; Hager, A.; Turk, R.; Stocker-Worgotter, E.; Grube, M. An isolated lichenicolous fungus forms lichenoid structures when co-cultured with various coccoid algae. *Symbiosis* **2007**, *44*, 127–136.

51. Gueidan, C.; Ruibal, C.; de Hoog, G.S.; Gorbushina, A.; Untereiner, W.A.; Lutzoni, F. A rock-inhabiting ancestor for mutualistic and pathogen-rich fungal lineages. *Stud. Mycol.* **2008**, *61*, 111–119.

Fungal Diversity in a Dark Oligotrophic Volcanic Ecosystem (DOVE) on Mount Erebus, Antarctica

Laurie Connell and Hubert Staudigel

Abstract: Fumarolic Ice caves on Antarctica's Mt. Erebus contain a dark oligotrophic volcanic ecosystem (DOVE) and represent a deep biosphere habitat that can provide insight into microbial communities that utilize energy sources other than photosynthesis. The community assembly and role of fungi in these environments remains largely unknown. However, these habitats could be relatively easily contaminated during human visits. Sixty-one species of fungi were identified from soil clone libraries originating from Warren Cave, a DOVE on Mt. Erebus. The species diversity was greater than has been found in the nearby McMurdo Dry Valleys oligotrophic soil. A relatively large proportion of the clones represented *Malassezia* species (37% of Basidomycota identified). These fungi are associated with skin surfaces of animals and require high lipid content for growth, indicating that contamination may have occurred through the few and episodic human visits in this particular cave. These findings highlight the importance of fungi to DOVE environments as well as their potential use for identifying contamination by humans. The latter offers compelling evidence suggesting more strict management of these valuable research areas.

Reprinted from *Biology*. Cite as: Connell, L.; Staudigel, H. Fungal Diversity in a Dark Oligotrophic Volcanic Ecosystem (DOVE) on Mount Erebus, Antarctica. *Biology* **2013**, *2*, 798-809.

1. Introduction

The subsurface biosphere has been among the most exciting, and rapidly evolving research ecosystem types in biogeosciences of the past 20 years [1]. Trace fossils of microbial dissolution in seafloor volcanic rocks suggest the presence of a Dark Oligotrophic Volcanic Ecosystem (DOVE) at least for the upper 500 m and extending back to first appearance of life on Planet Earth [2]. DOVEs take a special role in the study of the subsurface biosphere. They are volumetrically very significant and they contain abundant energy donors from the earth's interior, in the form of minerals and glasses that are highly reactive in low temperature hydrous environments. DOVEs commonly have active hydrothermal systems that readily circulate surface water and atmospheric gases through the interior of these volcanoes. The combination of surface-derived fluids and volcanic rocks from the interior of the earth providing abundant and effective combinations of electron acceptors and donors that can facilitate chemolithoautotrophic conditions for microbial communities to thrive without photosynthesis. Conditions of (near-) atmospheric oxygenation exist in very large fractions of DOVEs where water or air circulates relatively freely and recharges hydrothermal systems in active systems. These factors combine to make DOVEs a significant component of the subsurface biosphere and open up the possibility that DOVEs might provide biomass to the earth's surface offering a "rock bottom" for the food web.

The southernmost active volcano in the world, Mt. Erebus (3,795 m), has been the focus of research for decades [3]. Recently, ice caves and fumarolic ice towers near the summit have been recognized as a unique environment devoid of light and/or having a moist warm environment [4]. These sub-glacial fumaroles issue gases that are dominated by air with 80–100% humidity and up to 2% CO_2 [3,4]. CO_2 is one of the few sources of carbon available to these microbial communities, although small amounts of organic carbon may enter the cave through melt water from the surface during the summer months that can contain algae or wind delivered carbon sources.

Some of the Mt. Erebus DOVEs are now visited more frequently yet others still remain pristine environments and there is a need to determine which caves are best suited for future microbiological research through a microbial community survey to determine if there has been anthropogenic intrusion on these communities. Since the earliest days of Mt. Erebus cave research, ice caves were thought of as naturally protected environments to be used for the placement of experiments, storing gear and even food. There is even the possibly that members of Antarctic Heroic Age explorers from either the Nimrod Expedition (1907–1909) with the first ascent in 1908 or the Terra Nova Expedition (1910–1913) with the highest camp up to that time, left food or materials near the caves. Establishment of McMurdo Station in 1956 by the US Navy helped increase the access to Mt. Erebus and continuous research began in the early 1970's. One prominent Mt. Erebus cave, Warren Cave, is located in a logistically particularly strategic location, on a straight line between a recently discovered location of a camp by the Terra Nova Expedition (December 1912) and the summit of Mt. Erebus. This site is very close to Lower Erebus Hut, the operational base for the bulk of current research on Mt. Erebus.

The role of fungi in the DOVE communities is a new field of research and this is the first report of a fungal community associated with an Antarctic fumarole DOVE habitat.

2. Experimental Section

Sample site: Warren Cave on Mt. Erebus, Antarctica (77° 31.003 S; 167° 09.884 E) (Figure 1) has been visited by researchers annually over the past decade for the study of volcanic CO_2 emissions and temperature fluctuations by the Mount Erebus Volcano Observatory (MEVO) [4]. Warren Cave maintains a remarkably constant temperature. Temperatures in the fumarole studied were 18.5 °C inside the soil and 14.5 °C above the soil.

Sample collection: Soil substrate collected in 2010 from Warren Cave was taken from within a fumarolic vent issuing warm gas from beneath a protruding rock consisting of a patch of soil made up largely from coarse sand and fine gravel sized rock fragments (location labeled "GV 1" in [4]). The sample (Identification number: 10G439-WC) consisted of several pooled 10–15 g scoops collected aseptically in a sterile 50 mL tube. The sample was transported to the US at −20 °C. Soil temperature at the time of collection was 18.5 °C, the air temperature was 14.6 °C and soil pH was 5.2.

Soil analysis: Both soil moisture and carbon analyses were conducted by the University of Maine Analytical and Soil Testing Laboratory (Orono, ME, USA). Soil moisture was determined by gravimetric method at the time of soil drying for carbon analysis. Both total carbon and organic carbon were determined using the dry combustion method [5] with an Leco CN-2000 Carbon/Nitrogen analyzer.

Figure 1. Warren Cave near Mt. Erebus Summit. The map shows the upper section of Mt. Erebus and the location of Warren Cave. The insert photo shows the site where the sample was taken.

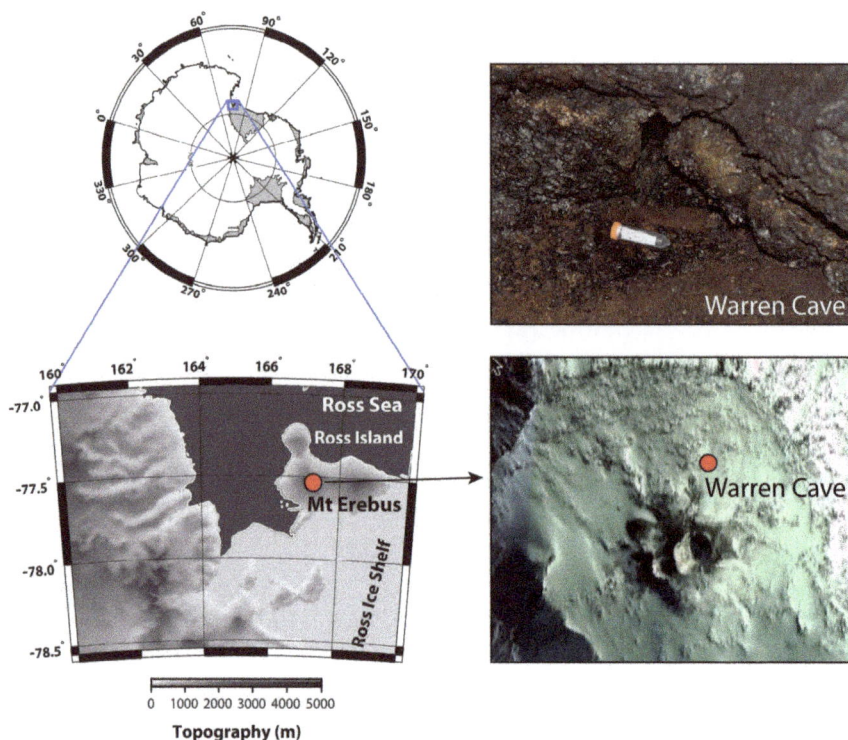

Nucleic acid extraction and analysis: DNA from 1.62 g of the soil pellet was extracted using a ZR Soil Microbe Midi kit (Zymo Research, Irvine, CA, USA). The soil pellet was extracted 2× and the DNA was pooled. Three pellets were extracted for each clone library. Specific ITS region amplicons were produced by PCR (100 ng/reaction) in 25 μL reactions using Illustra PuReTaq Ready-To-Go™ PCR Beads (GE Lifesciences, Piscataway, NJ, USA). PCR primer set ITS5/ITS4 [6] was used to target the ITS region for clone library construction. Initial denaturation was for 2 min at 95 °C and 35 cycles with a PTC-200 thermal cycler (MJ Research, Watertown, MA, USA) under the following conditions: 30 s at 95 °C, 30 s at 52.3 °C, 1 min at 72 °C with a final 72 °C 10 min extension. The resulting PCR products were cleaned prior to cloning using Promega SV Gel and PCR Clean-up System (Promega, Madison, WI, USA). Three PCR reactions were produced from each soil pellet DNA aliquot and pooled prior to library construction. Two ITS clone libraries ligations were produced. PCR products from multiple (4–8) clean PCR reactions (*i.e.*, showing only bands within expected ITS size ranges) were pooled, and purified using a PCR purification kit (QIAGEN, Valencia, CA, USA). Libraries were generated using a TOPO TA cloning kit and chemically competent *Escherichia coli* TOP10F cells (Invitrogen, Carlsbad, CA, USA). The High Throughput Genomic Unit (HTGU—University of Washington, Seattle, WA, USA) preformed the transformations, clone selection and sequencing using the vector T7 primer. The resulting data were screened for (1) poor quality sequence (below 80% quality) and (2) short sequences (shorter than 400 bp) using Sequencher v 5.0.1 (Gene Codes Corp.) and these sequences were eliminated from further analysis. Sequences that passed the first two steps were then screened for chimeric sequences using Chimera Checker [7] and chimeric sequences were eliminated from further analysis. Each remaining sequence was reviewed by hand and compared with NCIB database (BLAST and Tree Builder) for taxonomic assignments. Based on suggestion of Fell and coworkers [8] sequence homology of >98% were considered to be the same species. GenBank accession numbers are shown in Table 1 and can be accessed through GenBank. The closest match isolates selected for Table 1 are comprised only of fungi that have been cultured.

Table 1. Clones identified from Warren Cave with GenBank accession numbers and closest matches. Asterisks identify clones that fall below the 98% species identification threshold.

Warren Cave Clone Species	GenBank Accession Number	GenBank Closest Match	% match	GenBank Closest Match Species
Acremonium implicatum	KC785536	JQ692168	99%	Acremonium implicatum
*Acremonium sp.	KC785537	AB540571	94%	Acremonium cereale
Alternaria alternata	KC785538	AF218791	99%	Alternaria alternata
Aspergillus penicillioides	KC785539	HQ914939	99%	Aspergillus penicillioides
Aureobasidium pullulans	KC785542	FN868454	99%	Aureobasidium pullulans
Aureobasidium sp.	KC785543	HQ631013	99%	Aureobasidium sp. TMS-2011
Candida zeylanoides	KC785544	EF687774	100%	Candida zeylanoides
Cladosporium sp. 1	KC785545	HQ631003	100%	Cladosporium sp. TMS-2011 voucher
Cladosporium grevilleae	KC785546	JF770450	99%	Cladosporium grevilleae
Cladosporium sphaerospermum	KC785547	JQ776537	98%	Cladosporium sphaerospermum
Clavispora lusitaniae	KC785548	EU149777	99%	Clavispora lusitaniae
Cochliobolus lunatus	KC785549	HQ607915	100%	Cochliobolus lunatus
Cyphellophora laciniata	KC785550	EU035416	99%	Cyphellophora laciniata
Epicoccum nigrum	KC785551	HQ607859	100%	Epicoccum nigrum
Erysiphe polygoni	KC785552	AF011308	99%	Erysiphe polygoni
Gibellulopsis nigrescens	KC785553	KC156644	99%	Gibellulopsis nigrescens
*Hansfordia sp.	KC785554	HQ914948	96%	Hansfordia sp.
Lewia infectoria	KC785556	AY154718	99%	Lewia infectoria
Myrothecium verrucaria	KC785557	FJ235085	99%	Myrothecium verrucaria
Penicillium oxalicum	KC785558	JX231003	99%	Penicillium oxalicum
*Pezizomycotina sp.	KC785559	EU167561	96%	Pleiochaeta ghindensis
Phaeococcomyces nigricans	KC785560	AY843154	99%	Phaeococcomyces nigricans
Phaeosphaeria sp.	KC785561	HQ631018	99%	Phaeosphaeria sp. 1 TMS-2011 voucher
*Phialosimplex sp.	KC785562	GQ169326	93%	Phialosimplex chlamydosporus
Pleosporales sp.	KC785563	HQ207041	100%	Pleosporales sp. 24 PH
Saccharomyces cerevisiae	KC785564	AY939814	99%	Saccharomyces cerevisiae
Tetracladium sp. 1	KC785565	JF911760	98%	Tetracladium sp. QH32
Tetracladium sp. 2	KC785555	AB776690	99%	Tetracladium sp. SMU-1
Toxicocladosporium irritans	KC785566	EU040243	99%	Toxicocladosporium irritans
Verticillium dahliae	KC785567	HQ839784	99%	Verticillium dahliae
Volutella colletotrichoides	KC785568	AJ301962	100%	Volutella colletotrichoides

Table 1. *Cont.*

Warren Cave Clone Species	GenBank Accession Number	GenBank Closest Match	% match	GenBank Closest Match Species
Ceriporiopsis subvermispora	KC785569	FJ713106	99%	*Ceriporiopsis subvermispora*
Cryptococcus wieringae	KC785570	FN824493	99%	*Cryptococcus wieringae*
Cystofilobasidium macerans	KC785572	AF444317	100%	*Cystofilobasidium macerans*
Endophyte sp.	KC785573	EU977202	99%	*Fungal Endophyte* sp. P807B
Exidia glandulosa	KC785574	AY509555	99%	*Exidia glandulosa*
Exobasidium sp.	KC785575	EU784219	92%	*Exobasidium rhododendri*
Filobasidium floriforme	KC785576	AF190007	99%	*Filobasidium floriforme*
Ganoderma applanatum	KC785577	JX501311	99%	*Ganoderma applanatum*
Glaciozyma watsonii	KC785578	AY040660	99%	*Glaciozyma watsonii*
Hymenochaete sp.	KC785579	JN230420	97%	*Hymenochaete corrugata*
Hyphodontia rimosissima	KC785580	DQ873627	99%	*Hyphodontia rimosissima*
Irpex lacteus	KC785581	EU273517	99%	*Irpex lacteus*
Malassezia globosa	KC785582	KC152884	99%	*Malassezia globosa*
Malassezia restricta	KC785583	EU400587	99%	*Malassezia restricta*
Malassezia sp.	KC785585	KC141977	82%	*Malassezia sympodialis*
Mycena sp.	KC785587	JQ272379	99%	*Mycena* sp. 1 RB-2011
Peniophora lycii	KC785588	JX046435	99%	*Peniophora lycii*
Phanerochaete sp.	KC785589	GU934592	96%	*Phanerochaete* sp. 853
Polyporus sp.	KC785590	AF516599	97%	*Polyporus tuberaster*
Resinicium bicolor	KC785591	DQ826534	99%	*Resinicium bicolor*
Rhodotorula mucilaginosa	KC785592	HQ702343	99%	*Rhodotorula mucilaginosa*
Sistotrema brinkmanii	KC785594	DQ899095	99%	*Sistotrema brinkmannii*
Skeletocutis chrysella	KC785595	FN907916	99%	*Skeletocutis chrysella*
Sporobolomyces sp.	KC785596	EU002899	99%	*Sporobolomyces* sp.
Stereum sanguinolentum	KC785597	AY089730	99%	*Stereum sanguinolentum*
Stereum sp.	KC785598	FN539049	91%	*Stereum rugosum*
Trametes cubensis	KC785599	JN164923	99%	*Trametes cubensis*
Trichaptum sp.	KC785600	U63473	95%	*Trichaptum biforme*
Ustilago sp.	KC785601	AY740170	97%	*Ustilago drakensbergiana*
Ustilago tritici	KC785602	JN114419	99%	*Ustilago tritici*

Community analysis: The resulting passed sequences were classified into groups based on their phyla. Each group of sequences was aligned using MUSCLE web server alignment [9]. The alignments were used to create phylogenetic trees through the *Seaview* software program, version 4.3.1. A rooted neighbor-joining distance tree was generated, for each phyla separately (Ascomycetes and Basidiomycetes), based on nucleotide positions of the ITS region of the 5.8S gene. Bootstrap values were based on 100 replicates. GenBank accession numbers were listed for the outgroup sequences.

3. Results and Discussion

We investigated the fungal diversity in Warren Cave though clone libraries. Soil substrate extraction was used to concentrate the fungal portion of the community prior to total DNA extraction. The habitat was highly oligotrophic with only 126 μg/g organic carbon (151 μg/g total carbon) and relatively moist with 50% soil moisture. Overall fungal diversity was moderate [10] with 266 fungal ITS clone sequences representing a total of 61 species. All were within the Ascomycota (Figure 2) and Basidiomycota (Figure 3) phyla, with no Chytridiomycota or Zygomycota represented. Near equal distribution of Ascomycota and Basidiomycota taxa were represented in the Warren Cave clone libraries (31 Ascomycota/30 Basidiomycota) unlike in nearby McMurdo Dry Valley habitats where Basidiomycota dominate, especially in arid habitats [11–13]. The data in Table 1 show that a vast majority of clones found in Warren Cave can be identified to the species level (80%). Most of the remainder are relatively close matches with only one, a potential *Malassezia* species, with the closest match below 90% (82% identity with *Malassezia sympodialis*). The clones shown in this work are those that passed the criteria listed in the methods, yet within some of the chimeric clones (and not included in this analysis) DNA fragments of other organisms were found. These fragments were from organisms associated with humans. The most abundant of these were most closely identified as cabbage, soybeans, cereal grains and buckwheat. Interestingly, one of the dominant organisms found in the McMurdo Dry Valley soils, nematodes [14] were absent from these clone libraries, even in fragments.

Data on fungal communities in Antarctica still remain quite incomplete, however, it is possible to draw comparisons between our data from a Mt. Erebus DOVE to one other extreme environment in the McMurdo area, the nearby McMurdo Dry Valleys soils. The McMurdo Dry Valley soils are also highly oligotrophic, with some of the lowest organic carbon levels reported [15] and similar to those reported here. Their soil communities experience low temperature and rapid temperate swings, high UV radiation and desiccation [16], while the Mt. Erebus DOVEs have relatively moderate and constant temperature, no light, thus no UV radiation and a moist environment. In addition, many of the soils in the McMurdo Dry Valleys are basic with pH ranging up to pH10 [11], while this Warren Cave site was slightly acidic (pH 5.2). Therefore it is not surprising that the fungal communities are different. The typical number of fungal species found in any one soil community isolated from the McMurdo Dry Valleys is low, often below ten [11,12] whereas the fungal diversity found in Warren Cave was found to be much higher (61 species). The relative higher number of Ascomycota taxa found in Warren Cave compared with other studies in the McMurdo Dry Valleys may reflect the more stressful condition found in the latter. Yeast species in studies of the McMurdo Dry Valleys were dominated by basidiomycetous species (89%), most particularly those from the genus *Cryptococcus* (33%) [12]. The dominance of *Cryptococcus* species in soil, particularly arid soil, has been ascribed to their ability to produce polysaccharide capsules [17]. In contrast, only two *Cryptococcus* species were found in Warren Cave. Further, *Glaciozyma watsonii* has been isolated numerous times from soil from Continental Antarctic soil [12,18,19] and can be an abundant member of the McMurdo Dry Valley soil community, but are represented by only six clones (4.9%) in Warren Cave. *Rhodotorula mucilaginosa* has also been cultured from some of the most dry and cold locations in Antarctica, such as Sponsors Peak (03SP24) and a peak above Niebelungen Valley, in the Asgard Range (03NB35) [12] yet was represented by only 2 clones (1.6%) of the Warren Cave libraries.

Figure 2. Phylogenetic tree of Warren Cave species belonging to the phylum Ascomycota, obtained by neighbor-joining analysis of the Internal Transcribed Spacer (ITS) region of the 5.8S rDNA gene, with 100 full heuristic replications. Bootstrap values are as indicated on the tree. *Glaciozyma watsonii* sequence obtained from GenBank was used as the outgroup, with the GenBank accession number listed. *Candida zeylanoides, Saccharomyces cerevisiae. Clavispora lusitaniae, Cyphellophora laciniata, Acremonium implicatum, Aureobasidium pullulans* and *Erysiphe polygoni* sequences obtained from GenBank were used as closest relative reference sequences, with GenBank accession numbers listed.

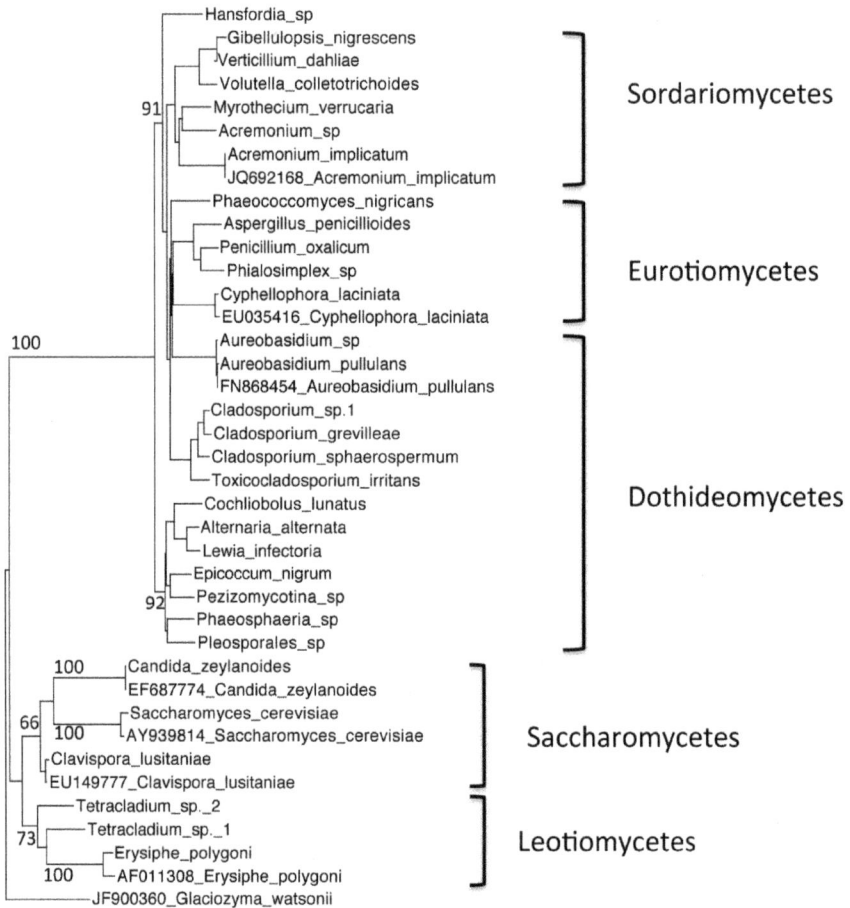

Figure 3. Phylogenetic tree of Warren Cave species belonging to the phylum Basidiomycota, obtained by neighbor-joining analysis of the Internal Transcribed Spacer (ITS) region of the 5.8S rDNA gene, with 100 full heuristic replications. Bootstrap values are as indicated on the tree. *Acremonium sp.* sequence obtained from GenBank was used as the outgroup, with the GenBank accession numbers listed. *Ustilago tritici, Glaciozyma watsonii, Rhodotorula mucilaginosa, Ganoderma applanatum, Malassezia globosa, Cystofilobasidium macerans* and *Filobasidium floriforme* sequences obtained from GenBank were used as closest relative reference sequences, with GenBank accession numbers listed.

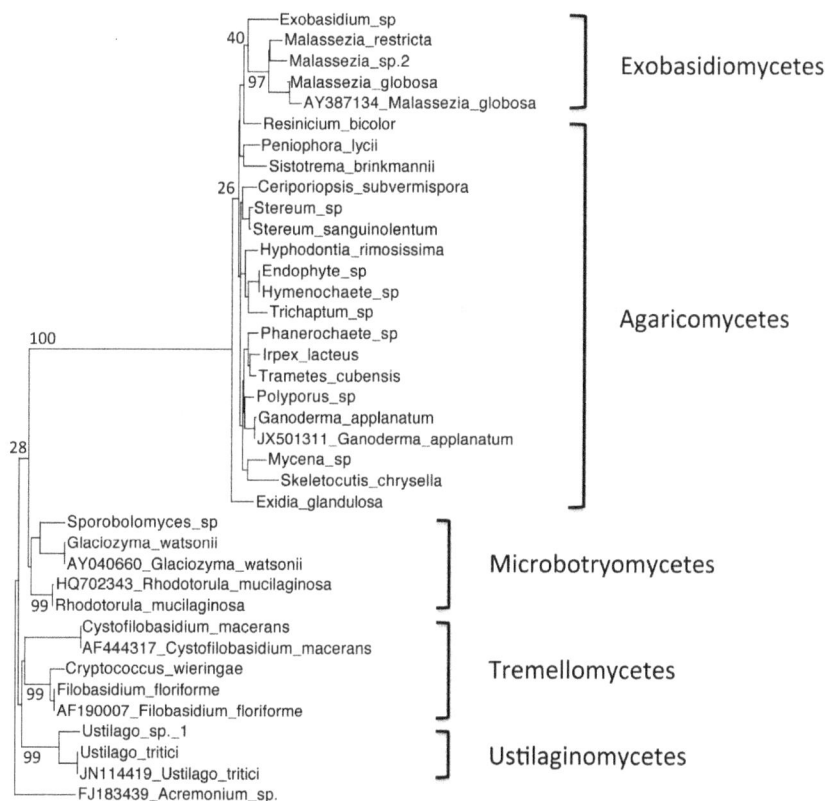

Aureobasidium pullulans a yeast-like fungus, *Aspergillus penicillioides* and *Alternaria alternata* were found to be the dominating members of the Ascomycota (Figure 4a) from the Warren Cave fungal community. All of these species are cosmopolitan and have been isolated numerous times from the Antarctic [20]. The most dominant Basidiomycota taxa found (Figure 4b) were *Malassezia* sp. (37% of the clones), yeasts most typically found associated with animals. Although *Malassezia* has also been identified from a clone library originating from McMurdo Dry Valleys desert soils of Taylor Valley, Antarctica (near a highly used pathway) [13] two of the species found in this study (*M. globosa* and *M. restricta*) are know to require lipids for growth and are common in human dandruff and seborrheic dermatitis [21]. This high proportion of human associated yeasts represented in the Warren Cave clone libraries suggests human contamination of the site. The second most abundant taxa of the Basidomycota was a *Peniophora* species, a member of a genus most known for wood rot [22].

Figure 4. The proportion of the number of fungal clones in this study by class belonging to (**a**) Ascomycota and (**b**) Basidiomycota.

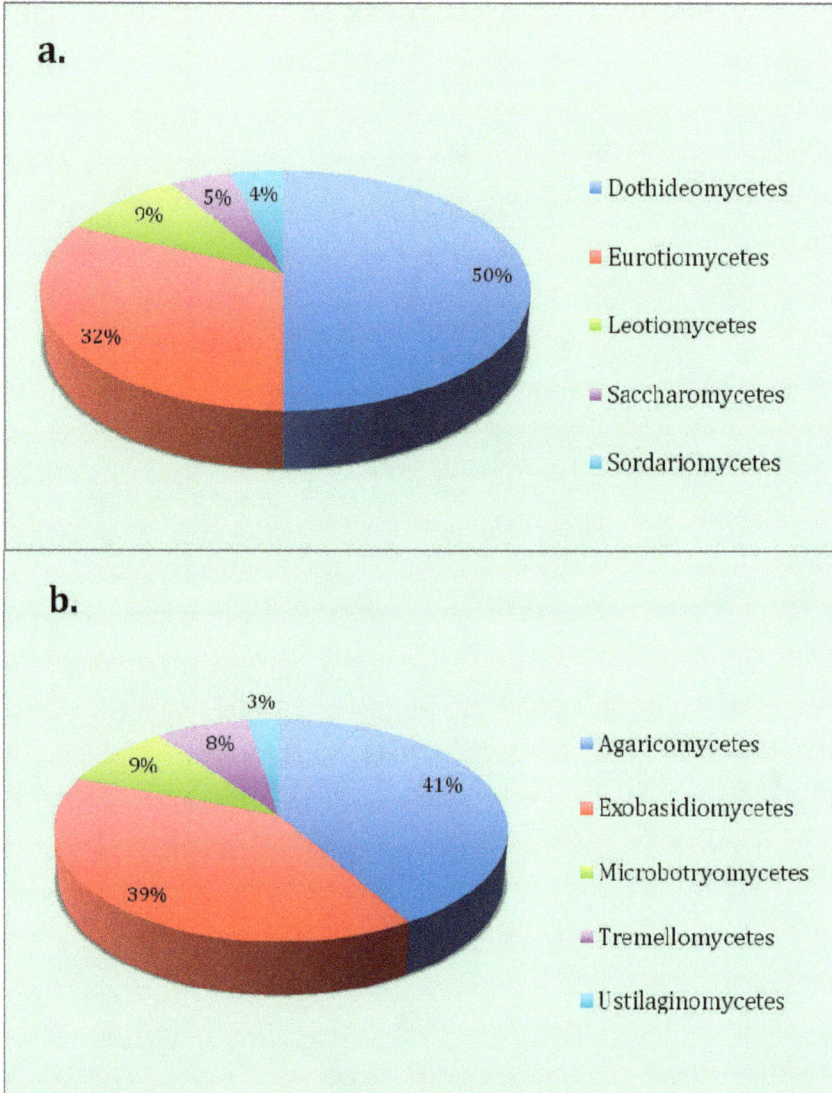

Although several fungal species found in Warren Cave are cosmopolitan, some have been shown to be capable of efficient colonization on minerals and sterile soil [23], abilities that enhance the probability that these organisms are active members of the DOVE community. For example *Irpex lactecus*, a white rot fungus, has been suggested as a potential bioremediation agent precisely because of its ability to utilize minerals and thrive in sterile soil. [24]. *Aureobasidium pullulans* has been isolated often from Antarctica, including McMurdo Dry Valley soil and has been found on the inner part of the Chernobyl inner containment system [25]. Both of these species may be early colonizers of oligotrophic habitats and potentially native Antarctic DOVE inhabitants.

4. Conclusions

The Warren Cave fungal community appears to have been influenced by humans. The cave has been visited throughout the years by both researchers and casual visitors. Evidence of human involvement in Warren Cave is supported by the findings of several *Malassezia* species. Finding species in a clone library alone does not necessary mean that they are active members of the community and more research will have to be carried out to determine their role, if any, in the Warren Cave soil community. As a result, it may not be surprising that Warren Cave is somewhat biologically compromised, and currently contains items that may impact and alter cave microbial communities (e.g., metal or bamboo stakes, instruments such as temperature recorders, batteries metal wires, *etc.*). It is clear that some of these practices profoundly impact the local microbial communities.

Even though Warren Cave has had human impact on its microbial community, other caves around the summit of Mt. Erebus remain untouched and therefore are potentially very valuable natural research laboratories for the study of DOVEs. The U.S. National Science Foundation has begun a procedure to develop a code of conduct for entering ice caves on Mt. Erebus. One of the first steps in this effort should be to determine which caves remain pristine and therefore are suitable candidates for microbial DOVE research.

Our data show that microbial communities in Mt. Erebus Ice Cave DOVEs contain diverse and specialized fungal communities that are likely to form a complex microbial foodweb that is independent from photosynthesis and primarily uses energy from chemolithotrophic metabolic processes. Very little is known about their function in these ecosystems, but several species are known to grow on mineral substrates in sterile soil supporting an active role. In addition, our study shows that a survey of fungi is a very sensitive indicator identifying the potential of human disturbance of these environments.

Acknowledgments

The authors would like to thank project co-PI Brad Tebo for insightful discussions and Rick Davis, Katherine Earle, Caleb Slemmons and Sarah Turner for their laboratory assistance, as well as Raytheon Polar Support Service, and PHI for logistical and laboratory support while in Antarctica. In addition, the authors give sincere thanks to the anonymous reviews for help in improving this manuscript. Funding was provided for this project by NSF Office of Polar Programs award #ANT-0739696 to the authors.

References

1. Edwards, K.J.; Becker, K.; Colwell, F. The deep, dark energy biosphere: Intraterrestrial life on earth. *Ann. Rev. Earth Planet. Sci.* **2012**, *40*, 551–568.
2. Staudigel, H.; Furnes, H.; McLoughlin, N.; Banerjee, N.R.; Connell, L.B.; Templeton, A.S. 3.5 billion years of glass bioalteration: Volcanic rocks as a basis for microbial life? *Earth Sci. Rev.* **2008**, *89*, 156–176.
3. Oppenheimer, C.; Kyle, P. Volcanology of erebus volcano, antarctica. *Spec. Issue J. Volcanol. Geotherm. Res.* **2008**, *177*, v–vii.
4. Curtis, A.; Kyle, P. Geothermal point sources identified in a fumarolic ice cave on erebus volcano, antarctica using fiber optic distributed temperature sensing. *Geophys. Res. Lett.* **2011**, *38*, doi:10.1029/2011GL048272.
5. Zimmerman, C.F.; Keefe, C.W.; Bshe, J. *Determination of Carbon and Nitrogen in Sediments and Particulates of Estuarine/Coastal Waters Using Elemental Analysis (Method 440.0) Revision 1.4*; National Exposure Research Laboratory: Cincinnati, OH, USA, 1997.
6. White, T.J.; Bruns, T.; Lee, S.; Taylor, J. Amplification and Direct Sequencing of Fungal Ribosomal RNA Genes for Phylogenetics. In *PCR Protocols: A Guide to Methods and Applications*; Innis, M., Gelfand, J., Sninsky, J., White, T.J., Eds.; Academic Press: Orlando, FL, USA, 1990; pp. 315–322.
7. Nilsson, R.H.; Abarenkov, K.; Veldre, V.; Nylinder, S.; de Wit, P.; Brosché, S.; Alfredsson, J.F.; Ryberg, M.; Kristiansson, E. An open source chimera checker for the fungal its region. *Mol. Ecol. Resour.* **2010**, *10*, 1076–1081.
8. Fell, J.W.; Boekhout, T.; Fonseca, A.; Scorzetti, G.; Statzell-Tallman, A.C. Biodiversity and systematics of basidiomycetous yeasts as determined by large-subunit rdna d1/d2 domain sequence analysis. *Int. J. Syst. Evolut. Microbiol.* **2000**, *50*, 1351–1371.
9. Edgar, R.C. Muscle: Multiple sequence alignment with high accuracy and high throughput. *Nucleic Acids Res.* **2004**, *32*, 1792–1797.
10. Rousk, J.; Bååth, E.; Brookes, P.C.; Lauber, C.L.; Lozupone, C.; Caporaso, J.G.; Knight, R.; Fierer, N. Soil bacterial and fungal communities across a ph gradient in an arable soil. *ISME J.* **2010**, *4*, 1340–1151.
11. Connell, L.B.; Redman, R.S.; Craig, S.D.; Rodriguez, R.J. Distribution and abundance of fungi in the soils of taylor valley, antarctica. *Soil Biol. Biochem.* **2006**, *38*, 3083–3094.
12. Connell, L.B.; Redman, R.S.; Craig, S.D.; Scorzetti, G.; Iszard, M.; Rodriguez, R.J. Diversity of soil yeasts isolated from south victoria land, antarctica. *Microb. Ecol.* **2008**, *56*, 448–459.
13. Fell, J.W.; Scorzetti, G.; Connell, L.B.; Craig, S.D. Biodiversity of micro-eukaryotes in antarctic dry valley soil with <5% soil moisture. *Soil Biol. Biochem.* **2006**, *38*, 3107–3119.
14. Adams, B.; Wall, D.; Gozel, U.; Dillman, A.R.; Chaston, J.M.; Hogg, I.D. The southernmost worm, *scottnema lindsayae* (nematoda): Diversity, dispersal and ecological stability. *Polar Biol.* **2006**, *30*, 809–815.
15. Fritsen, C.H.; Grue, A.; Priscu, J.C. Distribution of organic carbon and nitrogen in surface soils in the mcmurdo dry valleys, antarctica. *Polar Biol.* **2000**, *23*, 121–128.

16. Onofri, S.; Selbmann, L.; de Hoog, G.S.; Grube, M.; Barreca, D.; Ruisi, S.; Zucconi, L. Evolution and adaptation of fungi at boundries of life. *Adv. Space Res.* **2007**, *40*, 1657–16664.

17. Vishniac, H.S. A multivariate analysis of soil yeasts isolated from a latitudinal gradient. *Microb. Ecol.* **2006**, *52*, 90–103.

18. Thomas-Hall, S.R. Phylogenetic studies of fungi: Part a. Physiological and biochemical analysis of novel yeast species from antarctica. Ph.D. thesis, University of New England, Armidale, New South Wales, Australia, 2004.

19. Turchetti, B.; Thomas-Hall, S.R.; Connell, L.B.; Branda, E.; Buzzini, P.; Theelen, B.; Müller, W.H.; Boekhou, T. Psychrophilic yeasts from antarctica and european glaciers. Description of *glaciozyma* gen. Nov., *glaciozyma martinii* sp. Nov and *glaciozyma watsonii* sp. Nov. *Extremophiles* **2011**, *15*, 573–586.

20. Onofri, S.; Zucconi, L.; Tosi, S. *Continental Antarctic Fungi*; IHW-Verlag: Müchen, Germany, 2007; p. 247.

21. Batra, R.; Boekhout, T.; Gueho, E.; Cabanes, F.J.; Dawson, T.L., Jr.; Gupta, A.K. Malassezia baillon, emerging clinical yeasts. *FEMS Yeast Res.* **2005**, *5*, 1101–1113.

22. Kirk, P.M.; Cannon, P.F.; Minter, D.W.; Stalpers, J.A. *Dictionary of the Fungi*, 10th ed.; CABI Europe: Wallingford, UK, 2008; p. 507.

23. Gueidan, C.; Ruibal, C.; de Hoog, G.S.; Schneider, H. Rock-inhabiting fungi originated during periods of dry climate in the late devonian and middle triassic. *Fungal Biol.* **2011**, *115*, 987–996.

24. Novotny, C.; Erbanova, P.; Cajthaml, T.; Rothschild, N.; Dosoretz, C.; Sasek, V. *Irpex lacteus*, a white rot fungus applicable to water and soil bioremediation. *Appl. Microbiol. Biotechnol.* **2000**, *54*, 850–853.

25. Zhdanova, N.N.; Zakharchenko, V.A.; Vember, V.V.; Nakonechnaya, L.T. Fungi from chernobyl: Mycobiota of the inner regions of the containment structures of the damaged nuclear reactor. *Mycol. Res.* **2000**, *104*, 1421–1426.

Characterizing Microbial Diversity and the Potential for Metabolic Function at −15 °C in the Basal Ice of Taylor Glacier, Antarctica

Shawn M. Doyle, Scott N. Montross, Mark L. Skidmore and Brent C. Christner

Abstract: Measurement of gases entrapped in clean ice from basal portions of the Taylor Glacier, Antarctica, revealed that CO_2 ranged from 229 to 328 ppmv and O_2 was near 20% of the gas volume. In contrast, vertically adjacent sections of the sediment laden basal ice contained much higher concentrations of CO_2 (60,000 to 325,000 ppmv), whereas O_2 represented 4 to 18% of the total gas volume. The deviation in gas composition from atmospheric values occurred concurrently with increased microbial cell concentrations in the basal ice profile, suggesting that *in situ* microbial processes (*i.e.*, aerobic respiration) may have altered the entrapped gas composition. Molecular characterization of 16S rRNA genes amplified from samples of the basal ice indicated a low diversity of bacteria, and most of the sequences characterized (87%) were affiliated with the phylum, Firmicutes. The most abundant phylotypes in libraries from ice horizons with elevated CO_2 and depleted O_2 concentrations were related to the genus *Paenisporosarcina*, and 28 isolates from this genus were obtained by enrichment culturing. Metabolic experiments with *Paenisporosarcina* sp. TG14 revealed its capacity to conduct macromolecular synthesis when frozen in water derived from melted basal ice samples and incubated at −15 °C. The results support the hypothesis that the basal ice of glaciers and ice sheets are cryospheric habitats harboring bacteria with the physiological capacity to remain metabolically active and biogeochemically cycle elements within the subglacial environment.

Reprinted from *Biology*. Cite as: Doyle, S.M.; Montross, S.N.; Skidmore, M.L.; Christner, B.C. Characterizing Microbial Diversity and the Potential for Metabolic Function at −15 °C in the Basal Ice of Taylor Glacier, Antarctica. *Biology* **2013**, *2*, 1034-1053.

1. Introduction

During freezing, soluble and insoluble impurities (solutes, microbes, particles and gases) are physically excluded from the ice crystal lattice and concentrated into saline veins of liquid water found at the interface between ice crystals [1]. Despite the presence of liquid water, ice veins are environments in which microorganisms must endure physiochemical stresses such as low water activity, low pH, and ice recrystallization, as well as the biochemical challenges associated with low temperatures (e.g., reduced enzymatic activity and decreased membrane fluidity) [2]. There are two mechanisms by which microorganisms can be incorporated in glacial ice: aeolian deposition at the surface and entrainment of sediments in the basal zone [3,4]. Once entrapped in the ice, the long-term survival of a microbial population is constrained by their capability to endure the genetic and cellular damage that would accumulate in the absence of a functional metabolism. Damage to cellular macromolecules can be caused by a variety of physical and chemical mechanisms, including natural background ionizing radiation (e.g., produced from the decay of ^{40}K, ^{232}Th and

[238]U), L-amino acid racemization and spontaneous hydrolysis or oxidation of DNA [2,5]. As such, metabolically dormant microbial populations that remain frozen for extended periods of time would eventually accumulate a lethal amount of damage [5]. However, microorganisms with the capability to maintain a low level of metabolism requisite for mitigating genetic and cellular damage could theoretically persist, as long as suitable redox couples and nutrients were available to support their metabolic activity. Hence, the discovery of viable microbes persisting in ancient ice and permafrost [6–12] gives credence to the hypothesis that certain microorganisms are actively maintaining their cellular integrity under these conditions.

Analysis of gases entrapped in ice cores of glacial and basal ice from Antarctica (Siple Dome, Vostok), Greenland (North Greenland Ice Core Project, Greenland Ice Core Project) and South America (Sajama ice cap, Bolivia) have found concentrations of N_2O, CO_2 and CH_4 and stable isotopic compositions of N_2O and CH_4 [13–18] that do not correspond to atmospheric values. Microbial processes, such as nitrification and methanogenesis, are plausible explanations for the low $\delta^{18}O$-N_2O and $\delta^{13}C$-CH_4 values, respectively [17,19,20]. In support of this, laboratory studies have shown that microorganisms remain metabolically active at subzero temperatures, including respiration at temperatures of -33 °C [21] and -39 °C [22] and macromolecular synthesis at -15 °C [5,23,24]. Nevertheless, there are few data on the nature and constraints of *in situ* microbial activity in natural icy systems, and knowledge of subzero microbial physiology and its role in subglacial biogeochemical cycling is limited. Here, we present results from an investigation of microbial assemblages within basal ice horizons of Taylor Glacier, Antarctica. Basal ice is found in the deepest layers of a glacier and has a chemistry and physical structure that is directly affected by its proximity to the glacier bed [25]. Sedimentary debris becomes entrained in the ice at the basal zone, together with viable microorganisms and substrates suitable as energy and nutrient sources, which may create unique habitats within the ice [12,26–28]. The specific aim of this research was to investigate the potential for basal ice to serve as a microbial habitat, with the implication that microorganisms are ultimately responsible for the unusual concentration of gasses (e.g., CO_2 and O_2) found entrapped in these icy environments. Our data on active biogeochemical processes in the basal zone of Taylor Glacier is discussed in the broader context of polar ice sheets and potential habitats for life in icy extraterrestrial frozen environments.

2. Methods

2.1. Site Information and Field Sampling

Taylor Glacier is a 54 km outflow glacier of the East Antarctic Ice Sheet and is located at the western end of Taylor Valley in the McMurdo Dry Valleys of Victoria Land, terminating on the western shore of Lake Bonney (Figure 1A). During the austral summers of 2007 and 2009, two tunnels were excavated into the northern margin of Taylor Glacier to directly access a stratigraphic sequence of basal ice that was largely free of folding or distortions found in horizons at the margin. The tunnels were initiated on fresh ice aprons and extended 7–9 m in from the ice margin. In 2007, a vertical shaft (~5 m) was constructed at the end of the tunnel, and a 4 m vertical profile of basal ice was sampled. Three distinct basal ice facies were identified using the nomenclature of

Hubbard *et al.* [29]: (i) clean ice, containing <1 g L^{-1} debris; (ii) banded dispersed ice, containing debris up to 38% w/v; and (iii) solid ice, which is heavily debris laden (up to 60% w/v). The 4 m vertical profile collected contained ice from all three facies (Figure 1B), and the top of the sample profile (*i.e.*, access tunnel floor) was designated as the zero depth. Sample ice blocks measuring approximately 20 × 20 × 10 cm were cut using electric chainsaws with carbide tipped chains. During the 2009 season, a new access tunnel was excavated to directly intersect a layer of debris-rich banded dispersed ice (Figure 1C), and 27 large (40 × 30 × 15 cm) blocks of banded ice were collected. Temperature loggers deployed in the basal ice during the 2007 season indicated an ice temperature of −15 °C. All ice samples were shipped frozen to Montana State University and Louisiana State University and stored at −20 °C. Gas measurements for the ice samples are described in Montross [30] and Montross *et al.* [31].

2.2. Ice Decontamination and Sampling

The debris-free ice was subsequently cut using a band saw, and samples of the sediment-laden ice were cut using a masonry saw equipped with a diamond blade. The ice samples were handled with sterile stainless steel forceps and decontaminated in a class 100 laminar flow hood housed within a −5 °C freezer. The surface contaminated outer portion of the ice was removed based on a method developed by Christner *et al.* [32] for sampling deep ice cores recovered in boreholes containing hydrocarbon-based drilling fluids. The outermost surface of the ice sample was washed with 0.22 μm filtered 95% ethanol that was equilibrated to −5 °C. Samples were then rinsed with ice-cold 0.22 μm filtered, twice-autoclaved deionized water until an estimated minimum of ten millimeters of the outer sample surface had been removed. Sterile forceps were used to hold the samples during washing and were exchanged frequently to prevent carryover contamination. All samples were weighed before and after decontamination, and the decontamination method reduced the total ice mass of each sample by 15% to 25%. The cleaned samples were placed in sterile containers and melted at 4 °C (typically 16 h to 24 h).

2.3. Microbial Cell Density

Sections of both the clean ice and banded dispersed ice from the 2007 sample profile were selected for microscopic cell counts. Within a −5 °C freezer, a profile of the clean ice (depth 60–80 cm) was cut and sampled at a vertical resolution of 5 cm; the banded dispersed ice (depth 220–240 cm) was sampled at a vertical resolution of 2.5 cm. The ice samples were subsequently decontaminated and melted as described above (Section 2.2).

Figure 1. (**A**) Map and aerial photograph of Taylor Glacier, located in the McMurdo Dry Valleys of Victoria Land, Antarctica. The location of the 2007 and 2009 access tunnels are indicated. (**B**) Schematic of the 4 m deep basal ice profile sampled from Taylor Glacier in 2007. The top of the profile (designated 0 cm) was located in debris-poor clean ice, which was underlain by several layers of both debris-rich banded dispersed and laminated solid ice with a thick layer of basal solid ice as the lowermost unit. (**C**) Schematic of the debris-rich banded dispersed basal ice horizon sampled in 2009.

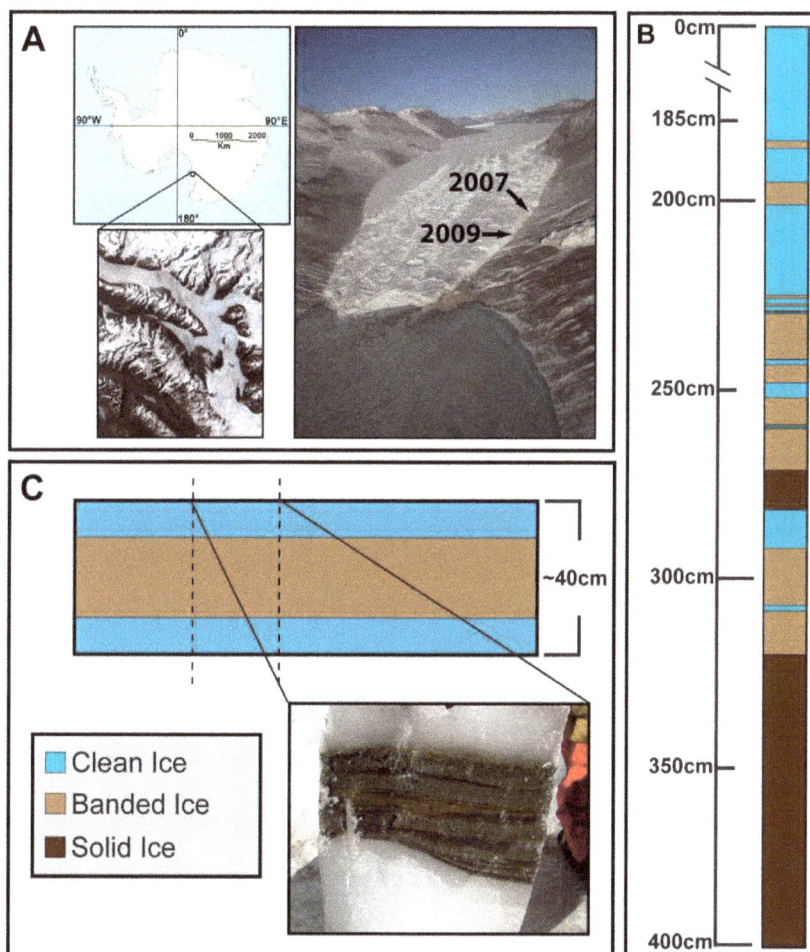

For enumeration, microbial cells attached to sediment particles were liberated from the solid phase with a modification of the method described by Trevors and Cook [33]. Nine milliliters of the sediment-melt water slurry was amended with 1 mL of a 1% (w/v) solution of $Na_4P_2O_7$ (pH 7.0), shaken at 200 rpm for 1 h at 4 °C and allowed to settle for 30 min. The supernatant was collected, and the cells within were fixed with sodium borate-buffered formalin (5% final concentration), stained with 2× SYBR Gold (Invitrogen) and filtered onto black polycarbonate 0.22 μm pore filters

(GE Water & Process Technologies). Identical samples that did not contain the formalin fixative were also prepared, stained with Baclight (Invitrogen) and filtered within 6 h after melting. The filters were mounted on glass slides with a glass coverslip using two drops of antifade solution and stored in the dark at 4 °C until counted. The antifade solution consisted of 90 mM p-phenylenediamine and 45% glycerol in phosphate buffered saline and was filtered through a 0.45 μm filter. Fifty random fields (field of view: 41,500 μm²) were counted using an Olympus BX51 epifluorescence microscope and a FITC filter cube (excitation from 455 to 500 nm and emission from 510 to 560 nm). Cell density estimates were calculated based on the average number of cells per field and normalized per gram of ice. The sediment content for each sample was determined by measuring the dry weight of sediment per gram of basal ice.

2.4. Enrichment and Isolate Culturing

Meltwater from the debris-rich banded dispersed ice in the 2007 sample profile (depth 195 cm to 200 cm; Figure 1) was vortexed for 1 min, and 100 μL of the slurry was spread plated on R2A (Difco), 10% R2A, 1% R2A, marine agar 2216 (Difco) and M9 minimal salts media (supplemented 20 mM glucose, acetate or pyruvate) in triplicate. The plates were incubated at 4, 10, 22 and 37 °C in the dark and examined daily for 60, 30, 15 and 7 days, respectively. Blank media controls were prepared and incubated in parallel with inoculated samples. Additional isolates from ice samples collected from a tunnel constructed at Taylor Glacier in 1999 were made available for this investigation; details of the tunnel location and physical and chemical properties of the ice are described in Samyn et al. [34]. Growth at 5, 15 and 22 °C was measured via optical density (620 nm) in marine broth 2216 (Difco) to determine the approximate optimal growth temperature of each isolate. Pasteurization of the melted ice was performed by heating at 80 °C for 10 min, followed by spread plating 100 μL of the sample on marine agar 2216 (Difco) in triplicate. The cultures were incubated aerobically at 22 °C, and the number of colony-forming units (CFU) was quantified and compared to control samples. Marine agar 2216 (Difco) consistently yielded the highest CFU mL^{-1} from samples, and therefore, was used for this assay.

Salt tolerance of select isolates was examined by culturing in marine broth 2216 (Difco) supplemented with up to 10% (w/v) of NaCl (intervals of 2% NaCl). Optical density (OD$_{620\ nm}$) of the cultures was monitored at 10 °C over two weeks using a NanoDrop spectrophotometer.

2.5. Molecular Analysis of Bacterial 16S rRNA Genes

Genomic DNA was extracted from the banded dispersed basal ice facies recovered in 2007 and 2009. For the sample from the 2007 profile, an entire sample block (20 × 20 × 10 cm, profile depth 220–240 cm (Figure 2)) was decontaminated and melted at 4 °C. The resulting meltwater was vigorously shaken (300 rpm) to achieve a homogenous sediment-meltwater slurry, 15 mL of which was centrifuged (4,500 × g; 10 min; 4 °C), and total DNA was extracted from 0.5 g of the sediment pellet using a MoBio PowerSoil DNA extraction kit, as per the manufacturer's instructions. For samples collected in 2009, ~132 kg of basal ice was selected for filter concentration prior to shipment back to the United States from Antarctica. After decontamination, the ice was placed at

4 °C in sterilized polypropylene containers and allowed to melt. Complete melting of this basal ice took place over a period of seven days, wherein the meltwater was concentrated onto filters. In order to remove larger sediment particles, the sediment-meltwater slurry was filtered consecutively through a series of five sterilized nylon monofilament filters of decreasing pore size (100, 75, 50, 25 and 10 µm) and, then, centrifuged at 700 × g (10 min; 4 °C) [35]. The supernatant (~90 L) was filtered at 4 °C under a 20 cm Hg vacuum onto eleven 90 mm, 0.22 µm Supor-200 filters (Pall Corporation). The filters were frozen at −80 °C and shipped to Louisiana State University for storage and analysis. DNA was extracted from one of the 90 mm filters through which 5.1 L of meltwater had been filtered using a MoBio PowerMax soil DNA extraction kit.

Figure 2. Analysis of the microbial cell density (**A**), concentrations of O_2 and CO_2 (**B**) and sediment content (**C**) throughout a vertical profile of banded dispersed basal ice (**D**). Error bars represent the standard error of the direct cell counts.

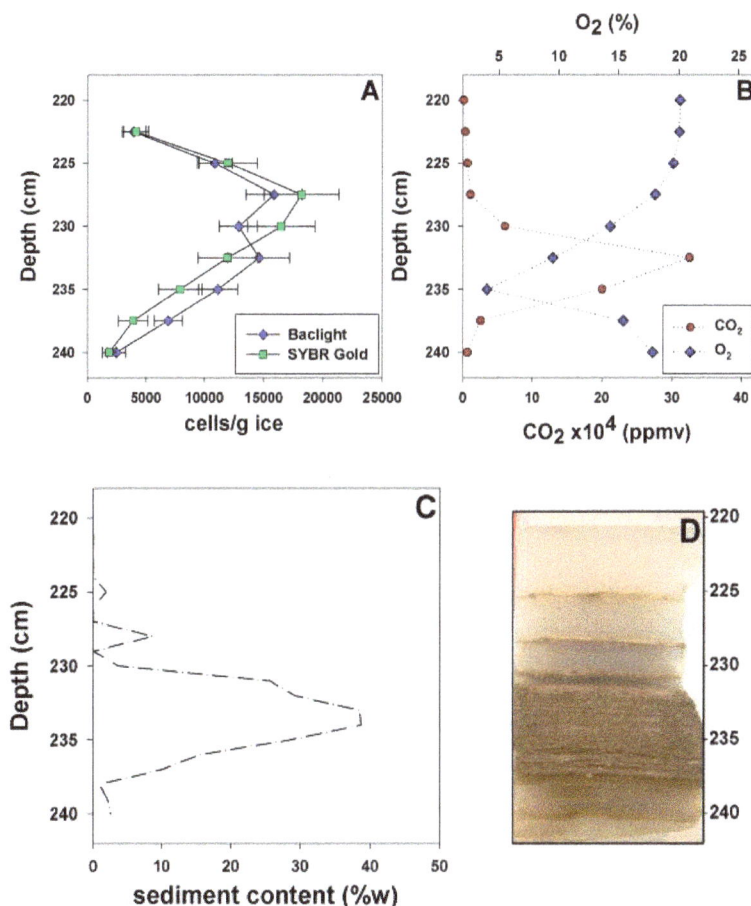

A portion of the 16S rRNA gene was amplified from the extracted genomic DNA samples using the primers, 27F (5'-AGAGTTTGATCCTGGCTCAG-3') and 1492R (5'-

GGTTACCTTGTTACGACTT-3') [36]. For the 2009 gDNA extracted from banded dispersed ice, the 50 μL PCR reaction contained 1.0 unit of *Taq* DNA polymerase (5PRIME), 1× Master*Taq* buffer, 1× *Taq*Master PCR enhancer, 1.5 mM Mg(C$_2$H$_3$O$_2$)$_2$, 15 pmol of each primer, 200 μM deoxynucleotide triphosphates (dNTPs), and ~100 pg of template DNA. Thirty cycles of PCR were done with a 45 s denaturation step at 94 °C, 60 s annealing step at 50.8 °C and extension at 72 °C for 60 s, followed by a final extension at 72 °C for 10 min. The 50 μL PCR reaction for the 2007 gDNA sample contained 5.0 units of AmpliTaq Gold DNA polymerase, LD (Invitrogen), 1× PCR Gold buffer, 3.5 mM Mg(C$_2$H$_3$O$_2$)$_2$, 30 pmol of each primer, 200 μM dNTPs and ~100 pg of template DNA. Forty-three cycles of time-release PCR were done with a 45 s denaturation step at 94 °C, 60 s annealing step at 50.8 °C and extension at 72 °C for 60 s, followed by a final extension at 72 °C for 10 min.

The PCR products obtained were examined by agarose gel electrophoresis, purified by ethanol precipitation and ligated into the pGEM T-Easy plasmid (Promega). Alpha complementation was used to identify clones containing inserts, and inserts of the predicted size were confirmed by PCR with primers that annealed to the flanking SP6 and T7 regions of the vector. Each clone was cultured in Luria Bertani medium amended with 100 μg mL^{-1} of ampicillin, and plasmid DNA was purified using the Qiagen MiniPrep kit, quantified on a Nanospec spectrophotometer and sequenced using BigDye Terminator (v. 3.1; Invitrogen) on an ABI 3130XL Genetic Analyzer (Applied Biosystems). The forward and reverse sequencing reads were manually trimmed to remove flanking vector and primer sequences and aligned with BioEdit software. The compiled sequences were aligned with SINA (v. 1.2.11) [37] using the SILVA reference database (release 113) [38]. Phylogenetic classification, diversity estimation and rarefaction were performed in MOTHUR [39]. For the diversity estimation, operational taxonomic units (OTUs) were identified at a genetic distance of 3%. The UCHIME algorithm [40] was used in MOTHUR to identify potential chimeric sequences, which were discarded from the analysis. Maximum likelihood phylogenetic trees were constructed in MEGA5 [41].

A total of 25 isolates were chosen for 16S rRNA gene sequencing based on differences in colony morphology, pigmentation, growth temperature and media of isolation (Table 1). Genomic DNA was extracted from each isolate using the UltraClean Microbial DNA isolation kit (MoBio Laboratories). Bacterial 16S rRNA genes were amplified from the genomic DNA by PCR using the primers, 27F and 1492R [36]. The 50 μL PCR reactions contained 1.0 units of MasterTaq DNA polymerase (5 PRIME), 1× Taq buffer, 1× TaqMaster PCR enhancer, 1.5 mM Mg (C$_2$H$_3$O$_2$)$_2$, 15 pmol of each primer, 200 μM dNTPs and ~300 ng of template DNA. Thirty cycles of amplification were performed with denaturation for 60 s at 96 °C, annealing at 50.8 °C for 1 min, extension at 72 °C for 2 min and a final extension for 10 min at 72 °C. PCR products of the expected length (≈1,500 bp) were purified by ethanol precipitation and sequenced as described above. Taxonomic assignments were performed using EzTaxon-E [42].

The DNA sequences obtained in this study were deposited in the GenBank database under accession numbers, KC777190 to KC777289.

Table 1. Phenotypic description of basal ice isolates and their phylogenetic relationships to cultured bacteria. N.D. = not determined.

Isolate	Closest Relative	% Identity	Sequence Length (bp)	Isolation Media	Isolation temp	Optimal temp (±5 °C)	Halotolerance (% NaCl)	Description
TG14	*Paenisporosarcina antarctica* N-05	99	1,379	R2A	10 °C	15 °C	~7.9% (w/v)	bright yellow, circular, flat
TG24	*Paenisporosarcina antarctica* N-05	99	1,363	marine	10 °C	15 °C	~9.9% (w/v)	tan, circular, convex
TG27	*Paenisporosarcina antarctica* N-05	99	1,366	M9 glucose	4 °C	15 °C	~7.9% (w/v)	off-white, shiny, convex
TG29	*Paenisporosarcina antarctica* N-05	99	1,364	M9 glucose	4 °C	22 °C	N.D.	yellow, shiny, convex
TG30	*Paenisporosarcina antarctica* N-05	99	1,369	M9 pyruvate	4 °C	22 °C	N.D.	white, shiny, convex
TG32	*Paenisporosarcina antarctica* N-05	99	1,364	10% R2A	4 °C	22 °C	N.D.	off-white, circular, shiny
TG25	*Paenisporosarcina antarctica* N-05	99	1,363	marine	10 °C	15 °C	N.D.	cream-yellow, shiny, convex
TG34	*Paenisporosarcina antarctica* N-05	100	1,368	1% R2A	4 °C	22 °C	N.D.	off-white, circular
TG21	*Paenisporosarcina macmurdoensis* CMS21w	99	1,367	marine	22 °C	22 °C	~7.9% (w/v)	bright yellow, rough
TG3	*Paenisporosarcina macmurdoensis* CMS21w	99	1,375	R2A	22 °C	22 °C	~5.9% (w/v)	off-white center, mucoid
TG6	*Paenisporosarcina macmurdoensis* CMS21w	99	1,382	1% R2A	22 °C	22 °C	~5.9% (w/v)	white-yellow, mucoid
TG7	*Paenisporosarcina macmurdoensis* CMS21w	99	1,387	1% R2A	22 °C	22 °C	N.D.	off-white, mucoid
TG18	*Paenisporosarcina macmurdoensis* CMS21w	99	1,349	R2A	22 °C	22 °C	N.D.	off-white, mucoid
TG11	*Paenisporosarcina macmurdoensis* CMS21w	99	1,373	R2A	22 °C	15 °C	N.D.	cream yellow, circular
TG2	*Paenisporosarcina macmurdoensis* CMS21w	99	1,382	R2A	22 °C	22 °C	N.D.	cream yellow, mucoid
TG15	*Paenisporosarcina macmurdoensis* CMS21w	99	1,376	R2A	10 °C	22 °C	N.D.	yellow, mucoid
TG17	*Paenisporosarcina macmurdoensis* CMS21w	99	1,377	R2A	10 °C	22 °C	N.D.	white-yellow, convex, mucoid
TG26	*Paenisporosarcina macmurdoensis* CMS21w	99	1,369	marine	10 °C	22 °C	N.D.	off-white, mucoid
TG19	*Paenisporosarcina indica*	99	1,369	marine	22 °C	22 °C	~7.9% (w/v)	dark brown, convex
TG9	*Paenisporosarcina indica*	99	1,372	R2A	22 °C	22 °C	~7.9% (w/v)	dull yellow, translucent, flat, mucoid
TG20	*Paenisporosarcina indica*	99	1,386	marine	22 °C	22 °C	N.D.	cream-yellow, convex
TG39	*Paenisporosarcina indica*	99	1,370	marine	4 °C	22 °C	N.D.	white, shiny, convex
TG10	*Paenisporosarcina quisquiliarum* SK 55	99	1,383	R2A	22 °C	22 °C	~5.9% (w/v)	off-white, shiny, convex
TG8	*Bacillus humi* LMG18435	97	1,367	1% R2A	22 °C	22 °C	~5.9% (w/v)	tan with brown center, circular
TG4	*Paraliobacillus quinghaiensis* YIMC158	99	1,418	R2A	22 °C	15 °C	~11.9% (w/v)	yellow, convex, rough

2.6. DNA and Protein Synthesis of Isolated Bacteria at −15 °C

Measurement of DNA and protein synthesis by cells frozen at −15 °C was carried out based on the procedure described by Christner [23]. Cultures (50 mL) of *Paenisporosarcina* sp. TG14 were grown aerobically (200 rpm) at 15 °C in marine broth 2216 (Difco). Cells were harvested from mid-exponential phase cultures by centrifugation (10 min, 4,500 × g). The harvested cells were then suspended in 50 mL of Taylor Glacier melt water, centrifuged, and suspended in 50 mL melt water at a concentration of 3.1 × 10^6 CFU mL^{-1}. Experiments were conducted using melt water from both debris-poor clean ice and debris-rich banded dispersed ice. The samples used for these experiments were taken adjacent to the main vertical sampling profile at sample depths of approximately ~100 cm depth for the clean ice and ~300 cm for the banded dispersed ice. Due to their opaque nature, sediment particles readily quench luminescence during liquid scintillation counting, and thus decrease measurement reliability. To mitigate this effect, coarse particles were allowed to settle from the samples for 18 h prior to harvesting melt water for these experiments. Aliquots (500 µL) of the cell suspension were amended with 1.7 µCi mL^{-1} of [^{3}H]-leucine (L-leucine [4,5-^{3}H], 84 Ci mmol^{-1} in ethanol:water 2:98; MP Biomedical) or 1.3 µCi mL^{-1} [^{3}H]-thymidine (thymidine [Methyl-^{3}H], 64 Ci mmol^{-1} in sterile water; MP Biomedical) to achieve a final concentration of 20 nM. Killed controls were amended with ice-cold 50% trichloroacetic acid (TCA) to a final concentration of 7% (w/v) 30 min prior to the addition of [^{3}H]-leucine or [^{3}H]-thymidine. All solutions were maintained on ice, and the samples were placed in a −80 °C freezer within 30s after the addition of either [^{3}H]-leucine or [^{3}H]-thymidine. After overnight (16 h) incubation at −80 °C, samples were transferred to a −15 °C thermally stable freezer (Revco ULT350-3-A32), which was designated as time zero. A HOBO U12 (Onset) data logger was used to log the temperature in the freezer every 10 min. Over the entire experimental time course, the mean temperature was −15.0 ± 0.5 °C. At each experimental time-point, frozen samples were removed from the freezer, immediately overlain with 100 µL of ice-cold 50% TCA, briefly centrifuged (10 s), and incubated at 4 °C to allow melting (final TCA concentration was 7%). After at least 30 min, the acid-insoluble macromolecules were pelleted by centrifugation at 17,000 × g for 15 min; the pellet was rinsed with 1 mL of ice-cold 5% TCA and centrifuged at 17,000 × g for 5 min. The residual was rinsed with 1 mL of ice-cold 70% ethanol, centrifuged at 17,000 × g for 5 min and the supernatant removed. The rinsed pellet was suspended in 1 mL of Cytoscint scintillation cocktail (Fisher, cat. no. BP458-4), and the radioactivity present was quantified using liquid scintillation spectrometry (Beckman LS6000IC scintillation counter). The number of disintegrations per minute (DPM) was calculated by determining counting efficiency using acetone-quenched standards of [^{3}H] toluene (American Radiolabeled Chemicals, cat# ARC182) in the Cytoscint cocktail. Incorporation rates per CFU were calculated over time and converted to molecules of substrate incorporated, per CFU, per day (molecules CFU^{-1} day^{-1}). Rate measurements were converted to grams of substrate carbon incorporated, per gram of cell carbon, per day (gC gC^{-1} day^{-1}) based on 65 fg C cell^{-1} [5]. A rectangular hyperbole (*i.e.*, one site saturation curve) was used as a best fit to model the data.

3. Results

3.1. Cell Concentration and Viability within Horizons of the Basal Ice

Direct counting using epifluorescent microscopy was used to quantify cell density in sampled horizons of the basal ice. The debris-poor clean ice horizons (2007 profile: 60 cm to 80 cm; selected as a representative piece of the clean ice) contained the lowest total cell concentrations of the entire basal ice profile: SYBR Gold and Baclight staining revealed counts ranging from 2.6 ± 0.2 to $4.9 \pm 0.4 \times 10^2$ cells g^{-1} ice (n = 4). A two-way ANOVA was conducted to test for differences in cell concentrations in the clean ice, and the data did not differ significantly between sampling depths within the 60–80 cm block [$F(3, 196) = 1.23$, $p = 0.30$]. In the debris-rich banded dispersed ice (profile depth 220 cm to 240 cm), total cell concentrations were approximately one to two orders of magnitude higher than the clean ice. SYBR Gold counts ranged from $1.8 \pm 0.1 \times 10^3$ cells g^{-1} to $1.8 \pm 0.4 \times 10^4$ cells g^{-1}, while Baclight counts ranged from $2.4 \pm 0.1 \times 10^3$ cells g^{-1} to $1.6 \pm 0.3 \times 10^4$ cells g^{-1} (Figure 2A); cell abundance estimates by the two methods were not significantly different ($\alpha = 0.05$). Cell concentration was positively correlated with sediment content [$r(14) = 0.60$, $p < 0.05$] and the concentration of CO_2 [$r(14) = 0.35$, $p < 0.10$; Figure 2]. In horizons with increased sediment content (profile depth 230–237.5 cm; Figure 2C), sediment content was correlated with CO_2 [$r(6) = 0.72$, $p < 0.025$] and O_2 [$r(6) = -0.79$, $p < 0.01$], whose concentrations were elevated and depleted, respectively, relative to atmospheric values [30,31]. Based on Baclight staining, cell viability in the banded dispersed ice was estimated at $73 \pm 9\%$, did not vary with sediment content, and was not significantly different ($\alpha = 0.05$) from the debris-poor clean ice ($78 \pm 5\%$).

3.2. Enrichment Culturing from the Basal Ice

Growth was observed on all media inoculated and incubated at 4, 10 and 22 °C, except M9 supplemented with acetate as a carbon source. No growth was observed at 37 °C. Enrichment culturing of heterotrophic bacteria at 22 °C with marine media (Difco 2216) in triplicate yielded an average of $9.7 \pm 1.5 \times 10^1$ CFU mL^{-1} in the banded dispersed ice melt-water from 195–200 cm, indicating that ca. 0.7% of the cells observed via SYBR Gold and Baclight staining were culturable under the conditions used. Pasteurization with duplicate melt-water samples was performed to determine if the cells cultured were in a vegetative state (cf., endospores) in the melted ice samples. Pasteurization of the meltwater reduced the recovery of viable cells by 96%. All of the isolates had an optimal growth temperature between 15 °C and 22 °C; moreover, all isolates were capable of growth at 4 °C (the lowest temperature tested). Of the isolates examined for halotolerance, all grew optimally in marine broth and had reduced growth rates when the marine broth was amended with additional NaCl. For the eleven isolates tested, the maximum salt concentration that supported growth was between ~5.9% and ~11.9% (w/v) NaCl (Table 1).

3.3. Phylogenetic Analysis of Clone Sequences and Isolates

Forty-three clones were sequenced from the clone library constructed from the 2009 banded dispersed basal ice sample, none of which possessed properties typical of chimeras. The 16S rRNA gene sequence clones in this library were phylogenetically related to the Firmicutes (77%) and the Gammaproteobacteria (Figure 3). The Firmicute-related clones were related to three genera: *Bacillus*, *Paenisporosarcina* and *Cohnella*, with *Bacillus*-related phylotypes being the most abundant (49%) in the library. The Gammaproteobacteria were represented by only two genera, *Acinetobacter* and *Psychrobacter*. For the 2007 banded dispersed ice gDNA, attempts at amplifying the 16S rRNA gene with MasterTaq as described above were not successful. However, following the manufacturer's recommendations for amplification of targets with low template concentrations, successful amplification of the 16S rRNA gene sequence was achieved using the AmpliTaq Gold DNA polymerase. Forty-one clones were sequenced from the 16S rRNA gene clone library constructed using DNA obtained from the 2007 banded dispersed basal ice sample (profile depth 220–240 cm), thirteen of which were discarded as potential chimeras or PCR artifacts. All the clones sequenced were related to members of four genera within the Firmicutes: *Paenibacillus*, *Paenisporosarcina*, *Bacillus* and *Jeotgalibacillus*. Sequences related to the *Paenisporosarcina* were the most commonly observed phylotype in the clone library, representing 65% of all the sequences characterized. Four clone sequences could not be classified beyond the phylum level and were, thus, listed as unclassified Firmicutes. Shannon diversity indices for both clone libraries were similar: 2007 H' = 1.81 ± 0.42; 2009 H' = 1.78 ± 0.22.

Figure 3. Taxonomic classification of partial 16S rRNA gene sequences amplified and cloned from the Taylor Glacier banded ice. Phylogenetic assignments were performed after 100 iterations using the naive Bayesian classifier [43] in MOTHUR with the SILVA database as a reference.

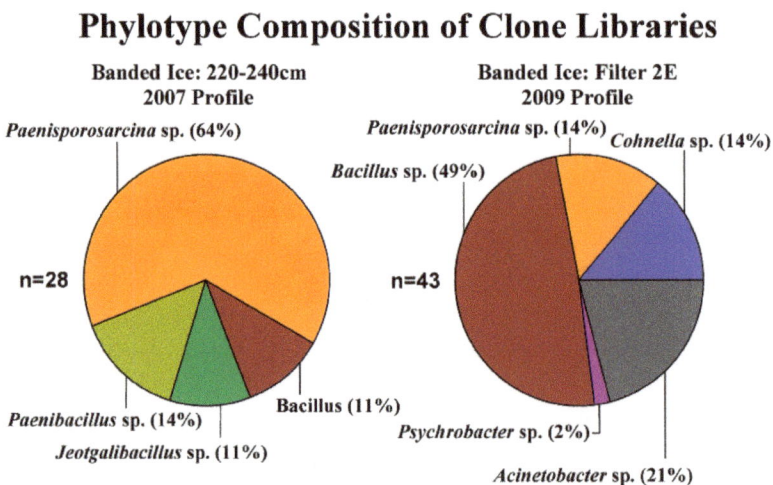

The cultured isolates belonged to three genera in the bacterial phylum, Firmicutes: *Paenisporosarcina*, *Bacillus* and *Paraliobacillus* (Table 1). Members of the *Paenisporosarcina* were the most abundant taxa cultured, representing 28 out of the 30 unique isolates characterized. The 16S rRNA gene sequences from the *Paenisporosarcina*-related isolates were similar ($\geq$97% identity) and clustered with sequences obtained in the clone libraries (Figure 4). Some of the clone sequences were 100% identical to those obtained from cultured isolates. Five of the isolates (shown in green in Figure 4) were recovered from Taylor Glacier basal ice samples obtained from the 1999 tunnel site [34].

3.4. Incorporation of Macromolecular Precursors in Melt water at −15 °C

Isolate *Paenisporosarcina* sp. TG14 had the fastest growth rate of the three isolates tested (TG14, TG19 and TG21) at 15 °C (3.0 h generation^{-1}) and was selected for a series of subzero metabolic assays to examine its physiological potential under conditions comparable to those in the basal ice. Cells of TG14 incorporated [^{3}H]-leucine and [^{3}H]-thymidine into acid-insoluble macromolecules when incubated at −15 °C in frozen melt water from either the clean or sediment-containing ice (Figure 5). During frozen incubation for 70 days in the clean ice melt water, TG14 incorporated an average of $1.5 \pm 0.2 \times 10^4$ and $4.0 \pm 0.2 \times 10^3$ molecules CFU^{-1} of [^{3}H]-leucine and [^{3}H]-thymidine, respectively. TG14 cells frozen in melt water from the sediment containing ice incorporated an average of $8.1 \pm 1.1 \times 10^3$ and $1.6 \pm 0.3 \times 10^3$ molecules CFU^{-1} of [^{3}H]-leucine and [^{3}H]-thymidine, respectively. For the clean ice samples, incorporation of [^{3}H]-leucine and [^{3}H]-thymidine was continuous and hyperbolic ($R^2 > 0.94$; $p < 0.0001$) over the entire time-course. For the banded dispersed ice melt water, incorporation of [^{3}H]-leucine was initially rapid, but high variability in the data resulted in a poor fit of the regression model ($\alpha = 0.05$); however, [^{3}H]-thymidine incorporation appeared to be partially hyperbolic ($R^2 = 0.41$; $p = 0.0455$; Figure 5). The maximum rates of [^{3}H]-leucine and [^{3}H]-thymidine incorporation in samples of the clean ice occurred during the first 90 h at $1.5 \pm 0.2 \times 10^3$ molecules CFU^{-1} day^{-1} and $2.3 \pm 0.9 \times 10^2$ molecules CFU^{-1} day^{-1}, respectively. For the banded dispersed ice meltwater incubation, the maximum rates of [^{3}H]-leucine and [^{3}H]-thymidine incorporation also occurred in the first 90 h at $2.5 \pm 0.9 \times 10^3$ molecules CFU^{-1} day^{-1} and $2.7 \pm 1.0 \times 10^2$ molecules cell^{-1} day^{-1}, respectively. There was not a significant difference between the maximum rates of [^{3}H]-leucine ($p > 0.38$) and [^{3}H]-thymidine ($p > 0.79$) incorporation measured in the clean ice and banded dispersed ice. However, cells in the clean ice incorporated significantly more [^{3}H]-leucine ($p = 0.049$) and [^{3}H]-thymidine ($p = 0.003$) than incubations conducted in the banded dispersed ice over 70 days.

372

Figure 4. Phylogenetic tree of *Paenisporosarcina*-related clone and isolate 16S rRNA gene sequences using a Maximum Likelihood method based on the Jukes-Cantor model. The partial 16S rRNA gene sequences, corresponding to nucleotides 108–1407 (*E. coli* numbering), were aligned to the SILVA reference database (v. 113). After filtering gaps, the final alignment was based on 1,288 nucleotides. Isolate sequences are shown in red and cloned sequences are shown in blue. Isolates recovered from the 1999 samples are shown in green. The bootstrap consensus tree inferred from 1,000 replicates is taken to represent the evolutionary history of the taxa analyzed [44]. Branches corresponding to partitions reproduced in less than 50% bootstrap replicates are collapsed. The percentage of replicate trees in which the associated taxa clustered together in the bootstrap test is shown next to the branches. The scale bar represents the number of changes per nucleotide position.

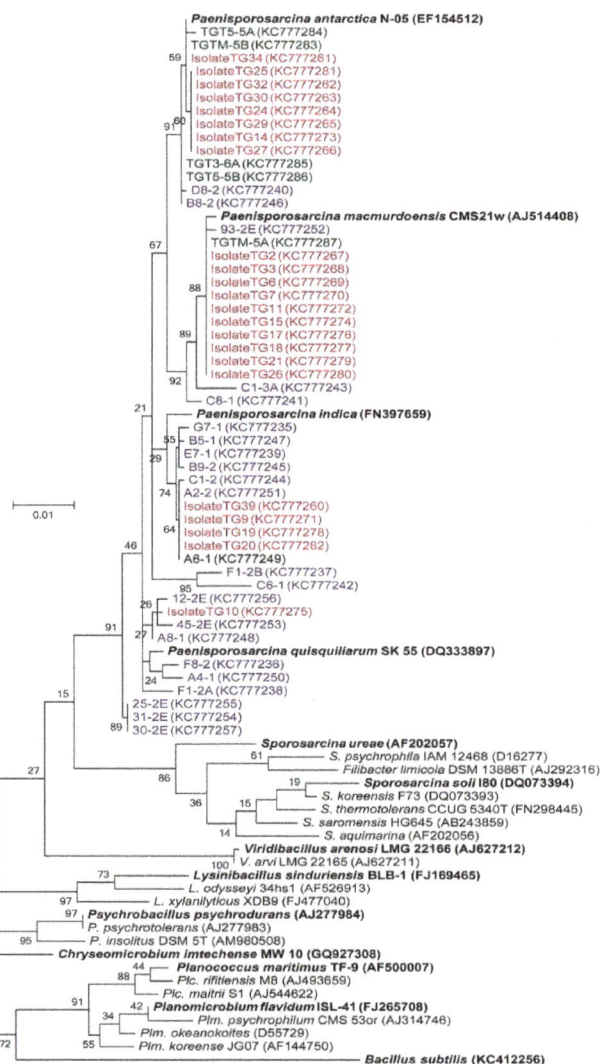

Figure 5. [^{3}H]-Leucine and [^{3}H]-thymidine incorporation into trichloroacetic acid (TCA) precipitable material at −15 °C by frozen cell suspensions of *Paenisporosarcina* sp. TG14 in clean ice melt water (white circles) and banded dispersed ice melt water (black triangles). The initial cell concentration of the TG14 cell suspensions was 3.1 × 10^6 colony-forming units (CFU) mL^{-1}. Error bars are the standard error of triplicate samples. Incorporation is reported as molecules of substrate incorporated per CFU on the left axis and grams of substrate carbon incorporated per gram of cell carbon on the right-hand axis. Where significant ($p < 0.05$), the best-fit regression curves (hyperbolic) are plotted as dashed lines.

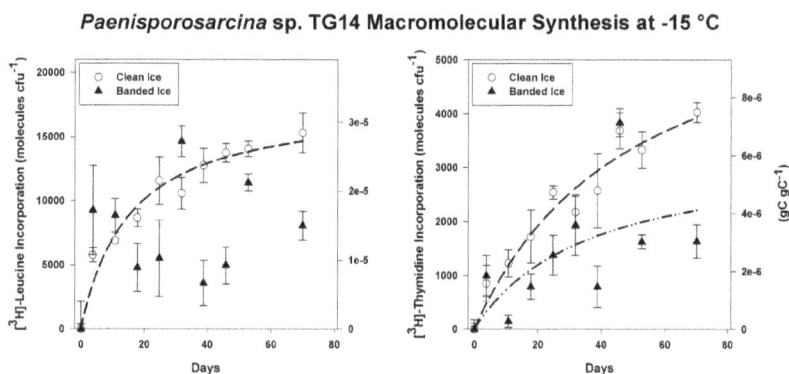

Paenisporosarcina sp. TG14 Macromolecular Synthesis at -15 °C

4. Discussion

Glaciers and ice sheets move under their own weight, entraining sediment particles and liquid water (*i.e.*, through regelation) from subglacial sources into the basal zone of the ice mass [12]. In some regions of East Antarctica, basal ice is estimated to constitute up to 50% of the total ice sheet thickness [45], and a globally significant pool of organic carbon (21,000 Pg) is hypothesized to be buried beneath the Antarctic Ice Sheet [46]. Although the long-term fate of this material is difficult to predict, geomicrobiological investigations of basal ice environments have produced strong lines of evidence that suggest microorganisms may be an important and active component of these systems [13–17,28]. Understanding the extent and nature of microbial processes in basal ice environments is critical for understanding biogeochemical cycling in the cryosphere and relevant to discussions regarding microbial life in extraterrestrial frozen environments, e.g., Mars [47].

Analysis of banded dispersed basal horizons from Taylor Glacier revealed a simultaneous increase of CO_2 (up to 325,000 ppmv) with a decrease of O_2 (as low as 4% of total gas volume) with respect to atmospheric concentrations (r (6) = −0.92, $p < 0.005$; Figure 2) [30,31,48]. These depths were associated with increased cell concentrations (Figure 2A), suggesting that the entrapped CO_2 and O_2 concentrations may have resulted from the aerobic respiration of organic matter. Indeed, the phylotypes identified in this study were related to predominantly aerobic microorganisms, and only two genera, *Bacillus* and *Paenibacillus*, are known to contain facultatively anaerobic species. These observations are similar to those in other basal ice environments, where similar gas anomalies have been reported. For instance, Souchez *et al.* [49]

reported elevated CO_2 (up to 130,000 ppmv) and CH_4 (up to 6,000 ppmv) concentrations in basal ice from the Greenland ice sheet. Likewise, Campen *et al.* [17] found excesses of CO_2, N_2O and CH_4 (32-, 240- and eight-times higher than atmospheric values, respectively) in ice from the Sajama glacier (Bolivia) and argued they resulted from microbial activity in the ice.

Molecular analysis of 16S rRNA genes amplified from Taylor Glacier basal ice revealed a low diversity of phylotypes from two bacterial phyla: the Firmicutes and Gammaproteobacteria. Although microorganisms attached to sediment particles in the basal ice may have been underrepresented in the 2009 samples, due to the sediment extraction procedure used, OTU diversity in the 2007 and 2009 was similar (H' = 1.81 and 1.78, respectively). The diversity of OTUs in the banded dispersed basal ice (H' = ~1.8) was significantly lower than that reported for Arctic permafrost (H' = 3.6; [50]) and similar to values found in Antarctic sea ice (H' = 1.1; [51]). In general, the molecular data are consistent with the low species richness observed in basal ice from the Bench Glacier (Alaska), the John Evans Glacier (Nunavut, Canada), and Greenland basal ice [26,52,53]. However, the subglacial environments of these glaciers appear to be largely dominated by members of the Betaproteobacteria (constituting between 25 to 51% of their clone libraries). From a phylogenetic perspective, our results are more similar to studies of permanent ground ice and permafrost in the Canadian Arctic, which found that the microbial assemblages inhabiting these frozen substrates were largely dominated by members of the Firmicutes (64% and 59%, respectively; [54,55]).

The most abundant phylotypes in the debris-rich ice horizons were affiliated with the genus, *Paenisporosarcina* (Figure 4). Enrichment culturing efforts were successful in obtaining a variety of psychrotolerant isolates with 16S rRNA gene identities >97% identical to the *Paenisporosarcina*-related phylotypes, as was also the case in previous culturing efforts on the Taylor Glacier basal ice [56]. Whether detected using molecular or culture-dependent approaches, members of the *Paenisporosarcina* have been documented in basal ice from three tunnel locations (~0.5 km apart) along the north margin of the Taylor Glacier. Some members of the Firmicutes, including species of *Paenisporosarcina* (e.g., *Ps. macmurdoensis*, *Ps. indica* and *Ps. quisquiliarum*), have the ability to differentiate into an endospore that is highly resistant to variety of environmental stresses. As such, their abundance in the ice could be attributed to selection, whereupon cells existing as environmentally durable spores would be more likely to persist in the ice after entrainment. However, results from pasteurization experiments indicated that 96% of the CFU were sensitive to heating at 80 °C, suggesting that the vast majority of the isolates obtained did not originate from heat resistant spores.

Members of the *Paenisporosarcina* have been documented in a variety of permanently cold environments, including permafrost [50,55], arctic saline springs [57], alpine glaciers [58] and the McMurdo Dry Valleys [59], implying this group may have specific adaptations to low temperature conditions. Due to the concentration of solutes into the premelt phase during freezing, microorganisms found in the unfrozen fraction of ice are subjected to considerable osmotic stress. For example, at −15 °C, the molarity of an ice vein environment is estimated to be approximately 4.5 M [2]. All of the *Paenisporosarcina*-related isolates from this study exhibited moderate halotolerance between 6% and 12% (w/v). These levels of halotolerance do not appear to be sufficient for the estimated salt concentrations occurring inside Taylor Glacier basal ice veins (I = 4.6 M; [2]); however, the cells may be able to sustain a low rate of maintenance-oriented metabolic activity exclusive of biomass production or grow at rates lower than those measurable after 80 days of incubation.

Experiments were conducted in an effort to examine bacterial metabolism under conditions that simulated those within the basal ice. To ensure minimal carry-over of nutrients from the marine media, the harvested cells were washed and suspended in Taylor Glacier melt water. Marine broth contains 1,800 µg L^{-1} PO$_4$-P, and if it is assumed that as much as 10 µL of residual supernatant remained in the tube between washes, the final solution would contain 72 pg L^{-1} PO$_4$-P; a value ~1,400-fold lower than the lowest concentration measured in Taylor Glacier clean ice [30]. *Paenisporosarcina* sp. TG14 incorporated at a maximum rate of approximately 1,500 to 2,500 molecules of [^{3}H]-leucine and 20 to 30 molecules of [^{3}H]-thymidine per day. Assuming an average protein length of 267 amino acids with 7.3% leucine content per protein [5], the rates observed could support the synthesis of approximately four proteins every hour and the replication of the *Paenisporosarcina* sp. TG14 genome (3.83 Mb) once every ~20 years. The annual N and P requirement to support this rate of DNA and protein synthesis in the TG14 cell population is estimated at approximately 2 ng N y^{-1} and 50 fg P y^{-1}. In terms of carbon turnover, these metabolic rates are extremely slow (0.81 to 2.1 × 10^{-7} gC gC^{-1} h^{-1}), but would be sufficient for a maintenance metabolism at −15 °C directed towards sustaining vital cellular processes (e.g. osmotic regulation, pH regulation, repair and/or turnover of macromolecules) [60].

From the melted debris-rich basal ice (230–235 cm depth), the highest measured concentrations of dissolved organic carbon, dissolved inorganic nitrogen and dissolved inorganic phosphorus were 63 mg L^{-1}, 0.68 mg L^{-1} and 5.5 µg L^{-1} respectively. These concentrations are approximately 10^2-fold higher than those measured in the debris-poor clean ice [30]. Nutrient concentrations in the banded dispersed ice melt water used in the −15 °C incubation experiments were not directly measured; however, by inference from other measurements as previously described [30], they would have been higher than in the clean ice meltwater. These higher nutrient concentrations in the banded dispersed ice melt water did not significantly increase the rate of macromolecular synthesis by *Paenisporosarcina* sp. TG14 at −15 °C, as compared with cells frozen in the clean ice (Figure 5). Further, more radiolabeled substrate was incorporated in the clean ice incubations over the entire time-course. There are several explanations for these data; first, the TG14 cells may have preferentially used substrates in the banded dispersed ice or endogenous sources of leucine or thymidine in the samples. Secondly, lower amounts of incorporation in the banded dispersed meltwater data could be due to scintillation interference from fine particulates in the clay size fraction. Thirdly, higher values of precursor incorporation may have been a result of differences in the [^{3}H]-leucine and [^{3}H] thymidine concentrations in the ice veins. Vein diameters, and, therefore, unfrozen water volume in ice formed from dilute solutions (*i.e.*, the clean ice) would be smaller than those in ice formed from more concentrated solutions (*i.e.*, the banded dispersed ice). The unfrozen water volume of the clean ice is significantly smaller than that of the banded dispersed ice based on bulk concentration differences (I = 0.1 and 160 mmol L^{-1}, respectively; [30]). As a result, 20 nmol L^{-1} of [^{3}H]-leucine or [^{3}H]-thymidine would exist at a higher concentration at −15°C in the liquid vein network in the clean ice. In support of this, micro-Raman spectroscopy of ice veins in clean Greenland glacial ice estimate that SO_4^{2-} and NO_3^- ions are $\sim 10^4$ more concentrated than they appear in the melted bulk phase [61]. Mathematical modeling of the Taylor Glacier basal ice chemistry predicts that at -15 °C, ice vein solutes are only ~ 35 times more concentrated than in the melted bulk phase [2]. It should be noted that these values are based on mathematical models of ideal solutions, assume thermodynamic equilibrium (*i.e.*, no supercooling), and do not account for heterogeneity across a volume of ice.

In summary, basal horizons of the Taylor Glacier contained elevated concentrations of microbial cells, which consisted largely of bacteria in the phylum, Firmicutes. Species from the genus, *Paenisporosarcina*, were a numerically abundant member of this assemblage and appear to have characteristics that would promote survival in ice (e.g., psychrotolerance, halotolerance, metabolic activity to temperatures as low as -33 °C [21] and a spore-based survival stage). Our data support the notion that basal ice environments are active microbial habitats. Moreover, the habitability of frozen environments may be more favorable than previously recognized, as very low concentrations of nutrients appear capable of supporting metabolism in the unfrozen liquid phase of ice. Given a suitable microbial inoculum, the basal ice environment is a potential environment for biogeochemical cycling beneath glaciers and ice sheets. Our results supporting the persistence of microbial metabolic function at frozen temperatures also suggest that frozen worlds, such as Mars, Europa or Enceladus, could harbor cryogenic habitats suitable for microbial life [62].

Acknowledgements

We thank Pierre Amato, Tim Brox, Amanda Achberger and Lindsay Knippenberg for assistance with field work in Antarctica. This research was supported by National Science Foundation awards, ANT-0636828 and 0636770, to B.C.C. and M.L.S., respectively, and grants to B.C.C. from the National Aeronautics and Space Administration (NNX10AN07A) and the Louisiana Board of Regents. Shawn Doyle and Scott Montross were partially supported by funding from the Louisiana Board of Regents Graduate Fellowship program and NSF-IGERT DGE 0654336, respectively. We are also grateful to S. Fitzsimons for guidance on tunnel construction and providing samples of basal ice from the Taylor Glacier.

Conflict of Interest

The authors declare no conflict of interest.

References

1. Price, P.B. A habitat for psychrophiles in deep Antarctic ice. *Proc. Natl. Acad. Sci. USA* **2000**, *97*, 1247–1251.
2. Doyle, S.M.; Dieser, M.; Broemsen, E.; Christner, B.C. General characteristics of cold-adapted microorganisms. In *Polar Microbiology: Life In a Deep Freeze*; Whyte, L., Miller, R.V., Eds.; ASM Press: Washington, DC, USA, 2012; pp. 103–125.

3. Miteva, V.I.; Sheridan, P.P.; Brenchley, J.E. Phylogenetic and physiological diversity of microorganisms isolated from a deep Greenland glacier ice core. *Appl. Environ. Microbiol.* **2004**, *70*, 202–213.

4. Xiang, S.-R.; Shang, T.-C.; Chen, Y.; Yao, T.-D. Deposition and postdeposition mechanisms as possible drivers of microbial population variability in glacier ice. *FEMS Microbiol. Ecol.* **2009**, *70*, 165–176.

5. Amato, P.; Doyle, S.M.; Battista, J.R.; Christner, B.C. Implications of subzero metabolic activity on long-term microbial survival in terrestrial and extraterrestrial permafrost. *Astrobiology* **2010**, *10*, 789–798.

6. Abyzov, S.S. Microorganisms in the Antarctic ice. *Antarct. Microbiol.* **1993**, *1*, 265–296.

7. Abyzov, S.S.; Lipenkov, V.Y.; Bobin, N.E.; Koudryashov, B.B. The microflora of the central Antarctic glacier and the control methods of the sterile isolation of the ice core for microbiological analysis. *Akad. Nauk SSSR Izv. Ser. Biol.* **1982**, *4*, 537–548.

8. Antony, R.; Krishnan, K.P.; Laluraj, C.M.; Thamban, M.; Dhakephalkar, P.K.; Engineer, A.S.; Shivaji, S. Diversity and physiology of culturable bacteria associated with a coastal Antarctic ice core. *Microbiol. Res.* **2012**, *167*, 372–380.

9. Christner, B.C.; Mosley-Thompson, E.; Thompson, L.G.; Reeve, J.N. Bacterial recovery from ancient ice. *Environ. Microbiol.* **2003**, *5*, 433–436.

10. Christner, B.C.; Mosley-Thompson, E.; Thompson, L.G.; Zagorodnov, V.; Sandman, K.; Reeve, J.N. Recovery and identification of viable bacteria immured in glacial ice. *Icarus* **2000**, *144*, 479–485.

11. Miteva, V.I.; Brenchley, J.E. Detection and isolation of ultrasmall microorganisms from a 120,000-year-old Greenland glacier ice core. *Appl. Environ. Microbiol.* **2005**, *71*, 7806–7818.

12. Skidmore, M.L.; Foght, J.M.; Sharp, M.J. Microbial life beneath a high Arctic glacier. *Appl. Environ. Microbiol.* **2000**, *66*, 3214–3220.

13. Ahn, J.; Wahlen, M.; Deck, B.L.; Brook, J.; Mayewski, P.A.; Taylor, K.C.; White, J.W.C. A record of atmospheric CO_2 during the last 40,000 years from the siple dome, Antarctica ice core. *J. Geophys. Res.* **2004**, *109*, doi:10.1029/2003JD004415.

14. Flückiger, J.; Blunier, T.; Stauffer, B.; Chappellaz, J.; Spahni, R.; Kawamura, K.; Schwander, J.; Stocker, T.F.; Dahl-Jensen, D. N_2O and CH_4 variations during the last glacial epoch: Insight into global processes. *Glob. Biogeochem. Cy.* **2004**, *18*, doi:10.1029/2003GB002122.

15. Souchez, R.; Janssens, L.; Lemmens, M.; Stauffer, B. Very low oxygen concentration in basal ice from summit, central Greenland. *Geophys. Res. Lett.* **1995**, *22*, 2001–2004.

16. Sowers, T. N_2O record spanning the penultimate deglaciation from the vostok ice core. *J. Geophys. Res.* **2001**, *106*, 31903–31914.

17. Campen, R.K.; Sowers, T.; Alley, R.B. Evidence of microbial consortia metabolizing within a low-latitude mountain glacier. *Geology* **2003**, *31*, 231–234.

18. Souchez, R.; Jouzel, J.; Landais, A.; Chappellaz, J.; Lorrain, R.; Tison, J.-L. Gas isotopes in ice reveal a vegetated central Greenland during ice sheet invasion. *Geophys. Res. Lett.* **2006**, *33*, doi:10.1029/2006GL028424.

19. Rohde, R.A.; Price, P.B.; Bay, R.C.; Bramall, N.E. *In situ* microbial metabolism as a cause of gas anomalies in ice. *Proc. Natl. Acad. Sci. USA* **2008**, *105*, 8667–8672.

20. Tung, H.C.; Bramall, N.E.; Price, P.B. Microbial origin of excess methane in glacial ice and implications for life on Mars. *Proc. Natl. Acad. Sci. USA* **2005**, *102*, 18292–18296.

21. Bakermans, C.; Skidmore, M. Microbial respiration in ice at subzero temperatures (−4 °C to −33 °C). *Environ. Microbiol. Rep.* **2011**, *3*, 774–782.

22. Panikov, N.; Flanagan, P.; Oechel, W.; Mastepanov, M.; Christensen, T. Microbial activity in soils frozen to below −39 °C. *Soil Biol. Biochem.* **2006**, *38*, 785-794.

23. Christner, B.C. Incorporation of DNA and protein precursors into macromolecules by bacteria at −15 °C. *Appl. Environ. Microbiol.* **2002**, *68*, 6435–6438.

24. Junge, K.; Eicken, H.; Swanson, B.D.; Deming, J.W. Bacterial incorporation of leucine into protein down to −20 °C with evidence for potential activity in sub-eutectic saline ice formations. *Cryobiology* **2006**, *52*, 417–429.

25. Knight, P.G. The basal ice layer of glaciers and ice sheets. *Quat. Sci. Rev.* **1997**, *16*, 975–993.

26. Skidmore, M.; Anderson, S.P.; Sharp, M.; Foght, J.; Lanoil, B.D. Comparison of microbial community compositions of two subglacial environments reveals a possible role for microbes in chemical weathering processes. *Appl. Environ. Microbiol.* **2005**, *71*, 6986–6997.

27. Sharp, M.; Parkes, J.; Cragg, B.; Fairchild, I.J.; Lamb, H.; Tranter, M. Widespread bacterial populations at glacier beds and their relationship to rock weathering and carbon cycling. *Geology* **1999**, *27*, 107–110.

28. Tung, H.C.; Price, P.B.; Bramall, N.E.; Vrdoljak, G. Microorganisms metabolizing on clay grains in 3-km-deep Greenland basal ice. *Astrobiology* **2006**, *6*, 69–86.

29. Hubbard, B.; Cook, S.; Coulson, H. Basal Ice facies: A review and unifying approach. *Quat. Sci. Rev.* **2009**, *28*, 1956–1969.

30. Montross, S.N. Biogeochemistry of Basal Ice from Taylor Glacier, Antarctica. Ph.D. Dissertation, Montana State University, Bozeman, MT, USA, 2012.

31. Montross, S.M.; Skidmore, M.; Christner, B.; Samyn, D.; Tison, J.L.; Lorrain, R.; Doyle, S.; Fitzsimons, S. Debris-rich basal ice as a microbial habitat, Taylor Glacier, Antarctica. *Geomicrobiol. J.* **2013**, doi:10.1080/01490451.2013.811316.

32. Christner, B.C.; Mikucki, J.A.; Foreman, C.M.; Denson, J.; Priscu, J.C. Glacial ice cores: A model system for developing extraterrestrial decontamination protocols. *Icarus* **2005**, *174*, 572–584.

33. Trevors, J.T.; Cook, S. A Comparison of plating media and diluents for enumeration of aerobic bacteria in a loam soil. *J. Microbiol. Meth.* **1992**, *14*, 271–275.

34. Samyn, D.; Svensson, A.; Fitzsimons, S. Dynamic implications of discontinuous recrystallization in cold basal ice: Taylor Glacier, Antarctica. *J. Geophys. Res.* **2008**, *113*, doi:10.1029/2006JF000600.

35. Robe, P.; Nalin, R.; Capellano, C.; Vogel, T.M.; Simonet, P. Extraction of DNA from soil. *Eur. J. Soil. Biol.* **2003**, *39*, 183–190.

36. Lane, D.J. 16S/23S rRNA sequencing. In *Nucleic Acid Techniques in Bacterial Systematics*; Stackebrandt, E., Goodfellow, M., Eds.; Wiley: New York, NY, USA, 1991; pp. 115–175.

37. Pruesse, E.; Peplies, J.; Glöckner, F.O. SINA: Accurate high-throughput multiple sequence alignment of ribosomal RNA genes. *Bioinformatics* **2012**, *28*, 1823–1829.

38. Pruesse, E.; Quast, C.; Knittel, K.; Fuchs, B.M.; Ludwig, W.; Peplies, J.; Glöckner, F.O. SILVA: A comprehensive online resource for quality checked and aligned ribosomal RNA sequence data compatible with ARB. *Nucleic Acids Res.* **2007**, *35*, 7188–7196.

39. Schloss, P.D.; Westcott, S.L.; Ryabin, T.; Hall, J.R.; Hartmann, M.; Hollister, E.B.; Lesniewski, R.A.; Oakley, B.B.; Parks, D.H.; Robinson, C.J.; *et al.* Introducing mothur: Open-source, platform-independent, community-supported software for describing and comparing microbial communities. *Appl. Environ. Microbiol.* **2009**, *75*, 7537–7541.

40. Edgar, R.C.; Haas, B.J.; Clemente, J.C.; Quince, C.; Knight, R. UCHIME improves sensitivity and speed of chimera detection. *Bioinformatics* **2011**, *27*, 2194–2200.

41. Tamura, K.; Peterson, D.; Peterson, N.; Stecher, G.; Nei, M.; Kumar, S. MEGA5: Molecular evolutionary genetics analysis using maximum likelihood, evolutionary distance, and maximum parsimony methods. *Mol. Biol. Evol.* **2011**, *28*, 2731–2739.

42. Kim, O.S.; Cho, Y.J.; Lee, K.; Yoon, S.H.; Kim, M.; Na, H.; Park, S.C.; Jeon, Y.S.; Lee, J.H.; Yi, H.; *et al.* Introducing EzTaxon-E: A prokaryotic 16S rRNA gene sequence database with phylotypes that represent uncultured species. *Int. J. Syst. Evol. Microbiol.* **2012**, *62*, 716–721.

43. Wang, Q.; Garrity, G.M.; Tiedje, J.M.; Cole, J.R. Naive bayesian classifier for rapid assignment of rRNA sequences into the new bacterial taxonomy. *Appl. Environ. Microbiol.* **2007**, *73*, 5261–5267.

44. Felsenstein, J. Confidence limits on phylogenies: An approach using the bootstrap. *Evolution* **1985**, *783–791.*

45. Bell, R.E.; Ferraccioli, F.; Creyts, T.T.; Braaten, D.; Corr, H.; Das, I.; Damaske, D.; Frearson, N.; Jordan, T.; Rose, K.; *et al.* Widespread persistent thickening of the east Antarctic ice sheet by freezing from the base. *Science* **2011**, *331*, 1592–1595.

46. Wadham, J.; Arndt, S.; Tulaczyk, S.; Stibal, M.; Tranter, M.; Telling, J.; Lis, G.; Lawson, E.; Ridgwell, A.; Dubnick, A. Potential methane reservoirs beneath Antarctica. *Nature* **2012**, *488*, 633–637.

47. Byrne, S.; Dundas, C.M.; Kennedy, M.R.; Mellon, M.T.; McEwen, A.S.; Cull, S.C.; Daubar, I.J.; Shean, D.E.; Seelos, K.D.; Murchie, S.L. Distribution of mid-latitude ground ice on Mars from new impact craters. *Science* **2009**, *325*, 1674–1676.

48. Samyn, D.; Fitzsimons, S.J.; Lorrain, R.D. Strain-induced phase changes within cold basal ice from Taylor Glacier, Antarctica, indicated by textural and gas analyses. *J. Glaciol.* **2005**, *51*, 611–619.

49. Souchez, R.; Lemmens, M.; Chappellaz, J. Flow-induced mixing in the GRIP basal ice deduced from the CO_2 and CH_4 records. *Geophys. Res. Lett.* **1995**, *22*, 41–44.

50. Steven, B.; Briggs, G.; McKay, C.P.; Pollard, W.H.; Greer, C.W.; Whyte, L.G. Characterization of the microbial diversity in a permafrost sample from the Canadian High Arctic using culture-dependent and culture-independent methods. *FEMS Microbiol. Ecol.* **2007**, *59*, 513–523.

51. Brinkmeyer, R.; Knittel, K.; Jürgens, J.; Weyland, H.; Amann, R.; Helmke, E. Diversity and structure of bacterial communities in Arctic versus Antarctic pack ice. *Appl. Environ. Microbiol.* **2003**, *69*, 6610–6619.

52. Cheng, S.M.; Foght, J.M. Cultivation-independent and -dependent characterization of bacteria resident beneath John Evans Glacier. *FEMS Microbiol. Ecol.* **2007**, *59*, 318–330.

53. Yde, J.C.; Finster, K.W.; Raiswell, R.; Steffensen, J.P.; Heinemeier, J.; Olsen, J.; Gunnlaugsson, H.P.; Nielsen, O.B. Basal ice microbiology at the margin of the Greenland ice sheet. *Ann. Glaciol.* **2010**, *51*, 71–79.

54. Lacelle, D.; Radtke, K.; Clark, I.D.; Fisher, D.; Lauriol, B.; Utting, N.; Whyte, L.G. Geomicrobiology and occluded O_2-CO_2-Ar gas analyses provide evidence of microbial respiration in ancient terrestrial ground ice. *Earth Planet. Sci. Lett.* **2011**, *306*, 46–54.

55. Steven, B.; Pollard, W.H.; Greer, C.W.; Whyte, L.G. Microbial diversity and activity through a permafrost/ground ice core profile from the Canadian High Arctic. *Environ. Microbiol.* **2008**, *10*, 3388–3403.

56. Christner, B.C.; Skidmore, M.L. Montana State University, Bozeman, MT, USA, Unpublished work, 2005.

57. Perreault, N.N.; Greer, C.W.; Andersen, D.T.; Tille, S.; Lacrampe-Couloume, G.; Lollar, B.S.; Whyte, L.G. Heterotrophic and autotrophic microbial populations in cold perennial springs of the high Arctic. *Appl. Environ. Microbiol.* **2008**, *74*, 6898–6907.

58. Reddy, G.; Manasa, B.P.; Singh, S.K.; Shivaji, S. *Paenisporosarcina indica* sp. nov., a psychrophilic bacterium from pindari glacier of the Himalayan Mountain Ranges and reclassification of *sporosarcina Antarctica* Yu *et al.*, 2008 as *Paenisporosarcina antarctica* comb. nov. and emended description of the genus *Paenisporosarcina*. *Int. J. Syst. Evol. Microbiol.* **2013**, in press.

59. Reddy, G.; Matsumoto, G.; Shivaji, S. *Sporosarcina macmurdoensis* sp. nov., from a cyanobacterial mat sample from a pond in the McMurdo Dry Valleys, Antarctica. *Int. J. Syst. Evol. Microbiol.* **2003**, *53*, 1363–1367.

60. Price, P.B.; Sowers, T. Temperature dependence of metabolic rates for microbial growth, maintenance, and survival. *Proc. Natl. Acad. Sci. USA* **2004**, *101*, 4631–4636.

61. Barletta, R.E.; Priscu, J.C.; Mader, H.M.; Jones, W.L.; Roe, C.H. Chemical analysis of ice vein microenvironments: II. analysis of glacial samples from Greenland and Antarctica. *J. Glaciol.* **2012**, *58*, 1109–1118.

62. Priscu, J.C.; Hand, K.P. Microbial habitability of icy worlds. *Microbe* **2012**, *7*, 167–172.

Contrasting Responses to Nutrient Enrichment of Prokaryotic Communities Collected from Deep Sea Sites in the Southern Ocean

David M. McCarthy, David A. Pearce, John W. Patching and Gerard T. A. Fleming

Abstract: Deep water samples (*ca.* 4,200 m) were taken from two hydrologically-similar sites around the Crozet islands with highly contrasting surface water productivities. Site M5 was characteristic of high productivity waters (high chlorophyll) whilst site M6 was subject to a low productivity regime (low chlorophyll) in the overlying waters. Samples were incubated for three weeks at 4 °C at *in-situ* and surface pressures, with and without added nutrients. Prokaryotic abundance increased by at least two-fold for all nutrient-supplemented incubations of water from M5 with little difference in abundance between incubations carried out at atmospheric and *in-situ* pressures. Abundance only increased for incubations of M6 waters (1.6-fold) when they were carried out at *in-situ* pressures and with added nutrients. Changes in community structure as a result of incubation and enrichment (as measured by DGGE banding profiles and phylogenetic analysis) showed that diversity increased for incubations of M5 waters but decreased for those with M6 waters. *Moritella* spp. came to dominate incubations carried out under *in-situ* pressure whilst the Archaeal community was dominated by *Crenarchaea* in all incubations. Comparisons between atmospheric and *in situ* pressure incubations demonstrated that community composition was significantly altered and community structure changes in unsuspplemented incubations at *in situ* pressure was indicative of the loss of functional taxa as a result of depressurisation during sampling. The use of enrichment incubations under *in-situ* conditions has contributed to understanding the different roles played by microorganisms in deep sea ecosystems in regions of low and high productivity.

Reprinted from *Biology*. Cite as: McCarthy, D.M.; Pearce, D.A.; Patching, J.W.; Fleming, G.T.A. Contrasting Responses to Nutrient Enrichment of Prokaryotic Communities Collected from Deep Sea Sites in the Southern Ocean. *Biology* **2013**, *2*, 1165-1188.

1. Introduction

Deep-sea prokaryotic communities play an important role in the recycling of organic matter and are thought to be responsible for at least 50% of global net mineralisation of organic matter in marine ecosystems [1]. Many of the long-standing challenges facing deep-sea studies have been overcome in recent years and our knowledge of prokaryotic community structure has been greatly advanced by the widespread application of rapid molecular "fingerprinting" techniques [1,2]. Several studies have shown a direct link between local organic enrichment and changes in community structure in natural and disturbed aquatic environments [3–5] and it has been shown that major shifts in community structure can occur over short time scales (days) in response to nutrient input [6,7]. The influence of nutrient in promoting changes in patterns of diversity and the distribution of functional and taxonomic grouping of prokaryotic assemblages warrants further investigation and in particular, the effects of local episodic nutrient inputs on deep-sea prokaryotic community structure remains poorly understood [8]. This is due in part to the difficulty in recovering samples from the deep sea and also to the problems associated with maintaining *in situ* conditions during incubation studies. For these and other reasons the effects of pressure on deep-sea microbial communities have also remained poorly defined [9].

The present study seeks to characterise how the dominant members of the bacterial and archaeal communities responded to nutrient amendment when incubated at *in situ* and atmospheric pressures. Furthermore, response to experimental nutrient addition was investigated at two different sites subject to contrasting nutrient deposition regimes. In this respect, the authors believe that this study represents a unique attempt to understand the dynamics of both bacterial and archaeal microbial community structures in deep polar seas. The work presented here was undertaken as part of a large multidisciplinary project (CROZEX) whose purpose was to determine whether the nature of organics deposited with phytodetritus on the deep-sea floor influences the organisms living there. It focused on the bloom that occurs annually north of the Crozet Islands and Plateau (Crozet) which were surveyed and compared with a high-nutrient low-chlorophyll (HNLC) region south of Crozet (as described in a special issue of Deep Sea Research [10]). The study sites M5 (high chlorophyll) and M6 (low chlorophyll) shown in Figure 1 were located in deep water (*ca.* 4,200 m) close to the Crozet islands (46.4°S, 52.0°E), lying within the Antarctic Polar Frontal Zone, east of the Southwest Indian Ocean ridge [11]. The surrounding area has been characterised as an oligotrophic "High Nutrient Low Chlorophyll" (HNLC) zone [12], a condition arising where phytoplankton numbers are low irrespective of high macro-nutrient concentrations. HNLC is considered to be caused by the scarcity of certain limiting nutrient, particularly iron and high grazing rates by micro-zooplankton [13].

Figure 1. Map showing positions of sites M5 and M6 respectively in the vicinity of the Crozet islands. Inset shows the location of the Crozet Plateau east of the Southwest Indian Ridge.

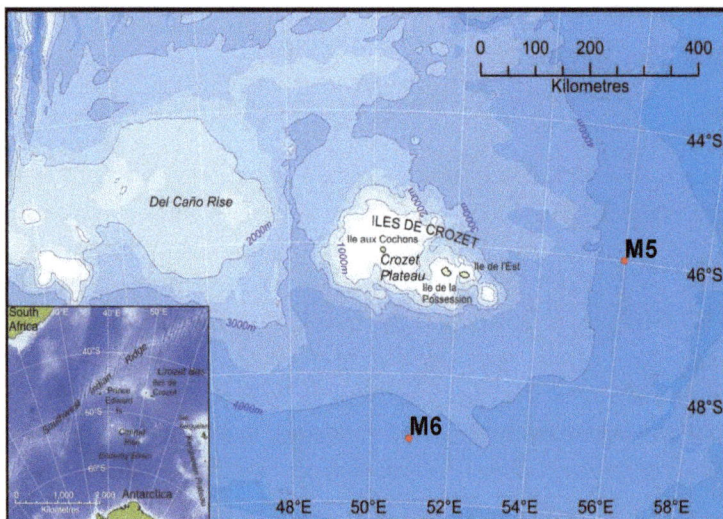

Site M5 was within an area to the northeast and downstream of the islands characterized by high chlorophyll concentrations during the Austral spring bloom period. The bloom is believed to result from an island-mass effect mediated by iron fertilisation from the islands and associated in-shore sediments [14]. The extent and circulation of the annual bloom is constrained by local seafloor topography, currents and winds [15,16]. Pollard and co-workers [17] have shown that the blooms in this region are seasonal and occur consistently over the same area and with the same temporal evolution.

Site M6 was in an area isolated from the bloom event. Hughes *et al.* [18] report an export productivity to 100 m depth of 40 g C m^{-2} y^{-1} at M5, compared to 10 g C m^{-2} y^{-1} at M6, and estimate the flux to the seafloor at M5 and M6 to be 2.0 g C m^{-2} y^{-1} and 0.5 g C m^{-2} y^{-1} respectively based on the export model assumptions of Martin *et al.*, [18,19]. The two sites are in close proximity and have almost identical environmental characteristics (Table 1) [18,20]. As such, the two sites are distinguished by differences in the primary productivity regime of the overlying waters. Whilst near bottom waters maybe strongly influenced by the physics of sheer caused by bottom currents, differences in microbial community structure between the two sites may be considered to result largely from the observed differences in flux of organic material to the deep sea.

Table 1. Bottom water properties, Austral summer 2004–2005. HC: high-chlorophyll; LC: low-chlorophyll. Data reproduced from [18] and [20].

	M5 (HC)	M6 (LC)
Depth (m)	4,224	4,212
Temperature (°C)	−0.22	−0.21
Salinity	34.67	34.67
Oxygen (μmol L^{-1})	230.9	231.0
Nitrate + Nitrite (μmol L^{-1})	31.90	32.32
Silicate (μmol L^{-1})	154.4	155.3
Phosphate (μmol L^{-1})	2.19	2.19
C flux to 100 m (g C m^{-2} y^{-1})	**40**	**10**
C flux to Seafloor (g C m^{-2} y^{-1})	**2**	**0.5**

Waters from sites M5 (high chlorophyll) and M6 (low chlorophyll) represented an excellent opportunity to directly compare the dominant microbial community structure arising out of different naturally-occurring nutrient regimes. Furthermore, this study focused on free-living assemblages in the benthic boundary layer, the extended layer of water directly above the sediment-water interface. This zone (along with the sediments) is of considerable interest because it represents the ultimate sink of surface-derived material and the site of resuspension of sediment material [21], and is therefore particularly important for understanding benthic-pelagic coupling.

A parallel study by Jamieson and co-workers [22] investigated the relationship between surface-derived particulate organic matter (POM) and bacterial abundance, community structure and composition in two sediment layers at M5 (high chlorophyll site) and M6 (low chlorophyll site). Results from this study demonstrated that these parameters were remarkably similar despite contrasting organic input in overlying waters.

In the present study, water samples were obtained from the water column approximately 10 m above these same sediments. Incubations were carried out at atmospheric and *in-situ* pressure (42 MPa) in order to compare the effects of incubation pressures on prokaryotic community responses to nutrient. It is worth noting that with some exceptions (e.g., [7]), investigations using waters from depth are usually carried out under conditions of atmospheric pressure [23,24] even when bacteria from extreme hadal depths are under study [25]. Deep-sea microorganisms may be characterised as belonging to particular groups depending on their growth response to hydrostatic pressure. Those that grow optimally at elevated pressures (which are characteristic of the deep sea) are defined as barophilic (piezophilic) while mesophiles grow preferentially at atmospheric pressures and are likely to be surface-derived. Some of these surface-derived microorganisms can survive at pressures that are encountered in the deep sea [26,27]. We hypothesise that deep-sea microbial populations undergo changes in community structure as a result of incubation at surface pressure. It has been shown elsewhere that incubations of deep-sea communities at surface pressures come to be dominated by surface-derived organisms that would not have been favored under *in-situ* conditions [7,28].

2. Results and Discussion

The four incubations were designed to represent a gradient of incubation conditions: with and without nutrients at *in-situ* pressure in order to assess the community response to nutrient, and with identical nutrient amendments at atmospheric pressure in order to compare the effects of incubation pressure on community structure and response to nutrient. The community response was determined for each incubation condition (±added nutrients, surface and *in-situ* pressure) by measuring changes in total cell number (abundance) and the number/differences in DGGE bands from communities obtained from the two sites. Denaturing gradient gel electrophoresis [29] has been widely used in conjunction with statistical analysis of "fingerprint" patterns [30] to elucidate the effects of nutrient enrichments on community structure [7,31], and relate such changes to environmental parameters [32,33] and thus appears in the literature as a useful method for visualising the major members of a microbial community, but there are certain limitations in this regard. The brightest bands in a DGGE profile are often assumed to represent the dominant members of the community [27]. However, the biases associated with PCR could cause relative under- or over-representation of a given taxon in the DGGE profile. Consequently, quantitative inferences about species richness must be confined to general statements about species predominance (for example, [34–36]). On the other end of the spectrum of abundance, the limit of resolution of this method seems to be about 1% of the community population, that is, only DNA from organisms comprising 1% or more of the community sample can be visualised [29,37]. In addition, this method can be difficult to apply to extremely complex communities that produce hundreds of bands on a DGGE profile, which become difficult to visualise individually [38]. The key to the approach taken in the present study is differences (*i.e.*, changes in community structure in response to the environmental change) compared with no difference (*i.e.*, no response). DGGE gels are shown in Figure 2.

Figure 2. Denaturing gradient gels for Bacteria and Archaea. Left panel, Bacteria: (**1**) M5, time zero; (**2**) M5, no nutrients/42 MPa; (**3**) M5, nutrients/42 MPa; (**4**) M5, no nutrients/1 atm; (**5**) M5, nutrients/1 atm; (M) DGGE Marker lane; (**6**) M6, time zero; (**7**) M6, no nutrients/42 MPa; (**8**) M6, nutrients/42 MPa; (**9**) M6, no nutrients/1 atm; (**10**) M6, nutrients/1 atm. Right panel, Archaea: (**a**) M5, time zero; (**b**) M5, no mutrients/42 MPa; (**c**) M5, nutrients/42 MPa; (**d**) M5 no nutrients/1 atm; (**e**) M5, nutrients/1 atm; (**f**) M6, time zero; (**g**) M6, no mutrients/42 MPa; (**h**) M6, nutrients/42 MPa; (**i**) M6, no nutrients/1 atm; (**j**) M6, nutrients/1 atm.

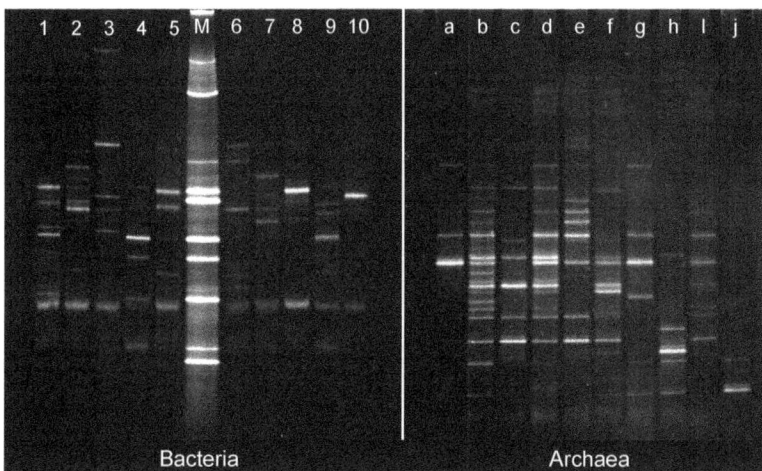

Obligate barophiles (piezophiles or hyperpiezophiles) require pressure for optimal growth [26]. There is evidence that the growth or metabolic rates of pressure-adapted microorganisms are not adversely affected by a reduction in hydrostatic pressure [39]. Decompression of the water sample was unavoidable during sample recovery and subsequent transfer to pressure vessels. Ideally samples would have been taken and transferred to pressure vessels using a specialised pressure-retaining sampling device to maintain conditions of *in-situ* pressure and temperature. It has been shown however that decompression itself does not necessarily cause lethal damage to deep-sea adapted bacteria [40,41]. Park & Clark [42] have shown that cell lysis of a deep-sea obligate barophile was avoided by ensuring a sufficiently slow rate of decompression (8.6 dbar s^{-1}, equivalent to 86 kPa s^{-1}). Seawater samples taken here were decompressed at a far lower rate of approximately 1 dbar s^{-1} (10 kPa s^{-1}). It has been argued that exposure to increases in temperature has a greater detrimental effect than changes in hydrostatic pressure [43,44] and may be responsible for many of the artifacts thought to be caused by decompression. Isothermal decompression at low temperature has been shown to preserve the viability of obligate barophiles (piezophiles), with ultrastructural damage only appearing to manifest over the course of several days incubation at atmospheric pressure [41]. Furthermore, others have indicated that maintenance of low temperatures can compensate for depressurisation in relation to the activity of barophiles (piezophiles) in deep sea sediments [45]. Great care was taken in the present study to ensure that sample temperature did not exceed 4 °C and that *in-situ* pressures were re-established within one hour of arriving on deck (except in the case of incubations at atmospheric pressure). The temperature of the water column during sampling did not exceed 3 °C until the last 200 m approaching the surface, where the temperature increased to between 4 °C and 6 °C. Samples were then immediately transferred to a temperature controlled laboratory and maintained at 4 °C. The bottom temperature of −0.22 °C could not be replicated aboard ship due to technical constraints.

2.1. Pre-Incubation Prokaryotic Abundance

Total (direct) counts and diversity indices for incubations are shown in Table 2. The total count for site M5 was three-fold lower than at that obtained for M6 site at the beginning of the incubation period (time zero) and this differential remained under all incubation conditions.

It is believed that prokaryotic biomass in the bathypelagic is regulated primarily by the availability of organic carbon, an example of a "bottom up" control. [46]. However, in this case the lower counts observed for waters taken from the high chlorophyll site M5 (pre-incubation) are inconsistent with such a model. This finding might be explained by increased viral lysis or protistan grazing [47,48] in which bacterial growth is initially stimulated by organic material from the phytoplankton bloom, but as the bloom evolves the enhanced biomass is diminished by increased protistan grazing [49]. Coincidental with the present study, Hughes and co-workers [18] reported the elevated abundance of benthic foraminifera at M5 which is reflective of the different nutrient regimes. They also suggest that only a few foraminiferal groups responded directly to the deposited organics and that non-phytodetritus associated species may still have benefited nutritionally by feeding on enhanced bacterial populations. Although these findings refer specifically to benthic foraminifera in the sediment, planktonic communities of bacteriovores such as ciliates and flagellates are known to respond in a similar fashion and can contribute to a transient reduction in bacterial numbers as part of a normal predator-prey interaction. Previous studies have observed similar sharp reductions in bacterial abundance immediately following a phytoplankton bloom [50] and a temporarily reduced bacterial abundance inside a bloom compared to adjacent waters [49]. In other cases, the reduction in bacterial abundance was accompanied by increased protist abundance and elevated Chlorophyll-a concentrations [51]. Interactions with protists and viruses are an important structuring factor for prokaryotic communities [52]. An evaluation of bacterial mortality and other food web interactions was not possible in the present study due to sampling limitations associated with incubations under pressure.

Table 2. Total counts (N), diversity (H) and dominance (c) for prokaryotic communities. Values refer to bacteria and archaea combined. Diversity indices were calculated from matrices of DGGE band peak heights. N: Total prokaryotic abundance ($n \times 10^4$ cells mL^{-1}), parenthetical values represent the standard error of the mean; H: Shannon's index of general diversity, c: Simpson's index of dominance. N+ denotes incubation with added nutrients; N− denotes incubation without nutrient addition.

	M5 (High Chlorophyll site)			M6 (Low Chlorophyll site)		
	N	H	c	N	H	c
Time Zero	1.19 (±0.14)	2.593	0.104	3.95 (±0.08)	3.021	0.064
Atm, N+	2.77 (±0.08)	2.789	0.08	4.52 (±0.13)	1.307	0.368
42 MPa, N+	2.64 (±0.10)	2.71	0.088	6.44 (±0.10)	2.118	0.182
Atm, N−	2.05 (±0.10)	2.845	0.082	3.56 (±0.14)	2.674	0.092
42 MPa, N−	1.82 (±0.16)	3.12	0.058	3.85 (±0.15)	2.761	0.087

2.2. Changes in Abundance and Community Structure after Incubations

The organic loading used during incubations is highly labile and untypical of organic material that usually reaches the deep sea. Similar substrates have been previously used to promote growth during the course of incubations of deep sea waters so that changes in community structure could be measured as a result of growth and enrichment [7]. Whilst the incubation pressure and nutrient additions were conditions applied by us, other potential changes would have been introduced by the process of sample recovery and the incubation technique. These are the perturbations acting on the enrichment carried out under *in situ* pressure and in the absence of added nutrients. Under these conditions there was a 1.6-fold increase in abundance at M5 (high chlorophyll site) but only a negligible change for the low-chlorophyll (M6) waters (Figure 3). Egan and co-authors [7] have previously considered that an increase in abundance during unsupplemented incubation might have resulted from the release of additional nutrients from cells damaged during recovery. Another possible explanation is the "wall effect", originally proposed by Zobell & Anderson as early as 1936 [53] in which growth is stimulated by the concentration of nutrients and microorganisms on the solid surfaces of an enclosed incubation [54].

It is likely that those microorganisms which come to dominance in incubations without added nutrients and *in situ* pressures are the active components of the deep sea microbial community. Of particular interest is the decrease in diversity as a result of incubations without added nutrient but under *in-situ* pressure (Table 3). It is probable that this was brought about by the loss of true or obligate barophiles as a result of decompression during sample recovery [55,56]. Those surviving decompression and subsequent recompression are barophiles which are also "atmospheric pressure tolerant". The increase in abundance for M5 waters (incubations with no added nutrients and at atmospheric pressure) was probably brought about by a positive growth response upon depressurisation of surface-derived organisms that sedimented on phytodetritus material.

Figure 3. Change in total prokaryotic abundance (*N*-fold increase) relative to time zero. Atm: Atmospheric pressure; M5: High chlorophyll (red); M6: Low chlorophyll (blue). Error bars represent the standard error of the mean.

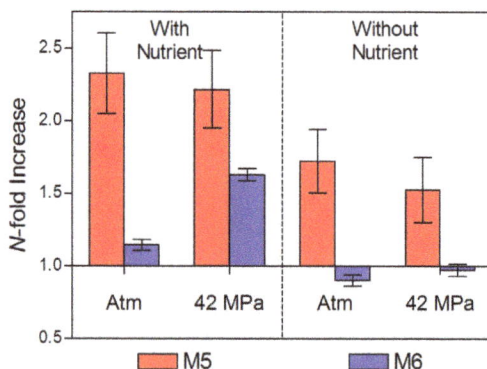

Table 3. Bacteria and Archaea: Shannon's diversity index (H). Separate values are presented for bacterial and archaeal populations and were calculated using DGGE band peak heights. N+ denotes incubation with added nutrients; N− denotes incubation without nutrient addition.

	HC (M5)		LC (M6)	
	Bact	Arch	Bact	Arch
Time Zero	2.34	1.122	2.459	2.203
Atm, N+	1.963	2.224	0.446	1.127
42 MPa, N+	2.127	1.895	1.365	1.595
Atm, N−	1.852	2.449	1.99	2.161
42 MPa, N−	1.84	2.851	2.276	1.819

The largest increase in abundance was found for M5 water with added nutrients. Incubations of M5 waters at atmospheric and *in-situ* pressures underwent similar increases in abundance (2.3-fold and 2.2-fold: Figure 3). Abundance increased with nutrient- supplementation and at *in-situ* pressure (1.6-fold increase) for incubations with M6 waters (low chlorophyll site) but remained relatively unchanged (1.1-fold increase) when incubations were carried out at atmospheric pressure. Site M6 may be dominated by an active community of obligate barophiles (piezophiles) which responded to nutrient input at *in-situ* pressure but did not do so when incubated at atmospheric pressure. Sequences from DGGE bands showed that the *in-situ* pressure incubations were dominated by three members of the genus Moritella (HQ731657, HQ731668 and HQ731671) closely affiliated with *M. abyssi* and *M. profunda* (Figure 4). Conversely, incubations of waters from site M5 (high chlorophyll) showed increases in abundance that were greater than those observed for M6 (low chlorophyll) waters irrespective of nutrient amendment or not. Post-incubation changes in abundance were similar in magnitude at atmospheric and *in-situ* pressures. This is indicative of a more evenly mixed community of piezophilic (barophilic) and piezosensitive types in the water column because of the deposition of surface-derived mesophiles associated with sedimenting phytodetritus at the high chlorophyll (M5) site. It is known that surface-derived microorganisms can survive in the deep ocean and respond to increased hydrostatic pressure through changes in cell shape and structure [28]. The increase in abundance observed for incubations with M5 water without added nutrients (Figure 3) may have been due to elevated background levels of organic nutrient caused by the arrival of phytodetritus material on the seafloor at M5 [18]. Alternatively, a "wall effect" [53,54] may have contributed to the increase in abundance. However, the same increase was not evident for M6 incubations (low chlorophyll site) without added nutrients.

The number of OTUs and their relative abundances were estimated from the number and relative intensity of DGGE bands. Shannon's general diversity index (H') and Simpson's index of dominance (c) were calculated using peak height values from densitometric scans of DGGE profiles. Although we recognise the limitations associated with using such indices for the purposes of defining diversity (in particular species richness) in the context of a limited number of DGGE-generated OTUs [57], it was beyond the scope of the present investigation to ascertain how rare taxa within the microbial community responded to perturbations. Recent advances in assessing biodiversity, such as high throughput sequencing, have demonstrated the presence of large numbers of rare (low abundance) taxa in natural environments that cannot be detected by cruder methods such as DGGE [2,58]. However, it is thought that this "rare biosphere" represents a largely dormant seed-bank of biodiversity, not directly involved in ecosystem functions such as nutrient cycling [58,59]. This study makes the assumption that taxa found to be abundant are involved in ecosystem functions and that rare taxa are not. If a rare and previously undetected taxa increases in abundance under incubation conditions to the level of detection afforded by the PCR/DGGE method then it can be assumed to be active under the given conditions. In other words this study provides a narrow view focused on those organisms that were abundant and therefore active at the two study sites and those that came to dominate the incubation experiments. Those organisms that remained below the level of detection are assumed to be irrelevant within the context of this study.

Even though nested PCR-DGGE is a powerful molecular fingerprinting method for detecting hierarchical taxonomic groups in microbial communities, it has been criticised for quantitative assessments, as the proportionality between initial template amounts and amplicon concentration can be lost through amplification cycling [60,61]. However, previous work has demonstrated that reliable quantifications can be performed as long as stringent optimisation of the PCR conditions is carried out to limit any potential bias associated with nested PCR-DGGE [7,62,63].

Figure 4. Evolutionary relationships of sequenced bacterial DGGE bands. Neighbor-Joining consensus tree of 16S rRNA V3 gene sequences obtained from DGGE analyses of bacterial communities in enrichment cultures. Bootstrap values were calculated from 2,000 replicates and only values >50% are shown. Scale bar represents the number of base substitutions per site.

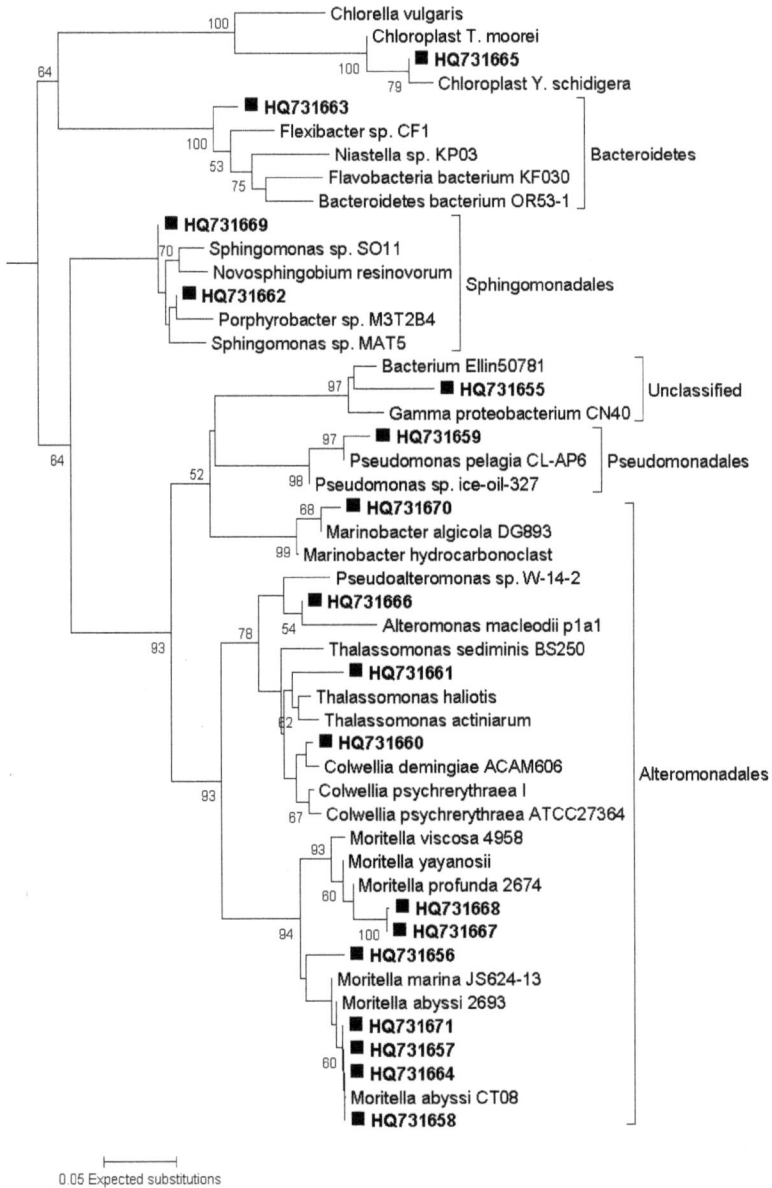

Diversity and dominance indices are shown in Tables 2 and 3. The Shannon index for the low chlorophyll site M6 (H' = 3.02) was somewhat greater than that of M5 (H' = 2.59) in the pre-incubation (time zero) samples. The Shannon index was higher for the high-chlorophyll site (M5) for all post-incubation samples. The changes in total prokaryotic diversity (bacteria and archaea combined) associated with each incubation condition (relative to the pre-incubation community) are shown in Figure 5. Diversity decreased for all incubations with water from site M6 (low chlorophyll waters). The extent of the decrease was greater in the presence of nutrients, and greatest for incubations with nutrients and at atmospheric pressure. The environmental conditions found here (incubation with added nutrients at surface pressure) might be considered to be the most different from the *in-situ* conditions found at M6 and may have favored the selection of a rare sub-population. The converse is evident for incubations of M5 waters (high chlorophyll site) which underwent minimal increases in diversity. The largest increase was measured for incubations at *in-situ* pressure (42 MPa) without added nutrients (Figure 5). These incubation conditions simulated the *in-situ* environment so changes in community structure were expected to be small.

Figure 5. Change in total prokaryotic community diversity (Shannon index, H) relative to time zero. Atm: Atmospheric pressure; M5: High chlorophyll (red); M6: Low chlorophyll (blue).

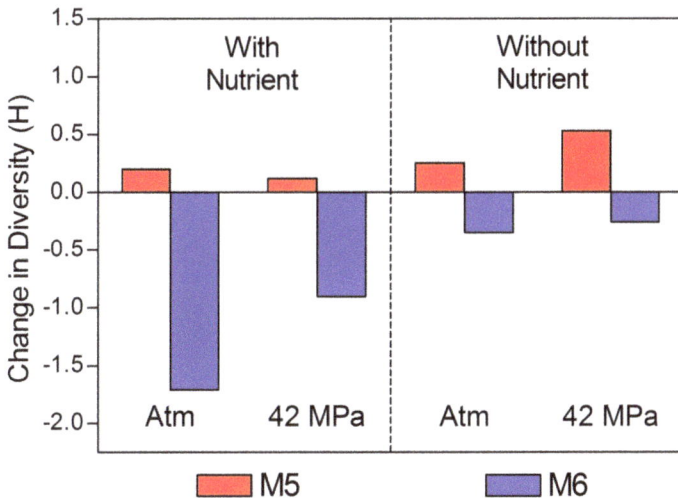

Those microorganisms that were present in the pre-incubation community at abundances below the DGGE detection threshold would have increased in number with the general increase in population size. These would have been detectable at the end of the incubation period thus giving an apparent increase in diversity. Furthermore, this may also be taken as evidence that the community structure was not greatly perturbed by decompression during sample recovery. If this had been the case, a greater reduction in diversity would have been expected due to the loss of decompression-sensitive members of the population.

Changes in abundance as a result of incubation (Figure 3) were analysed by paired two-tailed t-tests. There was no significant difference ($p = 0.706$) between the change in abundance due to incubation pressures (atmospheric compared with incubations at 420 MPa) and nutrient treatments (without and with added nutrients, $p = 0.215$). The change in abundance was found however to be significantly different between incubations of waters from the high chlorophyll region M5 and the low chlorophyll site M6 ($p = 0.012$). Regarding diversity data, no significant differences were found for incubations carried out at atmospheric and *in-situ* pressures ($p = 0.252$) or incubations without and with added nutrients ($p = 0.116$), but the overall population diversity for the high chlorophyll region (M5) was significantly different from that of the low chlorophyll region M6 ($p = 0.033$).

Whilst the presence of a functionally significant piezophile community in the deep-sea has been well established [26,44,45,64,65], it is still common for estimates of bacterial activity in the deep-sea to be carried out at atmospheric pressure [23–25]. Our results suggest that the changes that occur in deep sea communities incubated at atmospheric and *in-situ* pressures are similar in terms of population density and diversity indices. Community composition is strongly affected by incubation pressure and as such the results of incubations at atmospheric pressure may not provide a sufficiently accurate representation of the natural community. This may be particularly relevant for a low chlorophyll site like M6 which is more broadly representative of typical deep-sea environments than M5 (high chlorophyll site). For example, the bacterial community at the low chlorophyll site (M6) underwent a considerable degree of restructuring in response to nutrient amendment at atmospheric pressure. Analysis of M6 community structure incorporating sequence data (Figure 6) indicate that incubations carried out at atmospheric pressure and with added nutrients became dominated by four *Moritella* species. Two of these represented 50% of the post-incubation population and were not detected by DGGE in the pre-incubation community.

Conversely for the site M5 (high chlorophyll), all the OTUs detected after incubation at atmospheric pressure with added nutrients were already present in the pre-incubation community. A different community composition however was established when they were incubated at *in-situ* pressure. Again it should be noted that in this regard M5 and M6 communities exhibited different patterns of response to enrichments (Figures 6 and 7). Figure 7 shows the composition of the Archaeal incubation community at M5 (high chlorophyll site) was almost exclusively dominated (82%) by just two *Crenarchaeota* (accession numbers HQ731672 and HQ731679), but they comprised only 14% of the community at M6 (low chlorophyll site).

Figure 6. Changes in bacterial community structure in enrichment cultures. The percent contribution of sequenced DGGE bands to the total community. Relative abundance is derived from individual band intensities relative to the total sum intensity of all bands in the relevant lane. The large central pie represents the pre-incubation community. P+: 42 MPa with nutrient; P−: 42 MPa without nutrient; A+: Atmospheric pressure with nutrient; A−: Atmospheric pressure without nutrient.

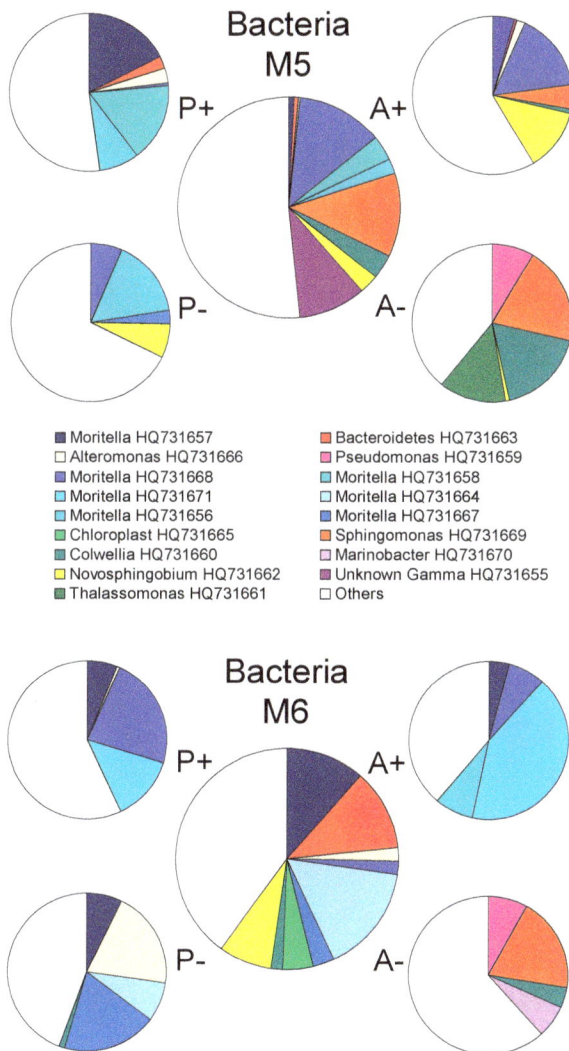

Bacteria M5

Moritella HQ731657
Alteromonas HQ731666
Moritella HQ731668
Moritella HQ731671
Moritella HQ731656
Chloroplast HQ731665
Colwellia HQ731660
Novosphingobium HQ731662
Thalassomonas HQ731661

Bacteroidetes HQ731663
Pseudomonas HQ731659
Moritella HQ731658
Moritella HQ731664
Moritella HQ731667
Sphingomonas HQ731669
Marinobacter HQ731670
Unknown Gamma HQ731655
Others

Bacteria M6

Figure 7. Changes in archaeal community structure in enrichment cultures. The percent contribution of sequenced DGGE bands to the total community. Relative abundance is derived from individual band intensities relative to the total sum intensity of all bands in the relevant lane The large central pie represents the pre-incubation community. P+: 42 MPa with nutrient; P−: 42 MPa without nutrient; A+: Atmospheric pressure with nutrient; A−: Atmospheric pressure without nutrient. No bands were sequenced from the M6 A+ incubation.

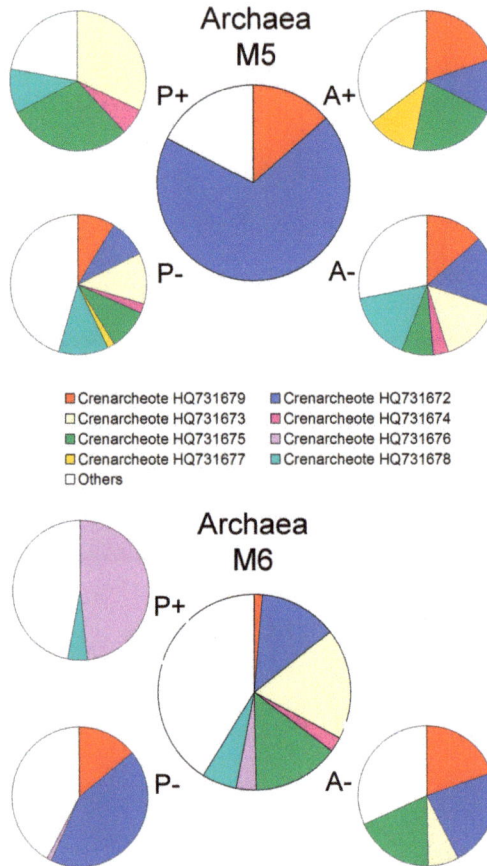

These two *Crenarchaeota* were prevalent in almost all incubations except for *in-situ* pressure with nutrients and were phylogenetically related to ammonia-oxidising chemolithoautotrophs *Nitrosopumilus maritimus* and *Cenarchaeum symbiosum* (Figure 8). Indeed, all the archaeal bands that were sequenced in this study were *Crenarchaeota*, This finding suggests a prevalent autotrophic role for the Archaeal assemblages at both sites, which would be in broad general agreement with earlier studies that have shown autotrophy to be the dominant archaeal metabolism in deep oceanic waters [66,67].

Figure 8. Evolutionary relationships of sequenced archaeal DGGE bands. Neighbour-Joining consensus tree of 16S rRNA V3 gene sequences obtained from DGGE analyses of archaeal communities in enrichment cultures. Bootstrap values were calculated from 2,000 replicates and only values >40% are shown. Scale bar represents the number of base substitutions per site.

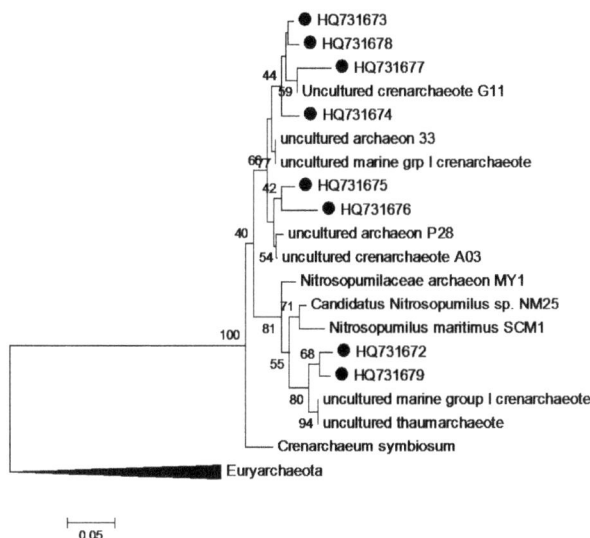

M6 (low chlorophyll site) was subject to a constant low nutrient status typical of deep sea habitats while M5 (high chlorophyll site) experienced relatively large episodic inputs of nutrient from seasonal surface productivity. The microbial community at M5 was found to be dominated by a piezophillic phenotype and was shown to be less capable of adapting to the changed conditions in the incubations, perhaps as a result of conditioning to allochtonous inputs. Furthermore, population size and diversity in the M5 community changed to a lesser degree (relative to the t0) in response to different incubation conditions. This may be due to the persistence of surface-adapted microorganisms transported to the deep on sedimenting particles. The growth rates of such piezosensitive microorganisms would be retarded by hydrostatic pressures encountered at depth but this does not normally result in cell death [68].

3. Experimental Section

3.1. Study Area and Sampling Sites

Study sites M5 (high-chlorophyll) and M6 (low-chlorophyll) were located 525 km apart in the Crozet archipelago region of the Southern Indian Ocean. M5 was situated approximately 350 km east (45°54.27'S, 56°24.83'E) of the Crozet Islands in an area influenced by an annual bloom of primary production [10]. M6 was located approximately 300 km south (49°04.7'S; 51°13.0'E) of the islands in an area considered to be relatively isolated from the bloom event. Sampling was carried out aboard the RRS Discovery (cruise D300) between December 2005 and January 2006, following termination of the local spring phytoplankton bloom.

3.2. Water Sampling

Water column samples were collected from 12 m (±2 m) above the seafloor using 10 L Niskin bottles mounted on a CTD frame. Upon recovery of the CTD, water samples (10 L) were transferred to chilled (4 °C) 25 L aseptic plastic containers and immediately removed to a constant temperature laboratory (4 °C) aboard ship.

3.3. Enrichment Incubations

Water samples (10 L) were shaken before aliquots (100 mL) of seawater were placed in sterile polyethylene sample bags (155 mm × 76 mm), with or without the addition of organic loading (2.5 mg mL^{-1} peptone, 0.5 mg mL^{-1} yeast extract and 1.285 mg mL^{-1} N-acetyl-d-glucosamine). Parallel samples (50 mL) were preserved as Time-zero controls for epifluorescence microscopy by the addition of 1% (v/v) particle-free buffered formalin. Bags (100 mL) were sealed aseptically with an impulse heat sealer (Packer, Basildon, Essex, UK) taking care to exclude all air. Bags were placed in water-filled titanium pressure vessels and incubated in darkness at 42 MPa and at 4 °C (±1 °C). Bags were also incubated at ambient atmospheric pressure (1 atm.) in darkness and at 4 °C. Samples taken from the high chlorophyll site (M5) site were incubated for 21 days, those from the low chlorophyll site (M6) for 18 days. Upon termination of the incubation experiments aliquots (10 mL) were preserved for epifluorescence microscopy by the addition of 1% (v/v) formalin, and stored at 4 °C. Epifluorescence microscopy was performed in accordance with the method of Parkes and co-workers [69] with the exception that Sybr-gold® (Invitrogen, Paisley, UK) was used instead of Acridine Orange. Appropriate volumes (1–5 mL) of formalin-fixed samples were filtered onto 25 mm (diameter) black Isopore® filter membranes of 0.2 μm pore size (Millipore, Cork, Ireland). Bacteria were counted on fifteen to thirty microscope fields for each sample, using a Nikon Optiphot-2 UV Microscope with B-2A excitation filter.

3.4. DNA Extraction

Microbial biomass was collected and concentrated by filtration of 90 mL of the enrichment broth through 0.22 μm pore Sterivex filter cartridges (Millipore, Cork, Ireland) using a peristaltic pump (Watson Marlow, Cornwall UK). Sterivex cartridges were stored at −80 °C until extraction of community genomic DNA was carried out using the QIAamp DNA mini kit (Qiagen, Hilden, Germany). The kit was used in accordance with the manufacturers protocol for the isolation of DNA from bacterial cultures (Qiagen DNA Mini-kit Protocol D, Hilden, Germany) with the following modifications aimed at scaling up the volume of reagents for use in a Sterivex cartridge:

Sterivex filter cartridges were sealed at both ends with Luer-lok caps and incubated with 500 μL of lysozyme buffer (20 mg mL^{-1} lysozyme, 20 mM Tris, 2 mM EDTA, 1.2% Triton® X-100) for one hour at 37 °C. Qiagen Buffer AL (450 μL) and 50 μL Proteinase K (Qiagen) were then added and the cartridge was incubated for a further 1–3 h at 56 °C with brief vortexing every twenty minutes.

The lysate was collected in a sterile 2.0 mL microcentrifuge tube, incubated at 90 °C for four minutes and flash cooled on ice. Pure Ethanol (500 µL) was added and the mixture was vortexed for 10 s. The mixture was then passed through the binding membrane of the Qiagen column in two 750 µL volumes by centrifugation at 10,000 rcf in a Sigma model 1–15 microcentrifuge. The binding, washing and elution steps were carried out as described in the manufacturers protocol without any further modification. DNA was eluted in 200 µL of Qiagen buffer AE and stored at −20 °C.

DNA yield and quality was estimated visually on a 1% Agarose gel by comparison to a DNA molecular weight marker (GelPilot 100 bp Plus Ladder, QIAGEN, Hilden, Germany).

3.5. PCR and DGGE

Community genomic DNA was used as template for PCR (Polymerase Chain Reaction) amplification of bacterial and archaeal 16S rRNA genes. Amplification yield and specificity were optimised by employing nested PCR primers and a "touchdown" amplification protocol [70]. The bacterial domain specific primer pair 27f and 1492r [71], and Archaeal domain specific primers 21f and 958r [72] were used to amplify 16S rRNA gene sequences from genomic DNA template. The reactions (50 µL) were prepared on ice and contained 250 µM each of the deoxynucleoside triphosphates, 1× PCR buffer, 7.5 picomoles of each oligonucleotide primer, 2.0–2.5 mM $MgCl_2$ and 1.5 units of Taq DNA polymerase (Bioline, London, UK) in 18 MΩ analytical grade water (Sigma, Dorset, UK). PCR was performed in a G-Storm GS1 thermal cycler (Labtech, East Sussex, UK) using the following programs:

Bacterial 16S rRNA gene: Denaturation for 4 min at 94 °C, 10 cycles of denaturation (30 s at 94 °C), annealing (1 min at 58 °C) and extension (1 min at 72 °C), followed by 22 cycles of denaturation (30 s at 92 °C), annealing (1 min at 58 °C) and extension (1 min at 72 °C), with a final extension for 10 min at 72 °C.

Archaeal 16S rRNA gene. Denatured for 4 min at 95 °C, 30 cycles of denaturation (1 min at 94 °C), annealing (1 min at 56 °C) and extension (1 min at 74 °C), with a final extension for 10 min at 72 °C.

Amplicons from this first round of amplification were used as templates for a second PCR in order to obtain fragments suitable for DGGE, using primers directed towards the internal V3 region of the 16S gene. Bacterial templates were amplified using the primers 341f and 517r [29], Archaeal templates were amplified with the primers 340f and 519r [73]. Forward primers were 5'-labelled with a 40-mer GC nucleotide "clamp" sequence [74]. The reactions (50 µL) contained 250 µM of each deoxynucleoside triphosphate, 1× PCR buffer, 3.75 picomoles of each oligonucleotide primer, 2.0 mM $MgCl_2$ and 0.75 units of Taq DNA polymerase. The PCR programs were as follows:

Bacterial 16S rRNA gene (V3 region): Denaturation for 4 min at 94 °C, a "touchdown" step consisting of 10 cycles of denaturation (30 s at 94 °C), annealing (1 min at 67 °C reducing by 1 °C each cycle) and extension (1 min at 72°C), followed by 20 cycles of denaturation (30 s at 92 °C), annealing (1 min at 57 °C) and extension (1 min at 72 °C), with a final extension for 20 min at 72 °C.

Archaeal 16S rRNA gene (V3 region): Denaturation for 4 min at 95 °C, a "touchdown" step consisting of 10 cycles of denaturation (1 min at 94 °C), annealing (1 min at 60.5 °C reducing by 0.5 °C each cycle) and extension (1 min at 72 °C), followed by 25 cycles of denaturation (1 min at 92 °C), annealing (1 min at 55.5 °C) and extension (1 min at 72 °C), with a final extension for 20 min at 72 °C.

PCR product sizes were verified by agarose gel electrophoresis with a molecular weight marker.

DGGE was performed with a D-Code Mutation Detection System (Bio-Rad Laboratories, Philadelphia, PA, USA). An 8% (w/v) polyacrylamide gel was used, incorporating a linear concentration gradient of denaturant ranging from 40% to 60% [100% denaturant = 7 M urea, 40% (v/v) formamide]. Gels were cast using a Model 475 Gradient Delivery System (Bio-Rad Laboratories, Philadelphia, PA, USA) in accordance with the manufacturers protocol except that ammonium persulphate (APS) and N,N,N',N'-tetramethylethylenediamine (TEMED) were used at 0.8% and 0.08% (v/v) respectively, and glycerol was incorporated into the gel at a final concentration of 2% (v/v). Electrophoresis was carried out in 1× TAE (40 mM Tris, 0.02 M sodium acetate, 1 mM EDTA) buffer at 60 °C for 16.5 h at 60 V. Gels were stained with ethidium bromide (% w/v) and photographed under UV illumination (300 nm).

3.6. Quantitative Analysis of DGGE Fingerprints

Digital images of DGGE gels were analysed using Total Lab TL120 software (Nonlinear Dynamics, Newcastle upon Tyne, UK). A densitometric scan of each lane was created and background noise was subtracted using a rolling disc algorithm. For each gel a matrix was constructed using the peak height values of bands in each densitometric profile. Principal coordinate analysis based on Euclidean distances was carried out using the MVSP Multivariate Statistics Package V3.13p [75]. The diversity of bacterial and archaeal communities was examined by the Shannon index of general diversity, H' and Simpsons index of dominance, c [76]. The equations used were:

$$H' = -\sum_{i=1}^{s} (n_i / N) \log(n_i / N) \qquad [77]$$

$$c = 1 - \sum_{i=1}^{s} (n_i / N)^2 \qquad [39]$$

where n_i is the height of individual band peaks and N is the sum of all band peak heights.

Matrices of peak height values derived from densitometric scans of separate archaeal and bacterial DGGE gels were combined for calculations of total community diversity.

3.7. Sequencing of DGGE Bands

A sterile 1 mL pipette tip was used to remove four small cores of gel across the centre of selected bands. The gel cores were collected in 50 μL of gel elution buffer (0.3 M NaCl; 3 mM EDTA; 30 mM Tris, pH 7.6) and DNA was eluted by incubation at 50 °C for twenty minutes followed by a further incubation for 18 h at room temperature. A 1 μL aliquot of the eluate was used as template for PCR. Primers (341f/517r for Bacteria and 340f/519r for Archaea) were used to re-amplify the recovered DNA and were screened by DGGE to ensure that all were resolved as discrete bands. The reactions (50 μL) contained 250 μM of each deoxynucleoside triphosphate, 1× PCR buffer, 3.75 picomoles of each oligonucleotide primer, 2.0 mM MgCl$_2$ and 0.75 units of Taq DNA polymerase. The PCR programs consisted of an initial denaturation step for 4 min at 94 °C, followed by 30 cycles of denaturation (30 s at 94 °C), annealing (45 s at 55 °C) and extension (40 s at 72 °C), with a final extension for 10 min at 72 °C.

For DNA sequencing the PCR was performed using primers without GC clamps. PCR products were purified using the QIAGEN QIAquick® PCR Purification kit and were submitted to GATC Biotech (Konstanz, Germany) for direct single strand sequencing using an ABI 3730 XL Illumina Genome Analyzer (Life Technologies, Grand Island, NY, USA). Twenty five sequences were recovered and were assigned accession numbers HQ731657 to HQ731679.

3.8. Phylogenetic Analysis of Sequence Data

The recovered partial 16S rRNA gene sequences were submitted to the Basic Local Alignment Search Tool [78] web portal maintained by the National Centre for Biotechnology Information [79] in order to identify database sequences with highest similarity. Sequences were aligned with ClustalW2 [80] and evolutionary analyses were conducted in MEGA5 [81]. Phylogenetic position was inferred using the Neighbour-Joining method [82]. The bootstrap consensus trees (Figures 4 and 8) were inferred from 2000 replicates [83] and are taken to represent the evolutionary history of the analysed sequences. The percentage of replicate trees in which the associated sequences clustered together in the bootstrap test are shown next to the branches [83]. The trees are drawn to scale, with branch lengths in the same units as those of the evolutionary distances used to infer the phylogenetic tree. The evolutionary distances were computed using the Kimura 2-parameter method [84] and are presented in units of the number of base substitutions per site.

4. Conclusions

This study demonstrated contrasting ecological responses to nutrient enrichment by the prokaryotic communities at the two hydrologically similar deep-sea sites which differed in the nutrient regime of the overlying waters. The community from the high chlorophyll site (M5) was better able to maintain population size and diversity under incubation conditions. This may be due to the persistence of surface-adapted microorganisms transported on sedimenting particles, or to the presence of an enhanced reservoir of diversity in the sediments and benthic boundary layer at M5. The dominant taxa within the community at M5 were found to belong to groups representing a greater range of metabolic diversity (compared to the low chlorophyll site M6), at least partly as a result of allochtonous input, and possibly as a result of community structure adaptation to cyclical disturbances. Furthermore, a contrasting response to organic nutrient was observed between bacterial and archaeal assemblages at the two sites which may suggest different functional roles in response to nutrient input. There were also important differences observed in community response when samples from either site were incubated under atmospheric and *in situ* pressures. The identity of the organisms which dominated M5 (high chlorophyll site) incubations at atmospheric pressure indicated that they may have been surface-derived. Their absence from pressure incubations would suggest that these species were disadvantaged *in situ* and were therefore not major contributors to deep-sea ecosystem functions. Changes in diversity and abundance associated with incubation of unsupplemented sea water under pressure would indicate the loss of obligate barophiles as a result of decompression during sampling.

Findings presented here and elsewhere [7,9] supports the tenet that studies which involve growing deep-sea microorganisms to examine community responses to perturbations should be carried out under conditions of *in situ* pressure. It is hoped that this study should stimulate further research into this area, particularly the effects of pressure on microbial community structure and the differences in community structure dynamics that arise as a result of conditioning. Future studies need to address the issue of sample depressurisation in order for results to be meaningful. Tamburini and co-authors [9] have recently reviewed prokaryotic responses to hydrostatic pressure in the ocean and placed some emphasis on the importance for the development of pressure retaining sampling systems. These will finally enable us to determine the role played by obligate barophiles in the deep ocean.

Acknowledgments

The authors would like to acknowledge the officers and crew of the RRS Discovery for their invaluable assistance.

Conflicts of Interest

The authors declare no conflict of interest.

References

1. Yokokawa, T.; Nagata, T. Linking bacterial community structure to carbon fluxes in MARINE environments. *J. Oceanogr.* **2010**, *66*, 1–12.
2. Pedros-Alio, C. Ecology: Dipping into the rare biosphere. *Science* **2007**, *315*, 192–193.
3. Dumestre, J.F.; Casamayor, E.O.; Massana, R.; Pedros-Alio, C. Changes in bacterial and archaeal assemblages in an equatorial river induced by the water eutrophication of Petit Saut dam reservoir (French Guiana). *Aquat. Microb. Ecol.* **2002**, *26*, 209–221.
4. Lebaron, P.; Servais, P.; Troussellier, M.; Courties, C.; Vives-Rego, J.; Muyzer, G.; Bernard, L.; Guindulain, T.; Schafer, H.; Stackebrandt, E. Changes in bacterial community structure in seawater mesocosms differing in their nutrient status. *Aquat. Microb. Ecol.* **1999**, *19*, 255–267.
5. Leflaive, J.; Danger, M.; Lacroix, G.; Lyautey, E.; Oumarou, C.; Ten-Hage, L. Nutrient effects on the genetic and functional diversity of aquatic bacterial communities. *FEMS Microbiol. Ecol.* **2008**, *66*, 379–390.
6. Riemann, L.; Steward, G.F.; Azam, F. Dynamics of bacterial community composition and activity during a mesocosm diatom bloom. *Appl. Environ. Microbiol.* **2000**, *66*, 578–587.

7. Egan, S.T.; McCarthy, D.M.; Patching, J.W.; Fleming, G.T.A. An investigation of the physiology and potential role of components of the deep ocean bacterial community (of the NE Atlantic) by enrichments carried out under minimal environmental change. *Deep Sea Res. Part I Oceanogr. Res. Pap.* **2012**, *61*, 11–20.

8. Jorgensen, B.B.; Boetius, A. Feast and famine: Microbial life in the deep-sea bed. *Nat. Rev. Microbiol.* **2007**, *5*, 770–781.

9. Tamburini, C.; Boutrif, M.; Garel, M.; Colwell, R.R.; Deming, J.W. Prokaryotic responses to hydrostatic pressure in the ocean—A review. *Environ. Microbiol.* **2013**, *15*, 1262–1274.

10. Pollard, R.T.; Venables, H.J.; Read, J.F.; Allen, J.T. Large-scale circulation around the Crozet Plateau controls an annual phytoplankton bloom in the Crozet Basin. *Deep Sea Res. Part II Top. Stud. Oceanogr.* **2007**, *54*, 1915–1929.

11. Pollard, R.; Sanders, R.; Lucas, M.; Statham, P. The Crozet natural Iron Bloom and Export Experiment (CROZEX). *Deep Sea Res. Part II Top. Stud. Oceanogr.* **2007**, *54*, 1905–1914.

12. Treguer, P.; Jacques, G. Dynamics of nutrients and phytoplankton, and fluxes of carbon, nitrogen and silicon in the Antarctic Ocean. *Polar Biol.* **1992**, *12*, 149–162.

13. Pitchford, J.; Brindley, J. Iron limitation, grazing pressure and oceanic High Nutrient-Low Chlorophyll (HNLC) regions. *J. Plankton Res.* **1999**, *21*, 525–547.

14. Planquette, H.; Statham, P.J.; Fones, G.R.; Charette, M.A.; Moore, C.M.; Salter, I.; Nedelec, F.H.; Taylor, S.L.; French, M.; Baker, A.R.; *et al.* Dissolved iron in the vicinity of the Crozet Islands, Southern Ocean. *Deep Sea Res. Part II Top. Stud. Oceanogr.* **2007**, *54*, 1999–2019.

15. Pollard, R.; Read, J. Circulation pathways and transports of the Southern Ocean in the vicinity of the Southwest Indian Ridge. *J. Geophys. Res.* **2001**, *106*, 2881–2898.

16. Young-Hyang, P.; Gamberoni, L.; Charriaud, E. Frontal structure, water masses, and circulation in the Crozet Basin. *J. Geophys. Res.* **1993**, *98*, 361–385.

17. Pollard, R.T.; Lucas, M.I.; Read, J.F. Physical controls on biogeochemical zonation in the Southern Ocean. *Deep Sea Res. Part II Top. Stud. Oceanogr.* **2002**, *49*, 3289–3305.

18. Hughes, J.A.; Smith, T.; Chaillan, F.; Bett, B.J.; Billett, D.S.M.; Boorman, B.; Fischer, E.H.; Frenz, M.; Wolff, G.A. Two abyssal sites in the Southern Ocean influenced by different organic matter inputs: Environmental characterisation and preliminary observations on the benthic foraminifera. *Deep Sea Res. Part II Top. Stud. Oceanogr.* **2007**, *54*, 2275–2290.

19. Martin, J.H.; Knauer, G.A.; Karl, D.M.; Broenkow, W.W. VERTEX: Carbon cycling in the northeast Pacific. *Deep Sea Res. Part I Oceanogr. Res. Pap.* **1987**, *34*, 267–285.

20. Sanders, R.; Morris, P.J.; Stinchcombe, M.; Seeyave, S.; Venables, H.; Lucas, M. New production and the f ratio around the Crozet Plateau in austral summer 2004–2005 diagnosed from seasonal changes in inorganic nutrient levels. *Deep Sea Res. Part II Top. Stud. Oceanogr.* **2007**, *54*, 2191–2207.

21. Turley, C. Bacteria in the cold deep-sea benthic boundary layer and sediment-water interface of the NE Atlantic. *FEMS Microbiol. Ecol.* **2000**, *33*, 89–99.

22. Jamieson, R.E.; Rogers, A.D.; Billett, D.S.M.; Pearce, D.A. Bacterial biodiversity in deep-sea sediments from two regions of contrasting surface water productivity near the Crozet Islands, Southern Ocean. *Deep Sea Res. Part A Oceanogr. Res. Pap.* **2013**, in press.

23. Danovaro, R.; Corinaldesi, C.; Luna, G.M.; Magagnini, M.; Manini, E.; Pusceddu, A. Prokaryote diversity and viral production in deep-sea sediments and seamounts. *Deep Sea Res. Part II Top. Stud. Oceanogr.* **2009**, *56*, 738–747.

24. Bendtsen, J.; Lundsgaard, C.; Middelboe, M.; Archer, D. Influence of bacterial uptake on deep-ocean dissolved organic carbon. *Glob. Biogeochem. Cycle* **2002**, *16*, 74:1–74:12.

25. Danovaro, R.; della Croce, N.; Dell'Anno, A.; Pusceddu, A. A depocenter of organic matter at 7800 m depth in the SE Pacific Ocean. *Deep Sea Res. Part I Oceanogr. Res. Pap.* **2003**, *50*, 1411–1420.

26. Fang, J.; Zhang, L.; Bazylinski, D.A. Deep-sea piezosphere and piezophiles: Geomicrobiology and biogeochemistry. *Trends Microbiol.* **2010**, *18*, 413–422.

27. Forney, L.J.; Zhou, X.; Brown, C.J. Molecular microbial ecology: Land of the one-eyed king. *Curr. Opin. Microbiol.* **2004**, *7*, 210–220.

28. Grossart, H.P.; Gust, G. Hydrostatic pressure affects physiology and community structure of marine bacteria during settling to 4000 m: An experimental approach. *Mar. Ecol. Prog. Ser.* **2009**, *390*, 97–104.

29. Muyzer, G.; de Waal, E.C.; Uitterlinden, A.G. Profiling of complex microbial populations by denaturing gradient gel electrophoresis analysis of polymerase chain reaction-amplified genes coding for 16S rRNA. *Appl. Environ. Microbiol.* **1993**, *59*, 695–700.

30. Fromin, N.; Hamelin, J.; Tarnawski, S.; Roesti, D.; Jourdain-Miserez, K.; Forestier, N.; Teyssier-Cuvelle, S.; Gillet, F.; Aragno, M.; Rossi, P. Statistical analysis of denaturing gel electrophoresis (DGE) fingerprinting patterns. *Environ. Microbiol.* **2002**, *4*, 634–643.

31. Rink, B.; Seeberger, S.; Martens, T.; Duerselen, C.D.; Simon, M.; Brinkhoff, T. Effects of phytoplankton bloom in a coastal ecosystem on the composition of bacterial communities. *Aquat. Microb. Ecol.* **2007**, *48*, 47–60.

32. Yoshida, A.; Nishimura, M.; Kogure, K. Bacterial community structure in the Sulu Sea and adjacent areas. *Deep Sea Res. Part II Top. Stud. Oceanogr.* **2007**, *54*, 103–113.

33. Wu, L.; Yu, Y.; Zhang, T.; Feng, W.; Zhang, X.; Li, W. PCR-DGGE fingerprinting analysis of plankton communities and its relationship to lake trophic status. *Int. Rev. Hydrobiol.* **2009**, *94*, 528–541.

34. Bonin, P.C.; Michotey, V.D.; Mouzdahir, A.; Rontani, J.-F. Anaerobic biodegradation of squalene: Using DGGE to monitor the isolation of denitrifying Bacteria taken from enrichment cultures. *FEMS Microbiol. Ecol.* **2002**, *42*, 37–49.

35. Nakagawa, T.; Fukui, M. Phylogenetic characterization of microbial mats and streamers from a Japanese alkaline hot spring with a thermal gradient. *J. Gen. Appl. Microbiol.* **2002**, *48*, 211–222.

36. Stephen, J.R.; Chang, Y.-J.; Gan, Y.D.; Peacock, A.; Pfiffner, S.M.; Barcelona, M.J.; White, D.C.; Macnaughton, S.J. Microbial characterization of a JP-4 fuel-contaminated site using a combined lipid biomarker/polymerase chain reaction-denaturing gradient gel electrophoresis (PCR-DGGE)-based approach. *Environ. Microbiol.* **1999**, *1*, 231–241.

37. Murray, A.E.; Preston, C.M.; Massana, R.; Taylor, L.T.; Blakis, A.; Wu, K.; DeLong, E.F. Seasonal and spatial variability of bacterial and archaeal assemblages in the coastal waters near Anvers Island, Antarctica. *Appl. Environ. Microbiol.* **1998**, *64*, 2585–2595.

38. Spiegelman, D.; Whissell, G.; Greer, C.W. A survey of the methods for the characterization of microbial consortia and communities. *Can. J. Microbiol.* **2005**, *51*, 355–386.

39. Grossi, V.; Yakimov, M.M.; Al Ali, B.; Tapilatu, Y.; Cuny, P.; Goutx, M.; La Cono, V.; Giuliano, L.; Tamburini, C. Hydrostatic pressure affects membrane and storage lipid compositions of the piezotolerant hydrocarbon-degrading Marinobacter hydrocarbonoclasticus strain #5. *Environ. Microbiol.* **2010**, *12*, 2020–2033.

40. Yayanos, A.A.; Dietz, A.S. Death of a hadal deep-sea bacterium after decompression. *Science* **1983**, *220*, 497–498.

41. Chastain, R.A.; Yayanos, A.A. Ultrastructural changes in an obligately barophilic marine bacterium after decompression. *Appl. Environ. Microbiol.* **1991**, *57*, 1489–1497.

42. Park, C.B.; Clark, D.S. Rupture of the cell envelope by decompression of the deep-sea methanogen methanococcus jannaschii. *Appl. Environ. Microbiol.* **2002**, *68*, 1458–1463.

43. Yayanos, A.A.; Dietz, A.S. Thermal inactivation of a deep-sea barophilic bacterium, isolate CNPT-3. *Appl. Environ. Microbiol.* **1982**, *43*, 1481–1489.

44. Patching, J.W.; Eardly, D. Bacterial biomass and activity in the deep waters of the eastern Atlantic—Evidence of a barophilic community. *Deep Sea Res. Part I Oceanogr. Res. Pap.* **1997**, *44*, 1655–1670.

45. Eardly, D.F.; Carton, M.W.; Gallagher, J.M.; Patching, J.W. Bacterial abundance and activity in deep-sea sediments from the eastern North Atlantic. *Prog. Oceanog.* **2001**, *50*, 245–259.

46. Nagata, T.; Fukuda, H.; Fukuda, R.; Koike, I. Bacterioplankton distribution and production in deep Pacific waters: Large-Scale geographic variations and possible coupling with sinking particle fluxes. *Limnol. Oceanogr.* **2000**, *45*, 426–435.

47. Thingstad, T.F. Elements of a theory for the mechanisms controlling abundance, diversity and biogeochemical role of lytic viruses in aquatic systems. *Limnol. Oceanogr.* **2000**, *45*, 1320–1328.

48. Pernthaler, J. Predation on prokaryotes in the water column and its ecological implications. *Nat. Rev. Microbiol.* **2005**, *3*, 537–536.

49. Hyun, J.H.; Kim, K.H. Bacterial abundance and production during the unique spring phytoplankton bloom in the central Yellow Sea. *Mar. Ecol. Prog. Ser.* **2003**, *252*, 77–88.

50. Guixa-Boixereu, N.; Lysnes, K.; Pedros-Alio, C. Viral lysis and bacterivory during a phytoplankton bloom in a coastal water microcosm. *Appl. Environ. Microbiol.* **1999**, *65*, 1949–1958.

51. Pinhassi, J.; Sala, M.M.; Havskum, H.; Peters, F.; Guadayol, O.; Malits, A.; Marrase, C. Changes in bacterioplankton composition under different phytoplankton regimens. *Appl. Environ. Microbiol.* **2004**, *70*, 6753–6766.

52. Jurgens, K.; Matz, C. Predation as a shaping force for the phenotypic and genotypic composition of planktonic bacteria. *Antonie Leeuwenhoek* **2002**, *81*, 413–434.

53. ZoBell, C.E.; Anderson, D.Q. Observations on the multiplication of bacteria in different volumes of sea water and the influence of oxygen tension and solid surfaces. *Biol. Bull. (Woods Hole)* **1936**, *71*, 324–342.

54. Zobell, C.E. The effect of solid surfaces upon bacterial activity. *J. Bacteriol.* **1943**, *46*, 39–56.

55. Kato, C.; Bartlett, D.H. The molecular biology of barophilic bacteria. *Extremophiles* **1997**, *1*, 111–116.

56. Jannasch, H.W.; Taylor, C.D. Deep-sea microbiology. *Annu. Rev. Microbiol.* **1984**, *38*, 487–514.

57. Bent, S.J.; Forney, L.J. The tragedy of the uncommon: Understanding limitations in the analysis of microbial diversity. *ISME J.* **2008**, *2*, 689–695.

58. Sogin, M.L.; Morrison, H.G.; Huber, J.A.; Welch, D.M.; Huse, S.M.; Neal, P.R.; Arrieta, J.M.; Herndl, G.J. Microbial diversity in the deep sea and the underexplored "rare biosphere". *Proc. Natl. Acad. Sci. USA* **2006**, *103*, 12115–12120.

59. Fuhrman, J.A. Microbial community structure and its functional implications. *Nature* **2009**, *459*, 193–199.

60. Muylaert, K.; VanDerGucht, K.; Vloemans, N.; Meester, L.; Gillis, M.; Vyverman, W. Relationship between bacterial community composition and bottom-up versus top-down variables in four eutrophic shallow lakes. *Appl. Environ. Microbiol.* **2002**, *68*, 4740–4750.

61. Zwart, G.; Kamst-van Agterveld, M.P.; van der Werff-Staverman, I.; Hagen, F.; Hoogveld, H.L.; Gons, H.J. Molecular characterization of cyanobacterial diversity in a shallow eutrophic lake. *Environ. Microbiol.* **2005**, *7*, 365–377.

62. Park, J.W.; Crowley, D.E. Nested PCR bias: A case study of Pseudomonas spp. in soil microcosms. *J. Environ. Monit.* **1039**, *12*, 985–988.

63. Touzet, N.; McCarthy, D.; Fleming, G.T. Molecular fingerprinting of lacustrian cyanobacterial communities: Regional patterns in summer diversity. *FEMS Microbiol. Ecol.* **2013**, doi:10.1111/1574-6941.12172.

64. Bianchi, A.; Garcin, J. Bacterial response to hydrostatic-pressure in seawater samples collected in mixed-water and stratified-water conditions. *Mar. Ecol. Prog. Ser.* **1994**, *111*, 137–141.

65. Jannasch, H.W.; Wirsen, C.O.; Taylor, C.D. Deep-Sea Bacteria: Isolation in the Absence of Decompression. *Science* **1982**, *216*, 1315–1317.

66. Herndl, G.J.; Reinthaler, T.; Teira, E.; van Aken, H.; Veth, C.; Pernthaler, A.; Pernthaler, J. Contribution of archaea to total prokaryotic production in the deep Atlantic Ocean. *Appl. Environ. Microbiol.* **2005**, *71*, 2303–2309.

67. Ingalls, A.E.; Shah, S.R.; Hansman, R.L.; Aluwihere, L.I.; Santos, G.R.; Druffel, E.R.M.; Pearson, A. Quantifying archaeal community autotrophy in the mesopelagic ocean using natural radiocarbon. *Proc. Natl. Acad. Sci. USA* **2006**, *103*, 6442–6447.

68. Oger, P.M.; Jebbar, M. The many ways of coping with pressure. *Res. Microbiol.* **2010**, *161*, 799–809.

69. Parkes, R.J.; Cragg, B.A.; Getliff, J.M.; Harvey, S.M.; Fry, J.C.; Lewis, C.A.; Rowland, S.J. A quantitative study of microbial decomposition of biopolymers in Recent sediments from the Peru Margin. *Mar. Geol.* **1993**, *113*, 55–66.

70. Don, R.H.; Cox, P.T.; Wainwright, B.J.; Baker, K.; Mattick, J.S. "Touchdown" PCR to circumvent spurious priming during gene amplification. *Nucleic Acids Res.* **1991**, doi:10.1093/nar/19.14.4008.

71. Achenbach, L.; Woese, C.R. 16S and 23S rRNA-like Primers. In *Archaea—A Laboratory Manual*; Sower, K.R., Schreier, H.J., Eds.; Cold Spring Harbor Laboratory Press: Cold Spring Harbor, NY, USA, 1995; pp. 521–523.

72. DeLong, E.F. Archaea in coastal marine environments. *Proc. Natl. Acad. Sci. USA* **1992**, *89*, 5685–5689.

73. Ovreas, L.; Forney, L.; Daae, F.; Torsvik, V. Distribution of bacterioplankton in meromictic Lake Saelenvannet, as determined by denaturing gradient gel electrophoresis of PCR-amplified gene fragments coding for 16S rRNA. *Appl. Environ. Microbiol.* **1997**, *63*, 3367–3373.

74. Sheffield, V.C.; Cox, D.R.; Lerman, L.S.; Myers, R.M. Attachment of a 40 base-pair G+C-rich sequence to genomic DNA fragments by the polymerase chain reaction results in improved detection of single base changes. *Proc. Natl. Acad. Sci. USA* **1989**, *86*, 232–236.

75. Kovach, W. *MVSP-A Multivariate Statistical Package for Windows, ver. 3.13q*; Kovach Computing Services: Pentraeth, Wales, UK, 1998.

76. Simpson, E.H. Measurement of diversity. *Nature* **1949**, *163*, 688.

77. Shannon, C.E.; Weaver, W. *The Mathematical Theory of Communication*; University of Illinois Press: Urbana, IL, USA, 1963; p. 29.

78. Altschul, S.F.; Madden, T.L.; Schaffer, A.A.; Zhang, J.; Zhang, Z.; Miller, W.; Lipman, D.J. Gapped BLAST and PSI-BLAST: A new generation of protein database search programs. *Nucleic Acids Res.* **1997**, *25*, 3389–3402.

79. Johnson, M.; Zaretskaya, I.; Raytselis, Y.; Merezhuk, Y.; McGinnis, S.; Madden, T.L. NCBI BLAST: A better web interface. *Nucleic Acids Res.* **2008**, *36*, W5–W9.

80. Chenna, R.; Sugawara, H.; Koike, T.; Lopez, R.; Gibson, T.J.; Higgins, D.G.; Thompson, J.D. Multiple sequence alignment with the Clustal series of programs. *Nucleic Acids Res.* **2003**, *31*, 3497–3500.

81. Tamura, K.; Peterson, D.; Peterson, N.; Stecher, G.; Nei, M.; Kumar, S. MEGA5: Molecular Evolutionary Genetics Analysis using Maximum Likelihood, Evolutionary Distance, and Maximum Parsimony Methods. *Mol. Biol. Evol.* **2011**, *28*, 2731–2739.

82. Saitou, N.; Nei, N. The neighbor-joining method: A new method for reconstructing phylogenetic trees. *Mol. Biol. Evol.* **1987**, *4*, 406–425.

83. Felsenstein, J. Confidence limits on phylogenies: An approach using the bootstrap. *Evolution* **1985**, *39*, 783–791.

84. Kimura, M. A simple method for estimating evolutionary rate of base substitutions through comparative studies of nucleotide sequences. *J. Mol. Evol.* **1980**, *16*, 111–120.

Polar Microalgae: New Approaches towards Understanding Adaptations to an Extreme and Changing Environment

Barbara R. Lyon and Thomas Mock

Abstract: Polar Regions are unique and highly prolific ecosystems characterized by extreme environmental gradients. Photosynthetic autotrophs, the base of the food web, have had to adapt physiological mechanisms to maintain growth, reproduction and metabolic activity despite environmental conditions that would shut-down cellular processes in most organisms. High latitudes are characterized by temperatures below the freezing point, complete darkness in winter and continuous light and high UV in the summer. Additionally, sea-ice, an ecological niche exploited by microbes during the long winter seasons when the ocean and land freezes over, is characterized by large salinity fluctuations, limited gas exchange, and highly oxic conditions. The last decade has been an exciting period of insights into the molecular mechanisms behind adaptation of microalgae to the cryosphere facilitated by the advancement of new scientific tools, particularly "omics" techniques. We review recent insights derived from genomics, transcriptomics, and proteomics studies. Genes, proteins and pathways identified from these highly adaptable polar microbes have far-reaching biotechnological applications. Furthermore, they may provide insights into life outside this planet, as well as glimpses into the past. High latitude regions also have disproportionately large inputs into global biogeochemical cycles and are the region most sensitive to climate change.

Reprinted from *Biology*. Cite as: Lyon, B.R.; Mock, T. Polar Microalgae: New Approaches towards Understanding Adaptations to an Extreme and Changing Environment. *Biology* **2014**, *3*, 56-80.

1. Introduction

Low-temperature environments represent probably the largest untouched biological resource on our planet because the largest proportion of the Earth's biomass exists in low temperate environments, largely marine. Polar microalgae, which form the base of a largely bottom-up controlled polar food web [1] have successfully adapted to the extreme and oscillating polar environmental gradients. In addition to freezing temperatures, these cold environments coincide with a host of other environmental challenges including solar, osmotic, oxidative and nutrient stress which have been well described in previous reviews [2–4]. The ephemeral nature of one of polar microalgae's major niches, sea-ice, makes it one of the most dynamic of the extreme environments on earth. The semi-enclosed sea-ice habitat harbours a very diverse community of organisms interacting on a very small scale, continually acclimating and adapting to strong and oscillating environmental conditions [5]. This promotes fast evolution through horizontal exchange and recombination of genetic material. Thus, these organisms represent a resource for identification of new species, new physiological mechanisms of adaptation and new genes. However, global warming due to increased atmospheric carbon dioxide concentrations has begun to seriously threaten the coldest environments on our planet, polar ecosystems. This could mean a loss of a vast pool of genetic diversity yet to be uncovered. And only through advances in our understanding of molecular mechanisms driving metabolism and community structure of polar microalgae will scientists be able to better estimate the current and future biogeochemical inputs and niche adaptability of polar microalgae.

The focus of this review is to describe the physiological mechanisms involved in microalgae adaptations to cryospheric conditions, emphasizing insights "omics" techniques have recently provided. The later section, "Using systems biology to understand a changing world," highlights gaps in knowledge and suggested priorities for future research with particular emphasis on "omics" applications. We also refer readers to previous reviews that have given insights into microalgae adaptations to polar environments, including several which have gone into great detail on sea-ice structure, biodiversity, primary production and niche adaptation [5–7], and those that have focused on polar lake microalgae [8,9] and cyanobacteria [10,11], as well as reviews of polar macroalgae [12] and bacteria [13–17]. There have also been very informative reviews of the metabolic and biogeochemical insights derived from the first two temperate microalgae genomes sequenced [18,19]. More general reviews of metagenomic [20], proteomic [21], and metabolomic [22] environmental applications are also available.

2. Polar Significance

Polar regions are unique, prolific ecosystems despite their inhospitable appearance [23]. The Antarctic continent covers an area of 14 million km^2 and amidst the world's largest desert are numerous perennial frozen lakes. It is surrounded by the Southern Ocean, a high nutrient low chlorophyll (HNLC) region, which is covered by seasonal sea-ice that can extend up to 20 million km^2, covering ~40% of the Southern Ocean during austral winters [24]. Southern polar glaciation arose ~40 million years ago coinciding with the separation of the Antarctic and South American continents and formation of the Antarctic Circumpolar Current [25]. In the northern hemisphere, regions of North America, Europe and Asia continents all lie within the Arctic Circle (66°33'N) and surround a shallow Arctic basin covered by up to 16 million km^2 of perennial and seasonal sea-ice [26]. However, the most recent northern glaciation event did not occur until ~3.5 million years ago, making this a much younger cryospheric region [25].

Together, these polar regions account for a large proportion of the Earth's surface area and have great impacts on global biogeochemical cycles. Due to increased CO_2 solubility at low water temperatures, deep water formation in the Southern Ocean sequesters large amounts of carbon (*i.e.*, 30% of global uptake despite accounting for only 10% of the surface area; (i.e., 30% of global uptake despite accounting for only 10% of the surface area; [27]). High river inputs and the largely refractory nature of the terrestrial dissolved organic matter entering the Arctic basin also leads to large CO_2 sequestration [28]. However, warming Arctic temperatures are mobilizing frozen methane deposits, a greenhouse gas that can further exacerbate climate warming [29]. Polar regions have also been shown to produce large biogenic sulfur fluxes to the atmosphere through the breakdown of the phytoplankton metabolite dimethylsulfoniopropionate (DMSP) to dimethylsulfide (DMS), a volatile gas [30]. Oxidized sulfate particles from DMS, in turn, help seed clouds which mitigate climate warming [31]. The qualitative and quantitative inputs polar ecosystems have on the various biogeochemical cycles and food webs are largely dependent on the microbial populations that control primary production and remineralization processes. While perennial permafrost severely limits terrestrial photoautotrophs, algae and cyanobacteria have successfully adapted to polar marine and/or fresh water niches. They form the base of these productive food webs, converting light energy and nutrients into chemical energy despite a physiologically challenging environment.

3. Microalgal Mechanisms to Thrive

3.1. Membrane Fluidity

Cell membranes control transport of nutrients and metabolic waste products in and out of the cells and are integral to the electron transport chains of cellular metabolism. Therefore, maintaining their fluidity under freezing temperatures is of utmost importance. Increases in unsaturated bonds promote a looser packing of lipids and decreased temperature of solidification. Increased concentrations of polyunsaturated fatty acids (PUFAs) is one of the most well-documented cold tolerance mechanisms and has been shown in polar diatoms [32], dinoflagellates [33], and chlorophytes [34–36]. Polar microalgae not only increase PUFA concentrations in cell membrane phospholipids, but perhaps more importantly, in the galactolipids integral to the chloroplast membrane [37–40]. The recent publication of the genome of a psychrotolerant green algae, Coccomyxa subellipsoidea, found amongst the most enriched gene families FA synthases, elongases, lipases, and desaturases [41], highlighting the importance of lipid metabolism under polar conditions. Desaturase enzymes are responsible for inserting double bonds into FAs at specific carbon locations and differential regulation of desaturases indicates locations of double bonds are tightly controlled [42,43]. Upregulation in response to salt stress indicates they are also likely involved in more than just temperature acclimation [44,45]. Unlike bacterial desaturases, de novo PUFA synthesis by eukaryotic desaturases can function independent of growth, which is important for rapid acclimation [9]. However, the sensory and signal pathways involved in PUFA synthesis remain to be elucidated in eukaryotic phytoplankton.

3.2. Enzyme Kinetics

Microbes are poikilothermic (i.e., they are in thermal equilibrium with their surrounding environment). Thus, bioenergetic demands of the cell must overcome the inhibiting effects of a low kinetic environment, most notably the freezing of molecules and decreased rates of catalysis. Recently, thanks to the sequencing of >30 prokaryotic genomes, specific protein structural changes promoting cryospheric enzyme flexibility were identified and statistically validated [14]. They include amino acid substitutions which decrease hydrophobic interactions, H-bonds and salt-bridges, particularly around the active site, which in turn can increase reaction rates by requiring less energy than induced fit mechanisms. A detailed review of enzymes kinetics in polar prokaryotes is given by Gerday and colleagues in this issue [15]. While the availability of only two polar microalgae genomes does not enable statistical investigations into amino acid substitutions, physiological and molecular techniques have shown the temperature hardiness of polar microalgae metabolic enzymes utilizes various mechanisms depending on the enzyme. For example, studies of ice diatoms found enzymes involved in nitrate, ammonium and carbon uptake have optimal temperature ranges at near freezing temperatures, while other metabolic enzymes such as nitrate reductase (NR) have more moderate optimal temperatures but less sensitivity to temperature changes [46]. Comparisons between psychrophilic and mesophilic chlorophytes also showed shifts to higher and more stable activity of psychrophilic metabolic enzymes at low temperatures [47,48]. Interestingly, NR from a psychrophilic chlorophyte was capable of utilizing both NADPH and NADH as energetic reductants, whereas the mesophilic chlorophyte could only use NADH [47]. Thus, in response to polar conditions which can reduce metabolic production of energy units, enzymes from polar microalgae seem to have evolved to be more flexible in their energy source. This is supported by the fact that measurements of NR:NADH activity showed maximal activities at higher temperatures than *in vivo* assays; Ferrara and colleagues [49] suggest this can be explained by the *in vivo* presence of enzymes such as glucose 6-phosphate dehydrogenase whose synthesize of NADPH improves other enzymes (*i.e.*, NR) cold activity. Another example of polar microalgae kinetic adaptations was the elevation of pyrophosphate-dependent phospho-fructo-kinase detected in the polar diatom *Fragilariopsis cylindrus* during salinity acclimation [50]. This ATP-independent form of an important glycolysis enzyme saves chemical energy units for other processes, such as osmolyte synthesis. A gene encoding a rhodopsin has also been found in *F. cylindrus* which has been suggested to serve as a trace-metal independent method for additional ATP synthesis [51,52]. Surprisingly, ribulose-1,6-bisphosphate carboxylase (RUBISCO), which is fundamental to phototrophic carbon fixation, shows an opposite trend of greater decreases in activity under cold temperatures when psychrophilic chlorophytes where compared to their temperate counterparts [53]. However, the cellular concentrations of RUBISCO enzymes in the polar species were twice that found in temperate algae [53]. A similar mechanism was observed in molecular studies in *Chlamydomonas subcaudata* which showed increased concentrations of ATP synthase proteins within the chloroplast [54]. Elevations in this enzyme's abundance may partly explain the ability for psychrophiles, but not mesophiles, to increase ATP production following cold-shock [55]. Most recently ribosomal proteins have also been found to be significantly increased at colder temperatures, presumable to maintain adequate translation within a low kinetics environment [56].

3.3. Compatible Solutes and Cryoprotectants

Cellular compatible solutes, typically associated with their role in salinity acclimation, also serve as freeze protection molecules and contribute to the high *in vivo* enzyme activities within psychrophilic organisms. These compounds reduce the intracellular freezing point and help maintain enzyme hydration spheres to stabilize catalytic activity [57]. There is a wide array of compatible solutes including: sugars, polyols, amino acids and their derivatives such as betaine and DMSP. The amino acid proline is an abundant compatible solute in many cryophilic microalgae, including the model ice diatom *F. cylindrus*. Genes for proline synthesis were strongly represented in *F. cylindrus* cold and salt stress EST libraries [58]. Proteomics studies further validated the importance of proline synthesis within this ice diatom with seven of the 36 proteins that increased in relative abundance in response to high salinity involved in the amino acid synthesis pathway of proline ([50]; Figure 1). Two isoforms of a homologue to the bacterial/archaeal glycine betaine methyltransferase protein were also elevated. This is an alternate enzyme for synthesis of the compatible solute betaine that, in contrast to the more typical eukaryotic choline oxidase pathway, produces large betaine concentrations without damaging H_2O_2 as a by-product [59]. High concentrations of DMSP, a compatible solute with important feedbacks in climate and biogeochemical cycles, are found in ice-diatom communities [60] in contrast to typically low levels in temperate diatoms [61]. This compound has been shown to stabilize enzymes against cold-induced denaturation [62]. Proteomics following the salinity shift of *F. cylindrus* identified candidate genes for all four steps of the proposed synthesis pathway (from previous radiolabeling metabolite studies by [63]) demonstrating the utility of this "omics" technique to identify candidate enzymes for poorly characterized metabolic pathways [50].

Figure 1. Functional groups for elevated (**A**) and reduced (**B**) proteins in *F. cylindrus* following a shift to high salinity. This is an example of the major functional groups associated with polar algae acclimation processes. Note that most proteins resolved on 2DE gels had predicted functions in contrast to the much larger percentage of unknown proteins in genomic and transcriptomic datasets, most likely due to gel bias towards highly abundant metabolic proteins.

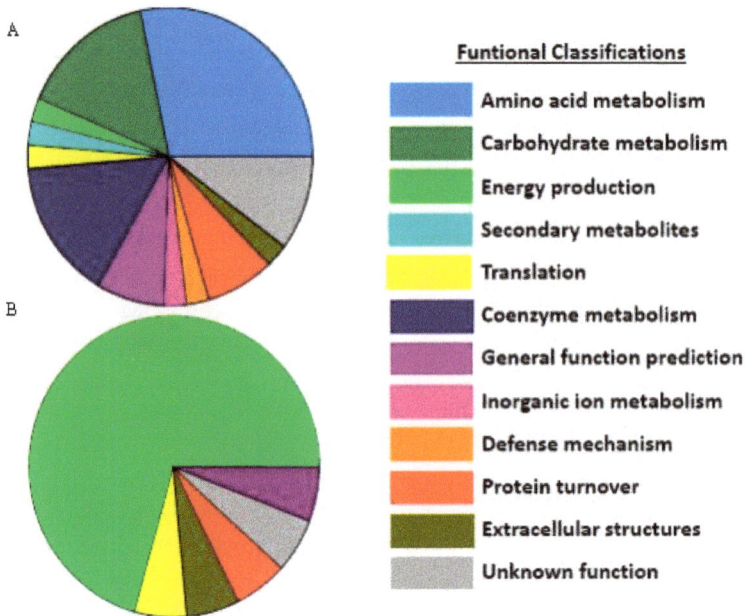

Late embryogenesis abundant (LEA) proteins are also considered multi-faceted cryoprotectants believed to work through hydrogen-bond stabilizing effects on enzymes [64]. LEA genes were found to be highly expressed in clone libraries from the polar chlorophyte *Chlorella vulgaris* (which was recently reclassified as *C. subellipsoidea*) and associated with cold-hardiness [65].

Additional novel LEA proteins were later isolated from this species by suppression subtractive hybridization and shown to protect lactate dehydrogenase activity from freeze inactivation [66].

A category of "cold-shock response" proteins has also been identified and is present across all taxa, from microbes to mammals [14]. These encompass a family of highly conserved small molecular weight proteins which bind to cold shock domains of single-stranded nucleic acids, as well as a number of RNA helicases. Cold-shock proteins are highly abundant in *F. cylindrus* two-dimensional protein gels (equivalent intensity to the highly abundant light harvesting complex proteins) and were significantly elevated following shifts from 4 °C to 0 °C [67]. Furthermore, six DNA/RNA helicases were identified within the *F. cylindrus* cold stress EST library [68]. It is believed that they work together to promote replication, transcription and translation under low temperature conditions by minimizing cold denaturation which otherwise forms kinks, coils and secondary structures that impede such processes.

3.4. Extracellular Compounds

Under freezing conditions it is also important for phytoplankton to maintain an aqueous external environment. To this end various extracellular modifiers are produced by microalgae within the ice such as ice-binding proteins (IBPs) which are excreted from the cells and inhibit ice growth and recrystallization and enhance brine retention through changes in ice channel structure [69]. IBPs were first identified in ice diatoms simultaneously in the salt shock EST library of *F. cylindrus* and through protein isolation and mass spectrometry from the ice-diatom *Navicula glaciei* [70]. The presence of these genes in all ice-algae tested to date (diatoms, prymnesiophytes, prasinophytes and chlorophytes), and complete absence from temperate species, suggests that IBPs play an important role in sea-ice adaptation [71]. Furthermore, phylogenetic analysis strongly suggests the genes were acquired through horizontal gene transfer (HGT) of bacterial IBPs [71,72]. Exopolymeric substances (EPS), composed of polysaccharides, amino acids, and proteins, are also highly abundant within the sea-ice [73]. Diatom EPS also helps retain salts, increase liquid brine fraction and thus create microhabitats within sea-ice [74]. Polysaccharide and cell wall metabolism gene families, which likely includes enzymes for synthesis of EPS and antifreeze glycoproteins, were enriched in the polar green algae *C. subellipsoidea* compared to temperate chlorophytes [41]. The role of EPS in polar adaptation is nicely reviewed by Ewert and Deming [7] within this special issue.

3.5. Light Acclimation

Light availability is highly variable in the polar environment and polar microalgae must both avoid photodamage under periods of high light and adapt to low light levels. Ice-diatoms are well adapted to low light levels. High photosynthetic efficiencies enable them to reach saturated growth at 20 μE m^{-2} s^{-1} and active photosynthesis has been observed at irradiances < 0.5 μE m^{-2} s^{-1}, 0.01% of incident irradiance [75,76]. Physiological studies revealed steady-state *F. cylindrus* cultures growing at 2 μE m^{-2} s^{-1} *versus* those grown at 15 μE m^{-2} s^{-1} exhibited increases in specific chloroplast PUFAs that enhance the fluidity of the thylakoid membrane and thus the flow of electrons, and this was associated with a near doubling of pigment concentrations and 50% reduction in carbohydrate concentration [38]. Fucoxanthin-chlorophyll binding proteins (FCPs), complex families of proteins that appear to have different specializations (e.g., light harvesting *versus* dissipation of excess energy), were the most redundant genes identified in *F. cylindrus* and *C. neogracile* EST libraries [68,77] but it has yet to be tested whether elevated FCP concentrations are also part of the high shade adaptability of polar diatoms. However, it appears that in ice diatoms a dense packaging of pigments and their binding proteins in conjunction with enhanced thylakoid fluidity enable high photosynthetic efficiencies at very low light, while carbohydrate utilization and alternate energy sources help offset any energetic deficiencies. The chlorophyte *Chlamydomonas raudensis*, which dominates the highly saline bottom layers of permanently ice-covered Antarctic lakes, is also highly shade adapted, but utilizes quite unique mechanisms (reviewed in [9]). It has lost many conserved short-term and long-term photoacclimation mechanisms such as non-photochemical quenching (NPQ), light harvesting complex state transitions between PSI and PSII, and alterations in pigment concentrations. Instead, *C. raudensis* has an extremely high PSII to PSI stoichiometry to maximize harvesting of low levels of blue light. Furthermore, in response to increased irradiance (up to 10-fold higher than natural conditions) *C. raudensis* increases growth rates to dissipate increased energy rather than exhibiting photoinhibition. In contrast, the chlorophyte species (Chlorella BI) isolated from an Antarctic pond, with higher and more fluctuating light levels, has maintained its ability for state transitions to balance PSI and PSII light absorption and thus maintain optimal photosynthetic activity under changing light conditions [78].

Despite their adaption to low light levels, ice diatoms are still capable of acclimating to high light (>350 μE m^{-2} s^{-1}) at temperatures down to -5 °C [79]. Like diatoms from other habitats, they utilize NPQ mechanisms, such as the diatoxanthin - diadinoxanthin xanthophyll cycle, to dissipate excess energy and prevent photoinhibition and cellular damage [80]. Xanthophyll cycle pigments can bind to the LHCx family of FCPs associated with the dissipation of excess energy [81]. The genome of *F. cylindrus* revealed a large expansion in the LHCx gene family compared to the two sequenced temperate diatom species [82]. Microarray studies with the polar diatom *Chaetoceros neogracile* found shifts from 20 to 600 μE m^{-2} s^{-1} resulted in significant elevations in LHCx proteins and antioxidant proteins, while those associated with light harvesting were significantly reduced [83]. The fact that growth rate was only 20-35% reduced over a 10 day period at this very high light level illustrates the photoacclimation capabilities of this species. Interestingly, work from various labs has shown low temperatures to elicit photoacclimation responses similar to high light, such as increased NPQ, PSII proteins, and photoprotective pigments [35,79,84]. Presumably these changes were a result of increased excitation pressures caused by low temperatures (*i.e.*, decreased enzyme kinetics inhibits efficiency of metabolic electron sinks leading to a build-up of reduced plastoquinone). The redox status of the plastoquinone pool in turn triggers phosphorylation cascades which initiate photoacclimation mechanisms [81,85]. Salinity shifts can also elicit photoacclimation mechanisms in polar chlorophytes and diatoms [67,86], again likely through the common mechanism of changes in excitation pressures. Thus, multiple environmental pressures require robust photoacclimation mechanisms for microalgae to thrive within the cryosphere.

3.6. Antioxidants

Photosynthesis creates an oxic environment intracellularly that can be exacerbated by reactive oxygen species (ROS) formation caused by low temperature and other stress induced metabolic imbalances. This is further magnified by the increased solubility of oxygen at low temperatures and restricted diffusion within sea-ice and permanently ice-covered lakes leading to hyperoxic extracellular environments. Thus, robust antioxidant systems are important for polar microalgae to cope with ROS. Polar diatoms [87], chlorophytes [88] and dinoflagellates [89] are resistant to UV damage, indicating highly effective antioxidant systems. High catalase activity in the sea-ice diatom *Entomoneis kufferathii* following exposure to high light and low temperatures [90] is one of many antioxidant systems that can help protect cells from oxidative damage. Studies with the polar diatom *Chaetoceros brevis* showed elevated levels in superoxide dismutase activity, in addition to xanthophyll cycling, also to be important for dissipating ROS brought on by irradiance shifts [91]. A survey of antioxidant systems showed polar algae also utilize ascorbate peroxidase and glutathione reductase as antioxidant systems; however, the activity levels following light stress did not show a clear geographic trend between polar and temperate species but rather indicated species specific responses that could be due to differences in photoacclimation capabilities and utilization of alternate antioxidant systems [92]. In fact a microarray study with the polar diatom *Chaetoceros neogracile* found amongst thermal stress response genes (following shift from 4 °C to 10 °C) a spectrum of antioxidant enzymes, including monoascorbate reductase, glutaredoxin, glutathione peroxidase, glutathione S-transferase (GST), and alternative oxidase, illustrating the diverse suite of ROS defense enzymes this psychrophile can utilize to mitigate oxidative damage [93]. Various compatible solutes such as proline and DMSP also have secondary benefits as antioxidants [94,95] and their high concentrations in polar microalgae likely contribute to protection from oxidative damage. Furthermore, symbiotic relationships between ice-diatom and epiphytic bacteria who scavenge ROS have been described [96].

Given the importance of PUFAs to cryoprotection and their sensitivity to oxidation, protection against ROS damage to these lipids is of utmost importance. GST conjugates reduced glutathione to electrophilic centers, particularly abundant within PUFAs, and protects them from more damaging oxidation such as lipid peroxidation and other reactions with H_2O_2. In addition to the GST transcript elevations in response to elevated temperatures mentioned above, proteomics studies have found significantly elevated GST protein levels to be involved in low temperature acclimation of the sea-ice chlorophyte, *Chlamydomonas sp.* [97] and high salinity acclimation of the polar diatom, *F. cylindrus* [50], further emphasizing its importance in environmental stress tolerance across diverse taxa of polar microalgae. Another important enzyme for protecting cellular proteins from oxidative damage is methionine sulfoxide reductase (Msr) which reduces methionine residues that become oxidized by ROS. MsrB has been shown to be a cold response protein in the *Arabidopsis* plant and knock-down mutants show decreased cold tolerance in the form of increased methionine oxidation, H_2O_2 formation, and electrolyte leakage [98]. Fourteen Msr homologues are present in the *F. cylindrus* genome (*versus* seven and ten in the temperate diatom genomes of *Thalassiosira pseudonana* and *Phaeodactylum tricornutum*, respectively) supporting an important role for Msr enzymes in cold and oxidative stress acclimation of polar microalgae. Comparisons between the genome of the psychrotolerant chlorophyte *C. subellipsoidea* and its temperate counterpart found enrichment in the polar species of a family of short-chain dehydrogenase/reductase enzymes whose substrates vary from alcohols, sugars, and steroids to xenobiotics [41]. Most proteins in this family have oxidoreductase activity and enrichment could indicate an important role in maintaining a balanced cellular redox state under polar conditions, perhaps complementary to aldehyde dehydrogenases mitigation of chemically reactive oxidized lipid aldehydes that can accumulate when environmental stress perturbs metabolic balance as shown in plants [99]. Interestingly, an aldehyde dehydrogenase protein was found to be significantly elevated in *F. cylindrus* during high salinity acclimation at 0 °C [50]. While there is still much work to be done to understand more thoroughly the molecular mechanisms enabling oxidative stress tolerance in polar microalgae, studies thus far indicate a diverse suite of mechanisms to prevent and mitigate ROS damage.

3.7. Dark Adaptation

Survival through winter's extended periods of darkness is key to photoautotrophic success in polar regions. Dark adaptation is also important in controlling microalgae seasonal and spatial distributions [100]. Incubation experiments found temperate diatoms to survive 21–35 days of darkness [100], while Antarctic species survived 4-9 month periods in the dark [101]. However, little is known of physiological mechanisms behind polar overwintering as logistics severely limits austral winter field studies. It is known that carbohydrate storage molecules such as glucan in diatoms and starch in chlorophytes are accumulated within polar algae and utilized during periods of darkness [9,102]. Furthermore, polar microalgae can also uptake dissolved organic material such as sugars and starches for energetic breakdown [103]. A recent study of polar pelagic algae also described a high plasticity in regards to inorganic carbon uptake in Southern Ocean phytoplankton [104]. Comparison of the recent polar chlorophyte genome *C. subellipsoidea* to its temperate counterparts showed gene enrichment in amino acid transporters and permeases which would promote enhanced uptake of organic nutrient sources [41]. Additionally, a number of carbohydrate metabolism gene families were present in *C. subellipsoidea* that did not have homologs in temperate chlorophyte species but instead appeared to have HGT origins. In diatoms, the presence of the urea cycle has been proposed as a means for recovering carbon and nitrogen depleted during photorespiration [18], while the FA β-oxidation pathway means lipids can be used as metabolic intermediates and for ATP synthesis [105]. This metabolic flexibility being revealed through diatom genomes is likely fundamental to their ability to thrive within the extreme and highly-variable environmental conditions which characterize polar regions.

Ice-algae cells can go into a winter resting stage during which time minimal changes in carbon and chlorophyll concentrations occur [106]. Low temperature shift in *F. cylindrus* (5 °C to −1.8 °C) were associated with reductions in photosynthesis and carbon fixation genes, and this was hypothesized to indicate a preparation towards a winter resting stage cued by decreasing temperatures [107]. *Xanthonema*, a class of heterokont snow algae, has been shown to disassemble PSII but keep LHC proteins intact in response to extended dark adaptation [108]. Thus, it appears they maintain thylakoid structural proteins primed to quickly reassembly PS in order to utilize the first short periods of irradiance in austral spring. Extended dark adaption studies in the chlorophyte *Koliella antarctica* showed similar PS changes, but also detected hallmarks of programmed cell death (PCD) in a subpopulation of cells [109]. However, return of cultures to low light after 60 days of darkness showed rapid recovery of growth and photosynthesis, clearly demonstrating viability of cells that did not undergo PCD and raising the question of PCD as an adaptive benefit to unicellular communities. There is still much debate to whether shifts to heterotrophic metabolism also play a role in adaption to extended periods of darkness. Transformation of *P. tricornutum* with a single gene, a glucose transporter, enabled this diatom to switch from photoautotrophic to heterotrophic growth [110], illustrating the ease for such a shift to occur given the high degree of HGT believed to occur within the sea-ice environment. Indeed, the chlorophyte *Chlorella* BI is capable of switching between autotrophic growth and heterotrophic growth, with highest growth rates achieved during mixotrophic growth in light with a glucose carbon source [78]. Many dinoflagellate species are also capable of heterotrophic growth, and shifts to mixotrophic sea-ice communities have been detected from late winter ice cores [111]. However, more research is needed to understand metabolic shifts occurring within species and through changes in community composition during seasonal periods of extended darkness.

4. Using Systems Biology to Understand a Changing World

Improving our understanding of the molecular mechanisms behind environmental acclimation and adaptation processes is key to predicting ecological and biogeochemical inputs of polar primary producers (Figure 2). Our molecular tool box has greatly advanced over the past couple decades, and the next hurdle is linking this knowledge to processes on the global scale with ever increasing resolution. While some mechanisms may be unique to polar environments, many are likely utilized in other environments to overcome similar bioenergetic pressures that may arise from common or quite distinct stressors. Temperature, light, nutrients, allelopathic and anthropogenic compounds, and chemical-physical processes (e.g., stratification, oxygen minimum zones, carbonate saturation depth) collectively control temporal and spatial taxonomic distributions depending on the biological potential of organisms (*i.e.*, genetic adaptability). Different evolutionary histories (e.g., endosymbiotic events, HGT, gene loss/expansion during niche specialization) provide different suites of genes which result in metabolic diversity. Differences in metabolism between species, in turn, ultimately result in different impacts on biogeochemical cycles. A systems biology approach to understand the complex genomic, transcriptomic, proteomic, and metabolomic interactions is required if we are to achieve a holistic understanding of feedbacks between organisms and their environment and eventually develop mathematical models capable of representing past, current, and predicting future biological inputs on various ecosystem parameters, such as climate, biodiversity, and nutrient availability (Figure 2). Measuring the effects of various environmental variables, individually and in combinations, on the vast array of phytoplankton is a laborious and logistically difficult matrix of experiments. Alternatively, environmental "omic" approaches reveals the importance of genes in the natural environment and can be correlated with environmental metadata to tease out significant relationships and important drivers of biodiversity, gene expression, and biogeochemical inputs. But in order to move beyond the current low resolution, broad taxonomic group models there are some important gaps (highlighted in the following paragraphs) that must be addressed, particularly with regard to polar primary producers due to their significance in global biogeochemical processes and the sensitivity of this region to climate change.

Figure 2. An array of advanced, high-throughput molecular techniques is enabling a global effort to link environmental parameters to biological distributions, physiological capabilities, protein expression, metabolic functions, and biogeochemical cycles. The ultimate goal is the ability to make high resolution predictions of biogeochemical and ecosystem inputs under current and future climates. The light blue bubbles represent areas to focus future research to improve our understanding and modeling efforts. OMZ–oxygen minimum zone, PTM – post-translational modification, TF–transcription factor.

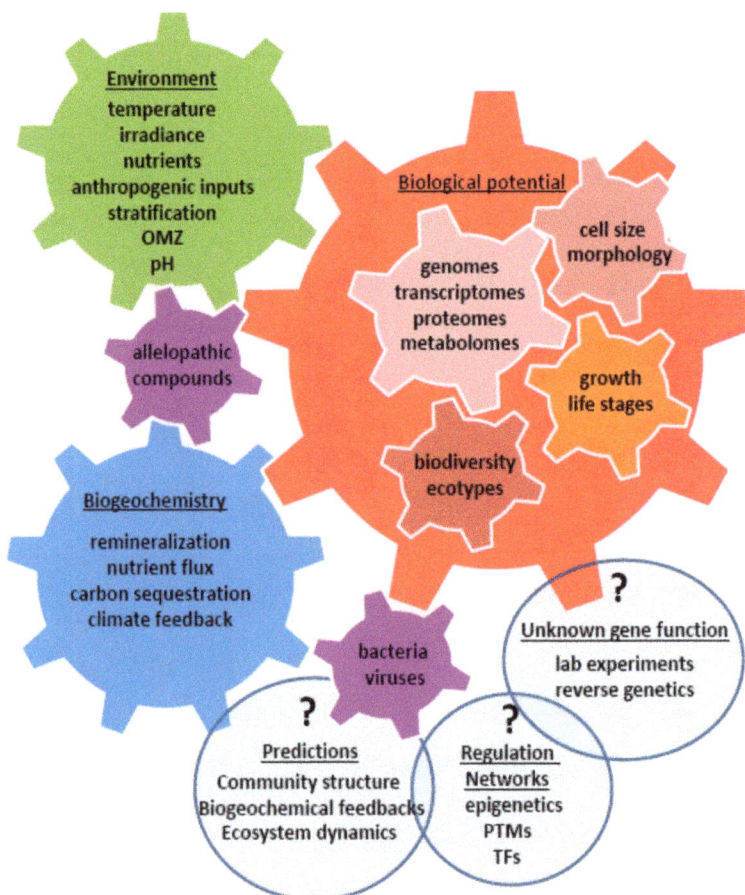

Polar microalgae genomes provide valuable information on their genetic repertoire, as well as promoter and intergenic regions whose diverse roles in gene regulation we are still just beginning to understand. Polar microalgae genomes also can provide novel proteins and pathways for biotechnological applications and insights into cryosphere bioenergetics that may even help us understand possibilities for life outside of this planet [112]. The *F. cylindrus* genome was the first psychrophilic microalgae genome to become publicly available [113] and insights from its annotation should be published in the near future. Recently the genome of a psychrotolerant green algae, *C. subellipsoidea*, was published [41]. Sequencing of a dominant polar haptophyte, *P. antarctica*, has also been started by the U.S. Department of Energy Joint Genome Institute in 2010 and likely will become public in 2014 (Arrigo, personal communication). Thanks to the rapid advancement of sequencing and annotation techniques, genome sequencing costs and time has been greatly reduced and the number of sequencing facilities has increased substantially. Thus science is now capable of an important next step to understanding polar microalgae physiology and global biogeochemical inputs: expand reference genomes and/or transcriptomes to include multiple species from various taxa representing different niche specialists, and importantly environmental isolates rather than clones that have been cultured for decades in the laboratory. Comparisons between ecotypes of the same species (*i.e.*, cosmopolitan species like *C. neogracile* and *C. raudensis* with polar and warm water strains) will also be incredibly valuable to understanding genes fundamental to cryospheric life. To help fill this gap The Gordon and Betty Moore Foundation has undertaken the Marine Microbial Eukaryote Transcriptome Sequencing Project [114] to sequence the transcriptomes of 750 samples from a diverse range of habitats, including many more polar species. This will enable us to better understand competitive advantages between species and ecotypes and differences in their metabolic potentials. Importantly, it will also substantially increase our database of reference genomes for identifying the taxonomic source of environmental metatranscriptome and metaproteome sequences and prevent misassignment due to limited reference libraries. Furthermore, a new technique of single-cell sequencing can now be used to provide genetic information on non-cultured environmental samples and detect differences and interactions at the individual cell level [115].

Determining which species are present and contributing to biogeochemical fluxes is not a simple task. Since traditional microscopy methods are laborious, molecular techniques have been gaining prominence. Hypervariable regions of the eukaryotic 18s small subunit ribosomal RNA (rRNA) sequence have been used to define eukaryotic phytoplankton taxonomic units at the level of species supergroups (*i.e.*, division or family level) and this has greatly expanded our appreciation of polar microalgae diversity, particularly with regards to the more fragile single-celled organisms [116,117]. Using 18s rRNA Charvet and colleagues [118] found the diversity of Antarctic lake flagellates was significantly underestimated by traditional microscopy and HPLC pigment methods. Similarly, 18s rRNA studies of Antarctic sea-ice communities under different irradiance regimes identified a very diverse eukaryotic community and identified three seasonal stages in community structure: mixed, dinoflagellate-dominated, and diatom-dominated [119]. However, it is important to note that 18s rRNA methods have their own biases towards particular species depending on the primer pairs used [120]. Furthermore, 18s rRNA copy number varies substantially between species [121]. And many 18s rRNA environmental sequences cannot be assigned to known taxa, as evidenced by preliminary analysis of the recent TARA expedition which could not assign 31% of 18s rRNA v9 sequences to known supergroups [122]. So while molecular taxonomic techniques increase sensitivity to species richness and provide a high-throughput means for gathering phenotypic information to link with metatranscriptome and metaproteome data, to apply 18s rRNA metagenomic techniques beyond a presence/absence assessment will require a greatly expanded reference database and increased efforts to quantify and normalize inherent sequencing biases. Furthermore, to truly improve our understanding of who is where and doing what will require the development of molecular markers able to define microalgae taxa with ever increasing specificity, eventually at the species and ecotype level.

Genes of unknown function made up six of the 10 most abundant genes in the *F. cylindrus* EST libraries [68] and 45% of the 1700 unique ESTs in *C. neogracile* [77]. While genes of unknown function are a common feature of all genomes, there are great possibilities for metabolic ingenuity and novel functions amongst these genes in polar microalgae. Environmental expression and functional characterization studies will be key to understanding their effect on an organisms biological potential and biogeochemical feedbacks (Figure 2). Even the assignment of genes to putative functions (*i.e.*, aldehyde dehydrogenase mentioned earlier) are usually only based on sequence similarity and conserved domains that place them in broad enzyme classes with a diverse array of cellular functions, which in a eukaryotic cell are also dependent on subcellular localization. This is further compounded by redundant enzymes and paralogs within a genome. Thus, while comparative "omics" using natural and laboratory controlled conditions provides candidate genes for important and novel regulatory and metabolic processes, a true understanding of function and activity will require protein characterization studies using over-expression and silencing transformations within appropriate host organisms. Such work is easy and well-established in bacterial systems, but these hosts lack eukaryotic organelles, chaperones, post-translational systems and the upstream and downstream signaling networks native to that gene. Gene overexpression and silencing techniques have been developed in temperate diatom species *P. tricornutum* and *T. pseudonana* [110,123,124]. However, the establishment of methods to transform polar diatoms and other phytoplankton taxa would greatly enhance our ability to functionally characterize novel polar genes. New techniques such as viral promoters and transcription activator-like effector nucleases are promising approaches to expand this technology into polar microalgae host models.

Various cell organelles and metabolic systems are involved in polar acclimation processes and interact through a complex network of sensory, signaling, and regulatory mechanisms. Understanding and predicting biogeochemical feedbacks requires untangling and ultimately quantifying this collection of synergistic and antagonistic intracellular interactions. In contrast to the large proportion of putative genes with unknown functions found in *F. cylindrus* transcript studies, only three of the 56 proteins identified in an *F. cylindrus* proteomics study were of unknown function (Figure 1; [50]). Protein gels are biased towards highly abundant and soluble proteins. Perhaps these strongly differentially-regulated gene transcripts of unknown function code for unique signaling and transcription factor proteins which do not need to be expressed at the same concentrations as metabolic enzymes to have immense physiological effects, hence their limited presence in differentially expressed 2D gel proteins. A high priority for understanding gene expression on a systems biology level must be deciphering gene networks, transcription factors and promoter binding regions, sensory/signaling pathways and post-translation regulation. Such studies have been applied to plants and cyanobacteria but are still in their infancy within temperate microalgae and have yet to be applied to polar counterparts.

The ability of polar microalgae to acclimate to a wide range of environmental gradients and their broader range in photosynthetic and metabolic responses compared to temperate counterparts [125] has led to the conclusion that they possess a high degree of phenotypic plasticity. Phenotypic plasticity is like a "cellular memory" that responds to environment and may be selected for in environments with multiple stressors and steep or rapidly changing gradients [19,126]. Epigenetic modifications, such as cytosine methylations and histone modifications can serve as "soft" heritable changes which effect transcription. Recently, a high level of methylation of transposable elements (TEs) and a subset of genes which tended to be involved in important metabolic activities of nutrient resource management were identified in the "methylome" of the diatom *P. tricornutum* [127]. Environmental cues, such as nitrate limitation, decreased methylation of specific genes and triggered an increase in transcript levels. On the other hand, TEs are mobile genetic units and activation by environmental stressors (likely through changes in methylation) can lead to genetic rearrangements, another mechanism facilitating evolution and environmental adaptation at a rapid rate [128]. Small non-coding (silencing) RNAs are another method for controlling phenotypic plasticity in plants and animals [129] and cDNA libraries generated from *T. pseudonana* small RNAs indicate that these are also likely important transcription and translation regulators in microalgae [130]. Clearly, there are still many gaps in our understanding of the molecular mechanisms behind the high level of phenotypic plasticity within polar microalgae which must be addressed if we are to begin quantifying environmental regulation of cellular feedbacks into biogeochemical processes.

Microalgae community structure, genetic mobility, and nutrient availability are all regulated by bacteria and viruses [131] which are abundant in polar environments [132]. Recently a study from an Antarctic hypersaline lake described a virophage-virus-prasinophyte interaction whereby the virophage limited virus-induced mortality and increased phytoplankton blooms [133]. Viruses have also been shown to stimulate PCD pathways in phytoplankton [134]. On the other hand, bacteria-derived infochemicals may stimulate diatom EPS capsule formation [135] and species-specific interaction between a bacterium and diatom may enhance diatom growth, likely through a bacteria produced phytohormone [136]. Yet only a few of the potential bioactive secondary metabolites (aka allelopathic compounds or infochemicals) responsible for microbial intra/interspecies communication have been described so far. These include diatom derived aldehydes shown to trigger nitric oxide signalling and PCD in phytoplankton [137] and diatom derived oxylipins, formed from oxygenated PUFAs and shown to disrupt zooplankton reproduction and development [138]. Studies specific to polar microbial communities are still lacking. The role of intra/interspecies communication and predator-prey interactions in polar ecosystems, particularly in regard to viral and bacterial regulation of primary production, bioremineralization processes, community structure, and biogeochemical cycling, should be a high priority of future research.

For some time now polar regions have shown an amplified sensitivity to climate change [139]. Increased temperatures and winds have increased sea-ice retreat, upper ocean freshening and nutrient upwelling in the relatively shallow Arctic basin; this, in turn, has increased pelagic primary production but with a shift from nano to picoplankton populations [140]. Thinning sea-ice, allowing for increased light penetration, has also led to large under ice phytoplankton blooms in the Arctic [141]. In the Southern Ocean increased stratification, resulting in a shallower mixed layer with increased light, has been postulated to favor diatom growth, but at the same time decreases in upwelling also due to stratification have been predicted to more negatively impact large diatoms compared to small phytoplankton [142]. Meanwhile a region around the Antarctic Peninsula has seen overall summer phytoplankton abundance decrease by 12% over the past 30 years [143]. Warming temperatures may also alter adaptation to other polar conditions, such as darkness [144].

Microalgae play a key role in biogeochemical cycling. Species composition, abundance, cell size and life history all determine the drawdown of organic matter and C, N, and P are sequestered in different ratios depending on such factors [145]. Furthermore, changes in lipid composition and other metabolic shifts associated with adaptations of polar microalgae to altered niches [32,146] will have complex effects on food quality, community structure and biogeochemical processes [147]. Importantly, polar microalgae are key players in two major climate feedback loops that mitigate global warming trends: deep ocean carbon sequestration [27] and cloud condensation sulfate particles [148]. Changes in species composition and/or environmental variables significantly affect the fluxes within these feedback loops but to what extant is still largely uncertain. Clearly, the need to better understand feedbacks between climate, primary production, and biogeochemical loops in polar regions is paramount, yet models are ripe with uncertainties inherent in attempts to define function based on broad taxonomic classifications, such as nano *versus* picoplankton or diatom *versus* haptophyte [149]. "Omics" techniques continue to generate a wealth of data towards understanding acclimation potentials and metabolic fluxes, as well as elucidating niche separation and climate change adaptability. The next major hurdle will be advancing our ability to quantify and model different molecular/metabolic strategies to give a finer resolution on functional groups and biogeochemical fluxes, particularly in the face of new ecological pressures. This requires close collaborations between molecular biologists and modelers in order to develop a holistic approach based on genomic and biochemical data. Such an integrative systems ecology approach will provide mechanistic insights into how climate change will impact polar phytoplankton communities.

5. Conclusions

Only over the past decade have modern molecular genomic, transcriptomic, proteomic, and metabolomic tools been applied to polar microalgae. Although still in its infancy, great insights have already been made in regards to adaptation and acclimation mechanisms of polar microalgae using these new techniques. As we improve our understanding of polar bioenergetics, resource management, metabolic fluxes, and community composition, our ability to understand feedbacks of polar microalgae on global biogeochemical processes will become clearer. Furthermore, discovery of novel genes and pathways could have profound impacts on biotechnological applications.

Acknowledgments

The authors thank Jan Strauss for thoughtful comments on the manuscript. Funding for BR Lyon was provided by NSF OISE (Award Number 1159163).

Conflicts of Interest

The authors declare no conflict of interest.

References

1. Smith, W.O.; Lancelot, C. Bottom-up versus top-down control in phytoplankton of the Southern Ocean. *Antarct. Sci.* **2004**, *16*, 531–539.
2. Kirst, G.O.; Wiencke, C. Ecophysiology of polar algae. *J. Phycol.* **1995**, *31*, 181–199.
3. Thomas, D.N.; Dieckmann, G.S. Antarctic Sea ice—A habitat for extremophiles. *Science* **2002**, *295*, 641–644.
4. Mock, T.; Thomas, D.N. Microalgae in Polar Regions: Linking Functional Genomics and Physiology with Environmental Conditions. In *Psychrophiles: From Biodiversity to Biotechnology*; Margesin, R., Schinner, F., Marx, J.-C., Gerday, C., Eds.; Springer: Berlin/Heidelberg, Germany, 2008; pp. 285–312.
5. Arrigo, K.R.; Mock, T.; Lizotte, M.P. Primary production in sea ice. In *Sea Ice: An Introduction to Its Physics, Chemistry, Biology and Geology*; Thomas, D., Dieckmann, G., Eds.; Wiley-Blackwell: Malden, MA, USA, 2010; pp. 143–183.
6. Mock, T.; Thomas, D.N. Recent advances in sea-ice microbiology. *Environ. Microbiol.* **2005**, *7*, 605–619.
7. Ewert, M.; Deming, J. Sea ice microorganisms: Environmental constraints and extracellular responses. *Biology* **2013**, *2*, 603–628.
8. Dolhi, J.; Maxwell, D.; Morgan-Kiss, R. Review: The Antarctic *Chlamydomonas raudensis*: An emerging model for cold adaptation of photosynthesis. *Extremophiles* **2013**, *17*, 711–722.
9. Morgan-Kiss, R.M.; Priscu, J.C.; Pocock, T.; Gudynaite-Savitch, L.; Huner, N.P.A. Adaptation and acclimation of photosynthetic microorganisms to permanently cold environments. *Microbiol. Mol. Biol. Rev.* **2006**, *70*, 222–252.
10. Singh, S.M.; Elster, J. Cyanobacteria in Antarctic Lake Environments. In *Algae and Cyanobacteria in Extreme Environments*; Seckbach, J., Ed.; Springer: Dordrecht, The Netherlands, 2007; Volume 11, pp. 303–320.
11. Vincent, W. Cold Tolerance in Cyanobacteria and Life in the Cryosphere. In *Algae and Cyanobacteria in Extreme Environments*; Seckbach, J., Ed.; Springer: Dordrecht, The Netherlands, 2007; Volume 11, pp. 287–301.
12. Wiencke, C.; Amsler, C.D. Seaweeds and Their Communities in Polar Regions. In *Seaweed Biology*; Wiencke, C., Bischof, K., Eds.; Springer: Berlin/Heidelberg, Germany, 2012; Volume 219, pp. 265–291.
13. D'Amico, S. Psychrophilic microorganisms: Challenges for life. *EMBO Rep.* **2006**, *7*, 385–389.

14. Casanueva, A.; Tuffin, M.; Cary, C.; Cowan, D.A. Molecular adaptations to psychrophily: The impact of 'omic' technologies. *Trends Microbiol.* **2010**, *18*, 374–381.

15. Gerday, C. Psychrophily and catalysis. *Biology* **2013**, *2*, 719–741.

16. Koh, E.Y.; Martin, A.R.; McMinn, A.; Ryan, K.G. Recent Advances and Future Perspectives in Microbial Phototrophy in Antarctic Sea Ice. *Biology* **2012**, *1*, 542–556.

17. Wilkins, D.; Yau, S.; Williams, T.J.; Allen, M.A.; Brown, M.V.; DeMaere, M.Z.; Lauro, F.M.; Cavicchioli, R. Key microbial drivers in Antarctic aquatic environments. *FEMS Microbiol. Rev.* **2013**, *37*, 303–335.

18. Parker, M.S.; Mock, T.; Armbrust, E.V. Genomic insights into marine microalgae. *Annu. Rev. Genet.* **2008**, *42*, 619–645.

19. Bowler, C.; Vardi, A.; Allen, A.E. Oceanographic and biogeochemical insights from diatom genomes. *J. Rev. Mar. Sci.* **2010**, *2*, 333–365.

20. Allen, A.E.; LaRoche, J.; Maheswari, U.; Lommer, M.; Schauer, N.; Lopez, P.J.; Finazzi, G.; Fernie, A.R.; Bowler, C. Whole-cell response of the pennate diatom *Phaeodactylum tricornutum* to iron starvation. *PNAS* **2008**, *105*, 10438–10443.

21. Tomanek, L. Environmental proteomics: Changes in the proteome of marine organisms in response to environmental stress, pollutants, infection, symbiosis, and development. *J. Rev. Mar. Sci.* **2011**, *3*, 373–399.

22. Weber, A.P.M.; Horst, R.J.; Barbier, G.G.; Oesterhelt, C. Metabolism and metabolomics of eukaryotes living under extreme conditions. In *International Review of Cytology*; Kwang, W.J., Ed.; Academic Press: Salt Lake City, UT, USA, 2007; Volume 256, pp. 1–34.

23. Bluhm, B.A.; Gebruk, A.V.; Gradinger, R.; Hopcroft, R.R.; Huettmann, F.; Kosobokova, K.N.; Sirenko, B.I.; Weslawski, J.M. Arctic marine biodiversity: An update of species richness and examples of biodiversity change. *Oceanography* **2011**, *24*, 232.

24. Lizotte, M.P. The contributions of sea ice algae to Antarctic marine primary production. *Am. Zool.* **2001**, *41*, 57–73.

25. Zachos, J.; Pagani, M.; Sloan, L.; Thomas, E.; Billups, K. Trends, rhythms, and aberrations in global climate 65 Ma to present. *Science* **2001**, *292*, 686–693.

26. Comiso, J.C. Large-scale characteristics and variability of the global sea ice cover. In *Sea Ice: An Introduction to Its Physics, Chemistry, Biology, and Geology*; Thomas, D.N., Dieckmann, G.S., Eds.; Blackwell Science Ltd.: Oxford, UK, 2003; pp. 112–142.

27. Sabine, C.L.; Feely, R.A.; Gruber, N.; Key, R.M.; Lee, K.; Bullister, J.L.; Wanninkhof, R.; Wong, C.S.; Wallace, D.W.R.; Tilbrook, B.; Millero, F.J.; Peng, T.-H.; Kozyr, A.; Ono, T.; Rios, A.F. The Oceanic Sink for Anthropogenic $CO2$. *Science* **2004**, *305*, 367–371.

28. Dittmar, T.; Kattner, G. The biogeochemistry of the river and shelf ecosystem of the Arctic Ocean: a review. *Mar. Chem.* **2003**, *83*, 103–120.

29. Shakhova, N.; Semiletov, I.; Salyuk, A.; Yusupov, V.; Kosmach, D.; Gustafsson, Ö. Extensive methane venting to the atmosphere from sediments of the East Siberian Arctic Shelf. *Science* **2010**, *327*, 1246–1250.

30. Kiene, R.; Kieber, D.; Slezak, D.; Toole, D.; del Valle, D.; Bisgrove, J.; Brinkley, J.; Rellinger, A. Distribution and cycling of dimethylsulfide, dimethylsulfoniopropionate, and dimethylsulfoxide during spring and early summer in the Southern Ocean south of New Zealand. *Aquat. Sci.* **2007**, *69*, 305–319.

31. Gunson, J.R.; Spall, S.A.; Anderson, T.R.; Jones, A.; Totterdell, I.J.; Woodage, M.J. Climate sensitivity to ocean dimethylsulphide emissions. *Geophys. Res. Lett.* **2006**, *33*, doi:10.1029/2005GL024982.

32. Teoh, M.-L.; Phang, S.-M.; Chu, W.-L. Response of Antarctic, temperate, and tropical microalgae to temperature stress. *J. Appl. Phycol.* **2012**, *1*, 1–13.

33. Thomson, P.G.; Wright, S.W.; Bolch, C.J.S.; Nichols, P.D.; Skerratt, J.H.; McMinn, A. Antarctic distribution, pigment and lipid composition, and molecular identification of the brine dinoflagellate Polarella glacialis (Dinophyceae). *J. Phycol.* **2004**, *40*, 867–873.

34. Osipova, S.; Dudareva, L.; Bondarenko, N.; Nasarova, A.; Sokolova, N.; Obolkina, L.; Glyzina, O.; Timoshkin, O. Temporal variation in fatty acid composition of *Ulothrix Zonata* (Chlorophyta) from ice and benthic communities of Lake Baikal. *Phycologia* **2009**, *48*, 130–135.

35. Fogliano, V.; Andreoli, C.; Martello, A.; Caiazzo, M.; Lobosco, O.; Formisano, F.; Carlino, P.A.; Meca, G.; Graziani, G.; Rigano, V.D.M.; Vona, V.; Carfagna, S.; Rigano, C. Functional ingredients produced by culture of Koliella antarctica. *Aquaculture* **2010**, *299*, 115–120.

36. Chen, Z.; He, C.; Hu, H. Temperature responses of growth, photosynthesis, fatty acid and nitrate reductase in Antarctic and temperate *Stichococcus. Extremophiles* **2012**, *16*, 127–133.

37. Mock, T.; Kroon, B.M.A. Photosynthetic energy conversion under extreme conditions—I: Important role of lipids as structural modulators and energy sink under N-limited growth in Antarctic sea ice diatoms. *Phytochemistry* **2002**, *61*, 41–51.

38. Mock, T.; Kroon, B.M.A. Photosynthetic energy conversion under extreme conditions—II: The significance of lipids under light limited growth in Antarctic sea ice diatoms. *Phytochemistry* **2002**, *61*, 53–60.

39. Gray, C.G.; Lasiter, A.D.; Leblond, J.D. Mono- and digalactosyldiacylglycerol composition of dinoflagellates. III. Four cold-adapted, peridinin-containing taxa and the presence of trigalactosyldiacylglycerol as an additional glycolipid. *Eur. J. Phycol.* **2009**, *44*, 439–445.

40. Morgan-Kiss, R.; Ivanov, A.G.; Williams, J.; Mobashsher, K.; Huner, N.P.A. Differential thermal effects on the energy distribution between photosystem II and photosystem I in thylakoid membranes of a psychrophilic and a mesophilic alga. *Biochim. Biophys. Acta* **2002**, *1561*, 251–265.

41. Blanc, G.; Agarkova, I.; Grimwood, J.; Kuo, A.; Brueggeman, A.; Dunigan, D.D.; Gurnon, J.; Ladunga, I.; Lindquist, E.; Lucas, S.; Pangilinan, J.; Proschold, T.; Salamov, A.; Schmutz, J.; Weeks, D.; Yamada, T.; Lomsadze, A.; Borodovsky, M.; Claverie, J.M.; Grigoriev, I.V.; Van Etten, J.L. The genome of the polar eukaryotic microalga *Coccomyxa subellipsoidea* reveals traits of cold adaptation. *Genome Biol.* **2012**, *13*, doi:10.1186/gb-2012-13-5-r39.

42. Suga, K.; Honjoh, K.-I.; Furuya, N.; Shimizu, H.; Nishi, K.; Shinohara, F.; Hirabaru, Y.; Maruyama, I.; Miyamoto, T.; Hatano, S.; Iio, M. Two low-temperature-inducible *Chlorella* genes for Δ-12 and omega-3 fatty acid desaturase (FAD): Isolation of Δ-12 and omega-3 fad cDNA clones. *Biosci. Biotechnol. Biochem.* **2002**, *66*, 1314–1327.

43. An, M.; Mou, S.; Zhang, X.; Ye, N.; Zheng, Z.; Cao, S.; Xu, D.; Fan, X.; Wang, Y.; Miao, J. Temperature regulates fatty acid desaturases at a transcriptional level and modulates the fatty acid profile in the Antarctic microalga Chlamydomonas sp. ICE-L. *Bioresour. Technol.* **2013**, *134*, 151–157.

44. Zhang, P.; Liu, S.; Cong, B.; Wu, G.; Liu, C.; Lin, X.; Shen, J.; Huang, X. A novel omega-3 fatty acid desaturase involved in acclimation processes of polar condition from Antarctic ice algae Chlamydomonas sp. ICE-L. *Mar. Biotechnol.* **2011**, *13*, 393–401.

45. An, M.; Mou, S.; Zhang, X.; Zheng, Z.; Ye, N.; Wang, D.; Zhang, W.; Miao, J. Expression of fatty acid desaturase genes and fatty acid accumulation in Chlamydomonas sp. ICE-L under salt stress. *Bioresour. Technol.* **2013**, *149*, 77–83.

46. Priscu, J.; Palmisano, A.; Priscu, L.; Sullivan, C. Temperature dependence of inorganic nitrogen uptake and assimilation in Antarctic sea-ice microalgae. *Polar Biol.* **1989**, *9*, 443–446.

47. Di Martino Rigano, V.; Vona, V.; Lobosco, O.; Carillo, P.; Lunn, J.E.; Carfagna, S.; Esposito, S.; Caiazzo, M.; Rigano, C. Temperature dependence of nitrate reductase in the psychrophilic unicellular alga *Koliella antarctica* and the mesophilic alga *Chlorella sorokiniana*. *Plant Cell Environ.* **2006**, *29*, 1400–1409.

48. Vona, V.; Di Martino Rigano, V.; Lobosco, O.; Carfagna, S.; Esposito, S.; Rigano, C. Temperature responses of growth, photosynthesis, respiration and NADH: Nitrate reductase in cryophilic and mesophilic algae. *New Phytol.* **2004**, *163*, 325–331.

49. Ferrara, M.; Guerriero, G.; Cardi, M.; Esposito, S. Purification and biochemical characterisation of a glucose-6-phosphate dehydrogenase from the psychrophilic green alga *Koliella antarctica*. *Extremophiles* **2012**, *17*, 53–62.

50. Lyon, B.R.; Lee, P.A.; Bennett, J.M.; DiTullio, G.R.; Janech, M.G. Proteomic analysis of a sea-ice diatom: Salinity acclimation provides new insight into the dimethylsulfoniopropionate production pathway. *Plant Physiol.* **2011**, *157*, 1926–1941.

51. Strauss, J.; Gao, S.; Morrissey, J.; Bowler, C.; Nagel, G.; Mock, T. A light-driven rhodopsin proton pump from the psychrophilic diatom *Fragilariopsis cylindrus*. In Proceeding of EMBO Workshop: The Molecular Life of Diatoms, Paris, France, 25–28 June 2013.

52. Marchetti, A.; Schruth, D.M.; Durkin, C.A.; Parker, M.S.; Kodner, R.B.; Berthiaume, C.T.; Morales, R.; Allen, A.E.; Armbrust, E.V. Comparative metatranscriptomics identifies molecular bases for the physiological responses of phytoplankton to varying iron availability. *PNAS* **2012**, *109*, E317–E325.

53. Devos, N.; Ingouff, M.; Loppes, R.; Matagne, R.F. RUBISCO adaptation to low temperatures: A comparative study in psychrophilic and mesophilic unicellular algae. *J. Phycol.* **1998**, *34*, 655–660.

54. Morgan, R.M.; Ivanov, A.G.; Priscu, J.C.; Maxwell, D.P.; Huner, N.P.A. Structure and composition of the photochemical apparatus of the antarctic green alga, *Chlamydomonas subcaudata*. *Photosynth. Res.* **1998**, *56*, 303–314.

55. Napolitano, M.J.; Shain, D.H. Distinctions in adenylate metabolism among organisms inhabiting temperature extremes. *Extremophiles* **2005**, *9*, 93–98.

56. Toseland, A.D.S.J.; Clark, J.R.; Kirkham, A.; Strauss, J.; Uhlig, C.; Lenton, T.M.; Valentin, K.; Pearson, G.A.; Moulton, V.; Mock, T. The impact of temperature on marine phytoplankton resource allocation and metabolism. *Nat. Clim. Change* **2013**, *3*, 979–984.

57. Welsh, D.T. Ecological significance of compatible solute accumulation by micro-organisms: From single cells to global climate. *FEMS Microbiol. Rev.* **2000**, *24*, 263–290.

58. Krell, A. Salt stress tolerance in the psychrophilic diatom *Fragilariopsis cylindrus*. Ph.D. Thesis, University of Bremen, Bremen, Germany, 2006.

59. Waditee, R.; Bhuiyan, M.N.H.; Rai, V.; Aoki, K.; Tanaka, Y.; Hibino, T.; Suzuki, S.; Takano, J.; Jagendorf, A.T.; Takabe, T.; Takabe, T. Genes for direct methylation of glycine provide high levels of glycinebetaine and abiotic-stress tolerance in *Synechococcus* and *Arabidopsis*. *Proc. Natl. Acad. Sci. USA* **2005**, *102*, 1318–1323.

60. DiTullio, G.R.; Garrison, D.L.; Mathot, S. Dimethylsulfoniopropionate in sea ice algae from the Ross Sea polynya. In *Antarctic Sea Ice: Biological Processes, Interactions and Variability*; Arrigo, K.R., Lizotte, M.P., Eds.; American Geophysical Union: Washington, DC, USA, 1998; pp. 139–146.

61. Keller, M.D.; Bellows, W.K.; Guillard, R.R.L. Dimethyl sulfide production in marine phytoplankton. In *Biogenic Sulfur in the Environment*; Saltzman, E.S., Cooper, W.J., Eds.; American Chemical Society: Washington, DC, USA, 1989; pp. 131–142.

62. Nishiguchi, M.K.; Somero, G.N. Temperature- and concentration-dependence of compatibility of the organic osmolyte [beta]-dimethylsulfoniopropionate. *Cryobiology* **1992**, *29*, 118–124.

63. Gage, D.A.; Rhodes, D.; Nolte, K.D.; Hicks, W.A.; Leustek, T.; Cooper, A.J.L.; Hanson, A.D. A new route for synthesis of dimethylsulphoniopropionate in marine algae. *Nature* **1997**, *387*, 891–894.

64. Tunnacliffe, A.; Wise, M.J. The continuing conundrum of the LEA proteins. *Naturwissenschaften* **2007**, *94*, 791–812.

65. Honjoh, K.-I.; Yoshimoto, M.; Joh, T.; Kajiwara, T.; Miyamoto, T.; Hatano, S. Isolation and characterization of hardening-induced proteins in *Chlorella vulgaris* C-27: Identification of late embryogenesis abundant proteins. *Plant Cell Physiol.* **1995**, *36*, 1421–1430.

66. Liu, X.; Wang, Y.; Gao, H.; Xu, X. Identification and characterization of genes encoding two novel LEA proteins in Antarctic and temperate strains of *Chlorella vulgaris*. *Gene* **2011**, *482*, 51–58.

67. Lyon, B.R. Medical University of South Carolina-Hollings Marine Lab, Charleston, SC, USA. Unpublished work, 2011.

68. Mock, T.; Krell, A.; Glockner, G.; Kolukisaoglu, U.; Valentin, K. Analysis of expressed sequence tags (ESTs) from the polar diatom *Fragilariopsis cylindrus*. *J. Phycol.* **2005**, *42*, 78–85.

69. Raymond, J.A. Algal ice-binding proteins change the structure of sea ice. *PNAS* **2011**, *108*, E198.

70. Janech, M.G.; Krell, A.; Mock, T.; Kang, J.S.; Raymond, J.A. Ice-binding proteins from sea ice diatoms (Bacillariophyceae). *J. Phycol.* **2006**, *42*, 410–416.

71. Raymond, J.A.; Kim, H.J. Possible role of horizontal gene transfer in the colonization of sea ice by algae. *PLoS ONE* **2012**, *7*, e35968.

72. Raymond, J.A.; Morgan-Kiss, R. Separate origins of ice-binding proteins in Antarctic *Chlamydomonas* species. *PLoS ONE* **2013**, *8*, e59186.

73. Krembs, C.; Eicken, H.; Junge, K.; Deming, J.W. High concentrations of exopolymeric substances in Arctic winter sea ice: Implications for the polar ocean carbon cycle and cryoprotection of diatoms. *Deep Sea Res. Part I* **2002**, *49*, 2163–2181.

74. Krembs, C.; Eicken, H.; Deming, J.W. Exopolymer alteration of physical properties of sea ice and implications for ice habitability and biogeochemistry in a warmer Arctic. *PNAS* **2011**, *108*, 3653–3658.

75. Mock, T.; Gradinger, R. Determination of Arctic ice algal production with a new in situ incubation technique. *Mar. Ecol. Prog. Ser.* **1999**, *177*, 15–26.

76. Cota, G.F. Photoadaptation of high Arctic ice algae. *Nature* **1985**, *315*, 219–222.

77. Jung, G.; Lee, C.G.; Kang, S.H.; Jin, E. Annotation and expression profile analysis of cDNAs from the Antarctic diatom *Chaetoceros neogracile*. *J. Microbiol. Biotechnol.* **2007**, *17*, 1330–1337.

78. Morgan-Kiss, R.; Ivanov, A.; Modla, S.; Czymmek, K.; Hüner, N.; Priscu, J.; Lisle, J.; Hanson, T. Identity and physiology of a new psychrophilic eukaryotic green alga, *Chlorella* sp., strain BI, isolated from a transitory pond near Bratina Island, Antarctica. *Extremophiles* **2008**, *12*, 701–711.

79. Ralph, P.J.; McMinn, A.; Ryan, K.G.; Ashworth, C. Short-term effect of temperature on the photokinetics of microalgae from the surface layers of Antarctic pack ice. *J. Phycol.* **2005**, *41*, 763–769.

80. Robinson, D.; Kolber, Z.; Sullivan, C. Photophysiology and photoacclimation in surface sea ice algae from McMurdo Sound, Antarctica. *Mar. Ecol. Prog. Ser.* **1997**, *147*, 243–256.

81. Lepetit, B.; Sturm, S.; Rogato, A.; Gruber, A.; Sachse, M.; Falciatore, A.; Kroth, P.G.; Lavaud, J. High light acclimation in the secondary plastids containing diatom *Phaeodactylum tricornutum* is triggered by the redox state of the plastoquinone pool. *Plant. Physiol.* **2013**, *161*, 853–865.

82. Green, B.; Alami, M.; Zhu, S.; Guo, J.; Maldonado, M. The LHC superfamily and the complex roles of its members in photoacclimation. In Proceedings of EMBO Workshop: The Molecular Life of Diatoms, Paris, France, 25–28 June 2013.

83. Park, S.; Jung, G.; Hwang, Y.-s.; Jin, E. Dynamic response of the transcriptome of a psychrophilic diatom, *Chaetoceros neogracile*, to high irradiance. *Planta* **2010**, *231*, 349–360.

84. Mock, T.; Hoch, N. Long-term temperature acclimation of photosynthesis in steady-state cultures of the polar diatom *Fragilariopsis cylindrus*. *Photosynth. Res.* **2005**, *85*, 307–317.

85. Szyszka, B.; Ivanov, A.G.; Huner, N.P. Psychrophily is associated with differential energy partitioning, photosystem stoichiometry and polypeptide phosphorylation in *Chlamydomonas raudensis*. *Biochimica et Biophysica Acta* **2007**, *1767*, 789–800.

86. Takizawa, K.; Takahashi, S.; Huner, N.P.; Minagawa, J. Salinity affects the photoacclimation of *Chlamydomonas raudensis* Ettl UWO241. *Photosynth. Res.* **2009**, *99*, 195–203.

87. Ryan, K.G.; McMinn, A.; Hegseth, E.N.; Davy, S.K. The effects of ultraviolet-b radiation on Antarctic sea-ice algae. *J. Phycol.* **2012**, *48*, 74–84.

88. Miao, J.; Li, G.; Hou, X.; Zhang, Y.; Jiang, Y.; Wang, B.; Zhang, B. Study on induced synthesis of anti-UV substances in the Antarctic algae. *High. Tech. Lett.* **2002**, *6*, 179–183.

89. Obertegger, U.; Camin, F.; Guella, G.; Flaim, G. Adaptation of a psychrophilic freshwater dinoflagellate to ultraviolet radiation. *J. Phycol.* **2011**, *47*, 811–820.

90. Schriek, R. Effects of light and temperature on the enzymatic antioxidative defense systems in the Antarctic ice diatom *Entomoneis kufferathii* Manguin. *Rep. Polar Res.* **2000**, *349*, 1–130.

91. Janknegt, P.J.; Van De Poll, W.H.; Visser, R.J.W.; Rijstenbil, J.W.; Buma, A.G.J. Oxidative stress responses in the marine antarctic diatom *Chaetoceros brevis* (bacillariophyceae) during photoacclimation. *J. Phycol.* **2008**, *44*, 957–966.

92. Janknegt, P.J.; De Graaff, C.M.; Van De Poll, W.H.; Visser, R.J.W.; Rijstenbil, J.W.; Buma, A.G.J. Short-term antioxidative responses of 15 microalgae exposed to excessive irradiance including ultraviolet radiation. *Eur. J. Phycol.* **2009**, *44*, 525–539.

93. Hwang, Y.-S.; Jung, G.; Jin, E. Transcriptome analysis of acclimatory responses to thermal stress in Antarctic algae. *Biochem. Biophys. Res. Commun.* **2008**, *367*, 635–641.

94. Sunda, W.; Kieber, D.J.; Kiene, R.P.; Huntsman, S. An antioxidant function for DMSP and DMS in marine algae. *Nature* **2002**, *418*, 317–320.

95. Chen, C.; Dickman, M.B. Proline suppresses apoptosis in the fungal pathogen *Colletotrichum trifolii*. *Proc. Natl. Acad. Sci. USA* **2005**, *102*, 3459–3464.

96. Hünken, M.; Harder, J.; Kirst, G.O. Epiphytic bacteria on the Antarctic ice diatom *Amphiprora kufferathii* Manguin cleave hydrogen peroxide produced during algal photosynthesis. *Plant. Biol.* **2008**, *10*, 519–526.

97. Kan, G.-F.; Miao, J.-L.; Shi, C.-J.; Li, G.-Y. Proteomic alterations of Antarctic ice microalga *Chlamydomonas sp.* under low-temperature stress. *J. Integr. Plant Biol.* **2006**, *48*, 965–970.

98. Kwon, S.J.; Kwon, S.I.; Bae, M.S.; Cho, E.J.; Park, O.K. Role of the methionine sulfoxide reductase MsrB3 in cold acclimation in Arabidopsis. *Plant Cell. Physiol.* **2007**, *48*, 1713–1723.

99. Kirch, H.-H.; Bartels, D.; Wei, Y.; Schnable, P.S.; Wood, A.J. The ALDH gene superfamily of Arabidopsis. *Trends Plant Sci.* **2004**, *9*, 371–377.

100. Peters, E. Prolonged darkness and diatom mortality: II. Marine temperate species. *J. Exp. Mar. Biol. Ecol.* **1996**, *207*, 43–58.

101. Peters, E.; Thomas, D.N. Prolonged darkness and diatom mortality I: Marine Antarctic species. *J. Exp. Mar. Biol. Ecol.* **1996**, *207*, 25–41.

102. van Oijen, T.; Leeuwe, M.; Gieskes, W.C. Variation of particulate carbohydrate pools over time and depth in a diatom-dominated plankton community at the Antarctic Polar Front. *Polar Biol.* **2003**, *26*, 195–201.

103. Palmisano, A.; Garrison, D. Microorganisms in Antarctic sea ice. In *Antarctic Microbiology*; Friedmann, E., Ed.; Wiley-Liss: New York, NY, USA, 1993; pp. 167–218.

104. Neven, I.A.; Stefels, J.; van Heuven, S.M.A.C.; de Baar, H.J.W.; Elzenga, J.T.M. High plasticity in inorganic carbon uptake by Southern Ocean phytoplankton in response to ambient CO_2. *Deep Sea Res. Part II* **2011**, *58*, 2636–2646.

105. Armbrust, E.V.; Berges, J.A.; Bowler, C.; Green, B.R.; Martinez, D.; Putnam, N.H.; Zhou, S.; Allen, A.E.; Apt, K.E.; Bechner, M.; Brzezinski, M.A.; Chaal, B.K.; Chiovitti, A.; Davis, A.K.; Demarest, M.S.; Detter, J.C.; Glavina, T.; Goodstein, D.; Hadi, M.Z.; Hellsten, U.; Armbrust, E.V. The genome of the diatom *Thalassiosira pseudonana*: Ecology, evolution, and metabolism. *Science* **2004**, *306*, 79–86.

106. Doucette, G.J.; Fryxell, G.A. *Thalassiosira antarctica*: vegetative and resting stage chemical composition of an ice-related marine diatom. *Mar. Biol.* **1983**, *78*, 1–6.

107. Mock, T.; Valentin, K. Photosynthesis and cold acclimation: Molecular evidence from a polar diatom. *J. Phycol.* **2004**, *40*, 732–741.

108. Baldisserotto, C.; Ferroni, L.; Moro, I.; Fasulo, M.P.; Pancaldi, S. Modulations of the thylakoid system in snow xanthophycean alga cultured in the dark for two months: comparison between microspectrofluorimetric responses and morphological aspects. *Protoplasma* **2005**, *226*, 125–135.

109. Ferroni, L.; Baldisserotto, C.; Zennaro, V.; Soldani, C.; Fasulo, M.P.; Pancaldi, S. Acclimation to darkness in the marine chlorophyte *Koliella antarctica* cultured under low salinity: hypotheses on its origin in the polar environment. *Eur. J. Phycol.* **2007**, *42*, 91–104.

110. Zaslavskaia, L.A.; Lippmeier, J.C.; Kroth, P.G.; Grossman, A.R.; Apt, K.E. Transformation of the diatom *Phaeodactylum tricornutum* (Bacillariophyceae) with a variety of selectable marker and reporter genes. *J. Phycol.* **2000**, *36*, 379–386.

111. Bachy, C.; Lopez-Garcia, P.; Vereshchaka, A.; Moreira, D. Diversity and vertical distribution of microbial eukaryotes in the snow, sea ice and seawater near the North Pole at the end of the polar night. *Front. Microbiol.* **2011**, *2*, doi:10.3389/fmicb.2011.00106.

112. Duarte, R.T.D.; Nóbrega, F.; Nakayama, C.R.; Pellizari, V.H. Brazilian research on extremophiles in the context of astrobiology. *Int. J. Astrobiol.* **2012**, *11*, 325–333.

113. Homepage of *Fragilariopsis cylindrus* Genome. Available online: http://genome.jgi-psf.org/Fracy1/Fracy1.home.html (accessed on 7 September 2013).

114. Marine MIcrobial Eukaryote Transcriptome Sequencing Project. Available online: www.marinemicroeukaryotes.org (accessed on 7 September 2013).

115. Yoon, H.S.; Price, D.C.; Stepanauskas, R.; Rajah, V.D.; Sieracki, M.E.; Wilson, W.H.; Yang, E.C.; Duffy, S.; Bhattacharya, D. Single-cell genomics reveals organismal interactions in uncultivated marine protists. *Science* **2011**, *332*, 714–717.

116. Lovejoy, C.; Massana, R.; Pedros-Alio, C. Diversity and distribution of marine microbial eukaryotes in the Arctic Ocean and adjacent seas. *Appl. Environ. Microbiol.* **2006**, *72*, 3085–3095.

117. Amaral-Zettler, L.A.; McCliment, E.A.; Ducklow, H.W.; Huse, S.M. A method for studying protistan diversity using massively parallel sequencing of V9 hypervariable regions of small-subunit ribosomal RNA genes. *PLoS ONE* **2009**, *4*, e6372.

118. Charvet, S.; Vincent, W.; Lovejoy, C. Chrysophytes and other protists in High Arctic lakes: molecular gene surveys, pigment signatures and microscopy. *Polar Biol.* **2012**, *35*, 733–748.

119. Piquet, A.M.T.; Bolhuis, H.; Davidson, A.T.; Thomson, P.G.; Buma, A.G.J. Diversity and dynamics of Antarctic marine microbial eukaryotes under manipulated environmental UV radiation. *FEMS Microbiol. Ecol.* **2008**, *66*, 352–366.

120. Potvin, M.; Lovejoy, C. PCR-based diversity estimates of artificial and environmental 18S rRNA gene libraries. *J. Eukaryotic Microbiol.* **2009**, *56*, 174–181.

121. Zhu, F.; Massana, R.; Not, F.; Marie, D.; Vaulot, D. Mapping of picoeucaryotes in marine ecosystems with quantitative PCR of the 18S rRNA gene. *FEMS Microbiol. Ecol.* **2005**, *52*, 79–92.

122. Malviya, S.; Veluchamy, A.; Bittner, L.; Tanaka, A.; Bowler, C. Comprehensive biogeographical insights into the complexity of marine diatom communities. In EMBO Workshop: The Molecular Life of Diatoms, Paris, France, 25–28 June 2013.

123. Poulsen, N.; Chesley, P.M.; Kröger, N. Molecular genetic manipulation of the diatom *Thalassiosira pseudonana* (bacillariophyceae). *J. Phycol.* **2006**, *42*, 1059–1065.

124. De Riso, V.; Raniello, R.; Maumus, F.; Rogato, A.; Bowler, C.; Falciatore, A. Gene silencing in the marine diatom *Phaeodactylum tricornutum. Nucleic Acids Res.* **2009**, *37*, e96.

125. Pocock, T.; Vetterli, A.; Falk, S. Evidence for phenotypic plasticity in the Antarctic extremophile Chlamydomonas raudensis Ettl. UWO 241. *J. Exp. Bot.* **2011**, *62*, 1169–1177.

126. Konstantinidis, K.T.; Braff, J.; Karl, D.M.; DeLong, E.F. Comparative metagenomic analysis of a microbial community residing at a depth of 4,000 meters at station ALOHA in the North Pacific Subtropical Gyre. *Appl. Environ. Microbiol.* **2009**, *75*, 5345–5355.

127. Veluchamy, A.; Lin, X.; Maumus, F.; Rivarola, M.; Bhavsar, J.; Creasy, T.; O'Brien, K.; Sengamalay, N.A.; Tallon, L.J.; Smith, A.D.; Rayko, E.; Ahmed, I.; Crom, S.L.; Farrant, G.K.; Sgro, J.-Y.; Olson, S.A.; Bondurant, S.S.; Allen, A.; Rabinowicz, P.D.; Sussman, M.R.; Bowler, C.; Tirichine, L. Insights into the role of DNA methylation in diatoms by genome-wide profiling in Phaeodactylum tricornutum. *Nat. Commun.* **2013**, *4*, doi:10.1038/ncomms3091.

128. Maumus, F.; Allen, A.E.; Mhiri, C.; Hu, H.; Jabbari, K.; Vardi, A.; Grandbastien, M.A.; Bowler, C. Potential impact of stress activated retrotransposons on genome evolution in a marine diatom. *BMC Genomics* **2009**, *10*, 624.

129. Richards, E.J. Inherited epigenetic variation—Revisiting soft inheritance. *Nat. Rev. Genet.* **2006**, *7*, 395–401.

130. Norden-Krichmar, T.M.; Allen, A.E.; Gaasterland, T.; Hildebrand, M. Characterization of the small RNA transcriptome of the diatom, *Thalassiosira pseudonana*. *PLoS ONE* **2011**, *6*, e22870.

131. Anesio, A.M.; Bellas, C.M. Are low temperature habitats hot spots of microbial evolution driven by viruses? *Trends Microbiol.* **2011**, *19*, 52–57.

132. Pearce, I.; Davidson, A.T.; Bell, E.M.; Wright, S. Seasonal changes in the concentration and metabolic activity of bacteria and viruses at an Antarctic coastal site. *Aquat. Microb. Ecol.* **2007**, *47*, 11–23.

133. Yau, S.; Lauro, F.M.; DeMaere, M.Z.; Brown, M.V.; Thomas, T.; Raftery, M.J.; Andrews-Pfannkoch, C.; Lewis, M.; Hoffman, J.M.; Gibson, J.A.; Cavicchioli, R. Virophage control of antarctic algal host–virus dynamics. *PNAS* **2011**, *108*, 6163–6168.

134. Vardi, A.; Haramaty, L.; Van Mooy, B.A.; Fredricks, H.F.; Kimmance, S.A.; Larsen, A.; Bidle, K.D. Host-virus dynamics and subcellular controls of cell fate in a natural coccolithophore population. *Proc. Natl. Acad. Sci. USA* **2012**, *109*, 19327–19332.

135. Kroth, P.; Windler, M.; Leinweber, K.; Schulze, B.; Spiteller, D.; Buhmann, M. Interactions of diatoms and bacteria in biofilms. In Proceedings of EMBO Workshop: The Molecular Life of Diatoms, Paris, France, 25–28 June 2013.

136. Amin, S.; Hmelo, L.; Parsek, M.; Armbrust, E.V. Multiple complex interactions between a toxigenic diatom and an associated bacterium revealed using whole cell transcriptomics. In Proceedings of EMBO: The Molecular Life of Diatoms, Paris, France, 25–28 June 2013.

137. Vardi, A.; Bidle, K.D.; Kwityn, C.; Hirsh, D.J.; Thompson, S.M.; Callow, J.A.; Falkowski, P.; Bowler, C. A diatom gene regulating nitric-oxide signaling and susceptibility to diatom-derived aldehydes. *Curr. Biol.* **2008**, *18*, 895–899.

138. Caldwell, G.S. The influence of bioactive oxylipins from marine diatoms on invertebrate reproduction and development. *Mar. Drugs* **2009**, *7*, 367–400.

139. Smetacek, V.; Nicol, S. Polar ocean ecosystems in a changing world. *Nature* **2005**, *437*, 362–368.

140. Tremblay, J.-É.; Robert, D.; Varela, D.; Lovejoy, C.; Darnis, G.; Nelson, R.; Sastri, A. Current state and trends in Canadian Arctic marine ecosystems: I. Primary production. *Clim. Change* **2012**, *115*, 161–178.

141. Arrigo, K.R.; Perovich, D.K.; Pickart, R.S.; Brown, Z.W.; van Dijken, G.L.; Lowry, K.E.; Mills, M.M.; Palmer, M.A.; Balch, W.M.; Bahr, F.; Bates, N.R.; Benitez-Nelson, C.; Bowler, B.; Brownlee, E.; Ehn, J.K.; Frey, K.E.; Garley, R.; Laney, S.R.; Lubelczyk, L.; Mathis, J.; Matsuoka, A.; Mitchell, B.G.; Moore, G.W.K.; Ortega-Retuerta, E.; Pal, S.; Polashenski, C.M.; Reynolds, R.A.; Schieber, B.; Sosik, H.M.; Stephens, M.; Swift, J.H. Massive phytoplankton blooms under Arctic sea ice. *Science* **2012**, *336*, 1408.

142. Marinov, I.; Doney, S.C.; Lima, I.D. Response of ocean phytoplankton community structure to climate change over the 21st century: partitioning the effects of nutrients, temperature and light. *Biogeosciences* **2010**, *7*, 3941–3959.

143. Montes-Hugo, M.; Doney, S.C.; Ducklow, H.W.; Fraser, W.; Martinson, D.; Stammerjohn, S.E.; Schofield, O. Recent changes in phytoplankton communities associated with rapid regional climate change along the western Antarctic Peninsula. *Science* **2009**, *323*, 1470–1473.

144. Karsten, U.; Schlie, C.; Woelfel, J.; Becker, B. Benthic diatoms in Arctic waters -ecological functions and adaptations. *Polarforschung* **2012**, *81*, 77–84.

145. Dunbar, R.B.; Arrigo, K.R.; Lutz, M.; DiTullio, G.R.; Leventer, A.R.; Lizotte, M.P.; Van Woert, M.L.; Robinson, D.H. Non-Redfield production and export of marine organic matter: A recurrent part of the annual cycle in the Ross Sea, Antarctica. *Antarct. Sci. Ser.* **2003**, *78*, 179–195.

146. Lee, S.H.; Whitledge, T.E.; Kang, S.-H. Spring time production of bottom ice algae in the landfast sea ice zone at Barrow, Alaska. *J. Exp. Mar. Biol. Ecol.* **2008**, *367*, 204–212.

147. Comeau, A.M.; Li, W.K.W.; Tremblay, J.-É.; Carmack, E.C.; Lovejoy, C. Arctic Ocean microbial community structure before and after the 2007 record sea ice minimum. *PLoS ONE* **2011**, *6*, e27492.

148. Gabric, A.J.; Shephard, J.M.; Knight, J.M.; Jones, G.; Trevena, A.J. Correlations between the satellite-derived seasonal cycles of phytoplankton biomass and aerosol optical depth in the Southern Ocean: Evidence for the influence of sea ice. *Glob. Biogeochem. Cycle* **2005**, *19*, 1–10.

149. Boyd, P.W.; Strzepek, R.; Fu, F.; Hutchinsc, D.A. Environmental control of open-ocean phytoplankton groups: Now and in the future. *Limnol. Oceanogr.* **2010**, *55*, 1353–1376.

The Distribution and Identity of Edaphic Fungi in the McMurdo Dry Valleys

Lisa L. Dreesens, Charles K. Lee and S. Craig Cary

Abstract: Contrary to earlier assumptions, molecular evidence has demonstrated the presence of diverse and localized soil bacterial communities in the McMurdo Dry Valleys of Antarctica. Meanwhile, it remains unclear whether fungal signals so far detected in Dry Valley soils using both culture-based and molecular techniques represent adapted and ecologically active biomass or spores transported by wind. Through a systematic and quantitative molecular survey, we identified significant heterogeneities in soil fungal communities across the Dry Valleys that robustly correlate with heterogeneities in soil physicochemical properties. Community fingerprinting analysis and 454 pyrosequencing of the fungal ribosomal intergenic spacer region revealed different levels of heterogeneity in fungal diversity within individual Dry Valleys and a surprising abundance of Chytridiomycota species, whereas previous studies suggested that Dry Valley soils were dominated by Ascomycota and Basidiomycota. Critically, we identified significant differences in fungal community composition and structure of adjacent sites with no obvious barrier to aeolian transport between them. These findings suggest that edaphic fungi of the Antarctic Dry Valleys are adapted to local environments and represent an ecologically relevant (and possibly important) heterotrophic component of the ecosystem.

Reprinted from *Biology*. Cite as: Dreesens, L.L.; Lee, C.K.; Cary, S.C. The Distribution and Identity of Edaphic Fungi in the McMurdo Dry Valleys. *Biology* **2014**, *3*, 466-483.

1. Introduction

Located between the Polar Plateau and Ross Sea in Southern Victoria Land, the McMurdo Dry Valleys (hereinafter the Dry Valleys) are the largest contiguous ice-free area on the Antarctic continent. Dry Valley soils are known as some of the oldest, coldest, driest, and most oligotrophic soils on Earth [1]; consequently, the Dry Valley ecosystem is characterized by a lack of nutrients [2], low precipitation levels and biologically available water [3–5], high levels of salinity [6–8], large temperature fluctuations [5,9,10], steep chemical and biological gradients [11], and high incidence of UV-solar radiation [12–14]. Early studies suggested that Dry Valley soils contained very little microbial biota [1], but recent molecular evidence has demonstrated the presence of diverse and heterogeneous bacterial communities potentially driven by steep physicochemical gradients [1,10,15–19]. In contrast, comparatively limited molecular evidence exists on the distribution and drivers of fungal communities in Dry Valley soils [20–23].

Fungal identification in Dry Valley soils by means of a combination of culturing and molecular tools (*i.e.*, denaturing gradient gel electrophoresis and DNA sequencing) has detected primarily members of Dikarya (*i.e.*, Ascomycota and Basidiomycota), including both filamentous and non-filamentous species [24–27]. A survey of Dry Valley sites including Mt Flemming, Allan Hills, New Harbor, and Ross Island revealed the dominant free-living fungal genera in Dry Valley soils as *Cadophora* (Ascomycota), *Cryptococcus* (Basidiomycota), *Geomyces* (Ascomycota), and *Cladosporium* (Ascomycota) [22]. A study of cultivable fungi in Taylor Valley showed that filamentous fungi appeared to be associated with high soil pH and moisture, whereas yeasts and yeast-like fungi had wider distribution across habitats examined [23]. Basidiomycetous *Cryptococcus* and *Leucosporidium* species were the most frequently isolated genera in a regional survey of yeasts and yeast-like fungi in the Dry Valleys [20]. The diversity of yeasts and yeast-like fungi was positively correlated with soil pH and negatively with conductivity [20]. The same study also revealed apparent segregation of *Cryptococcus* clades found in Taylor Valley and the Labyrinths of Wright Valley [20], hinting at the presence of localized communities adapted to environmental conditions, as has been reported for soil bacteria in the Dry Valleys [15]. A culture-based study of soils taken from McKelvey Valley detected no fungal colony-forming units (CFUs) in most of the samples [21], and a molecular survey of McKelvey Valley also detected no fungal signals in the soils [18]. However, sequences affiliated with genera *Dothideomycetes* (Ascomycota), *Sordariomycetes* (Ascomycota), and *Cystobasidiomycetes* (Basidiomycota) were found in endolithic and chasmolithic communities in McKelvey Valley [18]. The evidence so far suggests that the cultivable components of Dry Valley fungal communities are dominated by ascomycetous and basidiomycetous species, although their biogeography and factors that shape their distribution in the Dry Valleys remain unclear due to the lack of systematic and culture-independent evidence. Furthermore, the ecological relevance of fungi in Dry Valley soils remains unknown since neither cultivation nor molecular techniques can effectively distinguish active fungal cells from dormant spores.

For this study, we carried out a molecular survey of Dry Valley soil fungi at six study sites (Battleship Promontory, Upper Wright Valley, Beacon Valley, Miers Valley, Alatna Valley, and University Valley) using terminal restriction fragment length polymorphism (tRFLP) and 454 pyrosequencing analyses of the fungal ribosomal intergenic spacer. Soil physicochemical properties were also characterized to examine potential environmental drivers of fungal diversity.

2. Experimental

2.1. Sample Collection

Soil was collected at six different sites in the McMurdo Dry Valleys (Table 1 and Figure 1) as described previously [15]. Briefly, sampling sites were all located on a south facing, 0–20° slope. An intersection was made by two 50 m transects, with the intersection in the middle being the central sampling point (X or C). Four sampling points around the central point were marked (A–D with A being the southernmost point and the remaining points in an anti-clockwise order, or N, E, S, W). Five scoops of the top 2 cm of soil were collected and homogenized at each identified (1 m²) sampling point after pavement pebbles were removed. Samples were stored in sterile Whirl-Pak (Nasco International, Fort Atkinson, WI, USA) at −20 °C until returned to New Zealand, where they were stored at −80 °C until analysis.

Table 1. List of sampling sites.

Valley	Coordinates	Elevation	Sampling Date
Miers Valley	78°05.486'S, 163°48.539'E	171 m	December 2006
Beacon Valley	77°52.321'S, 160°29.725'E	1376 m	December 2006
Upper Wright Valley	77°31.122'S, 160°45.813'E	947 m	January 2008
Battleship Promontory	76°54.694'S, 160°55.676'E	1028 m	January 2008
Alatna Valley	76°54.816'S, 161°02.213'E	1057 m	November 2010
University Valley	77°51.668'S, 160°42.736'E	1680 m	November 2010

Figure 1. Antarctica is presented in the lower right corner, with the McMurdo Dry Valleys marked in a blue rectangle. The locations of the sampling sites within the McMurdo Dry Valleys are displayed by red dots.

2.2. Soil Chemistry

Soil moisture content was determined by drying 6 g of soil at 35 °C until its weight stabilized and then at 105 °C until the sample reached constant weight. Soil pH and electrical conductivity were determined using the slurry technique, which is based on a 2:5 unground dried soil:de-ionized water mixture rehydrated overnight before measurement, using a Thermo Scientific Orion 4 STAR pH/Conductivity meter (Thermo Scientific, Beverly, MA, USA). For total and organic carbon and nitrogen contents, dried soils were ground to fine powders using an agate mortar and pestle and precisely weighed out to 100 mg. Samples were analyzed with an Elementar Isoprime 100 analyzer (Elementar Analysensysteme, Hanau, Germany). Sample preparation for elemental analysis was adapted from US EPA Analytical Methods 200.2 (Revision 2.8, 1994) and Lee *et al.* [15], in which ground dried soil samples were acid digested and analyzed using an E2 Instruments Inductively

Coupled Plasma Mass Spectrometer (ICP-MS) (Perkin-Elmer, Shelton, CT, USA) at the Waikato Mass Spectrometry Facility following manufacturer protocols [15]. For soil grain size, 0.3–0.4 g of 2-mm-sieved dried soil was incubated overnight with 10% hydrogen peroxide. A second excess of hydrogen peroxide was then added to the sample and heated on a hotplate. Finally, 10 mL of 10% Calgon was added to the sample and left overnight before being placed in an ultrasonic bath for 5 min. Measurements were taken on a Mastersizer 2000 (Malvern, Taren Point, NSW, Australia).

2.3. DNA Extraction

DNA was extracted from soils using a modified version of a previously published cetyl trimethylammonium bromide (CTAB) bead beating protocol designed for maximum recovery of DNA from low biomass soils [15,28] (Supplementary Material Text). DNA quantification was done using the QuBit-IT dsDNA HS Assay Kit (Invitrogen, Carlsbad, CA, USA).

2.4. Terminal Restriction Fragment Length Polymorphism Analysis

Terminal restriction fragment length polymorphism analysis (tRFLP) was utilized to identify fungal community structure and relative diversity by amplifying the intergenic spacer (ITS) between the 18S and the 28S genes of the fungal *rrn* operon. PCR was performed in triplicate and pooled together to reduce stochastic inter-reaction variability. PCR master mix included 1x PCR buffer (with 1.5 mM Mg^{2+}) (Invitrogen, Carlsbad, CA, USA), 0.2 mM dNTPs (Roche Applied Science, Branford, CT, USA), 0.02 U Platinum Taq (Invitrogen, Carlsbad, CA, USA), 0.25 µM of both forward and reverse primer (Custom Science, Auckland, New Zealand) (ITS1-F and 3126R; Table S1), and 0.02 mg/mL bovine serum albumin (Sigma Aldrich, St. Louis, MO, USA) and was treated with ethidium monoazide at a final concentration of 25 pg/µL to inhibit contaminating DNA in the reagents [29]. PCR was carried out using the following thermal cycling conditions: 94 °C for 3 min; 35 cycles of 94 °C for 20 s, 52 °C for 20 s, 72 °C for 1 min 15 s; and 72 °C for 5 min on a DNA Engine thermal cycler (Bio-Rad Laboratories, Hercules, CA, USA). Successful PCR was confirmed with 1% Tris-acetate-EDTA (TAE) agarose gels, and PCR products were cleaned using the Ultraclean 15 DNA Purification kit (MOBIO Laboratories, Carlsbad, CA, USA) according to manufacturer instructions. DNA was quantified using the QuBit-IT dsDNA HS Assay Kit. 40 ng of DNA was digested with 2 U of MspI and 1× restriction enzyme buffer (Roche Applied Science, Branford, CT, USA) according to manufacturer instructions and purified with Ultraclean 15 DNA Purification kit. Lengths of fluorescent-labeled PCR amplicons (*i.e.*, tRFLP fragments) were determined by capillary electrophoresis at the Waikato DNA Sequencing Facility using an ABI 3130 Genetic Analyzer (Life Technologies, Carlsbad, CA, USA) at 10 kV, a separation temperature of 44 °C for 2 h, and the GeneScan 1200 LIZ dye Size Standard (Life Technologies, Carlsbad, CA, USA).

2.5. 454 Pyrosequencing

PCR protocol for preparing amplicons for pyrosequencing was identical to that for tRFLP, except a different reverse primer (ITS4, Table S1) was used. PCR products were purified using gel extraction and the QuickClean 5M PCR Purification Kit (GenScript, Piscataway, NJ, USA). A second round of PCR using fusion primers containing adapters for 454 pyrosequencing was performed (Table S1). These products were purified using Agencourt AMPure XP Beads (Beckman Coulter, Inc., Brea, CA, USA) for PCR amplicon recovery and removal of unincorporated dNTPs, primers, primer dimmers, salts and other contaminants (Beckman Coulter, Beverly, MA, USA) according to manufacturer instructions. Quality of PCR amplicon libraries was checked using the Agilent High Sensitivity DNA Kit with a BioAnalyzer (Agilent 2100, Agilent Technologies, Santa Clara, CA, USA) and the Kapa Library Quantification Kit—454 Titanium (Kapa Biosystems, Wilmington, MA, USA). 454 pyrosequensing was performed using a Roche 454 Junior sequencer at the Waikato DNA Sequencing Facility following manufacturer protocols.

2.6. Data Analysis

Environmental variables were $\log(x + c)$ transformed, where c is the 1st percentile value for the variable (except [Ag] where c is the mean due to low values), prior to analysis; pH values were not transformed. A Euclidean distance matrix was calculated in PRIMER 6 (PRIMER-E Ltd., Ivybridge, UK) from the transformed environmental variables and used for downstream analyses. tRFLP traces were first processed using PeakScanner 1.0 (Life Technologies, Carlsbad, CA, USA) to export all peaks above 5 relative fluorescence units (RFU). The resulting profiles were further processed using an in-house collection of python and R scripts (available from authors upon request) to identify true signal peaks as well as binning peaks based on their sizes. Briefly, peaks outside the size range of 50–1200 bp were excluded from analysis, and only peaks whose heights are greater than the 99% confidence threshold (*i.e.*, alpha value of 0.01) within a log-normal distribution were considered to be non-noise. Additionally, peaks had to be greater than 50 RFU to be considered non-noise, and all peaks above 200 RFU were by default designated as non-noise peaks. Peaks were then binned to the nearest 1 bp, and only peaks whose relative abundance was greater than 0.1% were retained. The resulting matrix of peaks expressed as relative abundances was imported into PRIMER 6, and a Bray-Curtis similarity matrix was calculated for downstream analyses. Using these distance matrices, PRIMER 6 was used to generate non-metric multidimensional scaling (MDS) plots, perform group-average hierarchical clustering, and carry out one-way analysis of similarities (ANOSIM) and biota-environmental stepwise (BEST) analyses.

454 pyrosquencing flowgrams were denoised using AmpliconNoise v1.24 [30], including a SeqNoise step to remove PCR errors and a Perseus step to remove PCR chimeras [30]. Denoised reads were aligned pair-wise using ESPIRIT [31], which directly generated a distance matrix. Mothur 1.26 was used to cluster the sequences at 0.15 distance with nearest neighbor clustering [32], and the representative sequences for the resulting operational taxonomic units (OTUs) were checked (blastn with word size of 7) against the GenBank *nr* database to allow manual identification of fungal ITS sequences (>250 bp and >80% similarity to known fungal ITS sequences). The curated sequences were then re-clustered using average neighbor at 0.05 distance. OTUs with fewer than 9 reads were excluded from downstream analysis as an aggressive filter against spurious OTUs that arose from non-specific PCR amplification and sequencing errors.

3. Results and Discussion

3.1. Soil Geochemistry

Soils from six Dry Valleys were characterized as loamy sand or sand due to their low clay (<2%) and silt (<13%) contents (Table S2), which is congruent with Antarctica's known slow and primarily physical weathering processes [7]. The coarse soil texture likely resulted from low erosivity of cold-based glaciers and salt weathering, which causes comminution of coarse fragments and provides a steady supply of sandy grains to the soils [7]. Consequently, these soils lack significant aggregation and have poor moisture retention capacity, which is consistent with their low gravimetric water content (Table S2). Water availability has been suggested to be a major factor controlling biomass and diversity of Antarctic vegetation [33,34]. Among the six study sites, Miers Valley soils contained the lowest average moisture content (0.53%, ANOVA p-value = 0.002; Table S2). But due to its low elevation (elev. 171 m) and variable wind direction, temperatures in Miers Valley can reach above 0 °C in austral summers [35]. This likely leads to increased water availability from melt streams of Miers and Adams Glaciers, which can trigger rapid responses from local microorganisms [16,34]. Water availability in austral summers is also elevated in Alatna Valley and Battleship Promontory, where transient ponds are formed from snow melt. This is in contrast with the low moisture content and water availability in higher (elev. >1500 m) and more inland valleys (e.g., University Valley). The high altitude of University Valley results in colder air temperatures all year round, leading to a lower net ice loss rate when compared to Beacon Valley (*ca.* 450 m below University Valley) [36]. Soil salt content is a proxy for water availability [37], and Miers Valley, Alatna Valley, and Battleship Promontory soils showed relatively low conductivity. Soil physicochemical properties (Table S2) were significantly different among the sampling sites (ANOSIM global R = 0.963, p-value = 0.001) with each valley clearly forming its own clade. In a broader view, distinct grouping patterns emerged for Miers Valley in the MDS plot (Figure 2), possibly due to its alkaline pH reflective of greater influence from salts of marine origin [38] and its higher C/N ratio. Overall, geochemical analysis revealed a wide range of soil salinity (107–3920 μS), low moisture content (1%–3% w/v), low levels of organic carbon (<0.46%) and nitrogen (<0.12%).

3.2. Community Fingerprinting with tRFLP

DNA extractions from soils proved difficult, and DNA samples from Beacon, University, and Upper Wright Valleys were mostly below the detection limit of 0.05 ng/μL. The highest recovery yields were obtained from Miers Valley samples, followed by those from Battleship Promontory and Alatna Valley (Table 2). Fungal tRFLP analysis of extracted DNA returned positive signals for 12 of the 30 soil samples, with no polymorphic fragments (PFs) detected in any of the samples from University Valley. A total of 33 PFs were obtained (Table 3), whose lengths varied between 145 and 781 bp. Samples from Battleship Promontory collectively returned the highest diversity (13 PFs), followed by Alatna Valley (11 PFs) and Miers Valley (5 PFs). ANOSIM analysis of PF profiles demonstrated statistically significant differences among valleys (ANOSIM global

$R = 0.731$, p-value = 0.001), and there was no robust correlation between diversity (PF count) and biomass (averaged DNA yield from 1 gram of soil) ($R = 0.35$, p-value = 0.06).

Figure 2. Nonmetric multidimensional scaling (MDS) plot based on Euclidean distances between soil physicochemical profiles. Significant correlations (Pearson $R > 0.25$) between plot ordinations and soil physicochemical properties are represented as vectors in gray.

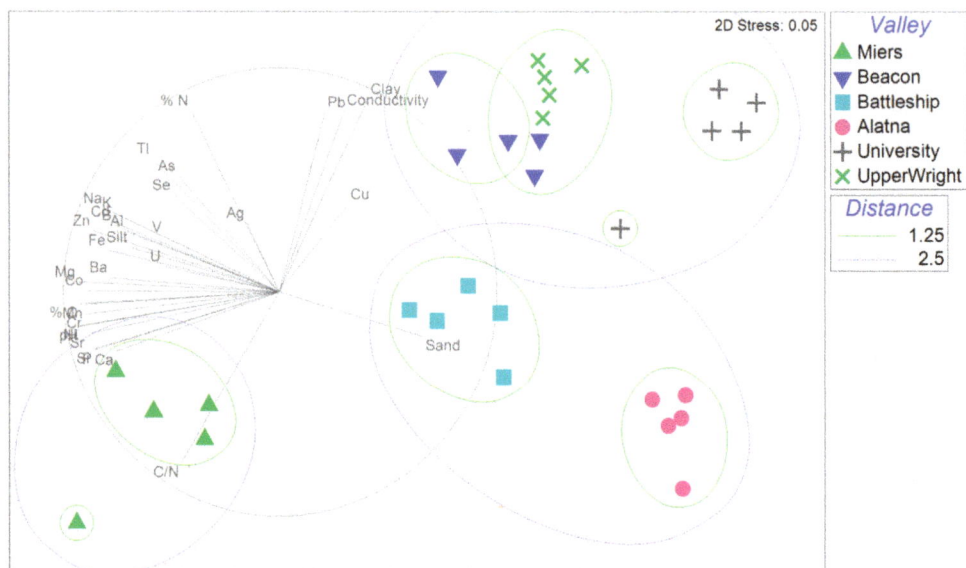

Table 2. Average concentrations of DNA extracted from 1 g of soil.

Valley	Average Concentration ± S.D.
Miers Valley	48.60 ± 27.79 ng/μL
Beacon Valley	0.48 ± 0.55 ng/μL
Battleship Promontory	20.87 ± 5.61 ng/μL
Upper Wright Valley	3.68 ± 7.57 ng/μL
Alatna Valley	15.84 ± 13.49 ng/μL
University Valley	0.05 ± 0.09 ng/μL

Table 3. Summary of terminal restriction fragment length polymorphism (tRFLP) polymorphic fragments (PF).

Valley	Total PF	Average PF ± S.D.
Miers Valley	5	1.0 ± 1.2
Beacon Valley	2	0.4 *
Battleship Valley	13	2.6 ± 1.5
Wright Valley	2	0.4 *
Alatna Valley	11	2.2 ± 3.2
University Valley	0	0

* S.D. not calculated.

Interestingly, a MDS plot of tRFLP data showed a clear separation of samples from Battleship Promontory and Alatna Valley (Figure 3), despite the fact that the two sampling sites are less than 5 km apart and within line-of-sight. This suggests that aeolian dispersal between these sites is very limited or outweighed by other environmental drivers that shape edaphic fungal diversity at these locations. There was only one sample each from Beacon and Upper Wright Valleys, but they were >50% similar to each other. Samples from Miers Valley were widely dispersed in the MDS plot, making Miers Valley a clear outlier.

Figure 3. Nonmetric multidimensional scaling (MDS) plot based on Bray-Curtis similarities of tRFLP profiles. Samples used for 454 PCR amplicon pyrosequencing are labeled by name. Significant correlations (Pearson $R > 0.25$) between plot ordinations and soil physicochemical properties are represented as vectors in gray.

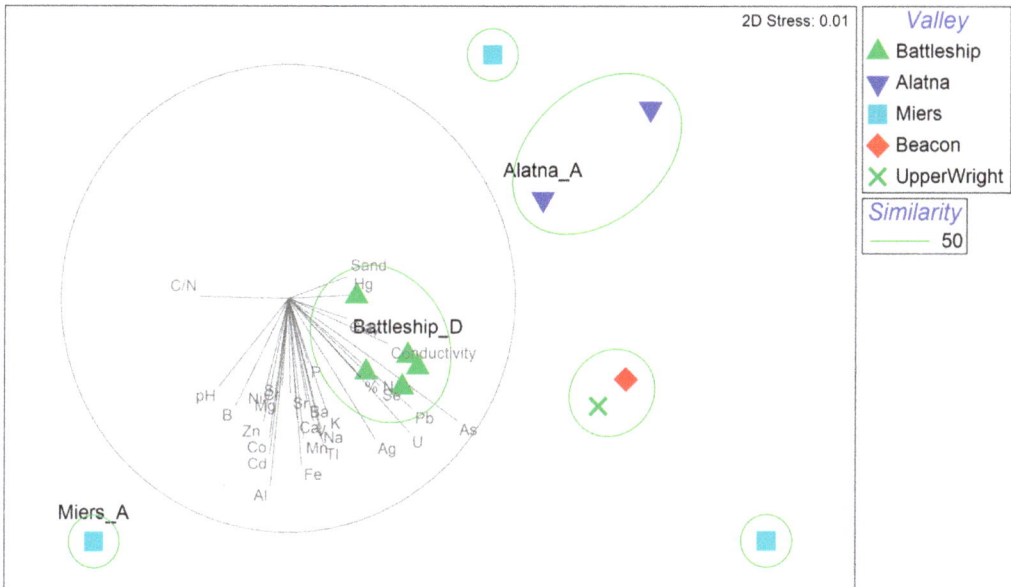

3.3. 454 Pyrosequencing

To identify the fungal species present, three samples that represented the greatest diversity based on results from tRFLP analysis were chosen for 454 PCR amplicon pyrosequencing. DNA extracted from Battleship Promontory sample D, referred to as Battleship_D, Alatna Valley sample N (Alatna_N), and Miers Valley sample A (Miers_A) appeared to be most representative of each major cluster (Figure 3). Fungal signals in Beacon and Upper Wright Valley were considered unsequenceable due to very low DNA extraction and amplification yields and therefore excluded from pyrosequencing. After filtering, denoising, chimera removal, and quality control, 262 fungal OTUs (from 21,101 reads) were obtained, of which 37 contained more than 9 reads (*i.e.*, >0.2% of the sample with fewest reads) and were used for downstream analysis. Species richness (Table 4) was highest in Miers Valley (31 OTUs from 1771 reads), followed by Battleship Promontory

(18 OTUs from 2091 reads), and Alatna Valley (17 OTUs from 5081 reads). A Venn diagram illustrates the distribution of OTUs among the three samples (Figure 4). Nine OTUs (representing 8943 reads) were found in all three Valleys (Figure 4), including the five most abundant OTUs.

Figure 4. Venn diagram of fungal OTUs.

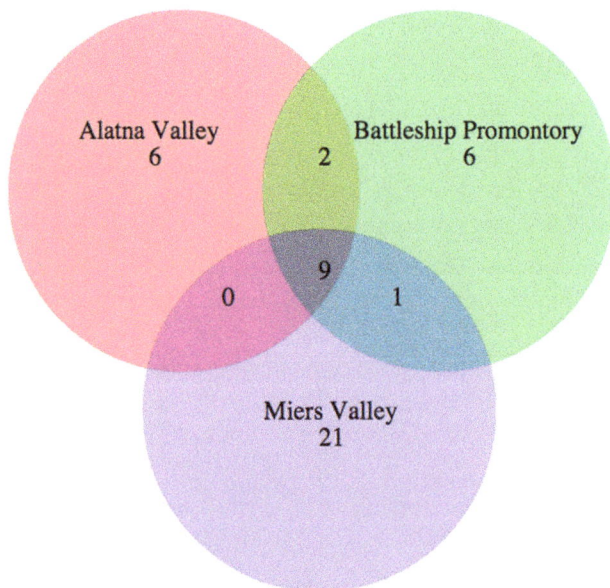

A significant number of OTUs were annotated as unclassified (Table 4 and Figure 5), which is likely reflective of the comparative lack of high quality annotated fungal ITS sequences in the GenBank *nr* database. Therefore, results that rely on classification of fungal sequences must be interpreted carefully. However, multiple studies identified Ascomycota and Basidiomycota as the dominant fungal phyla in the Dry Valleys [22,25,27,39], whereas our results showed an unexpected prominence of Chytridiomycota among all three valleys (Figure 5). It should be noted that Chytridiomycota were reported in a molecular survey on west Antarctic sites [40], including Signy Island, Mars Oasis, and Coal Nunatak, at significant abundances but not in the Dry Valleys.

Table 4. Overview of fungal OTUs from PCR amplicon pyrosequencing.

OTU #	Read Count			Total	Best Match in GenBank *nr* Database			
	AV_N	BP_D	MV_A		GenBank ID	Identity (%)	Phylum	Organism
3	1852	407	1	2283	AB032673	99	Basidiomycota	*Cryptococcus consortionis*
4	841	728	191	1760	EF432821	93	Chytridiomycota	*Lobulomycetales* sp. AF017
6	1058	122	369	1542	EF060799	99	Ascomycota	*Herpotrichiellaceae* sp. LM500
7	505	68	233	806	JF747078	99	Ascomycota	*Exophiala equina*
10	129	351	61	541	EU480339	93	Unknown	Uncultured clone
11	0	0	372	372	GQ250013	92	Ascomycota	*Cordyceps* sp. BCC22921
14	246	0	0	246	EF535204	90	Ascomycota	*Candelaria crawfordii* strain CHN265
16	179	0	0	179	FJ827708	90	Chytridiomycota	*Powellomyces* sp. PL 142
20	0	109	0	109	EU352772	93	Chytridiomycota	*Chytridiales* sp. JEL178
22	0	0	109	109	DQ457086	85	Unknown	Uncultured clone
24	0	83	0	83	AM901700	97	Ascomycota	*Ascomycete* sp. BF104
25	0	0	81	81	FJ827708	94	Chytridiomycota	*Powellomyces* sp. PL 142
26	80	0	0	80	GU184116	96	Ascomycota	*Acarospora rosulata* isolate ACABUL_USA2
28	36	30	0	66	KC222134	83	Ascomycota	*Trichoglossum octopartitum*
29	0	0	61	61	EF585664	83	Chytridiomycota	*Betamyces americaemeridionalis*
35	0	0	54	54	EU352770	92	Chytridiomycota	*Lobulomyces poculatus*
39	0	47	0	47	AF106527	91	Ascomycota	*Arthrobotrys arcuata* strain CBS 174.89
40	8	33	1	42	DQ494379	94	Ascomycota	*Vermispora fusarina*
41	12	27	3	42	JX171180	94	Basidiomycota	*Meira* sp. ANTCW08-165
45	34	1	5	40	FJ827741	96	Chytridiomycota	*Gaertneriomyces* sp. JEL 550
48	29	1	0	30	HQ634632	97	Ascomycota	*Chaetothyriales* sp. M-Cre1-2
49	29	0	0	29	JX124723	98	Ascomycota	*Taphrina* sp. CCFEE 5198
51	0	0	28	28	JX036093	93	Ascomycota	*Polysporina frigida*
54	0	10	17	27	EU352770	92	Chytridiomycota	*Lobulomyces poculatus*
56	0	0	25	25	JF809853	99	Chytridiomycota	*Betamyces* sp. PL 173

Table 4. *Cont.*

OTU #	Read Count				Best Match in GenBank *nr* Database			
	AV_N	BP_D	MV_A	Total	GenBank ID	Identity (%)	Phylum	Organism
59	0	0	23	23	AY373015	91	Unknown	*Olpidium brassicae*
60	0	22	0	22	JQ936330	99	Unknown	*Phaeosphaeriopsis* sp. CBP21E
61	0	0	22	22	JX219783	91	Ascomycota	*Cortinarius callisteus*
62	0	0	22	22	JN416510	89	Basidiomycota	*Basidiobolus* sp. BCU1
64	1	19	1	21	JX173100	99	Ascomycota	*Cladosporium* sp. AF13
67	18	0	0	18	AY781244	89	Unknown	*Ascomycete* sp. olrim401
68	0	18	0	18	AY394892	94	Ascomycota	*Mycorrhizal* sp. pkc11
72	0	0	17	17	EF634250	80	Chytridiomycota	*Coralloidiomyces digitatus*
78	0	15	0	15	EU480016	90	Unknown	Uncultured clone
101	0	0	11	11	JN882333	94	Chytridiomycota	*Monoblepharis hypogyna*
102	0	0	11	11	DQ485612	93	Chytridiomycota	*Rhizophydium carpophilum*
105	0	0	10	10	JQ711836	99	Basidiomycota	*Russula nigricans*

Abbreviations: OTU, operational taxonomic unit; AV_N, Alatna Valley sample N; BP_D, Battleship Promontory sample D; MV_A, Miers Valley sample A.

Figure 5. Phylum-level distribution of fungal OTUs.

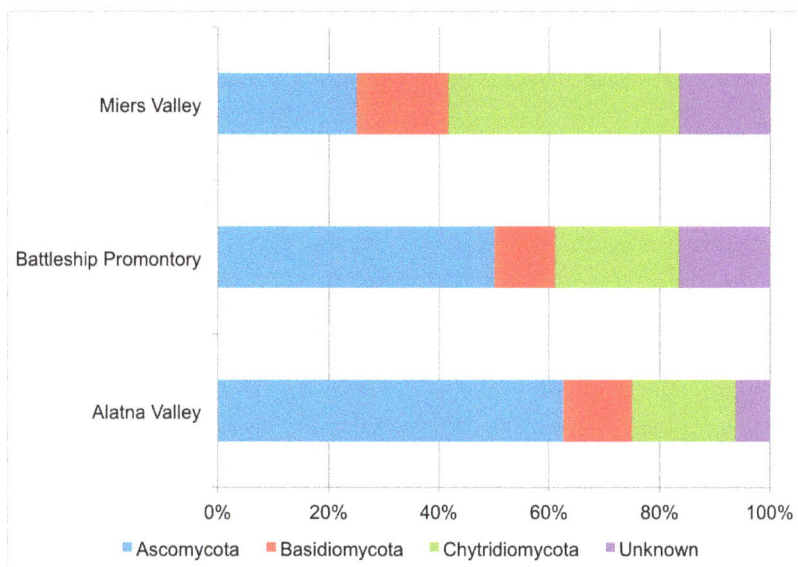

Contrary to fungal tRFLP results, PCR amplicon pyrosequencing analysis of the fungal ITS region identified Miers Valley as having the highest level of diversity of the three valleys (Figure 4), despite the lowest sequencing depth. In particular, Miers Valley appeared to harbor a limited presence of Ascomycota compared to the other two valleys, but also the highest number of Chytridiomycota OTUs (Figure 5).

The most abundant OTU (#3) was found in Alatna Valley (1875 reads), Battleship Promontory (407 reads), and Miers Valley (1 read) (Table 4). Its best match in GenBank (99% identity) was the psychrotolerant species *Cryptococcus consortionis* (Basidiomycota), which was previously observed and commonly found in Dry Valley soils [22,41]. *Cryptococcus consortionis* is characterized by the combination of amylase production and inability to utilize nitrate, cellobiose, D-galactose, *myo*-inositol, and mannitol [41]. The second most abundant OTU (#4) was also found in all three Dry Valleys (Table 4). Its best match in GenBank (93% similarity) was *Lobulomycetales* sp. AF017 (Chytridiomycota), which has been reported to occur in barren alpine soil in Peru [42]. Two other OTUs (#35 and #54) appeared to be affiliated with this genus as well.

Other abundant OTUs found in all three valleys (Table 4) were 99% similar to the species *Herpotrichiellaceae* sp. LM500 (Ascomycota) and 99.9% similar to *Exophiala equine* (Ascomycota), which was curiously reported to occur exclusively in waterborne cold-blooded animals [43]. Less abundant OTUs show similarity to fungal taxa described as Dry Valley lichen *Polysporina frigida* [44], *Meira* sp. ANTCW08-165 [45], and *Tetracladium* sp. ANTCW08-156 [45] which were previously detected in Antarctica. The genus *Cladosporium* has been reported as a dominant group by multiple studies [24,46,47] of pristine areas with little biotic influence [24,46], likely because of its prolific production of spores and high abundance in the air [24,47]. This is in contrast to our study, where *Cladosporium* species appear to be very rare (21 reads total). Notably, these fungi have been reported to survive repeated inoculations [24] and form spores, which can remain dormant for considerable periods of time [26]. It should be stressed that no conclusions can be drawn as to whether these fungi are active based on PCR amplicon pyrosequencing, as the method only detects the presence of DNA and does not indicate the viability of the organism [48,49].

3.4. Biogeography and Local Adaptation

The most important dispersal mechanisms for biomass in Antarctica have been suggested as aeolian transport [4,50,51]. If, as hypothesized previously [52], fungal species in the Dry Valleys are inactive spores that only respond to cultivation efforts and do not exhibit localized adaptations, neighboring valleys would be expected to harbor very similar fungal communities; for example, between Battleship Promontory and Alatna Valley and between Beacon and University Valley, which are located next to each other (<1 km) without any physical barrier. The tRFLP results indicated highly localized community structures, with Battleship Promontory and Alatna Valley forming statistically distinct clades (Figure 3). In addition, no fungal signals were detected in samples from University Valley while some were detected in Beacon Valley samples. Rao *et al.* previously hypothesized that the biogeography may be important for fungi in the Dry Valleys [52] and that fungal tolerance to saline conditions could confer selective advantage in high-elevation Dry Valleys [52]. Although the five most abundant OTUs reported here were found in all three samples sequenced, the relative abundances of individual OTUs were highly divergent. Since each of the sequenced samples can be considered representative of distinct diversity patterns found in the three Dry Valleys (Figure 3), the relative abundance patterns suggest that distinct fungal communities exist in each of these locations (Table 4). It should be noted that the limited spatial coverage in each Dry Valley and lack of replicates for sequencing analysis preclude definitive conclusions from being drawn, but these observations could indicate that aeolian transport plays a less important role than previous believed, or that Dry Valley fungal communities exhibit adaption to local conditions and thus are ecologically relevant.

3.5. Environmental Drivers of Fungal Distribution

Whether and how environmental factors shape fungal communities in Dry Valleys soils remains largely unexplored, but it has been suggested that both contemporary environmental conditions and historical contingencies play important roles in the distribution of fungal taxa in general [53]. It has been shown that abiotic factors play the most dominant role in extremely simplified food webs [5,11,54,55]. This makes the Dry Valleys soil ecosystem, with its extreme environmental stress, an excellent model for resolving the influence of abiotic factors on soil microbiota [19,55,56]. Miers Valley and Battleship Promontory, whose soils generally have a lower salinity, were reported to harbor greater bacterial and cyanobacterial diversity [15]. This study reveals similar trends for edaphic fungal diversity in these Dry Valleys as well as Alatna Valley; compared with Beacon Valley, University Valley, and Upper Wright Valley, where the lack of amplifiable fungal signal in extracted DNA could indicate potential limits of fungal growth and distribution.

Importantly, soil C/N ratios are higher in all three coastal and lower elevation valleys, which potentially indicate higher levels of primary productivity that can in turn sustain diverse populations of heterotrophic fungi [4,16,57]. Rao *et al.* suggested that substrate availability could limit diversity [52], since Dry Valley soils with higher carbon content harbored greater species richness [22,52]. Biota-environmental stepwise (BEST) analysis of soil physicochemical properties and tRFLP results supported this view, identifying C/N ratio as the most consistent differentiator of fungal community structure, followed by As and Ca (Supplementary Table S3). Calcium can be considered as a proxy for the mineral composition of underlying soils. The influence of arsenic on fungal populations is not clear since its concentrations are very low in our samples (Supplementary Table S2). The complete/near absence of detectable fungal signal in samples from University Valley and Beacon Valley is intriguing. Compared with other valleys, Beacon Valley and University Valley have higher elevations, resulting in lower average temperature and possibly less ice melting [36]. Therefore, contrary to an earlier hypothesis [52], lower temperature and water availability, combined with lower C/N ratio and higher salinity, may create conditions in these inland Dry Valleys that restrict fungal growth while permitting bacterial presence [15]. However, given that our samples were taken within comparatively small areas (2500 m^2) on south-facing slopes, the possibility that our observations are reflective of specific geographic features of the sampling sites cannot be ruled out. South-facing slopes of the Dry Valleys are generally colder due to the lack of solar radiation input [1] and possibly more oligotrophic (compared with north-facing slopes) [16], and as such may restrict the colonization and growth of fungi.

4. Conclusions

Soil physicochemical properties among the Dry Valley sites showed distinct grouping patterns, with each valley forming its own clade. tRFLP results revealed similar grouping patterns, with significant variations in relative abundances of fungal signals between sites. Miers Valley was identified as a clear outlier by geochemical and tRFLP analyses, which were corroborated by pyrosequencing results, showing that Miers Valley harbored the highest level of fungal diversity and an unexpected abundance of Chytridiomycota. This is in contrast with the relatively low abundance of Basidiomycota, which was previously reported as the most dominant fungal phyla in the Dry Valleys. In total, nine OTUs were found in all three valleys, including the five most abundant ones, indicating that a set of core fungal species is present throughout the Dry Valleys. However, the relative abundances of these dominant OTUs are notably different among the three sites, suggesting that there is significant biogeography for Dry Valley edaphic fungi and that they likely respond and adapt to local environmental conditions. This in turn implies that much of the fungal biomass in the Dry Valleys is biological active and ecologically relevant, rather than spores whose distribution pattern is largely dictated by aeolian transport. The comparative lack of fungal signals in the inland high elevation Dry Valleys suggests that environmental conditions at those locations may represent limits of fungal growth.

Acknowledgments

This research was supported by grants from the New Zealand Foundation for Research, Science and Technology (FRST) (UOWX0710) and the United States National Science Foundation (ANT-0944556, ANT-0944560) to S. Craig Cary. Charles K. Lee and S. Craig Cary were also supported by the New Zealand Marsden Fund (UOW0802 and UOW1003) and the New Zealand Antarctic Research Institute (NZARI2013-7). We would like to thank John Longmore of Waikato DNA Sequencing Facility, Anjana Rejendram of Waikato Stable Isotope Unit, Steve Cameron of Waikato Mass Spectrometry Facility, and Roanna Richards-Babbage and Eric Bottos of Thermopile Research Unit at University of Waikato for their support and assistance.

Author Contributions

Lisa L. Dreesens carried out DNA extraction and analysis. Lisa L. Dreesens and Charles K. Lee carried out data analyses and wrote the manuscript. Charles K. Lee and S. Craig Cary designed the study and carried out field sampling. Charles K. Lee and S. Craig Cary provided funding for the study and coordinated field expeditions.

Conflict of Interest

The authors declare no conflict of interest.

References

1. Cary, S.C.; McDonald, I.R.; Barrett, J.E.; Cowan, D.A. On the rocks: The microbiology of Antarctic Dry Valley soils. *Nat. Rev. Microbiol.* **2010**, *8*, 129–138.
2. Vishniac, H.S. The microbiology of Antarctic soils. In *Antarctic Microbiology*; Friedmann, E.I., Ed.; Wiley-Liss: New York, NY, USA, 1993; pp. 297–341.
3. Horowitz, N.H.; Cameron, R.E.; Hubbard, J.S. Microbiology of the Dry Valleys of Antarctica. *Science* **1972**, *176*, 242–245.
4. Wynn-Williams, D.D. Ecological aspects of Antarctic microbiology. In *Advances in Microbial Ecology*; Marshall, K.C., Ed.; Springer US: New York, NY, USA, 1990; Volume 11, pp. 71–146.
5. Doran, P.T.; Priscu, J.C.; Lyons, W.B.; Walsh, J.E.; Fountain, A.G.; McKnight, D.M.; Moorhead, D.L.; Virginia, R.A.; Wall, D.H.; Clow, G.D.; *et al.* Antarctic climate cooling and terrestrial ecosystem response. *Nature* **2002**, *415*, 517–520.
6. Claridge, G.G.C.; Campbell, I.B. The salts in Antarctic soils, their distribution and relationship to soil processes. *Soil Sci.* **1977**, *123*, 377–384.
7. Bockheim, J.G. Properties and classification of cold desert soils from Antarctica. *Soil Sci. Soc. Am. J.* **1997**, *61*, 224–231.
8. Treonis, A.M.; Wall, D.H.; Virginia, R.A. The use of anhydrobiosis by soil nematodes in the Antarctic Dry Valleys. *Funct. Ecol.* **2000**, *14*, 460–467.

9. Vincent, W.F. *Microbial Ecosystems of Antarctica*; Cambridge University Press: Cambridge, UK, 1988; p. 59.

10. Aislabie, J.M.; Chhour, K.L.; Saul, D.J.; Miyauchi, S.; Ayton, J.; Paetzold, R.F.; Balks, M. Dominant bacteria in soils of Marble Point and Wright Valley, Victoria Land, Antarctica. *Soil Biol. Biochem.* **2006**, *38*, 3041–3056.

11. Poage, M.A.; Barrett, J.E.; Virginia, R.A.; Wall, D.H. The influence of soil geochemistry on nematode distribution, McMurdo Dry Valleys, Antarctica. *Arct. Antarct. Alp. Res.* **2008**, *40*, 119–128.

12. Priscu, J.C. *Ecosystem Dynamics in A Polar Desert: The McMurdo Dry Valleys, Antarctica*, 1st ed.; American Geophysical Union: Washington, DC, USA, 1998; Volume 72, p. 369.

13. Smith, R.C.; Prezelin, B.B.; Baker, K.S.; Bidigare, R.R.; Boucher, N.P.; Coley, T.; Karentz, D.; MacIntyre, S.; Matlick, H.A.; Menzies, D.; *et al.* Ozone depletion: Ultraviolet radiation and phytoplankton biology in Antarctic waters. *Science* **1992**, *255*, 952–959.

14. Tosi, S.; Brusoni, M.; Zucconi, L.; Vishniac, H. Response of Antarctic soil fungal assemblages to experimental warming and reduction of UV radiation. *Polar Biol.* **2005**, *28*, 470–482.

15. Lee, C.K.; Barbier, B.A.; Bottos, E.M.; McDonald, I.R.; Cary, S.C. The inter-valley soil comparative survey: The ecology of Dry Valley edaphic microbial communities. *ISME J.* **2012**, *6*, 1046–1057.

16. Wood, S.A.; Rueckert, A.; Cowan, D.A.; Cary, S.C. Sources of edaphic cyanobacterial diversity in the Dry Valleys of eastern Antarctica. *ISME J.* **2008**, *2*, 308–320.

17. Niederberger, T.D.; McDonald, I.R.; Hacker, A.L.; Soo, R.M.; Barrett, J.E.; Wall, D.H.; Cary, S.C. Microbial community composition in soils of Northern Victoria Land, Antarctica. *Environ. Microbiol.* **2008**, *10*, 1713–1724.

18. Pointing, S.B.; Chan, Y.; Lacap, D.C.; Lau, M.C.; Jurgens, J.A.; Farrell, R.L. Highly specialized microbial diversity in hyper-arid polar desert. *Proc. Natl. Acad. Sci. USA* **2009**, *106*, 19964–19969.

19. Wall, D.H.; Virginia, R.A. Controls on soil biodiversity: Insights from extreme environments. *Appl. Soil. Ecol.* **1999**, *13*, 137–150.

20. Connell, L.; Redman, R.; Craig, S.; Scorzetti, G.; Iszard, M.; Rodriguez, R. Diversity of soil yeasts isolated from South Victoria Land, Antarctica. *Microb. Ecol.* **2008**, *56*, 448–459.

21. Arenz, B.E.; Blanchette, R.A. Distribution and abundance of soil fungi in Antarctica at sites on the Peninsula, Ross Sea Region and McMurdo Dry Valleys. *Soil Biol. Biochem.* **2011**, *43*, 308–315.

22. Arenz, B.E.; Held, B.W.; Jurgens, J.A.; Farrell, R.L.; Blanchette, R.A. Fungal diversity in soils and historic wood from the Ross Sea region of Antarctica. *Soil Biol. Biochem.* **2006**, *38*, 3057–3064.

23. Connell, L.; Redman, R.; Craig, S.; Rodriguez, R. Distribution and abundance of fungi in the soils of Taylor Valley, Antarctica. *Soil Biol. Biochem.* **2006**, *38*, 3083–3094.

24. Farrell, R.L.; Arenz, B.E.; Duncan, S.M.; Held, B.W.; Jurgens, J.A.; Blanchette, R.A. Introduced and indigenous fungi of the Ross Island historic huts and pristine areas of Antarctica. *Polar Biol.* **2011**, *34*, 1669–1677.

25. Blanchette, R.A.; Held, B.W.; Arenz, B.E.; Jurgens, J.A.; Baltes, N.J.; Duncan, S.M.; Farrell, R.L. An Antarctic hot spot for fungi at Shackleton's historic hut on Cape Royds. *Microb. Ecol.* **2010**, *60*, 29–38.

26. Duncan, S.M.; Farrell, R.L.; Jordan, N.; Jurgens, J.A.; Blanchette, R.A. Monitoring and identification of airborne fungi at historic locations on Ross Island, Antarctica. *Polar Sci.* **2010**, *4*, 275–283.

27. Selbmann, L.; de Hoog, G.S.; Mazzaglia, A.; Friedmann, E.I.; Onofri, S. Fungi at the edge of life: Cryptoendolithic black fungi from Antarctic desert. *Stud. Mycol.* **2005**, *51*, 1–32.

28. Coyne, K.J.; Hutchins, D.A.; Hare, C.E.; Cary, S.C. Assessing temporal and spatial variability in *Pfiesteria piscicida* distributions using molecular probing techniques. *Aquat. Microb. Ecol.* **2001**, *24*, 275–285.

29. Rueckert, A.; Morgan, H.W. Removal of contaminating DNA from polymerase chain reaction using ethidium monoazide. *J. Microbiol. Methods* **2007**, *68*, 596–600.

30. Quince, C.; Lanzen, A.; Davenport, R.J.; Turnbaugh, P.J. Removing noise from pyrosequenced amplicons. *BMC Bioinform.* **2011**, *12*, 1–18.

31. Sun, Y.; Cai, Y.; Liu, L.; Yu, F.; Farrell, M.L.; McKendree, W.; Farmerie, W. Esprit: Estimating species richness using large collections of 16s rRNA pyrosequences. *Nucleic Acids Res.* **2009**, *37*, e76.

32. Schloss, P.D.; Westcott, S.L.; Ryabin, T.; Hall, J.R.; Hartmann, M.; Hollister, E.B.; Lesniewski, R.A.; Oakley, B.B.; Parks, D.H.; Robinson, C.J.; *et al.* Introducing mothur: Open-source, platform-independent, community-supported software for describing and comparing microbial communities. *Appl. Environ. Microbiol.* **2009**, *75*, 7537–7541.

33. Kennedy, A.D. Water as a limiting factor in the Antarctic terrestrial environment: A biogeographical synthesis. *Arct. Alp. Res.* **1993**, *25*, 308–315.

34. McKnight, D.M.; Tate, C.M.; Andrews, E.D.; Niyogi, D.K.; Cozzetto, K.; Welch, K.; Lyons, W.B.; Capone, D.G. Reactivation of a cryptobiotic stream ecosystem in the McMurdo Dry Valleys, Antarctica: A long-term geomorphological experiment. *Geomorphology* **2007**, *89*, 186–204.

35. Katurji, M.; Zawar-Reza, P.; Zhong, S. Surface layer response to topographic solar shading in Antarctica's Dry Valleys. *J. Geophys. Res. Atmos.* **2013**, *118*, 12332–12344.

36. Pollard, W.H.; Lacelle, D.; Davila, A.F.; Andersen, D.; McKay, C.P.; Marinova, M.; Heldmann, J. Ground ice conditions in University Valley, McMurdo Dry Valleys, Antarctica. In Proceedings of the Tenth International Conference on Permafrost (TICOP), Salekhard, Russia, 25–29 June 2012; Volume 1, pp. 305–310.

37. Lamsal, K.; Paudyal, G.N.; Saeed, M. Model for assessing impact of salinity on soil water availability and crop yield. *Agric. Water Manag.* **1999**, *41*, 57–70.

38. Campbell, I.B.; Claridge, G.G.C. *Antarctica: Soils, Weathering Processes and Environment*; Elsevier Science Publishers: Amsterdam, The Netherlands, 1987; Volume 16, p. 406.

39. Duncan, S.M.; Farrell, R.L.; Thwaites, J.M.; Held, B.W.; Arenz, B.E.; Jurgens, J.A.; Blanchette, R.A. Endoglucanase-producing fungi isolated from Cape Evans historic expedition hut on Ross Island, Antarctica. *Environ. Microbiol.* **2006**, *8*, 1212–1219.

40. Lawley, B.; Ripley, S.; Bridge, P.; Convey, P. Molecular analysis of geographic patterns of eukaryotic diversity in Antarctic soils. *Appl. Environ. Microbiol.* **2004**, *70*, 5963–5972.

41. Vishniac, H.S. *Cryptococcus socialis* sp. nov. and *Cryptococcus consortionis* sp. nov., Antarctic Basidioblastomycetes. *Int. J. Syst. Bacteriol.* **1985**, *35*, 119–122.

42. Simmons, D.R.; James, T.Y.; Meyer, A.F.; Longcore, J.E. *Lobulomycetales*, a new order in the *Chytridiomycota. Mycol. Res.* **2009**, *113*, 450–460.

43. De Hoog, G.S.; Vicente, V.A.; Najafzadeh, M.J.; Harrak, M.J.; Badali, H.; Seyedmousavi, S. Waterborne *Exophiala* species causing disease in cold-blooded animals. *Persoonia* **2011**, *27*, 46–72.

44. Kantvilas, G.; Seppelt, R.D. *Polysporina frigida* sp. Nov. from Antarctica. *Lichenologist* **2006**, *38*, 109–113.

45. Slemmons, C.; Johnson, G.; Connell, L.B. Application of an automated ribosomal intergenic spacer analysis database for identification of cultured Antarctic fungi. *Antarct. Sci.* **2013**, *25*, 44–50.

46. Kerry, E. Effects of temperature on growth rates of fungi from Subantarctic Macquarie Island and Casey, Antarctica. *Polar Biol.* **1990**, *10*, 293–299.

47. Marshall, W.A. Seasonality in Antarctic airborne fungal spores. *Appl. Environ. Microbiol.* **1997**, *63*, 2240–2245.

48. Adams, B.J.; Bargett, R.D.; Ayres, E.; Wall, D.H.; Aislabie, J.; Bamforth, S.; Bargagli, R.; Cary, C.; Cavacini, P.; Conell, L.; *et al.* Diversity and distribution of Victoria Land biota. *Soil. Biol. Biochem.* **2006**, *38*, 3003–3018.

49. Fletcher, L.D.; Kerry, E.J.; Weste, G.M. Microfungi of MacRobertson and Enderby Iands, Antarctica. *Polar Biol.* **1985**, *4*, 81–88.

50. Vincent, W.F. Evolutionary origins of Antarctic microbiota: Invasion, selection and endemism. *Antarct. Sci.* **2000**, *12*, 374–385.

51. Marshall, W.A. Biological particles over Antarctica. *Nature* **1996**, *383*, 680.

52. Rao, S.; Chan, Y.; Lacap, D.C.; Hyde, K.D.; Pointing, S.B.; Farrell, R.L. Low-diversity fungal assemblage in an Antarctic Dry Valleys soil. *Polar Biol.* **2012**, *35*, 567–574.

53. Green, J.L.; Holmes, A.J.; Westoby, M.; Oliver, I.; Briscoe, D.; Dangerfield, M.; Gillings, M.; Beattie, A.J. Spatial scaling of microbial eukaryote diversity. *Nature* **2004**, *432*, 747–750.

54. Convey, P. The influence of envrionmental characteristics on life history attributes of Antarctic terrestrial biota. *Biol. Rev.* **1996**, *71*, 191–225.

55. Hogg, I.D.; Cary, C.S.; Convey, P.; Newsham, K.K.; O'Donnell, A.G.; Adams, B.J.; Aislabie, J.; Frati, F.; Stevens, M.I.; Wall, D.H. Biotic interactions in Antarctic terrestrial ecosystems: Are they a factor? *Soil. Biol. Biochem.* **2006**, *38*, 3035–3040.

56. Hopkins, D.W.; Sparrow, A.D.; Novis, P.M.; Gregorich, E.G.; Elberling, B.; Greenfield, L.G. Controls on the distribution of productivity and organic resources in Antarctic Dry Valley soils. *Proc. R. Soc. Sci. B* **2006**, *273*, 2687–2695.

57. Barrett, J.E.; Virginia, R.A.; Wall, D.H.; Cary, S.C.; Adams, B.J.; Hacker, A.L. Co-variation in soil biodiversity and biogeochemistry in Northern and Southern Victoria Land, Antarctica. *Antarct. Sci.* **2006**, *18*, 535–548.

MDPI AG
Klybeckstrasse 64
4057 Basel, Switzerland
Tel. +41 61 683 77 34
Fax +41 61 302 89 18
http://www.mdpi.com/

Biology Editorial Office
E-mail: biology@mdpi.com
http://www.mdpi.com/journal/biology